Solar Cooling Technologies

Energy Systems: From Design to Management

Series Editor: Vincenzo Bianco
Università di Genova, Italy

PUBLISHED TITLES:

Analysis of Energy Systems: Management, Planning and Policy
Vincenzo Bianco

Solar Cooling Technologies
Sotirios Karellas, Tryfon C. Roumpedakis, Nikolaos Tzouganatos, Konstantinos Braimakis

For more information about this series, please visit https://www.crcpress.com/Energy-Systems/book-series/CRCENESYSDESMAN

Solar Cooling Technologies

Sotirios Karellas
Tryfon C. Roumpedakis
Nikolaos Tzouganatos
Konstantinos Braimakis

CRC Press
Taylor & Francis Group
Boca Raton London New York

CRC Press is an imprint of the
Taylor & Francis Group, an **informa** business

CRC Press
Taylor & Francis Group
6000 Broken Sound Parkway NW, Suite 300
Boca Raton, FL 33487-2742

First issued in paperback 2020

ISBN 13: 978-0-367-73317-9 (pbk)
ISBN 13: 978-1-138-06017-3 (hbk)

Library of Congress Cataloging-in-Publication Data

Names: Karellas, Sotirios, author. | Roumpedakis, Tryfon, author. | Tzouganatos, Nikolaos, author. |
Braimakis, Konstantinos, author.
Title: Solar cooling technologies / Sotirios Karellas, Tryfon, Roumpedakis, Nikolaos Tzouganatos
and Konstantinos Braimakis.
Description: Boca Raton : Taylor & Francis, a CRC title, part of the Taylor & Francis imprint, a member of
the Taylor & Francis Group, the academic
division of T&F Informa, plc, 2018. | Includes bibliographical references and index.
Identifiers: LCCN 2018006650| ISBN 9781138060173 (hardback : alk. paper) | ISBN 9781315163178 (ebook)
Subjects: LCSH: Solar air conditioning.
Classification: LCC TH7687.9 .K37 2018 | DDC 697.9/3--dc23
LC record available at https://lccn.loc.gov/2018006650

Visit the Taylor & Francis Web site at
http://www.taylorandfrancis.com

and the CRC Press Web site at
http://www.crcpress.com

Contents

Preface

The growing need for renewable energy deployment due to the depletion of fossil fuel reserves and the increasing global concerns regarding the environmental impact of conventional methods of energy make the augmented integration of solar energy into the energy mix a top priority.

Up until the last few years, most of the relevant research and development endeavors as well as energy policy directions have aimed at the use of renewable energy sources for electricity generation and heating, which ultimately resulted in the installation of many systems worldwide, covering a wide range of applications. Cooling, on the contrary, is still primarily based on conventional methods reliant on fossil-fuel-derived electricity. Given that most of the existing refrigeration technologies have already been widely investigated and optimized, researchers have turned their interest on the utilization of renewable energy sources as a primary energy input to either produce the electricity required by refrigeration systems, or entirely substitute it.

In this context, solar energy technologies presumably represent the most promising candidate for providing renewable electricity and heat for driving sustainable heating and cooling systems. Solar-driven cooling applications are especially attractive as a means for peak electricity load shaving, due to the diurnal coincidence of the maximum solar radiation intensity with the peak cooling demand in buildings. This concurrence between the solar availability and the peak building demands turns the solar cooling and heating applications into the dominant sustainable option and a field of great scientific interest.

Solar cooling systems make use of a collector system to capture and transform solar radiation into electricity or useful heat, which is used to drive a refrigeration unit. It is a rather modern technology since, for the time being, there are not many installed applications worldwide. However, taking into account the total surface of solar collectors that have been globally installed for heating, it is easy to appreciate the great potential and opportunities of extending their use for the additional production of cooling. The ongoing development of solar collector systems and cooling installations is expected to lead to a substantial reduction of the high investment costs of the technology, which are at the moment one of its main barriers.

This book provides a detailed theoretical and technical overview of the available schemes and conversion pathways for the implementation and optimization of solar cooling technologies. After covering each technology separately, a comparison including basic process economics is held to enable the reader to better understand the advantages and weaknesses of each option and gain insights into their special characteristics and applicability.

Sotirios Karellas
Tryfon C. Roumpedakis
Nikolaos Tzouganatos
Konstantinos Braimakis

Authors

Sotirios Karellas is a professor at the School of Mechanical Engineering at the National Technical University of Athens (NTUA) and a visiting professor at the Technische Universität München and at the Universität Bayreuth in Germany. He is a specialist on energy systems, energy storage, solar-thermal energy, biomass, Organic Rankine Cycle (ORC) technology, decentralized energy systems, heat pumps, and trigeneration systems. He has more than 100 relevant publications in scientific journals and conferences and is currently supervising six PhD students at NTUA working in the field of energy production/conversion. He has participated in numerous of projects at NTUA (2006–present) and at Technische Universität München (2001–2006), assuming both technical and coordination responsibilities. He has significant industrial experience in power production plants, co/trigeneration systems, heat pumps, and chillers. He is a full-time member of five scientific journal editorial boards dealing with energy systems and renewable energy sources.

Tryfon C. Roumpedakis is an MSc Mechanical Engineer. He graduated from the School of Mechanical Engineering at the National Technical University of Athens, Greece, in 2014. In 2016, he earned his MS in Power Production and Management at the School of Electrical Engineering at the National Technical University of Athens. He is currently pursuing a second master's degree in Mechanical Engineering with a specialization in Sustainable Processes and Energy Technologies at Technische Universiteit (TU) Delft, Netherlands. During 2014–2015, he worked in the Laboratory of Steam Boilers and Thermal Plants at the National Technical University of Athens as a research engineer. He continues to collaborate with the lab as a project engineer in projects relevant to his specialization. His main fields of expertise are organic Rankine cycle and refrigeration technologies.

Nikolaos Tzouganatos earned his MEng degree from the Department of Mechanical Engineering at the National Technical University of Athens in 2010 and his MSc degree in Energy Science & Technology from ETH Zurich in 2012. In 2016, he earned his PhD from the Department of Mechanical and Process Engineering of ETH Zurich. He performed his doctoral research at the Solar Technology Laboratory of the Paul Scherrer Institute, Switzerland, in the field of solar thermochemistry and, particularly, on solar reactor development for zinc recycling and other metallurgical processes. Part of his work has been performed in close collaboration with the Commonwealth Scientific & Industrial Research Organisation (CSIRO) during his research visit at the National Solar Energy Centre in Newcastle, Australia in 2014. After working as a Postdoctoral Research Associate at the Professorship of Renewable Energy Carriers of ETH Zurich on the development of a parabolic dish solar collector-reactor system for the production of syngas by simultaneous splitting of H_2O and CO_2 via ceria redox reactions, he was appointed Project Leader at the Strategic Asset Management Department of the Zurich Municipal Electric Utility (ewz) in May 2017. His main activities include the development of a risk-based maintenance strategy and the prioritization of investments across the ewz hydropower plant portfolio as well as the analysis of wind turbine SCADA data for wind farm performance assessment and condition monitoring.

Konstantinos Braimakis is currently a PhD student at the Laboratory of Steam Boilers and Thermal Plants at the School of Mechanical Engineering of the National Technical University of Athens, Greece, in which he earned his diploma in 2013. His main focus is the exergy analysis as well as the thermodynamic and thermoeconomic optimization of Organic Rankine applications for cogeneration and trigeneration, heat transfer analysis and the investigation of zeotropic mixture working fluids.

1

Introduction

1.1 Global Energy Production and Resources

The world primary energy demand, estimated at $560 \cdot 10^{18}$ J in 2012, has been projected to increase 50% by 2040 under the International Energy Agency's (IEA's) Current Policies Scenario (International Energy Agency 2013), an increase primarily driven by rapid world population growth and the continuous economic development of emerging markets. China is the driving force of energy demand growth, accounting for almost 35% of the projected increase. However, despite the steady upward trajectory of energy demand, a slowdown in demand growth has been observed over the last few years, mainly as a result of energy efficiency gains and structural changes in the global economy that favor less energy-intensive activities. Comprehensive data for 2012 reveal a growth rate of 1.7%, which is slightly lower than the 1.9% growth rate obtained in 2011 (International Energy Agency 2013).

Today, approximately 82% of the rapidly expanding global energy demand is satisfied by fossil fuels. Despite the convenience of their use, fossil fuels represent a major source of anthropogenic CO_2 emissions and other types of pollution, all of which cause significant environmental impact and pose serious adverse effects to public health. Despite the CO_2-cutting measures announced by many counties after the 2015 United Nations Climate Change Conference, emissions are expected to rise from 31.6 Gt in 2012 to 38 Gt in 2040 according to the IEA's New Policies Scenario (International Energy Agency 2013). Furthermore, fossil fuel reserves and resources are finite and unevenly distributed across the globe, raising additional concerns about the security of energy supply.

Meeting the rapidly growing energy demand while curbing anthropogenic CO_2 emissions to meet currently established targets, avoiding the fast depletion of the fossil fuel reserves, and ensuring the security of energy supply poses a serious challenge to sustainable development and urges the need for the introduction of carbon-neutral fuels and cost-efficient deployment of renewable energy resources. Accounting for 21% of global electricity production, and predicted to further increase to 33% by 2040 according to the IEA's New Policies Scenario (International Energy Agency 2013), renewables have acquired an increasingly important role in the power generation sector and in the promotion and widespread integration of renewable energy sources across the global energy economy.

Unlike the power generation sector, heat generation for the buildings and industrial sectors is almost totally dependent on fossil fuels. More than 40% of the natural gas primary energy supply and approximately 20% of the coal primary supply and oil primary supply are used for heat production, thereby being responsible for ~39% of global energy-related CO_2 emissions. Despite the fact that heat demand accounts for almost 80% of the total energy demand in the buildings sector and that heat production processes remain a major source of anthropogenic CO_2 emissions, the share of renewables (including biomass) in global heat production reached only 25% in 2012, thus leaving tremendous potential for improvement. Solid biomass currently accounts for 68.7% of total renewable heat generation and is the most dominant energy source as it is widely used for heating and cooking in developing countries. However, concerns about local pollution from biomass use, low fuel conversion efficiency, and high capital costs pose serious challenges to further deployment in that sector. Despite these challenges, a significant growth of heat produced by modern renewables is foreseen by the IEA's New Policies Scenario. An average annual growth rate of 2.6% is expected between 2012–2040, leading to a 109% increase in the total amount of renewable heat produced by 2040 (International Energy Agency 2013).

1.2 Solar Energy

Solar energy is an attractive candidate for satisfying the electricity and heating and cooling demands of the industrial and buildings sectors since it is essentially inexhaustible and its utilization is ecologically benign. Roughly 0.1% of the earth's surface would suffice to supply the current annual global energy demand assuming a collection efficiency of merely 20% (Steinfeld and Meier 2004), thus it offers tremendous potential, as solar energy technology is constantly advancing and capital costs are decreasing. While the sun is the most powerful source of energy in our solar system, the dilute nature of solar radiation, as best indicated by the maximum energy flux (~1 kW·m^{-2}) reaching the earth's surface, limits the efficiency at which solar energy can be utilized. Furthermore, the high intermittency of solar radiation due to changing weather and atmospheric conditions, such as cloud, sand, or dust cover and snow interference, represent additional limitations to the efficient utilization of solar energy and to a continuous energy supply.

1.2.1 Solar-Generated Electricity

A series of important technological developments resulted in significant cost reductions that led, along with government support programs and subsidies, to a 50% expansion of solar electricity generation by photovoltaics (PV) over the last decade. In 2012, solar PV technology accounted for almost 1.4% of the world's renewable electricity generation, with a total installed capacity of 97.7 GW and a generation of 61 TWh (International Energy Agency 2013). While solar PVs already possess the highest growth rate among any renewable energy technology, ongoing technological improvements encourage the production of solar cells at even lower costs and higher efficiency. Thus, solar electricity generation is projected to grow by 9% annually over the next 30 years. Specifically, according to the IEA's New Policies Scenario, solar PV-based electricity is expected to rise to 950 TWh by 2035, reaching a 2.6% share of global electricity generation (International Energy Agency 2013). However, despite continuous research for the development of higher-efficiency solar PVs, the growth of PV technologies will continue to be closely linked to the provision of government subsidies in the near future, as generation costs are still high when compared to the average wholesale electricity price.

1.2.2 Solar Thermal Energy

A wide range of systems has also been developed to harness solar energy to generate thermal energy for use in the industrial, residential, and commercial sectors. Solar heating and cooling are the most widespread applications, and are promising candidates for achieving the goal of anthropogenic CO_2 emission reduction because they are capable of replacing conventional fossil-fuel-based technologies that deliver hot water, and hot and cold air. Additionally, the use of solar thermal energy systems to satisfy heating and cooling demands will also reduce electricity demand during peak load periods by replacing conventional electrically-powered heating and air conditioning systems.

Solar energy can be captured by a variety of solar thermal collector systems; each system is classified according to the temperature at which generated heat is supplied. At working temperatures below ~50°C, solar thermal collectors are mainly used for swimming pool heating and crop drying, while the most common applications of solar collectors with working temperatures up to ~120–150°C are water and space heating for the residential and industrial sectors. Although heating is the most common use of low-temperature solar thermal collectors, solar thermal energy can also be used to provide cooling and air conditioning for a building or a district cooling network through the utilization of solar thermal cooling technologies (e.g. solar-driven heat pumps, desiccant cooling systems, and absorption/adsorption chillers). Since all the aforementioned applications require low- to mid-range temperatures, the use of conventional flat-plate and evacuated tube collectors is sufficient for harnessing solar energy. However, fundamental limitations arise due to the dilute nature of solar radiation and the low solar energy conversion efficiency of these solar collectors when solar heat needs to be delivered at higher temperatures, mostly in the case of industrial applications. When we define the conversion efficiency of a perfectly insulated solar thermal collector receiver as:

$$\eta = \frac{Q_{in} - Q_{losses}}{Q_{in}} = 1 - \frac{Q_{losses}}{Q_{in}} = 1 - \frac{A_{receiver} \cdot \varepsilon \cdot \sigma \cdot T^4}{A_{receiver} \cdot q_{in}} = 1 - \frac{\varepsilon \cdot \sigma \cdot T^4}{q_{in}} \qquad (1.1)$$

where Q_{in} and Q_{losses} are the solar power input and the thermal losses from the solar collector receiver respectively, $A_{receiver}$ is the receiver area, ε and T are the emissivity and temperature of the solar receiver respectively, $\sigma = 5.67{\cdot}10^{-8}$ W·m^{-2}·K^{-4} is the Stefan Boltzmann constant, and q_{in} is the solar power input per unit receiver area, we note that the conversion efficiency can be increased by increasing the solar power input Q_{in} while holding the thermal losses Q_{losses} toward the surroundings constant. This can be achieved by concentrating sunlight incident from a large area onto a smaller area by means of solar concentrating mirrors. Concentrators are reflective optical devices that collect dilute low-flux solar radiation and direct it toward a solar receiver that is positioned at the reflector's focal point by deflecting and focusing the rays of light. Thus, the solar power input Q_{in} delivered to the receiver is augmented and a higher conversion efficiency is achieved according to Eq. (1.1). The capability of solar concentrators to collect and focus solar energy can be expressed in terms of their mean flux concentration ratio $\tilde{C}$:

$$\tilde{C} = \frac{Q_{in}}{I \cdot A_{receiver}} \qquad (1.2)$$

where I is the direct normal irradiance (DNI). High concentration ratios imply lower heat losses from smaller receivers and result in the conversion of sunlight into high-temperature process heat, which can be utilized for driving thermodynamic cycles for power generation or for process heat applications. Process heat can be delivered at low or moderate temperatures to heat gases and liquids used in various industrial processes, or at high temperatures to drive energy-intensive thermal and thermochemical processes such as the solar-driven production of metals (Vishnevetsky and Epstein 2015; Tzouganatos et al. 2013), minerals (Meier et al. 2005), and fuels (Marxer et al. 2015).

However, in contrast to most non-concentrating solar collectors, concentrating solar collectors can mainly only make efficient use of direct beam solar radiation as they consist of imaging optics, i.e. optical lenses aiming to form an image of the light source. Therefore, their application is restricted to areas with limited cloud and dust cover such as deserts and subtropical regions. To some extent, the flux concentration obtained with concentrating solar collectors can be further increased by mounting a non-imaging concentrator in tandem with the primary concentrating optical system. Non-imaging concentrators (Welford and Winston 1989) are capable of collecting and redirecting both beam and direct radiation to the receiver and have relatively low concentration ratios, i.e. in the single digits. Besides their application in tandem with imaging solar concentrator systems, non-imaging optics are often also used in combination with both flat-plate and evacuated tube solar collectors to increase their solar power input Q_{in} and conversion efficiency. Figure 1.1 gives an overview of the temperature ranges that are relevant to the different types of solar collector technologies and applications driven by solar heat.

Of all the aforementioned applications, solar hot water heating systems are considered the most mature technology. Between 2010 and 2011, the solar thermal collector total capacity increased from 195.8 GWth to 245 GWth worldwide (International Energy Agency 2012),

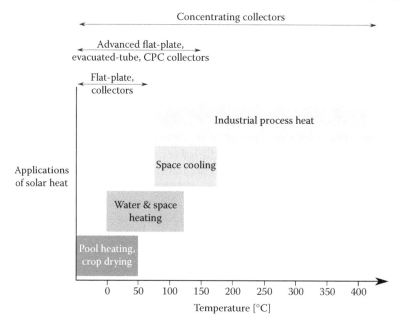

FIGURE 1.1
Solar collector technologies and their temperature ranges along with applications driven by solar heat. (Adapted from International Energy Agency, *Technology Roadmap Solar Heating and Cooling*, edited by International Energy Agency, Paris, France, 2012.)

indicating substantial growth rates in solar thermal technologies over the past few years, a trend that is expected to further increase in upcoming years. Solar water heaters in particular experience high growth rates as a result of their cost-effectiveness vis-à-vis electric and gas heaters. Specifically, the average annual cost of solar water heaters over their lifetime is estimated at 27 $/year, which is less than one-third of the annual cost of a gas (82 $/year) or an electric water (95 $/year) heater (International Energy Agency 2012). Solar district heating and low-temperature industrial process heat applications are in an advanced demonstration stage and close to commercialization. On the other hand, despite the 40–70% increase of small- to large-scale solar cooling installations between 2004 and 2011 (Mugnier and Jakob 2012), solar cooling systems are not yet economically viable and further research is necessary in order to achieve cost competitiveness and high levels of market adoption. Until 2011, 97% of the installed solar cooling systems made use of standardized flat-plate and evacuated tube collectors, but concentrating collectors that enable the production of high-temperature heat as required by highly-efficient thermally driven chillers have gained increasing scientific attention during the recent years (Ayadi et al. 2012). Overall, through continuous efforts by the industry and with governmental support, the IEA's Solar Heating and Cooling (SHC) Technology Roadmap envisages solar heating and cooling technologies to develop in such a way that solar energy may be able to meet more than 16–17% of the total final energy use for low-temperature heating and cooling by 2050, respectively (International Energy Agency 2012).

1.2.3 Cogeneration of Solar Electricity and Heat

The majority of solar collectors are currently designed for either electricity or heat generation. At times of peak solar radiation, the conversion efficiency of both solar PVs and thermal collectors acquires its maximum. However, under these conditions, high temperatures develop throughout PV cells with an adverse effect on their efficiency as the voltage output decreases linearly with increasing temperature. Operation at temperatures close to ~60–70°C may reduce cell efficiency by 10–20% (Carlson et al. 2000), while extreme increases in temperature can also result in damages to the PV module materials. To prevent temperature-driven production losses and high operation and maintenance (O&M) costs over the cell's lifetime, proper heat dissipation is required. In conventional PV technology, however, waste heat removed from the modules is not utilized, which leads to low exergy conversion efficiency. Thus, in order to recover the sensible heat of the cooling medium and achieve a higher overall system efficiency vis-à-vis PV modules or thermal collectors alone, hybrid PV-thermal solar collectors that combine PV modules with a heat extraction unit have been developed (Riffat and Cuce 2011). The resulting heat that is generated is then transferred through a contacting flow of cooling fluid (gas or liquid) to a heat exchanger in order to meet heat demand, thus achieving cogeneration of solar electricity and heat at a system efficiency of ~70%. Low-temperature heat extracted from PV cells is mostly applied for domestic water heating, while hightemperature heat can be used for electricity generation via organic Rankine cycles (Freeman et al. 2017) and thermoelectric generators, or for the production of solar-assisted heating and cooling (e.g. solar-driven heat pumps, desiccant coolers, adsorption/absorption cycles). High-temperature applications become increasingly efficient and economically viable with progressively higher temperatures. Because convective and radiation heat losses from the collector are strongly dependent on temperature, optimizing the design of hybrid PV-thermal collectors for efficient operation at high temperatures represents a major engineering challenge for this technology and leaves tremendous potential for improvement. In recent years, extensive research activities have been carried out to improve the performance of PV-thermal collectors and to reduce their cost. These research activities include (1) the evacuation of PV-thermal collectors to reduce convective heat losses and (2) the application of selective coatings to a solar collector's surface in order to achieve lower radiation heat losses.

1.2.4 Solar Energy Storage

Although the sun is the most abundant source of energy, exploitation of solar energy is restricted to daytime hours. Additionally, the intermittent nature of solar radiation due to unstable weather conditions introduces further limitations to the efficient use and conversion of solar energy. Therefore, the incorporation of energy storage mechanisms is indispensable for attaining high-efficiency, cost competitive solar energy systems and for enabling the continuous dispatchability of solar electricity and heat.

While battery banks represent the most widespread candidate for storing solar-generated electricity, thermal energy is predominantly stored in the form of sensible heat by changing the temperature of a storage medium. Typical sensible thermal energy storage media are liquids such as hot water and molten salts (Pacheco et al. 2002) or solid materials such as rocks (Zanganeh et al. 2012), sand, and metals. Hot water storage is the most common method of thermal storage for domestic heating, while other storage technologies are mostly used for higher-temperature commercial applications. To further reduce anthropogenic greenhouse gases (GHGs), emissions research activities are being carried out to investigate the feasibility of the valorization of waste materials (e.g. metal slags) produced by the extractive metallurgical industry as sensible heat storage media. Sensible heat storage systems offer important advantages such as simplicity of design and low material costs, but suffer from low thermal energy storage density and high temperature variations of the storage medium during charging and discharging phases. These factors directly affect the applicability of sensible storage systems since large volumes of storage material are required and they are not capable of providing heat at constant temperatures.

Smaller storage volumes can be achieved when storing solar thermal energy in the form of latent heat (Lane 1986) by changing the phase of the heat storage medium, also known as the phase change material (PCM). Latent heat storage systems are characterized by a higher storage density as the enthalpy changes that accompany phase transitions are considerably higher than sensible enthalpy changes, while the temperature of the storage medium during charging and discharging remains nearly constant. Most latent heat storage systems make use of solid-liquid PCMs but, depending on the phase change state, storage materials can be classified into solid-solid, liquid-gas, and solid-gas PCMs. Despite the major advantages of latent thermal energy storage systems, the technology is still in the development phase. Phase change materials are considered capital-intensive and their technical characteristics (low chemical stability and thermal conductivity which results in performance degradation after thermal cycling and inefficient power extraction from the storage medium during discharge) pose limitations to the commercialization of the technology. A third mechanism for efficiently storing intermittent solar energy is by means of reversible chemical reactions (Aydin et al. 2015). Thermochemical storage systems offer even higher energy storage density compared to PCMs due to the high amounts of heat absorbed/released in the endo-/exothermic chemical reactions taking place during charging and discharging phases, respectively. During the charging process, a reversible endothermic reaction proceeds by absorbing excess solar heat and reactant A is converted to products B and C according to:

$$A + solar\ heat \longleftrightarrow B + C \tag{1.3}$$

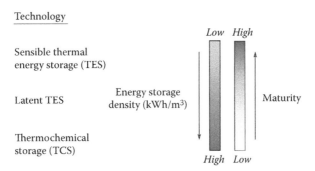

FIGURE 1.2
Energy storage density and maturity level of thermal storage technologies.

During discharge, the recombination of components B and C in the reverse reaction (1.3) is accompanied by the release of heat. Component A is regenerated for later use in the cycling process. Besides the high thermal storage density of thermochemical storage systems, these systems are capable of storing thermal energy with almost no thermal losses as the storage usually takes places at ambient temperature. Thus, the only thermal losses during the storage period are related to sensible thermal energy losses during the initial cooling of the products B and C from the reaction temperature (T_{rxn}) to the ambient temperature (T_{amb}). Any other losses are due to changes in the properties of the storage materials and material degradation after thermal cycling. However, despite the significant advantages of thermochemical energy storage, this technology is still in its early development stage. Figure 1.2 presents the superiority of various thermochemical energy storage systems in terms of energy storage density and maturity level.

1.3 Refrigeration Applications

In general, refrigeration refers to the extraction of heat from a low temperature heat source and its rejection to a higher temperature heat sink so as to maintain the temperature of the heat source below a desired level (Wang et al. 2000). The applications of refrigeration include commercial and industrial refrigeration and air conditioning (American Society of Heating and Air-Conditioning Engineers 2014). In terms of industrial refrigeration, one very important application area is the refrigeration and freezing of food products, which is used to preserve the quality of food products and enhance storage life (Stoecker 1998).

As will be discussed more thoroughly in the following chapters, refrigeration systems can be broadly categorized according to their production method of cooling effect:

- Vapor compression cooling: These systems are currently the most widely used in refrigeration applications. The basic cycle consists of a compressor that pressurizes the refrigerant up to a high pressure before it is condensed. Due to the mechanical compression used in such cycles, it is common to refer to vapor compression refrigeration as mechanical refrigeration. The latent heat released in the condenser is rejected to the environment or to a high temperature heat sink. The refrigerant is then throttled to return to the low pressure. The cooling load is produced in the evaporator of the system, which absorbs heat from the heat source that needs to be cooled down.

- Absorption cooling: The cooling effect is again produced in the evaporator of the system. After the low pressure evaporation, the refrigerant is absorbed in an aqueous absorbent. The main difference with mechanical refrigeration is that the phase change of the refrigerant in absorption cooling is achieved by adding heat, in the regenerator of the system, to heat the solution and cause the refrigerant to vaporize. The almost pure refrigerant is then condensed and throttled to return to the low pressure of the cycle and then reenters the evaporator for the restart of the cycle.

- Adsorption cooling: Adsorption is a similar method of cooling to absorption in the way that the phase change is thermally achieved. However, in adsorption, a solid material—adsorbent—is used to adsorb the refrigerant and increase its pressure. The adsorption is a highly exothermal reaction; thus, a secondary stream has to absorb the produced heat. The high pressure refrigerant is then condensed before entering the desorber. In the desorber, where heat is added to drive the desorption process, part of the previously adsorbed refrigerant is released in order for the adsorbent to be able to re-adsorb refrigerant during the adsorption phase of the next cycle. After the desorption de-pressurizes the refrigerant, it is led to the low-pressure evaporator to produce the cooling effect and then is led back to the adsorption bed for the restart of the cycle In its simplest form, an adsorption cycle consists of an evaporator, a condenser, and an adsorption bed, operating in two main phases per each cycle of operation. During the first half period of a cycle, the adsorption bed is in adsorption mode and connected to the condenser. During the second half period, the bed is in desorption mode and connected to the evaporator of the adsorption chiller.

- Desiccant cooling: Desiccant cooling is a special category that takes advantage of the heat required for the dehumidification of air. In most desiccant wheels, the desiccant material is silica gel. In a desiccant wheel using silica gel, the wheel turns and the silica gel passes in countercurrent flow to the "wet" air, adsorbing moisture. A regenerating zone ensures the drying of the desiccant for reuse in the cycle and the rejection of moisture. The air dehumidification, after proper designing of the cycle, is high enough to install an evaporator and produce the cooling effect.

1.3.1 Historical Overview

This section presents some milestones in the history of refrigeration technology.

- **1755** William Cullen produced ice-making temperatures by decreasing water pressure inside of a closed container with the use of an air pump.
- **1805** Oliver Evans proposed a closed loop refrigeration cycle.
- **1834** Jacob Perkins patented a closed cycle vapor compression refrigerator under the title "Apparatus and means for producing ice, and in cooling fluids."
- **1844** Dr. John Gorrie designed the first commercial reciprocating refrigeration machine.
- **1853** Alexander Twinning developed an ice-making system with a production capacity of approximately a ton per day.

- **1855** Charles Williams Siemens developed an ice-making machine with an aqueous solution of calcium chloride as working fluid, achieving a temperature reduction of 16 K (Reif-Acherman 2012).
- **1856** James Harrison patented a vapor compression refrigerator using ether, alcohol, or ammonia in Australia.
- **1860** Ferdinand Carré introduced the first ammonia absorption system to use water and sulfuric acid, which produced ice based on the chemical affinity of ammonia for water.
- **1862** Dr. Alexander Kirk developed a commercial refrigerator in Europe.
- **1864** Charles Tellier patented a refrigeration unit operating with dimethyl ether in France.
- **1868** Thaddeus Lowe designed a refrigerated ship using a compression system operating with carbon dioxide.
- **1872** David Boyler designed the first ammonia refrigerating compressor and then developed the first ammonia ice-making compression system in 1873.
- **1873** Paul Giffard developed an open-cycle refrigerating system consisting of two single-acting cylinders.
- **1875** Raoul Pictet developed a refrigerating machine with anhydrous sulfurous dioxide as working medium.
- **1876** Carl von Linde patented an improved method for gas liquefaction.
- **1886** Franz Wildhausen patented a carbon dioxide compressor.
- **1889** Everard Hesketh of J&E Hall developed a compound compressor to further enhance the performance of carbon dioxide systems.
- **1891** Eastman Kodak installed the first air conditioning system for the storage of photographic films in New York.
- **1894** The first domestic air conditioning system is installed in Frankfurt.
- **1903** Abbe Audiffren designed the first fully hermetic refrigeration unit.
- **1904** Willis Carrier designed a central air conditioning system using an air washer.
- **1911** General Electric (GE) released a domestic refrigeration unit powered by gas.
- **1918** Kelvinator released the first refrigerator with an automatic control.
- **1920** Edmund Copeland and Harry Edwards used isobutane in refrigerators.
- **1922** Baltzar von Platen and Carl Munters improved the principle of the absorption cooling cycle by using a three-fluid configuration, which was based on the pump-less operation of the cycle.
- **1926** Savage Arms developed a compressor that used a mercury column to pressurize the refrigerant gas.
- **1927** GE released the first electrically-powered refrigerator to use a hermetic compressor.
- **1928** Paul Crosley introduced a NH_3-H_2O absorption type refrigeration machine.
- **1931** Midgely and Hene discovered the refrigerant family called Freon.
- **1931** Electrolux released the first domestic refrigerator based on the absorption cooling principle proposed by von Platen and Munters.

- **1933** Miller and Fonda invented the rotary silica gel dehumidifier.
- **1938** The Trane Company introduced the first direct-drive, hermetic, centrifugal chiller.
- **1939** Copeland introduced the semi-hermetic, field-serviceable compressor.
- **1939** Introduction of the first dual temperature (freezer-refrigerator) domestic refrigerator.
- **1939** The first commercial heat pump is installed in Switzerland.
- **1945** Carrier introduced the first large (with a capacity of 352–2,460 kW) commercial $LiBr-H_2O$ absorption chillers.
- **1953** A solar refrigeration unit, using 10 m^2 of parabolic mirror, with a production capacity of 250 tons of ice per day, was installed in Tashkent, Uzbekistan.
- **1970s** Arkla Industries developed the first commercial absorption chiller for solar cooling.
- **1974** Mario Molina and F. Sherwood Rowland released a scientific report discussing the ozone layer depletion caused by CFCs.

1.4 Refrigerants

Refrigerant is defined as the working fluid that is used to produce the cooling effect of a refrigeration system. On the other hand, the fluid that is cooled down by the refrigeration system is referred to as the cooling medium of the refrigeration cycle.

During the early years of refrigeration applications, the most commonly used refrigerants were ammonia and sulfur dioxide. However, in modern domestic and industrial applications, synthetic freons are the dominant refrigerants. Midgley revolutionized the field of refrigeration with the invention of chlorofluorocarbon (CFC) R12 in the early 1930s (Hundy et al. 2008). The desirable thermodynamic properties of CFCs, as well as their oil miscibility and non-toxicity, established them as the most widely used refrigerants of the period. However, environmental concerns raised over the last 30 years have led to the replacement of these fluids with more environmentally friendly refrigerants.

The freons used for refrigeration applications are identified by an index "R" followed by a four-digit number that is used to identify the exact refrigerant. The first digit, which is omitted if it is equal to 0, refers to the number of unsaturated carbon-carbon bonds. The second digit is equal to the number of carbon (C) atoms minus one and is also omitted if it is equal to 0. The third digit indicates the number of hydrogen (H) atoms in the compound plus one. Finally, the last digit refers to the number of fluorine (F) atoms. In order to specify the number of chlorine (Cl) atoms, the sum of the fluorine and hydrogen atoms should be subtracted by the total number of atoms that can be connected to the carbon atoms. In many cases, several isomers for a given compound exist. Each isomer has different properties and thus a lowercase letter has been added to identify the different isomers. For example, the isomers of R142 are presented as follows:

- R142: (Chlorodifluoroethane) CH_2ClCHF_2
- R142a: (1-Chloro-1, 2-difluoroethane) $CH_2FCHClF$
- R142b: (1-Chloro-1, 1-difluoroethane) CH_3CClF_2

TABLE 1.1

Composition and Tolerances for R-407 Mixtures

	Component Composition (%)		
Refrigerant No.	R32	R125	R134a
R407A	20±2	40±2	40±2
R407B	10±2	70±2	20±2
R407C	23±2	25±2	52±2
R407D	15±2	15±2	70±2
R407E	25±2	15±2	60±2
R407F	30±2	30±2	40±2

The 400 and 500 series refer to mixtures and are designated based on the mass proportions of the refrigerants they consist of. The 400 series refers to zeotropic mixtures, while the 500 series refers to azeotropic mixtures. The refrigerant number designates which components are in the mixture, however, no data is provided for their amount within the mixture. For this reason, an uppercase letter is added as a suffix. In general, the numbers are in chronological order based on the time that the refrigerant was approved by ASHRAE. For example, Table 1.1 presents the R407 mixtures approved by ASHRAE and their corresponding compositions.

The 600 series consists of miscellaneous organic compounds (e.g. isopentane has been designated with the R601a, while N-pentane is R601). The 700 series consists of inorganic compounds including ammonia (R717), water (R718), and carbon dioxide (R744). The identification number for these compounds is formed by adding their relative molecular mass to 700. For instance, water's molecular mass is approximately 18, so the refrigerant number designated to water is R718.

1.4.1 Safety, Toxicity, and Flammability

In the early years of refrigeration technology, CFCs and HCFCs were extensively investigated and used. However, environmental concerns raised over their impact on the destruction of the ozone layer led to their phase out via the Montreal Protocol, as will be discussed later. These fluids include: R11, R12, R115 (CFCs), and R21, R123, R141b, R142b (HCFCs) (Bao and Zhao 2013).

Apart from the environmental impact of these working fluids, the safety issues regarding their use—specifically the toxicity and flammability of each working fluid—are also a matter of importance. For this reason, ASHRAE classified the refrigerants into six groups, based on the hazard of their use. Each group is designated with an uppercase letter and a number. The capital letter refers to the toxicity of the fluid and the allowable maximum exposure:

- Class A: Occupational exposure limit (OEL) of 400ppm or greater
- Class B: OEL is less than 400 ppm

The number classifies the refrigerant based on its flammability level at a temperature of 60 °C and a pressure of 1.013 bar. The four refrigerant groups based on flammability are classified as follows:

- Class 1: No flame propagation at the tested conditions
- Class 2: Flame propagation at the examined conditions, with a lower flammability limit (LFL) greater than 0.10 kg/m^3 at a temperature of 23 °C and a pressure

of 1.013 bar, and heat of combustion less than 19,000 kJ/kg (American Society of Heating and Air-Conditioning 2013)

- Class 2L: Subdivision of Class 2, with less flammable fluids that have a burning velocity of less than 100mm/s at a temperature of 23 °C and a pressure of 1.013 bar
- Class 3: Highly flammable refrigerants, with a LFL lower than 0.10 kg/m³ at a temperature of 23 °C and a pressure of 1.013 bar, and heat of combustion greater than 19,000 kJ/kg (American Society of Heating and Air-Conditioning 2013)

Based on these classifications, groups A3 and B3 are considered extremely hazardous and are not allowed to be used. Fluids in groups A2, A2L, B2, and B2L are used according to the specific regulations set out in each application area and under specific limitations. Finally, fluids in groups A1 and B1 are the least hazardous and on most occasions are not subject to restrictions (restrictions only apply to B1 fluids based on their specific toxicity). The hazard level of each of the six groups, as classified by ASHRAE, is summarized in Table 1.2.

1.4.2 Regulations and Phase Out

The destruction of the ozone layer and the impact of global warming have led to increasing concerns over the effects of several refrigerants. In the 1980s, certain measures and prohibitions were proposed as a response to these concerns, however only three main regulations have been signed since. These regulations include:

- The Montreal Protocol, in 1987, for the phase out of ozone-depleting substances (Welch et al. 2008)
- The Kyoto Protocol, in 1997, for the reduction of global warming (Philander 2008)
- The EU F-Gas Regulation, which came into effect July 2006, for regulation of the use of fluorinated gases (European Parliament 2013)

The Kyoto Protocol includes few HFC regulations as it mainly addresses the control of greenhouse gas emissions (especially CO_2). The first actual restrictions in the use of refrigerants were introduced by the Montreal Protocol, in 1987, which identified the fluids responsible for ozone depletion and announced a plan for their phase out. The Montreal Protocol came into effect on January 1, 1989 and was signed by 49 countries. In accordance with the plans set out by the Montreal Protocol, a freeze on the production of high ozone-depleting CFCs was enacted in 1996 and a subsequent freeze on the production of HCFCs took effect in 2013. European Union (EU) members agreed to freeze the production of CFCs after January 1, 1995 (Bryant 1997). Regarding the use of CFCs, the Montreal Protocol

TABLE 1.2

ASHRAE Safety Group Classification Limits

Increasing flammability →	Flammability	Chemical Formula	Molecular Mass [kg/kmol]
	Higher flammability	A3	B3
	Lower flammability	A2	B2
		A2L	B2L
	No flame propagation	A1	B1
		Lower toxicity	Higher toxicity

Source: American Society of Heating, Refrigerating, and Air-Conditioning Engineers, *2013 ASHRAE Handbook: Fundamentals*, Atlanta, GA: ASHRAE, 2013.

stated that consumption should be reduced at least 20% below that of 1986 levels by July 1, 1993, and that consumption should be reduced 50% below the base year (1986) by July 1, 1998. The phase-out procedure for HCFCs followed, with a reduction goal of 35% by January 1, 2004, and a reduction goal of up to 90% by January 1, 2015. A complete freeze on HCFCs production was scheduled for January 1, 2016 (Powell 2002).

A main contention of the Montreal Protocol was the fact that it focused on banning fluids with high ozone depletion potential (ODP). This allowed many low ODP refrigerants that were nevertheless harmful because of their high global warming potential (GWP) to continue being produced (Norman et al. 2008). Furthermore, despite the fact that annual production of ozone-depleting substances is down to less than 5% today, the ozone layer is not expected to recover until 2060 because of the long half-life of some of these substances (Woodcock 2009).

For this reason, the EU F-gas Regulation was introduced in Europe to reduce the use of fluorinated gases (e.g. HFCs) over time. The F-gas Regulation was first introduced in 2006 and prohibitions came into force gradually from 2007 to 2009. As it was adopted in 2015, the F-gas Regulation prohibited certain types of equipment. More specifically, domestic freezers using HFCs with a GWP greater than 150 were banned after January 1, 2015. A more gradual ban on commercial refrigerators and freezers utilizing harmful HFCs has been set in place. Freezers using HFCs with a GWP of 2,500 or greater will be banned after January 1, 2020, and those with a GWP of 150 or greater will be banned after January 1, 2022 (European Parliament 2013). Furthermore, under the F-Gas Regulation, certain bans were introduced for specific equipment utilizing fluorinated gases (e.g. fire protection equipment containing PFCs was banned as of July 4, 2007). Figure 1.3 presents the use of the different types of refrigerants per family within the EU from 1990 to, based on projections, 2030. As the figure shows, the use of CFCs and HCFCs has been mainly replaced by HFCs, according to aforementioned regulations.

According to the aforementioned regulations, Table 1.3 presents an overview of the main refrigerants to be replaced and their proposed short- and long-term replacements.

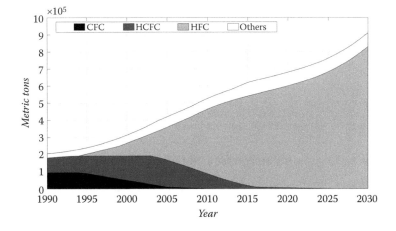

FIGURE 1.3
Overview of the total amount of refrigerant used per family in the EU from 1990 to 2030. (Adapted from Clodic, D., and S. Barrault, *1990 to 2010 Refrigerant Inventories for Europe. Previsions on Banks and Emissions from 2006 to 2030 for the European Union*, Brussels: Armines/ERIE, 2011.)

TABLE 1.3

Refrigerant Replacements According to Regulations

Type of System	Type of Application	Refrigerant to be Replaced	Short-Term Replacement	Long-Term Replacement
Refrigeration	Domestic	R12	R134a	R152a
		R409A	R437A	R290
				R600a
				R1234yf
	Commercial	R22	R407C	R32
		R502	R422	R152a
			R410A	R290
				R600a
	Industrial	R11	R123	R123
				R717
				R744
				Cascade R717/R744
A/C	Domestic and Commercial	R22	R407C	R32
			R410A	R152a
			R422D	R290
				R600a
				R1234yf
	Automotive	R12	R134a	R152a
			R437a	R744
				R1234yf
Freezing	Industrial	R11	R123	R123
				R717
				R744
				Cascade R717/R744

Source: Dinis Gaspar, Pedro, and Pedro Dinho da Silva, *Handbook of Research on Advances and Applications in Refrigeration Systems and Technologies*, Hershey, PA: Engineering Science Reference, 2015.

1.4.3 Overview of Common Refrigerants and Their Basic Properties

Table 1.4 lists some of the most widely used refrigerants and their basic properties to present readers with an overview of the environmental aspects of these fluids as well as to present the corresponding pressures of a typical compression cycle operating with these fluids.

Figure 1.4 presents the behavior of critical temperatures and pressures with respect to molecular weight for certain refrigerant categories. As the table shows, the critical temperature generally increases with the molecular mass within a category, while the critical pressure decreases with an increase in the molecular mass within the same fluid category.

TABLE 1.4

Overview of Common Refrigerants and Their Basic Properties (Evaporation Pressure at 10°C and Condensation Pressure at 40°C)

Working Fluid	Molecular Mass [kg/kmol]	Critical Pressure [bar]	Critical Temperature [°C]	Evaporation Pressure [bar][a]	Condensation Pressure [bar][b]	Compression Ratio	ODP	GWP 100yrs	ASHRAE Safety Group
HCs									
Ethane (R170)	30.07	48.72	32.13	16.30 [$T_e = -15°C$]	46.55 [$T_c = 30°C$]	2.857	0	5.5	A3
Propane (R290)	44.10	42.51	96.74	7.32	13.69	1.872	0	3.3	A3
Isobutane (R600a)	58.12	36.29	134.66	2.59	5.31	2.051	0	3.0	A3
Butane (R600)	58.12	37.96	151.98	1.76	3.79	2.149	0	4.0	A3
Isopentane (R601a)	72.15	33.78	187.20	0.64	1.52	2.383	0	4.0	A3
HFCs									
R23	70.01	48.32	26.14	16.27 [$T_e = -15°C$]	49.99 [$T_c = 25°C$]	2.889	0	14200	A1
R32	52.02	57.82	78.11	12.81	24.78	1.935	0	716	A2L
R41	34.03	58.97	44.13	30.15	53.88	1.787	0	92	–
R123	152.93	36.62	183.68	0.62	1.55	2.488	0.02	77	B1
R125	120.02	36.18	66.02	10.49	20.09	1.914	0.02	3420	A1
R134a	102.03	40.59	101.06	4.88	10.17	2.082	0	1370	A1
R143a	84.04	37.61	72.71	9.64	18.31	1.900	0	4180	A2L
R152a	66.05	45.17	113.26	4.39	9.09	2.073	0	133	A2
R161	48.06	50.91	102.15	6.97	13.66	1.959	0	12	–
R227ea	170.03	29.25	101.75	3.31	7.03	2.123	0	3580	A1
R236fa	152.04	32.00	124.92	1.92	4.38	2.277	0	9820	A1
R245fa	134.05	36.51	154.01	1.01	2.51	2.486	0	1050	B1
R404A	97.60	37.29	72.05	9.55	18.29	1.916	0	3700	A1
R405A	111.91	42.85	106.14	5.67	11.15	1.967	0.021	5300	A1
R407A	90.11	45.15	82.26	9.49	18.37	1.936	0	2100	A1

(*Continued*)

TABLE 1.4 (CONTINUED)

Overview of Common Refrigerants and Their Basic Properties (Evaporation Pressure at 10°C and Condensation Pressure at 40°C)

Working Fluid	Molecular Mass [kg/kmol]	Critical Pressure [bar]	Critical Temperature [°C]	Evaporation Pressure [bar][a]	Condensation Pressure [bar][b]	Compression Ratio	ODP	GWP 100yrs	ASHRAE Safety Group
R407C	86.20	46.29	86.03	8.99	17.49	1.944	0	1700	A1
R408A	87.01	42.95	83.14	8.88	17.01	1.916	0.019	3000	A1
R409B	97.43	46.99	109.26	6.05	11.82	1.954	0.037	1600	A1
R410A	72.59	49.02	71.35	12.58	24.26	1.928	0	2100	A1
R507A	98.86	37.05	70.62	9.77	18.70	1.914	0	3800	A1
PFCs									
R218	188.02	26.40	71.87	6.56	12.75	1.943	0	8830	A1
RC318	200.031	27.78	115.23	2.24	4.92	2.194	0	10300	A1
HFOs									
R1233zdE	130.50	36.24	166.45	n/a	n/a	n/a	0	1	–
R1234yf	114.04	33.82	94.70	5.10	10.18	1.996	0	4.4	A2L
R1234ze	114.04	36.36	109.37	3.64	7.67	2.105	0	6	A2L
R1336mzzZ	164.05	29.0	171.30	n/a	n/a	n/a	0	2	A1

Sources: Lemmon, E.W. et al., NIST Standard Reference Database 23: Reference Fluid Thermodynamic and Transport Properties—REFPROP. 9.0, 2010. Braimakis, Konstantinos et al., "Comparison of Environmentally Friendly Working Fluids for Organic Rankine Cycles", in *Advances in New Heat Transfer Fluids: From Numerical to Experimental Techniques*, edited by Alina Adriana Minea, Boca Raton: CRC Press, 2016; American Society of Heating, Refrigerating, and Air-Conditioning Engineers, *2013 ASHRAE Handbook: Fundamentals*, Atlanta, GA: ASHRAE, 2013.

[a] Evaporation pressure at 15°C, unless specified otherwise.

[b] Condensation pressure at 40°C, unless specified otherwise.

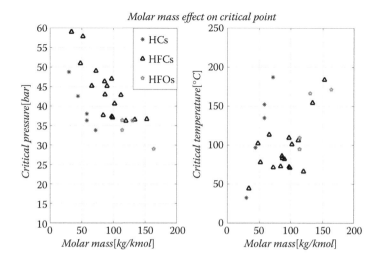

FIGURE 1.4
Critical temperatures and pressures with respect to molecular weight for certain refrigerant categories.

1.4.4 Optimal Properties of Refrigerants

When selecting a refrigerant as the working medium for a conventional refrigeration or heat pump system, desirable properties include:

- High critical temperature well outside the working range, because working at the affinity of the critical point may create problems with the efficient operation of the compressor and the condenser
- Low condensing pressure to allow for a construction of lighter compressors
- Low specific volume, which will lead to lower compressor and piping costs
- Low boiling point
- High latent heat value
- Non-toxicity and non-flammability based on the classification by ASHRAE
- Low ODP and low GWP
- Chemical stability and miscibility with lubricant oils in order to allow for the lubrication of the system compressor's moving parts (in the case of a vapor compression cycle)
- Non-corrosive properties, so as to protect against corrosion in the pipes and the components, as well as fouling in the heat exchangers of the system
- High thermal conductivity to allow for smaller-sized heat exchangers
- High dielectric strength
- Low cost and high commercial availability

Obviously, no singular fluid can combine all the aforementioned properties, thus there is always a compromise to be made based on the requirements of each specific case study.

1.5 Solar Cooling Status

Although solar cooling technologies were introduced much earlier, it was not until the late 1990s that this technology began to emerge in the market as a result of increasing oil, gas, and electricity prices.

According to the IEA, solar cooling will claim a market share of approximately 17% by 2050. However, at this time, the market has not yet considerably grown, with the predicted return on investment for such systems being measurable in 10–15 years—a time frame almost equal to their lifetime (United Nations Environment Programme 2014; Baldwin and Cruickshank 2012).

At the moment, only a few manufacturers provide a complete solar cooling unit. Such units consist of the solar collector, a hot water storage tank, a pump set, a chiller, a control unit, and a heat rejection unit. Some of the main manufacturers are listed in Table 1.5 below.

TABLE 1.5

List of Main Solar Cooling Unit Manufacturers

Manufacturer (Country)	Product Name	Cooling Capacity [kW$_c$]	Cooling Technology/ Working Pair
EDF Optimal Solutions (France)	Package system	17.5–210	Absorption (LiBr-H$_2$O)
Gasokol (Austria)	Absorption	15–200	Absorption (LiBr-H$_2$O)
	coolySun	8–15	Adsorption (silica gel-H$_2$O)
Hotspot Energy (United States)	ACDC12b	3.37	Conventional AC technology
Jiangsu Huineng (China)	Solar central air conditioning	11–175	Absorption (LiBr-H$_2$O)
Kloben (Italy)	SOLARTIK	17.5–105	Absorption (LiBr-H$_2$O)
Lucy solar (China)	Solar central air conditioning system	11.5–175	Absorption (LiBr-H$_2$O)
Sakura Corporation (Japan)	AC/DC Hybrid Solar A/C	2.6–7.0	Conventional AC technology
Schücko International KG (Germany)	LB cooling system	15–30	Absorption (LiBr-H$_2$O)
SK Sonnenklima GmbH (Germany)	Suninverse	10	Absorption (LiBr-H$_2$O)
Solarnext AG (Germany)	chillii® cooling kits	15–175	Absorption (LiBr-H$_2$O)
		19–100	Absorption (H$_2$O-NH$_3$)
		10	Adsorption (silica gel-H$_2$O)
		10–30	Adsorption (zeolite-H$_2$O)
Sol-ution (Austria)	EAW SE	15–54	Absorption (LiBr-H$_2$O)
	SOLACS	7.5–30	Adsorption (silica gel-H$_2$O)
Vicot (China)	Solar central air conditioning system	17.5–141	Absorption (LiBr-H$_2$O)

Sources: Meunier, Francis, and Daniel Mugnier. *La Climatisation Solaire: Thermique Ou Photovoltaïque*: Dunod, 2013; Stryi-Hipp, Gerhard, *Renewable Heating and Cooling: Technologies and Applications*, Amsterdam: Elsevier, 2016.

Apart from standardized solar cooling systems, there is a more mature market for the separate components of a solar cooling system, including the solar thermal collector (or the PV module for the exploitation of the solar energy) and the chiller (to be coupled with either the aforementioned solar thermal collector or the PV module) for the production of cooling power.

Regarding PV modules, their commercial availability is more than adequate on a global basis. Flat-plate collectors and evacuated tube collectors are also widespread due to their relatively easy implementation.

The market for thermally driven chillers, the second main component of a solar cooling unit, has also considerably grown. Such chillers are globally available and in a wide range of capacities.

Balaras et al. (2007) carried out a review of the status of solar air conditioning in Europe. Several projects were evaluated, including adsorption, absorption, steam jet, and desiccant cooling technologies. The capital costs for these systems were found to range between 1,286–8,420 €/kW, as shown also in Figure 1.5. As expected, lower costs were reported for larger cooling capacities, while the highest cost was reported for a small-scale research NH_3-H_2O absorption test rig.

Otanicar et al. (2012) carried out an economic and environmental analysis of the most common solar cooling schemes. The results of the analysis showed that the cost of solar thermal cooling systems will not decrease as much as the cost of PV cooling systems, unless the cost of refrigeration itself decreases significantly or through technological progress that allows COP values above 1 to be achieved. As shown in Figure 1.6, cost projections indicate a drop of approximately 40% by 2030 for the overall solar cooling system, as was also predicted by the IEA (Eicker 2014). Yet, the predicted costs by 2030 for the three solar thermal cooling options are still considerably higher than the predicted cost of approximately $30,000 for the respective PV cooling system. Henning et al. (2013) claimed that in order to expand the market for this technology, focus should be paid to the designing, commissioning, operation, and maintenance of such systems in order to allow for broader knowledge of the use of solar cooling systems.

Likewise, Hartmann et al. (2011) compared the economics of a solar thermal system to a conventional system for two different climates–Frankfurt, Germany, and Madrid, Spain–and found that the annual costs for the solar thermal system were considerably higher

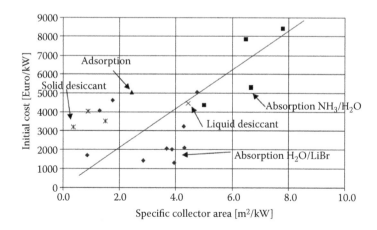

FIGURE 1.5
System capital cost as a function of the specific collector area. (Reproduced from Balaras, Constantinos A. et al., *Renewable and Sustainable Energy Reviews*, 11 (2): 299–314, 2007.)

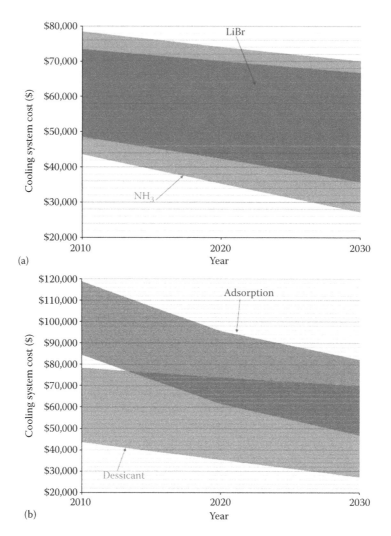

FIGURE 1.6
Cost projection for solar absorption (a) and solar-powered adsorption and desiccant cooling (b). (Reproduced from Otanicar, Todd et al., *Solar Energy*, 86 (5): 1287–1299, 2012.)

than the conventional system. More specifically, in the case of Frankfurt, the annual cost of a solar thermal system was 140% higher, while for Madrid the respective increase was 126% in comparison to the annual costs of a conventional air conditioning system.

According to Jakob (2016), the solar cooling market grew significantly between 2004 and 2014, as shown in Figure 1.7. By the end of 2014, there were approximately 1,200 solar cooling systems installed worldwide. Based on the current trend and the economy of scale, larger scale systems have higher market potential in comparison to domestic systems.

Figure 1.8 shows the behavior of the cost of solar cooling kits from 2007–2012. As the graph indicates, there is a general trend toward a reduction of cost as more systems are installed and adequate knowledge of their operational behavior becomes available. According to Figure 1.8, even though there is a considerable decrease in the cost for all scales, there is plenty of room yet for solar cooling systems to become competitive with the conventional chillers that are available at a cost range of 500–1,000 €/kW.

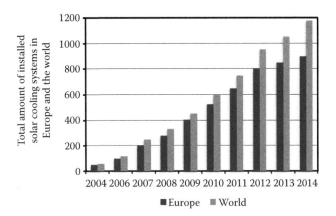

FIGURE. 1.7
Overview of the solar cooling market from 2004–2014. (Reproduced from Jakob, Uli, "Solar Cooling Technologies", in *Renewable Heating and Cooling*, 119–136, Woodhead Publishing, 2016.)

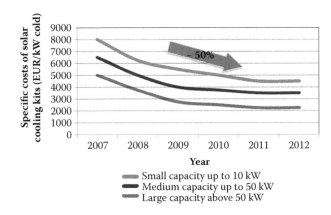

FIGURE 1.8
Cost trend for different scales of solar cooling systems from 2007–2012. (Reproduced from Jakob, Uli, "Solar Cooling Technologies", in *Renewable Heating and Cooling*, 119–136, Woodhead Publishing, 2016.)

Nomenclature

$A_{receiver}$	Receiver's surface	[m²]
$\bar{C}$	Mean flux concentration ratio	[–]
COP	Coefficient of performance	[–]
I	Direct normal irradiance (DNI)	[W/m²]
LFL	Lower flammability limit	[kg/m³]
OEL	Occupational exposure limit	[ppm]
$\dot{Q}$	Heat flux	[W]
T	Temperature	[K]

Greek Symbols

ε	Emissivity	[–]
η	Efficiency	[–]
σ	Stefan Boltzmann constant	$[\text{W}\cdot\text{m}^{-2}\cdot\text{K}^{-4}]$

Subscripts

in	Input
losses	Losses

Abbreviations

CFC	Chlorofluorocarbon
GWP	Global warming potential
HCFC	Hydrochlorofluorocarbon
HFC	Hydrofluorocarbons
ODP	Ozone depletion potential
PFC	Perfluorinated compound
PV	Photovoltaic

References

American Society of Heating, Refrigerating, and Air-Conditioning Engineers 2013. *2013 ASHRAE Handbook: Fundamentals.* Atlanta, GA: ASHRAE.

American Society of Heating, Refrigerating, and Inc Air-Conditioning Engineers. 2014. *2014 ASHRAE Handbook—Refrigeration (SI Edition).* American Society of Heating, Refrigerating and Air-Conditioning Engineers. Atlanta, GA: ASHRAE.

Ayadi, Osama, Marcello Aprile, and Mario Motta. 2012. "Solar Cooling Systems Utilizing Concentrating Solar Collectors—An Overview." *Energy Procedia* 30 (Supplement C): 875–883.

Aydin, Devrim, Sean P. Casey, and Saffa Riffat. 2015. "The Latest Advancements on Thermochemical Heat Storage Systems." *Renewable and Sustainable Energy Reviews* 41 (Supplement C): 356–367.

Balaras, Constantinos A., Gershon Grossman, Hans-Martin Henning, Carlos A. Infante Ferreira, Erich Podesser, Lei Wang, and Edo Wiemken. 2007. "Solar Air Conditioning in Europe—An Overview." *Renewable and Sustainable Energy Reviews* 11 (2): 299–314.

Baldwin, Christopher, and Cynthia A. Cruickshank. 2012. "A Review of Solar Cooling Technologies for Residential Applications in Canada." *Energy Procedia* 30 (Supplement C): 495–504.

Baniyounes, Ali M., Yazeed Yasin Ghadi, M. G. Rasul, Mohammad H. Alomari, and Adnan Manasreh. 2014. "An Overview of Solar Cooling Technologies Markets Development and its Managerial Aspects." *Energy Procedia* 61 (Supplement C): 1864–1869.

Bao, Junjiang, and Li Zhao. 2013. "A Review of Working Fluid and Expander Selections for Organic Rankine Cycle." *Renewable and Sustainable Energy Reviews* 24: 325–342.

Braimakis, Konstantinos, Tryfon Roumpedakis, Aris-Dimitrios Leontaritis, and Sotirios Karellas. 2016. "Comparison of Environmentally Friendly Working Fluids for Organic Rankine Cycles." In *Advances in New Heat Transfer Fluids: From Numerical to Experimental Techniques*, edited by Alina Adriana Minea. Boca Raton: CRC Press.

Bryant, A. C. 1997. *Refrigeration Equipment: A Servicing and Installation Handbook*. Burlington: Elsevier.

Carlson, D. E., G. Lin, and G. Ganguly. 2000. "Temperature Dependence of Amorphous Silicon Solar Cell PV Parameters." Photovoltaic Specialists Conference, 2000. Conference Record of the Twenty-Eighth IEEE.

Clodic, D., and S. Barrault. 2011. *1990 to 2010 Refrigerant Inventories for Europe. Previsions on Banks and Emissions from 2006 to 2030 for the European Union*. Brussels: Armines/ERIE.

Dincer, Ibrahim. 2017. "Refrigeration Systems and Applications." In. Newark: John Wiley & Sons, Incorporated.

Dincer, Ibrahim, and Tahir Abdul Hussain Ratlamwala. 2016. "Fundamentals of Absorption Refrigeration Systems." In *Integrated Absorption Refrigeration Systems: Comparative Energy and Exergy Analyses*, 1–25. Cham: Springer International Publishing.

Dinis Gaspar, Pedro, and Pedro Dinho da Silva. 2015. *Handbook of Research on Advances and Applications in Refrigeration Systems and Technologies*. Hershey, PA: Engineering Science Reference.

Eicker, Ursula. 2014. *Energy Efficient Buildings with Solar and Geothermal Resources*. Chichester, West Sussex, United Kingdom: John Wiley & Sons Inc.

Elvas, M. C., I. M. Peres, and S. Carvalho. 2010. "Making Science Cooler: Carré's Apparatus." 4th International Conference of the ESHS, Barcelona.

European Parliament. 2013. Fluorinated Greenhouse Gases P7_TA-PROV(2014)0223.

Freeman, J., I. Guarracino, S. A. Kalogirou, and C. N. Markides. 2017. "A Small-Scale Solar Organic Rankine Cycle Combined Heat And Power System with Integrated Thermal Energy Storage." *Applied Thermal Engineering* 127 (Supplement C): 1543–1554.

Granryd, Eric, högskolan Kungliga Tekniska, energiteknik Institutionen för, K. T. H. Industrial Engineering, Management, K. T. H. Royal Institute of Technology, and Technology Department of Energy. 2009. *Refrigerating Engineering*. Stockholm: Royal Institute of Technology, KTH, Department of Energy Technology, Division of Applied Thermodynamics and Refrigeration.

Grupp, Michael. 2012. *Time to Shine: Applications of Solar Energy Technology*. Salem, Massachusetts: Scrivener.

Hartmann, N., C. Glueck, and F. P. Schmidt. 2011. "Solar Cooling for Small Office Buildings: Comparison of Solar Thermal and Photovoltaic Options for Two Different European Climates." *Renewable Energy* 36 (5): 1329–1338.

Henning, Hans-Martin. 2013. *Solar Cooling Handbook: A Guide to Solar Assisted Cooling and Dehumidification Processes*: Birkhäuser.

Hundy, G. F., A. R. Trott, and T. C. Welch. 2008. "Refrigerants." In *Refrigeration and Air-Conditioning*, 4th ed. 30–40. Oxford: Butterworth-Heinemann.

Infante Ferreira, Carlos, and Dong-Seon Kim. 2014. "Techno-Economic Review of Solar Cooling Technologies Based on Location-Specific Data." *International Journal of Refrigeration* 39: 23–37.

International Energy Agency. 2012. *Technology Roadmap Solar Heating and Cooling*. edited by International Energy Agency. Paris, France.

International Energy Agency. 2013. *World Energy Outlook 2013*. edited by International Energy Agency. Paris, France.

Jakob, Uli. 2016. "Solar Cooling Technologies" In *Renewable Heating and Cooling*, 119–136. Woodhead Publishing.

Kalogirou, Soteris A. 2014. "Solar Energy Engineering: Processes and Systems." Amsterdam: Elsevier/Academic Press.

Kim, D. S., H. Van der Ree, and C. A. Infante Ferreira. 2007. "Solar Absorption Cooling." PhD dissertation, Korea University, Seoul. Retrieved from https://repository.tudelft.nl/islandora/object/uuid:290a8429-6316-433f-b052-7b639e2be62a?collection=research

Kiss, Anton Alexandru, and Carlos A. Infante Ferreira. 2017. *Heat Pumps in Chemical Process Industry*. Boca Raton: CRC Press.

Labouret, Anne, and Michel Villoz. 2010. *Solar Photovoltaic Energy*. Stevenage: Institution of Engineering and Technology.

Lane, George A. 1986. *Solar Heat Storage: Latent Heat Materials*, Vol. 2, *Technology*. Boca Raton: CRC Press.

Refprop database.

Lemmon, E. W. H., M. L., and McLinden, M. O. 2010. NIST Standard Reference Database 23: Reference Fluid Thermodynamic and Transport Properties - REFPROP. 9.0.

Marxer, Daniel, Philipp Furler, Jonathan Scheffe, Hans Geerlings, Christoph Falter, Valentin Batteiger, Andreas Sizmann, and Aldo Steinfeld. 2015. "Demonstration of the Entire Production Chain to Renewable Kerosene via Solar Thermochemical Splitting of H2O and CO2." *Energy & Fuels* 29 (5): 3241–3250.

Meier, Anton, Enrico Bonaldi, Gian Mario Cella, and Wojciech Lipinski. 2005. "Multitube Rotary Kiln for the Industrial Solar Production of Lime." *Journal of Solar Energy Engineering* 127 (3): 386–395.

Meunier, Francis, and Daniel Mugnier. 2013. *La Climatisation Solaire: Thermique Ou Photovoltaïque*: Dunod.

Miller, Rex, Edwin P. Anderson, and Mark R. Miller. 2004. *Refrigeration: Home and Commercial*. Indianapolis: Wiley.

Molina, Mario J., and F. S. Rowland. 1974. "Stratospheric Sink for Chlorofluoromethanes: Chlorine Atomc-Atalysed Destruction of Ozone." *Nature* 249 (5460): 810–812.

Mugnier, Daniel, and Uli Jakob. 2012. "Keeping Cool with the Sun." *International Sustainable Energy Review* 6 (1): 28–30.

Norman, Catherine, Stephen DeCanio, and Lin Fan. 2008. "The Montreal Protocol at 20: Ongoing Opportunities for Integration with Climate Protection." *Global Environmental Change* 18 (2): 330–340.

Norton, Brian. 2014. "Introduction." In *Harnessing Solar Heat*, 1–8. Dordrecht: Springer Netherlands.

Otanicar, Todd, Robert A. Taylor, and Patrick E. Phelan. 2012. "Prospects for Solar Cooling—An Economic and Environmental Assessment." *Solar Energy* 86 (5): 1287–1299.

Pacheco, James E., Steven K. Showalter, and William J. Kolb. 2002. "Development of a Molten-Salt Thermocline Thermal Storage System for Parabolic Trough Plants." *Transactions-American Society of Mechanical Engineers Journal of Solar Energy Engineering* 124 (2) :153–159.

Pearson, Andy. 2005. "Carbon Dioxide—New Uses for an Old Refrigerant." *International Journal of Refrigeration* 28 (8): 1140–1148.

Philander, S. George. 2008. *Encyclopedia of Global Warming and Climate Change*. Sage Publications.

Powell, Richard L. 2002. "CFC Phase-Out: Have We Met the Challenge?" *Journal of Fluorine Chemistry* 114 (2): 237–250.

Reif-Acherman, Simón. 2012. "The Early Ice Making Systems in the Nineteenth Century." *International Journal of Refrigeration* 35 (5): 1224–1252.

Riffat, Saffa B, and Erdem Cuce. 2011. "A Review on Hybrid Photovoltaic/Thermal Collectors and Systems." *International Journal of Low-Carbon Technologies* 6 (3): 212–241.

Sinnott, R. K. 2009. *Chemical Engineering Design: SI Edition*: Elsevier Science.

Srikhirin, Pongsid, Satha Aphornratana, and Supachart Chungpaibulpatana. 2001. "A Review of Absorption Refrigeration Technologies." *Renewable and Sustainable Energy Reviews* 5 (4): 343–372.

Steinfeld, Aldo, and Anton Meier. 2004. "Solar Fuels and Materials." *Encyclopedia of Energy* 5: 623–637.

Stoecker, W. F. 1998. *Industrial Refrigeration Handbook*. New York: McGraw-Hill.

Stryi-Hipp, Gerhard. 2016. *Renewable Heating and Cooling: Technologies and Applications*. Amsterdam: Elsevier.

Tzouganatos, N., R. Matter, C. Wieckert, J. Antrekowitsch, M. Gamroth, and A. Steinfeld. 2013. "Thermal Recycling of Waelz Oxide Using Concentrated Solar Energy." *JOM* 65 (12): 1733–1743.

United Nations Environment Programme. 2014. *Assessment on the Commercial Viability of Solar Cooling Technologies and Applications in the Arab Region*.

Vishnevetsky, I., and M. Epstein. 2015. "Solar Carbothermic Reduction of Alumina, Magnesia and Boria Under Vacuum." *Solar Energy* 111 (Supplement C): 236–251.

Wang, R. Z., and R. G. Oliveira. 2006. "Adsorption Refrigeration—An Efficient Way to Make Good Use of Waste Heat and Solar Energy." *Progress in Energy and Combustion Science* 32 (4): 424–458.

Wang, Shan K., Z. Lavan, and Paul Norton. 2000. *Air Conditioning and Refrigeration Engineering.* Boca Raton: CRC Press.

Welch, T., Helen Carwardine, and Ken Butcher. 2008. *Refrigeration.* London: Chartered Institution of Building Services Engineers.

Welford, W. T., and R. Winston. 1989. "Concentrators and Their Uses." In *High Collection Nonimaging Optics*, 1–7. Academic Press.

Whitman, William C. 2009. *Refrigeration and Air Conditioning Technology.* 6th ed. Clifton Park: Delmar.

Woodcock, Ashley. 2009. "The Montreal Protocol: Getting Over the Finishing Line?" *The Lancet* 373 (9665): 705–706.

Zanganeh, Giw, A. Pedretti, S. Zavattoni, M. Barbato, and A. Steinfeld. 2012. "Packed-bed Thermal Storage for Concentrated Solar Power—Pilot-Scale Demonstration and Industrial-Scale Design." *Solar Energy* 86 (10): 3084–3098.

2

Thermodynamic Cycles for Solar Cooling

2.1 Carnot Cycle for Refrigeration

In principle, during a continuous cooling application, heat absorbed at a low temperature should be released at a higher temperature environment, as shown in Figure 2.1. This is basically the reverse process of a thermal machine. In the ideal case, the process follows a reverse Carnot, which consists of two isothermal processes between two adiabatic processes. As shown in Figure 2.2, the reverse Carnot cycle consists of four phases:

- 1–2: Work is provided in the cycle to adiabatically compress the refrigerant and raise its temperature from the low temperature T_C to the high temperature of the cycle T_H.
- 2–3: The refrigerant isothermally rejects heat Q_H at a temperature T_H. The heat is rejected reversibly from the system by being in contact with a high temperature heat sink, with a temperature equal to or lower than T_H.
- 3–4: The refrigerant is adiabatically expanded to the low temperature T_C of the cycle.
- 4–1: The refrigerant evaporates, reversibly absorbing heat Q_C, at a constant temperature T_C from a cold reservoir. This heat, transferred from the cold reservoir to the system, is the cooling load of the cycle, which results in the decrease of the cold reservoir's temperature.

After the completion of process 4–1, the refrigerant is led to the compressor for the restart of the cycle.

Based on Figure 2.2, the heat that is rejected in the environment, $\dot{Q}_H$, is calculated as follows:

$$\dot{Q}_H = \dot{m} \cdot (h_2 - h_3) \tag{2.1}$$

The cooling capacity of the cycle, $\dot{Q}_C$, is calculated as follows:

$$\dot{Q}_C = \dot{m} \cdot (h_1 - h_4) \tag{2.2}$$

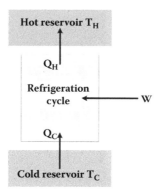

FIGURE 2.1
Working principle of a cooling machine.

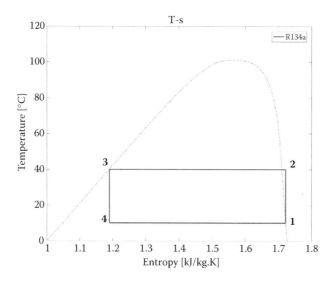

FIGURE 2.2
T-s diagram of the reverse Carnot process for refrigeration.

Finally, the work consumed at the compressor, $\dot{W}$, is calculated as follows:

$$\dot{W} = \dot{m} \cdot (h_2 - h_1) \tag{2.3}$$

The most common measure for the efficiency of a cooling system is the coefficient of performance (COP):

$$COP = \frac{\textit{heat absorbed at the low temperatur } T_C}{\textit{work consumed}} \tag{2.4}$$

Thus, the COP for the Carnot cycle is equal to:

$$COP = \frac{|\dot{Q}_C|}{|\dot{W}|} \qquad (2.5)$$

Applying the first law of thermodynamics, the change of internal energy in the cycle is derived to be zero, hence:

$$|\dot{W}| = |\dot{Q}_H| - |\dot{Q}_C| \qquad (2.6)$$

Then, if Eq. (2.6) is divided by $|\dot{Q}_C|$, the following is derived:

$$\left|\frac{\dot{W}}{\dot{Q}_C}\right| = \left|\frac{\dot{Q}_H}{\dot{Q}_C}\right| - 1$$

However, it is known for the Carnot cycle that:

$$\frac{|\dot{Q}_H|}{|\dot{Q}_C|} = \frac{T_H}{T_C}$$

Thus,

$$\left|\frac{\dot{W}}{\dot{Q}_C}\right| = \frac{T_H}{T_C} - 1 = \frac{T_H - T_C}{T_C}$$

Finally, it is concluded from Eq. (2.5) that the COP for the Carnot cycle for refrigeration (reversible COP) is equal to:

$$COP_{rev} = \frac{T_C}{T_H - T_C} \qquad (2.7)$$

In correspondence to the Carnot cycle for thermal machines, the COP for the reverse Carnot cycle gives the maximum value of the COP for every cooling device working between temperatures TH and TC. The fact that expansion and compression are considered adiabatic and isentropic in the Carnot cycle, as well as the fact that both evaporation and condensation take place without any losses, ensure that the Carnot cycle achieves the maximum cooling output for a given work input. Hence, the Carnot cycle is an idealized cycle and is used as the performance limit for every cooling device operating under the same conditions.

The reversible nature of the cycle indicates that it requires infinite time, thus resulting in zero average cooling output as well as zero average work input.

2.1.1 Tutorial on the Carnot Cycle

Assume a refrigeration cycle with a high temperature equal to 50°C (323.15 K) and an evaporation temperature of 10°C (283.15 K). Considering that the cooling capacity of the cycle is 8.25 kW and the power consumption of the compressor is equal to 3.75 kW, find the following:

i. The heat rejected to the environment in the condenser
ii. The COP of the cycle
iii. The minimum power input of the cycle (if a Carnot cycle was considered for the same cooling capacity)

Solution:

i. Using Eq. (2.6), it is possible to calculate the heat rejection as follows:

$$\left| \dot{Q}_H \right| = \left| \dot{W} \right| + \left| \dot{Q}_C \right| = 12 \ kW$$

ii. From Eq. (2.5), it is possible to determine the cycle's COP as follows:

$$COP = \frac{\left| \dot{Q}_C \right|}{\left| \dot{W} \right|} = \frac{8.25}{3.75} = 2.2$$

iii. From Eq. (2.7), the reversible COP can be determined as follows:

$$COP_{rev} = \frac{T_C}{T_H - T_C} = \frac{283.15}{323.15 - 283.15} = 7.079$$

Hence, the minimum power input of the cycle is equal to:

$$\dot{W}_{rev} = \frac{\dot{Q}_C}{COP_{rev}} = 1.165 \ kW$$

2.2 The Main Components of Mechanical Refrigeration

As already described in the previous chapter, a typical mechanical refrigeration system consists of a compressor, a condenser, a throttling device, and an evaporator. Before presenting the vapor compression cycle in detail, the following section presents an overview of some of the basic features of the technologies applied in the aforementioned components and their specific energy and exergy analyses.

2.2.1 Compressor

In a mechanical refrigeration system, the compressor is used to increase the pressure of the working fluid from the evaporating pressure up to the condensation pressure as well as to circulate the fluid in the cycle. Due the irreversibilities related to the compressor—even in a case where the outlet of the evaporator is saturated—superheated vapor conditions are at the outlet of the compressor. Based on their working principle, compressors are divided into two broad categories: displacement compressors and dynamic compressors. Figure 2.3 shows the main categories and subcategories of commercially available compressors.

The basic difference between dynamic compressors and positive displacement compressors is that, in positive displacement compressors, the work exchange takes place periodically by trapping parts of the flow in cavities and pressurizing them while also moving them from the inlet to the outlet of the compressor, while, in dynamic compressors, the compression process takes place continuously (Boyce 2012). For small-scale applications, positive displacement compressors are more favorable because of their lower design volume flow rates, their higher pressure ratios, and their smaller rotational speed when compared to dynamic compressors (Quoilin 2011; Hung et al. 1997).

The main subcategories of positive displacement compressors are rotary compressors and reciprocating compressors, as shown in Figure 2.3, with each subcategory further divided into smaller categories.

In order to analyze a thermodynamic cycle, some basic equations for each component should be introduced first. Assume use of the compressor shown in Figure 2.4.

Applying a mass balance across the compressor, the following equation is easily derived:

$$\dot{m}_i = \dot{m}_o \Rightarrow \rho_i \dot{V}_i = \rho_o \dot{V}_o \Rightarrow \frac{\dot{V}_i}{v_i} = \frac{\dot{V}_o}{v_o} \tag{2.8}$$

where $\dot{m}$ refers to the mass flow rate (kg/s), ρ is the density (kg/m³), $\dot{V}$ is the volumetric flow rate (m³/s) and v is the specific volume (m³/kg).

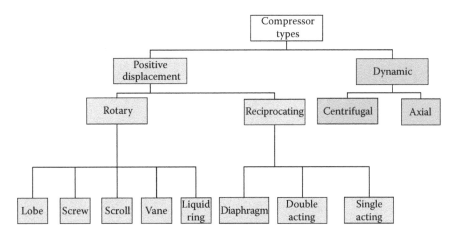

FIGURE 2.3
Main types of compressors used for industrial applications.

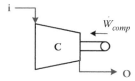

FIGURE 2.4
Schematic diagram of a compressor (i: inlet, o: outlet).

On the other hand, assuming there are no heat losses in the compressor, the application of an energy balance in the compressor results in the following equation for the calculation of the work consumption in the compressor:

$$\dot{W}_{comp} + \dot{m}_i h_i = \dot{m}_o h_o \Rightarrow$$
$$\dot{W}_{comp} = \dot{m}(h_o - h_i) \tag{2.9}$$

The performance of a compressor is influenced by several parameters. Among them, one of the most important parameters is the compression ratio, which is defined as shown in Eq. (2.10):

$$\pi_{comp} = \frac{P_o}{P_i} \tag{2.10}$$

where P_o refers to the discharge pressure and P_i refers to the suction pressure of the compressor.

To measure the performance of the compressor, there are three main indicators: the isentropic efficiency in terms of the energy losses; the volumetric efficiency, which reflects the losses affecting the flow in the compressor; and the exergetic efficiency, from the exergy point of view.

The isentropic efficiency of a compressor is defined as the ratio of the isentropic power required to achieve a certain pressure ratio divided by the actual power to compress the fluid up to the discharge pressure:

$$\eta_{is,comp} = \frac{\dot{W}_{is,comp}}{\dot{W}_{comp}} = \frac{\dot{m}(h_{o,is} - h_i)}{\dot{m}(h_o - h_i)} = \frac{h_{o,is} - h_i}{h_o - h_i} \tag{2.11}$$

where the term $h_{o,is}$ refers to the enthalpy of the vapor at a pressure equal to that of the discharge ($P_{o,s} = P_o$) and an entropy equal to that of the suction ($s_{o,s} = s_i$).

The volumetric efficiency of a compressor measures the losses that are affecting the flow in the compressor, and it is defined as the ratio of the actual volume in the suction divided by the geometrical suction volume:

$$\eta_{vol,comp} = \frac{\dot{V}_{i,actual}}{\dot{V}_{i,geom.}} \tag{2.12}$$

Exergy destruction, $\dot{Ex}_{loss}$, in an adiabatic compressor can be derived by applying an exergy balance in the compressor:

$$\dot{Ex}_i = \dot{Ex}_o + \dot{Ex}_{loss} \Rightarrow$$
$$\dot{Ex}_{loss} = \dot{W}_{comp} + \dot{Ex}_i - \dot{Ex}_o \Rightarrow \qquad (2.13)$$
$$\dot{Ex}_{loss} = \dot{W}_{comp} - \dot{m}[(h_o - h_i) - T_0(s_o - s_i)]$$

where the negative term on the left hand side is also called the reversible work of compression:

$$\dot{W}_{rev,comp} = \dot{m}\left[(h_o - h_i) - T_0(s_o - s_i)\right] \qquad (2.14)$$

thus, the exergetic efficiency of the compressor may be defined as the ratio of the reversible work of the compression process and the actual work of compression:

$$\eta_{ex,comp} = \frac{\dot{W}_{rev,comp}}{\dot{W}_{comp}} = 1 - \frac{\dot{Ex}_{loss}}{\dot{W}_{comp}} \qquad (2.15)$$

2.2.1.1 Tutorial on the Compressor

Consider a compressor that is used to pressurize 5 kg/s of R1234yf. The conditions at suction are the following: T_{suc} = 288.15 K and P_{suc} = 3 bar. The temperature at the discharge line is equal to T_{dis} = 320.236 K, and the pressure ratio of the compressor is π_{comp} = 2.47. Furthermore, it is a is given that the geometrical suction volume is equal to 354.91 L/s. Determine the following:

i. The compressor's actual power consumption
ii. The isentropic efficiency of the compressor
iii. The volumetric efficiency of the compressor and the volume displacement (in L/rev) if the compressor has a rotational speed of 1,500 rpm
iv. The exergetic efficiency of the compressor under investigation

Solution:

i. The properties of the used refrigerant at the suction line are derived from RefProp (Lemmon et al. 2010):

$$h_{suc}\left(T = T_{suc} = 288.15\,K, P = P_{suc} = 3\,bar\right) = 214.1531\,kJ/kg$$
$$s_{suc}\left(T = T_{suc} = 288.15\,K, P = P_{suc} = 3\,bar\right) = 0.79305\,kJ/kg.K$$
$$v_{suc}\left(T = T_{suc} = 288.15\,K, P = P_{suc} = 3\,bar\right) = 0.064594\,m^3/kg$$

Given the pressure ratio, it is possible to determine the discharge pressure:

$$\pi_{comp} = \frac{P_{dis}}{P_{suc}} \Rightarrow P_{dis} = \pi_{comp} \cdot P_{suc} = 7.41\ bar$$

Thus, with the temperature and the pressure of the discharge known, it is possible to determine the outlet properties:

$$h_{dis}\left(T = T_{dis} = 320.236\ K,\ P = P_{dis} = 7.41 bar\right) = 237.4022\ kJ/kg$$
$$s_{dis}\left(T = T_{dis} = 320.236\ K,\ P = P_{dis} = 7.41\ bar\right) = 0.81062\ kJ/kg.K$$

Finally, the power consumption of the compressor can be determined based on Eq. (2.9):

$$\dot{W}_{comp} = \dot{m}(h_{dis} - h_{suc}) = 5 \cdot (237.4022 - 214.1531) \Rightarrow$$
$$\dot{W}_{comp} = 116.2456\ kW$$

ii. To determine the isentropic efficiency, the enthalpy of the isentropic to the suction point must be calculated first:

$$h_{dis,is}(T = T_{dis} = 320.236\ K,\ s = s_{suc}) = 231.8224\ kJ/kg$$

Hence, given Eq. (2.11), the isentropic efficiency is determined as follows:

$$\eta_{is,comp} = \frac{h_{dis,is} - h_{suc}}{h_{dis} - h_{suc}} = \frac{231.8224 - 214.1531}{237.4022 - 214.1531} = 76\%$$

The actual compression process, as calculated above, is presented in Figure 2.5 in a T-s diagram.

iii. The actual suction volume is determined by the mass flow rate and the specific volume at the suction:

$$\dot{V}_{suc,actual} = \dot{m} \cdot v_{suc} = 5\ kg/s \cdot 0.064594\ m^3/kg = 0.32297\ m^3/s$$

On the other hand, it is a given that the geometrical suction volume is 0.35491 m³/kg. Thus, using Eq. (2.12), the volumetric efficiency can be determined as:

$$\eta_{vol,comp} = \frac{\dot{V}_{suc,actual}}{\dot{V}_{suc,geom.}} = \frac{0.32297\ m^3/s}{0.35491\ m^3/s} = 91\%$$

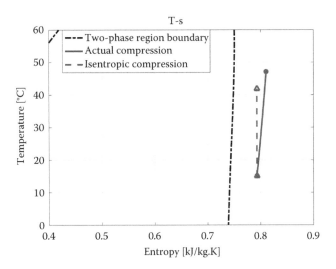

FIGURE 2.5
Overview of the compression process in a T-s diagram.

iv. To determine the exergetic efficiency, the compressor's reversible work, according to Eq. (2.14):

$$\dot{W}_{rev,comp} = \dot{m}[(h_{dis} - h_{suc}) - T_0(s_{dis} - s_{suc})] \Rightarrow$$
$$\dot{W}_{rev,comp} = 5 \cdot [(237.4022 - 214.1531) - 293.15 \cdot (0.81062 - 0.79305)] \Rightarrow$$
$$\dot{W}_{rev,comp} = 90.4922 \, kW$$

Hence, based on Eq. (2.15), the exergetic efficiency is equal to:

$$\eta_{ex,comp} = \frac{\dot{W}_{rev,comp}}{\dot{W}_{comp}} = \frac{90.4922}{116.2456} = 77.8457\%$$

2.2.2 Condenser

The main function of a condenser is to condense the working fluid by rejecting heat to the environment. Based on this fact, it is evident that during the design of a condenser, care should be taken not only to achieve the required condensation but also to respect the requirements stated for the secondary stream, to which heat is dumped.

The main types of condensers based on cooling medium include: water-cooled condensers, air-cooled condensers, and evaporative condensers. When it comes to water-cooled condensers in particular, different types of heat exchangers can be applied depending on the cooling capacity of the heat exchanger, the working fluid, the stream conditions (pressure and temperature at the inlet), and the amount of water required for the condensation process. Plate heat exchangers (PHEXs), tube-in-tube heat exchangers, and shell-and-tube heat exchangers (S&T HEXs) are the most widely used. In most applications, condensation

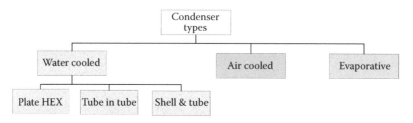

FIGURE 2.6
Main types of condensers for refrigeration applications.

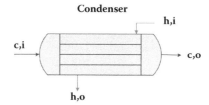

FIGURE 2.7
Schematic diagram of a condenser (H: hot side, C: cold side).

takes place on the shell side, thus the shell also serves as a receiver, collecting the condensate at the bottom of the heat exchanger. Figure 2.6 offers a visual representation of the most common commercial types of condensers.

To carry out an energy and exergy analysis, a condenser like the one in Figure 2.7 will be considered. Figure 2.8 presents the composite curve with the two streams involved in the heat transfer considered for the energy and exergy analysis of the condenser.

Applying an energy balance to the hot side of the condenser, between inlet (i) and outlet (o), the following equation is easily derived:

$$\dot{m}_h h_{h,i} = \dot{m}_h h_{h,o} + \dot{Q}_h \Rightarrow \dot{Q}_h = \dot{m}_h \cdot (h_{h,i} - h_{h,o}) \tag{2.16}$$

Assuming, furthermore, that the condenser is insulated and thus no heat transfer toward the surrounding environment takes place, the energy balance is simplified in the following form:

$$\dot{m}_h h_{h,i} + \dot{m}_c h_{c,i} = \dot{m}_h h_{h,o} + \dot{m}_c h_{c,o}$$
$$\dot{m}_h \cdot (h_{h,i} - h_{h,o}) = \dot{m}_c \cdot (h_{c,o} - h_{c,i}) \tag{2.17}$$

Hence, under these assumptions and based on Eq. (2.16) and Eq. (2.17), the heating load of the investigated condenser is equal to:

$$\dot{Q}_{cond} = \dot{m}_h \cdot \left(h_{h,i} - h_{h,o} \right) = \dot{m}_c \cdot \left(h_{c,o} - h_{c,i} \right) \tag{2.18}$$

In order to measure the performance of a heat exchanger, including the condenser, the most common method is to calculate its effectiveness. In this case, effectiveness is defined as the ratio of the actual heat transfer rate divided by the maximum heat transfer rate.

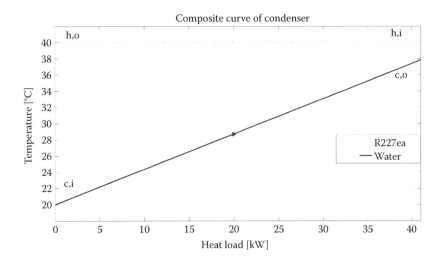

FIGURE 2.8
Composite curve of a condenser operating with R227ea rejecting heat to a water stream.

Before presenting the equation of effectiveness in terms of temperature, it is crucial to define the maximum heat transfer rate, $\dot{Q}_{max}$.

In order to specify the maximum heat transfer rate, the heat capacities, C_h and C_c, respectively, of the two streams need to be compared. If $C_h > C_c$, then the cold stream will experience the highest temperature change by being heated—in an ideal case—up to a temperature equal to the inlet of the hot side. Thus:

$$C_h = \dot{m}_h \cdot c_{p,h} > C_c = \dot{m}_c \cdot c_{p,c} : \dot{Q}_{max} = C_c \cdot (T_{h,i} - T_{c,i}) \tag{2.19}$$

On the other hand, if $C_h < C_c$, then the hot stream will experience the highest temperature change and will be cooled down to a temperature equal to the inlet of the cold side ($T_{c,i} = T_{h,o}$), hence:

$$C_h < C_c \ : \ \dot{Q}_{max} = C_h \cdot (T_{h,i} - T_{c,i}) \tag{2.20}$$

Hence, a heat exchanger's effectiveness can be defined as follows:

$$\varepsilon = \frac{\dot{Q}_{hex}}{\dot{Q}_{max}} = \begin{cases} \dfrac{C_c \cdot (T_{c,o} - T_{c,i})}{C_c \cdot (T_{h,i} - T_{c,i})} = \dfrac{T_{c,o} - T_{c,i}}{T_{h,i} - T_{c,i}}, & \text{if } C_h > C_c \\[2ex] \dfrac{C_h \cdot (T_{h,o} - T_{h,i})}{C_h \cdot (T_{h,i} - T_{c,i})} = \dfrac{T_{h,o} - T_{h,i}}{T_{h,i} - T_{c,i}}, & \text{if } C_h > C_c \end{cases} \tag{2.21}$$

However, in the case of a condenser, there is no temperature change on the hot side. Thus effectiveness is defined as follows:

$$\varepsilon_{cond} = \frac{\dot{Q}_{cond}}{\dot{Q}_{max}} = \frac{T_{c,o} - T_{c,i}}{T_{h,i} - T_{c,i}} \tag{2.22}$$

The entropy increase in the condenser can be determined by applying an entropy balance to the condenser:

$$\dot{S}_i = \dot{S}_o + \Delta\dot{S}_{cond} \Rightarrow$$

$$\Delta\dot{S}_{cond} = \dot{m}_h \cdot (s_{h,i} - s_{h,o}) + \frac{\dot{Q}_{cond}}{T_c} \tag{2.23}$$

Exergy losses in the condenser can be specified in a similar way, by applying an exergy balance

$$\dot{Ex}_i = \dot{Ex}_o + \dot{Ex}_{loss} \Rightarrow$$

$$\dot{Ex}_{loss} = (\dot{Ex}_{h,i} - \dot{Ex}_{h,o}) + (\dot{Ex}_{c,i} - \dot{Ex}_{c,o}) \Rightarrow$$

$$\dot{Ex}_{loss} = \dot{m}_h[(h_{h,i} - h_{h,o}) - T_0(s_{h,i} - s_{h,o})] + \dot{m}_c[(h_{c,i} - h_{c,o}) - T_0(s_{c,i} - s_{c,o})]$$

which, based on Eq. (2.18), can be simplified to the following:

$$\dot{Ex}_{loss} = T_0[\dot{m}_h \cdot (s_{h,o} - s_{h,i}) + \dot{m}_c \cdot (s_{c,o} - s_{c,i})] \tag{2.24}$$

Finally, the exergy efficiency of the condenser can be expressed as follows:

$$\eta_{ex,cond} = 1 - \frac{\dot{Ex}_{loss}}{\dot{Ex}_{h,i} - \dot{Ex}_{h,o}} \tag{2.25}$$

2.2.2.1 Tutorial on the Condenser

Assume a heat exchanger in which 0.4 kg/s of R227ea are being condensed at a temperature of 40°C (313.15 K). The heat is rejected to a water stream that enters the condenser with a mass flow rate of 0.55 kg/s, at a pressure of 3 bar and a temperature of 20°C (293.15 K). No pressure losses are considered. The inlet of R227ea is assumed to be saturated vapor and at the exit, in normal operation, the stream is saturated liquid. Calculate the following:

   i. The heat duty of the condenser
  ii. The effectiveness of the condenser
 iii. The entropy increase
  iv. The exergy efficiency of the condenser

Solution:

   i. For the saturated vapor of R227ea at a temperature of 313.15 K, it is possible, using RefProp (Lemmon et al. 2010), to estimate the thermodynamic properties at the inlet of the hot side:

$$h_{h,i}(T = 313.15 \ K) = 349.3158 \, kJ/kg$$
$$s_{h,i}(313.15 \ K) = 1.4859 \, kJ/kg.K$$

In a similar way, it is possible to determine the properties at the outlet (saturated liquid):

$$h_{h,o}(T = 313.15 \ K) = 246.7925 \, kJ/kg$$
$$s_{h,o}(313.15 \ K) = 1.1585 \, kJ/kg.K$$

The properties of the water at the inlet are equal to:

$$h_{c,i}\left(T = 293.15 \ K, \ P = 3 \ bar\right) = 84.1942 \, kJ/kg$$
$$s_{c,i}\left(T = 293.15 \ K, \ P = 3 \ bar\right) = 0.2964 \, kJ/kg$$

Thus, based on Eq. (2.18), the heat duty of the condenser is equal to:

$$\dot{Q}_{cond} = \dot{m}_h \cdot \left(h_{h,i} - h_{h,o}\right) = 0.4 \cdot \left(349.316 - 246.793\right) \Rightarrow$$
$$\dot{Q}_{cond} = 41.009 \ kW$$

ii. In order to determine the effectiveness of the condenser, previously, the water's outlet temperature must be specified. For this reason, an energy balance is applied in the condenser:

$$\dot{m}_h \cdot (h_{h,i} - h_{h,o}) = \dot{m}_c \cdot (h_{c,o} - h_{c,i}) \Rightarrow$$
$$h_{c,o} = h_{c,i} + \frac{\dot{m}_h}{\dot{m}_c} \cdot (h_{h,i} - h_{h,o}) = 84.1942 + \frac{0.4}{0.55}(349.316 - 246.793) \Rightarrow$$
$$h_{c,o} = 158.7566 \ kJ/kg$$

The corresponding temperature to the enthalpy $h_{c,o}$ and a pressure of 3 bar is:

$$T_{c,o}(h_{c,o} = 158.7566 \ kJ/kg, P = 3 \ bar) = 310.99 \ K(37.84°C)$$

Finally, based on Eq. (2.22), the effectiveness of the condenser can be determined:

$$\varepsilon_{cond} = \frac{T_{c,o} - T_{c,i}}{T_{h,i} - T_{c,i}} = \frac{310.99 - 293.15}{313.15 - 293.15} = 89.2\%$$

iii. The entropy at the outlet of the water stream is equal to:

$$s_{c,o}(h_{c,o} = 158.7566 \text{ kJ/kg}, P = 3 \text{ } bar) = 0.5433 \text{ } kJ/kg.K$$

The mean temperature of the cold (water) stream is equal to:

$$\bar{T}_c = \frac{T_{c,i} + T_{c,o}}{2} = 302.07 \text{ } K(28.92°C)$$

Hence, based on Eq. (2.23), the entropy increase is equal to:

$$\Delta\dot{S}_{cond} = \dot{m}_h \cdot \left(s_{h,i} - s_{h,o}\right) + \frac{\dot{Q}_{cond}}{\bar{T}_c} = 0.4 \cdot (1.4859 - 1.1585) + \frac{41009}{302.07} \Rightarrow$$

$$\Delta\dot{S}_{cond} = 0.2667 kJ/kg.K$$

iv. Based on Eq. (2.24), the exergy loss in the condenser is equal to:

$$\dot{Ex}_{loss} = T_0\left[\dot{m}_h \cdot \left(s_{h,o} - s_{h,i}\right) + \dot{m}_c \cdot \left(s_{c,o} - s_{c,i}\right)\right] \Rightarrow$$

$$\dot{Ex}_{loss} = 293.15\left[0.4 \cdot (1.1585 - 1.4859) + 0.55 \cdot (0.5433 - 0.2964)\right] \Rightarrow$$

$$\dot{Ex}_{loss} = 1.42 \text{ } kW$$

The exergy change in the hot stream is equal to:

$$\dot{Ex}_{h,i} - \dot{Ex}_{h,o} = \dot{m}_h[(h_{h,i} - h_{h,o}) - T_0(s_{h,i} - s_{h,o})] \Rightarrow$$

$$\dot{Ex}_{h,i} - \dot{Ex}_{h,o} = 0.4 \cdot [(349.316 - 246.793) - 293.15 \cdot (1.4859 - 1.1585)] \Rightarrow$$

$$\dot{Ex}_{h,i} - \dot{Ex}_{h,o} = 2.62 \text{ } kW$$

Thus, from Eq. (2.25), the exergy efficiency can be specified:

$$\eta_{ex,cond} = 1 - \frac{\dot{Ex}_{loss}}{\dot{Ex}_{h,i} - \dot{Ex}_{h,o}} = 45.78\%$$

2.2.3 Evaporator

The evaporator is the heat exchanger at which the cooling effect is produced by absorbing heat from a heat source in order to evaporate the working medium of the refrigeration cycle. There are two main types of evaporators used in practice, depending on the cooling medium: (1) evaporators that cool down a liquid, most commonly water, and (2) evaporators that cool air or other gas-phased media. In the liquid cooling evaporators, as in the case

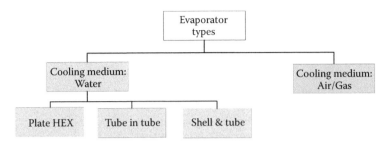

FIGURE 2.9
Main evaporator types used in refrigeration applications.

of water condensers, all kinds of heat exchangers can be used, with plate-heat exchangers, tube-in-tube, and shell-and-tube heat exchangers being the most widely applied, as shown in Figure 2.9. Especially in the case of shell-and-tube heat exchangers, fluid allocation is highly dependent on the specific application, with both cases of vaporization taking place on the shell side and cases in which the vaporizing fluid is flowing inside the tubes.

The energy and exergy analysis of an evaporator is similar to that of a condenser, so only the main equations will be presented below. Consider an evaporator similar to the one presented in Figure 2.10.

Under the assumption of being insulated from the surroundings and by applying an energy balance, the cooling load of the evaporator is found to be equal to:

$$\dot{Q}_{evap} = \dot{m}_h \cdot (h_{h,i} - h_{h,o}) = \dot{m}_c \cdot (h_{c,o} - h_{c,i}) \tag{2.26}$$

The effectiveness of the evaporator (no temperature change occurs in the cold stream, unless there is a certain level of superheating) is defined as follows:

$$\varepsilon_{evap} = \frac{\dot{Q}_{evap}}{\dot{Q}_{max}} = \frac{T_{h,i} - T_{h,o}}{T_{h,i} - T_{c,i}} \tag{2.27}$$

The entropy increase in the evaporator, in a similar way to the condenser, can be determined by applying an entropy balance:

$$\Delta \dot{S}_{evap} = \dot{m}_h \cdot (s_{c,o} - s_{c,i}) + \frac{\dot{Q}_{evap}}{T_h} \tag{2.28}$$

Finally, the exergy destruction and the exergy efficiency of the evaporator can be expressed as follows:

$$\dot{Ex}_{loss} = \dot{m}_h [T_0(s_{h,o} - s_{h,i})] + \dot{m}_c [T_0(s_{c,o} - s_{c,i})] \tag{2.29}$$

$$\eta_{ex,evap} = 1 - \frac{\dot{Ex}_{loss}}{\dot{Ex}_{h,i} - \dot{Ex}_{h,o}} \tag{2.30}$$

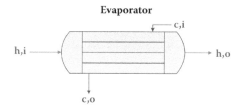

FIGURE 2.10
Schematic diagram of an evaporator.

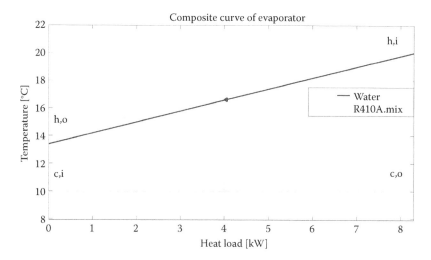

FIGURE 2.11
Composite curve of an evaporator operating with R245fa used to cool water.

Figure 2.11 presents the composite curve of one such evaporator, showing the outlet and inlet conditions of both streams.

2.2.4 Throttling Device

Throttling devices are used in refrigeration applications to reduce the pressure of the refrigerant from a high pressure at the outlet of the condenser to a low pressure of the evaporator's inlet (in the case of a single-stage vapor compression cycle) by throttling the flow. Through this process, a regulation of the flow is achieved so that proper conditions for the entrance of the fluid into the evaporator are achieved. In refrigeration, the most widely used throttling devices include thermostatic expansion valves, constant pressure expansion valves, float valves, and capillary tubes, as shown in Figure 2.12.

Consider the throttling device presented in Figure 2.13.

Applying an energy balance with the assumption of insulation from the surrounding environment and of no pressure or flow losses taking place, it is easily concluded that there is no enthalpy change in the throttling device:

$$\dot{m}_i \cdot h_i = \dot{m}_o \cdot h_o \xrightarrow{\dot{m}_i = \dot{m}_o} h_i = h_o \tag{2.31}$$

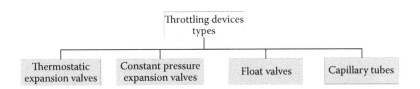

FIGURE 2.12
Types of throttling devices used in refrigeration applications.

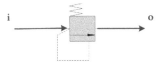

FIGURE 2.13
Schematic diagram of a throttling device.

The entropy change across the throttling device can be calculated by applying an entropy balance:

$$\Delta \dot{S}_{throttl.} = \dot{m}_i \cdot (s_o - s_i) \tag{2.32}$$

On the other hand, the exergy loss in the throttling device, as determined by an exergy balance, is equal to:

$$\dot{E}x_{loss} = -\dot{m}[(h_o - h_i) - T_0(s_o - s_i)] \tag{2.33}$$

And the corresponding exergy efficiency of the throttling device is equal to:

$$\eta_{ex,throttl.} = 1 - \frac{\dot{E}x_{loss}}{\dot{E}x_i - \dot{E}x_o} \tag{2.34}$$

2.3 The Vapor Compression Cycle

The most well-known method of cooling is based on the phase changes of a certain working medium and the heat that is transferred during these phase changes. The heat that is absorbed during the evaporation of the medium can provide the required heat QC for a cooling cycle. The heat that is released during the condensation stands for the QH of the cycle. The basic outline of one such system is presented in Figure 2.14.

In an ideal vapor compression cycle, the working medium undergoes the following processes (Figure 2.15):

- 1–2: Compression of the working vapor up to the condensation pressure. At the exit of the compressor, the refrigerant is superheated.

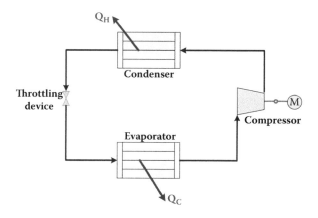

FIGURE 2.14
Simple vapor compression cycle system.

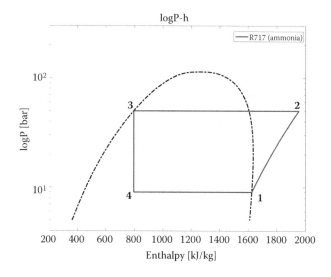

FIGURE 2.15
Ideal vapor compression cycle in logP-h diagram using refrigerant R134a.

- 2–3: Condensation of the working medium. At the exit of the condenser, in the ideal vapor compression cycle, the stream is saturated.
- 3–4: Throttling of the condensed stream until evaporation pressure. At the exit of the throttling device, the refrigerant is inside the two-phase region.
- 4–1: Evaporation of the working medium. At the exit of the evaporator, in the case of the ideal vapor compression cycle, the stream is saturated.

The vapor compression cycle is widely used for refrigeration applications covering a broad range of power outputs from a few W up to several MW per unit. As should be obvious, such systems require the existence of adequate mechanical energy to drive the compressor. A limiting factor for the selection of the operating pressures and temperatures is the ambient temperature at which heat is rejected via the condenser.

Given the fact that the components of a vapor compression cycle have been already analyzed in terms of their energy aspects as well as their exergy aspects, this section will discuss only the most key aspects regarding the performance of the cycle.

Based on the definition of the COP, as shown in Eq. (2.5), the COP is equal to:

$$COP = \frac{|\dot{Q}_C|}{|\dot{W}_{comp}|}$$

The cooling load, $\dot{Q}_C$, in a vapor compression cycle is equal to the cooling load of the evaporator (based on the points of Figure 2.14):

$$\dot{Q}_C = \dot{m} \cdot (h_1 - h_4) \tag{2.35}$$

On the other hand, the power consumed by the compressor is equal to:

$$\dot{W}_{comp} = \dot{m} \cdot (h_2 - h_1) \tag{2.36}$$

Hence, the COP for an ideal vapor compression cycle is equal to:

$$COP = \frac{|\dot{Q}_C|}{|\dot{W}_{comp}|} = \frac{\dot{m} \cdot (h_1 - h_4)}{\dot{m} \cdot (h_2 - h_1)} = \frac{h_1 - h_4}{h_2 - h_1} \tag{2.37}$$

The exergy efficiency of the cycle is easily determined by the ratio of the actual COP of the vapor compression cycle divided by the COP of the Carnot cycle operating with the same high and low temperatures:

$$\eta_{ex,vcc} = \frac{COP}{COP_{carnot}} \tag{2.38}$$

The exergy loss in the cycle is equal to the sum of all exergy losses of the components, as they were defined in their respective equations (2.13), (2.24), (2.29), and (2.33):

$$\dot{Ex}_{loss,vcc} = \dot{Ex}_{loss,comp} + \dot{Ex}_{loss,cond} + \dot{Ex}_{loss,evap} + \dot{Ex}_{loss,throttl.} \tag{2.39}$$

2.3.1 Actual Vapor Compression Cycle

In real-time applications, the vapor compression cycle deviates from the previously described behavior, mainly as a result of irreversibilities related to the compressor. Apart from deviation from isentropic compression, there are many other issues related to pressure drops across the components, resulting in the diagrams shown in Figure 2.16. Additionally, heat transfer toward and from the surroundings, should also be taken into consideration.

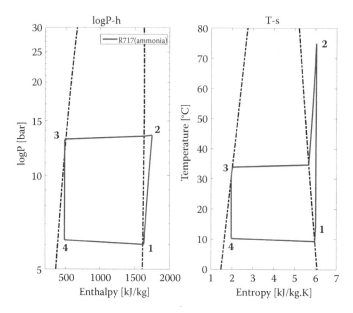

FIGURE 2.16
LogP-h and T-s diagrams of a real vapor compression cycle.

2.3.2 Subcooling and Superheating

Subcooling is the heat rejection that takes place at (or after) the condenser, resulting in the cooling of the stream from a saturated liquid to a subcooled liquid state. The main function of subcooling is to ensure that no vapor enters the throttling device. Vapor bubbles may interfere with the operation of the throttling device and disrupt flow regulation by not allowing the throttling device to expand the fluid up to the evaporation pressure properly, thus creating non-ideal conditions at the evaporator inlet.

On the other hand, superheating is additional heating that is supplied to the saturated vapor at the evaporator, turning the stream into superheated vapor. The main reason for the application of superheating is to ensure that there are no liquid droplets at the entrance of the compressor as this may lead to the malfunction of the compressor. Superheating is realized in most applications as additional heat supplied to the evaporator and thus higher cooling load. However, compression work is increasing at the same time, thus an optimization must be carried out based on the specific case study.

2.3.3 Multi-Stage and Cascade Vapor Compression Systems

Both multi-stage and cascade systems are defined as vapor compression systems that have more than one compression stage. The main difference is that cascade systems use more than one working fluid, while multi-stage systems operate with the same working fluid during all stages.

The main advantage of multi-stage systems compared to a single-stage vapor compression cycle is that multi-stage systems operate with lower pressure ratios, and thus have higher compression efficiencies, greater cooling effect, and higher flexibility. Consider a two-stage vapor compression cycle with P_{evap}, P_i, and P_{cond} as the three pressures in the cycle. In order to realize a two-stage system apart from the several compression stages, a flash receiver is used in most applications. The receiver is used for the cooling of the working fluid down to the saturated temperature of the intermediate stage while extracting the saturated vapor from the top of the tank in order to mix it with the exit stream of the low pressure compressor. A thermodynamic analysis carried out to define the optimum intermediate pressure to maximize the performance of the cycle easily determines that:

$$P_i = \sqrt{P_{cond} \cdot P_{evap}} \qquad (2.40)$$

In the case of an n-stage vapor compression cycle, the optimum pressure ratio of each stage is the following:

$$\pi_{stage} = \sqrt[n]{P_{cond} \cdot P_{evap}} \qquad (2.41)$$

2.3.4 Tutorial on the Vapor Compression Cycle

Consider an ideal vapor compression cycle (that is, no pressure losses in all components and isenthalpic expansion in the throttling device) operating with R404A with a mass flow of 0.5 kg/s. Evaporation takes place at a temperature of 10°C. The compressor has an operating pressure ratio of $\pi = 2.9$ and an isentropic efficiency of 70%. Furthermore, it is given that there is no subcooling and a superheating of 5°C. The inlet conditions of the cooling water in the condenser are as follows: There is a mass flow rate equal to 0.405 kg/s, a pressure of 2 bar, and a temperature of 20°C. On the other hand, the inlet conditions of the water in the evaporator are as follows: There is a mass flow rate equal to 3.5 kg/s, a pressure of 1 bar, and a temperature of 20°C. Determine the following:

i. The temperature, pressure, enthalpy, and entropy of each point in the vapor compression cycle
ii. The COP of the cycle
iii. The exergy efficiency
iv. The exergy losses in each component

Solution:

i. For R404A, the corresponding saturation pressure for an evaporation at 10°C is found to be equal to 8.1575 bar. It is given that the level of superheating is 5°C, thus the outlet temperature of the evaporator is 15°C. For the found pressure and temperature, it is easy to determine the enthalpy and entropy of point (1):

$$h_1 \left(T_1 = 288.15 \ K, \ P_1 = 8.1575 \ bar \right) = 374.3769 \ kJ/kg$$
$$s_1 \left(T_1 = 288.15 \ K, \ P_1 = 8.1575 \ bar \right) = 1.6859 \ kJ/kg.K$$

Based on the operating pressure ratio, the high pressure of the cycle is easily determined:

$$P_2 = \pi \cdot P_1 = 2.9 \cdot 8.1575 = 23.6567 \; bar$$

The enthalpy of point (2is) has the same entropy as point (1) and a pressure equal to P_2:

$$h_{2,is}(s_{2s} = s_1, P_{2s} = P_2) = 395.5366 \; kJ/kg$$

Hence, the enthalpy of point (2) can be derived from the isentropic efficiency of the compressor:

$$\eta_{is,comp} = \frac{h_{2,is} - h_1}{h_2 - h_1} \Rightarrow h_2 = h_1 + \frac{h_{2,is} - h_1}{\eta_{is,comp}} = 404.6051 \; kJ/kg$$

Thus the temperature and entropy at the condenser's inlet can be determined:

$$T_2 \left(P_2 = 23.6567 \; bar, \; h_2 = 404.6051 \; kJ/kg \; \right) = 340.63 \, K$$
$$s_2 \left(P_2 = 23.6567 \; bar, \; h_2 = 404.6051 \; kJ/kg \; \right) = 1.7128 \; kJ/kg.K$$

The condenser's outlet was given to be saturated liquid at 23.6567 bar, thus:

$$T_3 \left(P_3 = 23.6567 \; bar, sat.liquid \right) = 324.19 \; kJ/kg$$
$$h_3 \left(P_3 = 23.6567 \; bar, sat.liquid \right) = 277.9898 \; kJ/kg$$
$$s_3 \left(P_3 = 23.6567 \; bar, sat.liquid \right) = 1.3242 \; kJ/kg.K$$

The temperature in point (4) is known, as point (4) is within the two-phase region at 8.1575 bar, thus the temperature is 283.15K. The enthalpy at point (4) is the same as point (3) and the pressure is equal to the low pressure of the cycle. Thus:

$$s_4(P_4 = 8.1575 \; bar, h_2 = 277.9898 \; kJ/kg) = 1.3455 \; kJ/kg.K$$

Figure 2.17 shows the T-s diagram of the vapor compression cycle, based on the results of the aforementioned analysis.

ii. The power consumption of the compressor is easily determined based on Eq. (2.9):

$$\dot{W}_{comp} = \dot{m}(h_2 - h_1) = 15.1141 \; kW$$

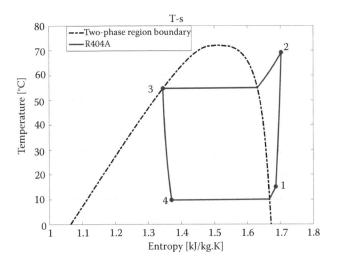

FIGURE 2.17
T-s diagram of the investigated vapor compression cycle.

The cooling power output of the cycle, based on Eq. (2.26), is equal to:

$$\dot{Q}_c = \dot{m} \cdot (h_1 - h_4) = 48.1936 \; kW$$

Hence, the COP of the cycle, based on Eq. (2.37), is equal to:

$$COP = \frac{|\dot{Q}_C|}{|\dot{W}_{comp}|} = 3.1886$$

iii. The Carnot cycle that is operating within the same high and low temperatures, based on Eq. (2.7), has a COP equal to:

$$COP_{rev} = \frac{T_C}{T_H - T_C} = \frac{T_4}{T_2 - T_4} = 4.8932$$

Hence, based on Eq. (2.38), the exergy efficiency of the vapor compression cycle is equal to:

$$\eta_{ex,vcc} = \frac{COP}{COP_{carnot}} = 65.1647\,\%$$

iv. Based on Eq. (2.13), the exergy loss in the compressor can be determined as follows ($T_0 = 298.15K$):

$$\dot{Ex}_{loss,1-2} = \dot{W}_{comp} - \dot{m}[(h_2 - h_1) - T_0(s_2 - s_1)] = 3.8726 \; kW$$

To calculate the exergy loss in the condenser the entropy at the inlet and outlet of the water stream, the following must be first be defined:

$$s_{c,i}\left(P_{cw} = 2bar, T_{c,i} = 293.15\ K\right) = 0.2964\ kJ/kg.K$$
$$h_{c,i}\left(P_{cw} = 2bar, T_{c,i} = 293.15\ K\right) = 84.100\ kJ/kg$$

The outlet enthalpy can be determined by an energy balance in the condenser:

$$\dot{m}_h \cdot \left(h_{h,i} - h_{h,o}\right) = \dot{m}_c \cdot \left(h_{c,o} - h_{c,i}\right) \Rightarrow$$
$$h_{c,o} = h_{c,i} + \frac{\dot{m}_h}{\dot{m}_c} \cdot \left(h_{h,i} - h_{h,o}\right) = 240.4153\ kJ/kg$$

And the corresponding entropy:

$$s_{c,o}(P_{cw} = 2\ bar, h_{c,o} = 240.4153\ kJ/kg) = 0.7983\ kJ/kg.K$$

Finally, based on Eq. (2.24), the exergy loss in the condenser is equal to $\left(T_0 = \dfrac{T_{c,o} + T_{c,i}}{2} = 311.85\ K\right)$:

$$\dot{Ex}_{loss,2-3} = T_0[\dot{m}_{R404A} \cdot (s_3 - s_2) + \dot{m}_{cw} \cdot (s_{c,o} - s_{c,i})] = 2.7922\ kW$$

In a similar way, for the evaporator the properties of the water stream must be defined as:

$$s_{h,i}\left(P_{hw} = 1\ bar,\ T_{h,i} = 293.15\ K\right) = 0.2964\ kJ/kg.K$$
$$h_{h,i}\left(P_{hw} = 1\ bar,\ T_{h,i} = 293.15\ K\right) = 84.006\ kJ/kg$$

The outlet enthalpy can be determined by an energy balance in the evaporator:

$$\dot{Q}_{evap} = \dot{m}_{hw} \cdot (h_{h,i} - h_{h,o}) \Rightarrow h_{h,o} = h_{h,i} - \frac{\dot{Q}_{evap}}{\dot{m}_{hw}} = 70.2364\ kJ/kg$$

Thus, the corresponding entropy at the water's outlet is equal to:

$$s_{h,o}(P_{hw} = 1\ bar, h_{h,o} = 70.2364\ kJ/kg) = 0.2492\ kJ/kg.K$$

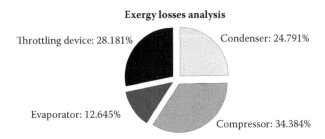

FIGURE 2.18
Contribution of each component to the total exergy losses in the cycle.

Finally, based on Eq. (2.29), the exergy loss in the evaporator is equal to $\left(T_0 = \dfrac{T_{h,o} + T_{h,i}}{2} = 291.51 \ K\right)$:

$$\dot{Ex}_{loss,4-1} = T_0[\dot{m}_{hw} \cdot (s_{h,o} - s_{h,i}) + \dot{m}_{R404A} \cdot (s_1 - s_4)] = 1.4242 \ kW$$

The exergy loss in the throttling device, based on Eq. (2.33), is equal to ($T_0 = 298.15 \ K$):

$$\dot{Ex}_{loss,3-4} = -\dot{m}[(h_4 - h_3) - T_0(s_4 - s_3)] = 3.1740 \ kW$$

Hence, the overall exergy loss in the investigated cycle is, based on Eq. (2.39), equal to:

$$\dot{Ex}_{loss,vcc} = \dot{Ex}_{loss,1-2} + \dot{Ex}_{loss,2-3} + \dot{Ex}_{loss,3-4} + \dot{Ex}_{loss,4-1} \Rightarrow$$
$$\dot{Ex}_{loss,vcc} = 11.2630 \ kW$$

Figure 2.18 presents the percentages of total exergy loss for each component in the investigated vapor compression cycle.

2.4 Absorption Cooling Cycle

Absorption cycles have similarities to vapor compression cycles in that both use an evaporator, a condenser, and an expansion device. The main difference lies in the fact that absorption machines are thermally driven, using heat to increase the pressure of the fluid exiting the evaporator and to deliver it to the condenser. Absorption chillers are commonly powered by waste heat, solar energy, geothermal sources, or the combustion of biomass or natural gas.

The cooling effect is, as in mechanically driven chillers, produced in the evaporator. Apart from the condenser and the expansion device, absorption chillers, in their simplest configuration, implement two more heat exchangers, the generator, and the absorber. By adding heat to the generator, the refrigerant, which is volatile and dissolved in a carrier, is separated from the carrier solution. The refrigerant is then redissolved in the carrier at the absorber. Absorption is an exothermic process, resulting in the need for heat rejection in the absorber. Between the absorber and the generator, as shown in Figure 2.19, a pump is installed to elevate the pressure of the solution. However, the impact of the pump from an energy point of view is minor, and in typical cases does not exceed the 1% of the chiller's nominal cooling output. On the other hand, to reduce the pressure of the solution exiting the generator before it enters the absorber, an expansion device is implemented. In order to enhance the system's energy utilization, a heat recovery unit is installed between the absorber and the generator to transfer heat from the hot (and weak in terms of concentration) solution exiting the generator to the cold (and strong in terms of concentration) solution exiting the absorber.

The commercial range of applications for absorption chillers extends from small domestic units (with a cooling output of few hundred W) to large industrial chillers (with cooling capacities of tens of MWs). The main advantages of absorption chillers are their long lifespans and their efficient part-load operation. Furthermore, their most common working pairs, water-lithium bromide and ammonia-water, which will be discussed also in Chapter 5, are both refrigerants with minimal ODP and GWP, especially when compared to the less environmentally friendly refrigerants used in mechanical compression cycles. In terms of cost consideration, the relatively high initial cost of absorption chillers is considered unattractive for wider application, yet the low operating costs offer potential for broader use of such systems. Their main drawbacks are the temperature restrictions relative to their working pairs, their heavier weight in comparison to vapor compression cooling systems, and their crystallization issues (Air Conditioning Refrigeration Institute 1998).

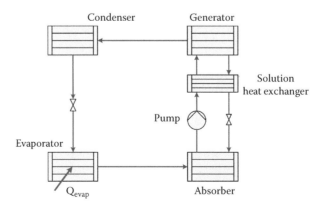

FIGURE 2.19
Schematic of a typical single-stage absorption cycle.

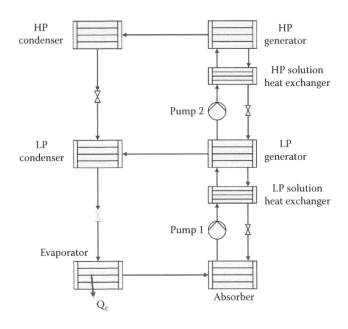

FIGURE 2.20
Schematic of a typical double-stage absorption cycle.

2.4.1 Multiple Stage Absorption Chillers

In order to exploit heat sources more efficiently and to enhance the absorption chiller's COP, multi-stage absorption chillers have been introduced. The number of stages is equivalent to the number of generator heat exchangers that are installed in the chiller at different temperature levels. However, due to their complexity and the increased cost of such systems, only double-stage absorption chillers have reached commercialization.

Double-stage absorption chillers generate almost twice the refrigerant vapor generated in the respective single-stage unit, and thus a significant increase in the COP is reported, allowing double-stage chillers to more efficiently exploit higher temperature heat sources (Gordon and Ng 2000). A schematic of a typical double-stage absorption chiller is presented in Figure 2.20. Apart from enhancing the COP, multi-stage configurations also allow for a higher temperature lift (the temperature difference between the condenser and the evaporator) (Granryd et al. 2009).

2.4.2 Energy Considerations for an Absorption Cycle

In order to carry out an energetic analysis of the absorption process, an enthalpy concentration diagram for the working binary mixture must be used, as shown in Figure 2.21. Use of one such diagram is explained practically in the absorption's cycle tutorial.

Figure 2.22 (a) and the corresponding ideal T-s diagram Figure 2.22 (b) present the ideal (and most simplified) layout of an absorption chiller. As shown in Figure 2.22 (b), line 4-1 represents the evaporation process, in which the cooling effect is produced. Cycle 1-2-3-4 is comparable with a vapor compression cycle, the main difference is that it is coupled with solution cycle 5-6-7-8. Lines 5-6 represents the desorption process, while lines 7-8 represent absorption.

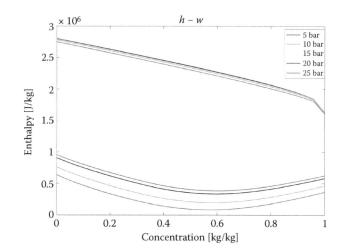

FIGURE 2.21
The enthalpy concentration diagram for ammonia water.

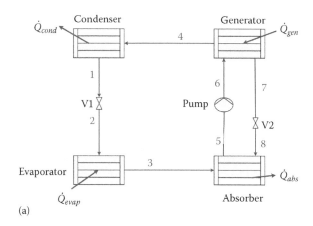

(a)

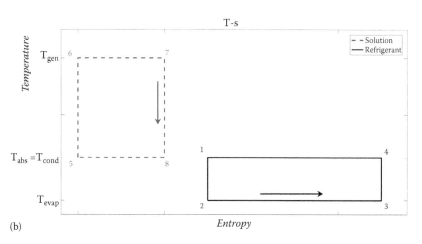

(b)

FIGURE 2.22
(a) Schematic of a single-stage absorption cycle and (b) its corresponding T-s diagram.

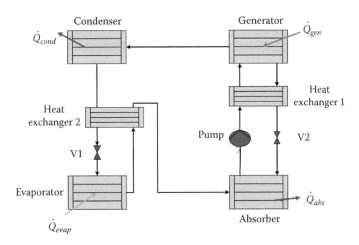

FIGURE 2.23
Single-stage absorption chiller with two internal heat exchangers.

In actual applications, two heat exchangers are implemented in the system to transfer heat internally and reduce external heat loads. The energy and exergy analysis presented below are based on the numbering found in the schematic in Figure 2.23.

Condenser:
$$\dot{m}_1 h_6 = \dot{Q}_{cond} + \dot{m}_1 h_1 \,(\text{energy balance}) \qquad (2.42)$$

Heat exchanger 2:
$$\dot{m}_1 h_1 + \dot{m}_1 h_4 = \dot{m}_1 h_2 + \dot{m}_1 h_5 \,\left(\text{energy balance}\right) \Rightarrow$$

$$h_1 + h_4 = h_2 + h_5 \qquad (2.43)$$

Exp. valve 1 (V1):
$$h_2 = h_3 \,(\text{energy balance}) \qquad (2.44)$$

Evaporator:
$$\dot{m}_1 h_3 + \dot{Q}_{evap} = \dot{m}_1 h_4 \,(\text{energy balance}) \qquad (2.45)$$

Absorber:
$$\dot{m}_1 h_5 + \dot{m}_{10} h_{12} = \dot{Q}_{abs} + \dot{m}_7 h_7 \,(\text{energy balance}) \qquad (2.46)$$

$$\dot{m}_1 + \dot{m}_{10} w_{ws} = \dot{m}_7 w_{ss} \,(\text{mass balance}) \qquad (2.47)$$

Heat exchanger 1:
$$\dot{m}_7 h_8 + \dot{m}_{10} h_{12} = \dot{m}_7 h_9 + \dot{m}_{10} h_{11} \,(\text{energy balnce}) \qquad (2.48)$$

Solution circulation ratio:

$$f = \frac{w_6 - w_{ss}}{w_{ss} - w_{ws}}$$

(2.49)

$$\xrightarrow{\text{(2.48),(2.49)}} fh_8 + (f-1)h_{12} = fh_9 + (f-1)h_{11}$$

(2.50)

Exp. valve 2 (V2):

$$h_{11} = h_{12} \,(\text{energy balance})$$

(2.51)

Pump:

$$\dot{m}_7 h_7 + \dot{W}_{pump} = \dot{m}_7 h_8 \,(\text{energy balnce})$$

(2.52)

Generator:

$$\dot{Q}_{gen} + \dot{m}_7 h_9 = \dot{m}_1 h_6 + \dot{m}_{10} h_{10} \,(\text{energy balnce})$$

(2.53)

Based on the above equations and the assumption of no external heat losses, the overall energy balance in the chiller can be expressed as follows:

$$\dot{Q}_{gen} + \dot{W}_{pump} + \dot{Q}_{evap} = \dot{Q}_{cond} + \dot{Q}_{abs}$$

(2.54)

In the case of absorption cooling, the system is thermally driven, as already mentioned. Thus, the COP for an absorption cycle, compared to the definition of the COP for mechanically driven refrigeration systems, is modified based on the following equation:

$$COP_{abs,refr.} = \frac{cooling\ output\ of\ the\ evaporator}{thermal\ input\ at\ the\ generator} = \frac{\dot{Q}_{evap}}{\dot{Q}_{gen}}$$

(2.55)

For heat pump mode, the COP of an absorption machine is defined as the following:

$$COP_{abs,HP} = \frac{heat\ rejection\ at\ the\ absorber\ and\ the\ condenser}{thermal\ input\ at\ the\ generator\ and\ the\ evaporator} \Rightarrow$$

$$COP_{abs,HP} = \frac{\dot{Q}_{abs} + \dot{Q}_{cond}}{\dot{Q}_{gen} + \dot{Q}_{evap}}$$

(2.56)

In general, the COP of absorption machines is lower than that of mechanical compression units. However, one big advantage of absorption is the potential to exploit decentralized thermal sources, such as waste heat sources, in a more environmentally friendly way. Furthermore, the fact that compression of the working fluid takes place in a liquid state reduces the impact of compression work to a minimum, while in mechanically driven refrigeration, energy consumption in the compressor is 20–50% of the net cooling output (Gordon and Ng 2000).

2.4.3 Exergy Considerations

As already mentioned, the analysis of exergy losses per component in the absorption chiller is expressed based on the numbering in Figure 2.23. At this point, it must be mentioned that the exergy losses in the expansion devices have been excluded from this analysis because they are negligible.

Condenser:

$$\Delta \dot{E}x_{cond} = \dot{m}_1 (ex_6 - ex_1) \tag{2.57}$$

Heat exchanger 2:

$$\Delta \dot{E}x_{hex,2} = \dot{m}_1 \left[(ex_1 - ex_2) + (ex_4 - ex_5) \right] \tag{2.58}$$

Evaporator:

$$\Delta \dot{E}x_{evap} = \dot{m}_1 (ex_3 - ex_4) + \dot{Q}_{evap} \left(1 - \frac{T_a}{T_{evap}} \right) \tag{2.59}$$

Absorber:

$$\Delta \dot{E}x_{abs} = \dot{m}_1 ex_5 + \dot{m}_{10} ex_{12} - \dot{m}_7 ex_7 \tag{2.60}$$

Heat exchanger 1:

$$\Delta \dot{E}x_{hex,1} = \dot{m}_7 (ex_8 - ex_9) + \dot{m}_{10} (ex_{10} - ex_{11}) \tag{2.61}$$

Pump:

$$\Delta \dot{E}x_{pump} = \dot{m}_7 (ex_7 - ex_8) + \dot{W}_{pump} \tag{2.62}$$

Generator:

$$\Delta \dot{E}x_{gen} = \dot{m}_7 ex_9 - \dot{m}_{10} ex_{10} - \dot{m}_1 ex_6 + \dot{Q}_{gen} \left(1 - \frac{T_a}{T_{gen}} \right) \tag{2.63}$$

Total:

$$\Delta \dot{E}x_{tot} = \sum \Delta \dot{E}x_i \tag{2.64}$$

In order to specify the Carnot efficiency of an absorption refrigeration machine, an energy flow balance is applied:

$$\dot{Q}_{evap} + \dot{Q}_{gen} = \dot{Q}_{abs} + \dot{Q}_{cond} \tag{2.65}$$

Furthermore, given the fact that in an ideal case the cycle is reversible, the following can be derived:

$$0 = \frac{\dot{Q}_{gen}}{T_{gen}} - \frac{\dot{Q}_{abs}}{T_{abs}} - \frac{\dot{Q}_{cond}}{T_{cond}} + \frac{\dot{Q}_{evap}}{T_{evap}} \xrightarrow{T_{abs}=T_{gen}}$$

$$-\frac{\dot{Q}_{gen}}{T_{gen}} = -\frac{\dot{Q}_{abs}}{T_{abs}} - \frac{\dot{Q}_{cond}}{T_{abs}} + \frac{\dot{Q}_{evap}}{T_{evap}} \xrightarrow{*T_{abs}}$$

$$-T_{abs}\frac{\dot{Q}_{gen}}{T_{gen}} = -\dot{Q}_{abs} - \dot{Q}_{cond} + T_{abs}\frac{\dot{Q}_{evap}}{T_{evap}} \xrightarrow{(2.63)}$$

$$-\dot{Q}_{gen}\frac{T_{abs}}{T_{gen}} = -\dot{Q}_{gen} - \dot{Q}_{evap} + \dot{Q}_{evap}\frac{T_{abs}}{T_{evap}} \Longrightarrow$$

$$\dot{Q}_{evap} = \frac{\dot{Q}_{gen}\left(1 - \dfrac{T_{abs}}{T_{gen}}\right)}{\dfrac{T_{abs}}{T_{evap}} - 1} \tag{2.66}$$

Thus, the Carnot COP for absorption refrigeration can be defined as follows:

$$COP_{Carnot,abs,refr} = \frac{\dot{Q}_{evap}}{\dot{Q}_{gen}} \xrightarrow{(2.45)}$$

$$COP_{Carnot,abs,refr} = \frac{\dfrac{1}{T_{abs}} - \dfrac{1}{T_{gen}}}{\dfrac{1}{T_{evap}} - \dfrac{1}{T_{abs}}} \tag{2.67}$$

Hence, the second law efficiency can easily be derived as follows:

$$\eta = \frac{COP_{abs,refr}}{COP_{Carnot,abs,refr}} \tag{2.68}$$

On the other hand, the Carnot efficiency for heat pump operation is equal to:

$$COP_{Carnot,abs,HP} = 1 + COP_{Carnot,abs,refr} \tag{2.69}$$

2.4.4 Tutorial on the Absorption Cycle

Assume the absorption cycle that is presented in Figure 2.23. The cycle is operating with a NH_3-H_2O working pair and the cooling capacity is equal to 100 kW. The evaporation pressure is 2.5 bar and the high pressure in the cycle is equal to 12 bar. The mass flow rate, $\dot{m}_1$, through the condenser is equal to 0.085 kg/s. The concentration of the strong solution is 0.5 kg/kg, and the respective value for the weak solution is a given equal to 0.35 kg/kg. The temperature of point 6 is 5 K higher than the temperature of the strong solution entering the generator. Pressure losses and other irreversibilities—including pump efficiency—are not considered in this tutorial. Determine the following:

i. Pressure, temperature, and enthalpy in every point
ii. The mass flow rates of the weak and the strong solutions
iii. The COP of the cycle

Solution:

i. Based on the fact that the condensation pressure is 12 bar, point 1 can be determined in both Figure 2.24 and Figure 2.25 (no subcooling is considered, thus point 1 is saturated liquid).

$$T_1 = T\left(p = 12 \ bar, \ w = 100\%\right) = 304.1 \ K$$
$$h_1 = h\left(p = 12 \ bar, \ w = 100\%\right) = 489.4 \ kJ/kg$$

At the exit of the evaporator, point 4, there is saturated vapor at a pressure of 2.5 bar, thus both the temperature and the enthalpy can be determined:

$$T_4 = T\left(p = 2.5 \ bar, \ w = 100\%\right) = 259.5 \ K$$
$$h_4 = h\left(p = 2.5 \ bar, \ w = 100\%\right) = 1589.3 \ kJ/kg$$

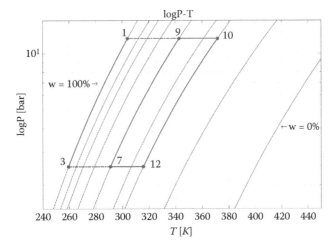

FIGURE 2.24
Logarithmic pressure temperature diagram for the investigated system.

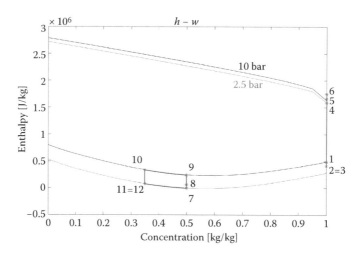

FIGURE 2.25
Enthalpy concentration diagram for the tutorial.

The cooling capacity was a given equal to 100 kW, hence:

$$Q_{evap} = \dot{m}_1(h_4 - h_3) \Rightarrow$$

$$h_3 = h_4 - \frac{Q_{evap}}{\dot{m}_1} = 412.8 \ kJ/kg$$

(The temperature of point 3 is equal to the temperature of point 4). Given the fact that through the expansion valve there is no enthalpy change considered, the enthalpy of point 2 is considered equal to point 3 ($h_3 = h_2$). For the determined enthalpy and pressure of point 2, the temperature can also be determined from the diagram in Figure 2.25.

$$T_2 = T\left(p = 12 \ bar, h_2 = 412.8 \ kJ/kg\right) = 288.0 \ K$$

At the exit of the evaporator, point 4, there is saturated vapor at a pressure of 2.5 bar, thus both the temperature and the enthalpy can be determined:

$$T_4 = T\left(p = 2.5 \ bar, w = 100\%\right) = 259.5 \ K$$
$$h_4 = h\left(p = 2.5 \ bar, w = 100\%\right) = 1589.3 \ kJ/kg$$

Using Eq. (2.43), the following can be determined:

$$h_1 + h_4 = h_2 + h_5 \Rightarrow$$
$$h_5 = h_1 + h_4 - h_2 = 1665.9 \ kJ/kg$$

Hence, the temperature of point 5 can also be determined:

$$T_5 = T\left(p = 2.5 \ bar, \ h_2 = 1665.9 \ kJ/kg\right) = 291.8 \ K$$

Based on the fact that the low pressure is 2.5 bar, point 7 can be determined in both Figure 2.24 and Figure 2.25 (point 7 is saturated liquid).

$$T_7 = T\left(p = 2.5 \ bar, \ w = 50\%\right) = 291.1 \ K$$
$$h_7 = h\left(p = 2.5 \ bar, \ w = 50\%\right) = -0.8 \ kJ/kg$$

And in a similar way for the weak solution of point 12:

$$T_{12} = T\left(p = 2.5 \ bar, \ w = 35\%\right) = 315.8 \ K$$
$$h_{12} = h\left(p = 2.5 \ bar, \ w = 35\%\right) = 80.2 \ kJ/kg$$

Through the expansion valve there is no enthalpy change considered, hence the enthalpy of point 11 is considered equal to point 12 ($h_{11} = h_{12}$). For the determined enthalpy and pressure of point 11, the temperature can also be determined from the diagram in Figure 2.25.

$$T_{11} = T\left(p = 12 \ bar, \ h_{11} = 80.2 \ kJ/kg\right) = 315.6 \ K$$

For point 9 (strong solution), the concentration and the pressure is given, hence its enthalpy and temperature can be easily determined (saturated liquid):

$$T_9 = T\left(p = 12 \ bar, \ w = 50\%\right) = 342.9 \ K$$
$$h_9 = h\left(p = 12 \ bar, \ w = 50\%\right) = 243.5 \ kJ/kg$$

In a similar way for the weak solution of point 10:

$$T_{10} = T\left(p = 12 \ bar, \ w = 35\%\right) = 371.7 \ K$$
$$h_{10} = h\left(p = 12 \ bar, \ w = 35\%\right) = 338.9 \ kJ/kg$$

The solution circulation ratio can be specified by the concentrations of the solutions:

$$f = \frac{w_6 - w_{ss}}{w_{ss} - w_{ws}} = 3.33$$

Hence, from Eq. (2.50), the enthalpy of point 8 can be determined:

$$fh_8 + (f-1)h_{12} = fh_9 + (f-1)h_{11} \Rightarrow$$
$$h_8 = 62.4 \ kJ/kg$$

Hence, the temperature of point 8 is equal to:

$$T_8 = T\left(p = 12 \ bar, \ h_2 = 62.4 \ kJ/kg \ \right) = 304.6 \ K$$

It is given that the temperature of point 6 is 5 K higher than the temperature of the strong solution entering the generator (point 9):

$$T_6 = T_9 + 5 = 347.9 \ K$$

Hence, the corresponding enthalpy of point 6 is equal to:

$$h_1 = h\left(p = 12 \ bar, \ T_6 = 347.9 \ K, \ w = 100\%\right) = 1755.3 \ kJ/kg$$

   The results of the aforementioned analysis are listed for every point of the cycle in Table 2.1.

ii. The mass flow of the strong solution can be easily determined from the solution circulation ratio as follows:

$$\dot{m}_7 = f \cdot \dot{m}_1 = 0.283 \ kg/s$$

The mass flow of the weak solution can be easily determined from the solution circulation ratio as follows:

$$\dot{m}_{10} = (f-1) \cdot \dot{m}_1 = 0.198 \ kg/s$$

TABLE 2.1

Overview of the Main Properties for Every Point in the Cycle

Point	Pressure (bar)	Temperature (K)	Enthalpy (kJ/kg)	Concentration (%)
1	12	304.1	489.4	100
2	12	288.0	412.8	100
3	2.5	259.5	412.8	100
4	2.5	259.5	1,589.3	100
5	2.5	291.8	1,665.9	100
6	12	347.9	1,755.3	100
7	2.5	291.1	−0.8	50
8	12	304.6	62.4	50
9	12	342.9	243.5	50
10	12	371.7	338.9	35
11	12	315.6	80.2	35
12	2.5	315.8	80.2	35

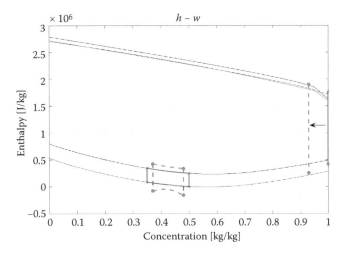

FIGURE 2.26
Overview of the cycle modification due to irreversibilities.

iii. To determine the cycle's COP, the thermal load of the generator must first be calculated using the energy balance from (2.53):

$$\dot{Q}_{gen} = \dot{m}_1 h_6 + \dot{m}_{10} h_{10} - \dot{m}_7 h_9 = 147.43 \; kW$$

The COP can then be determined using Eq. (2.55):

$$COP_{abs,refr.} = \frac{\dot{Q}_{evap}}{\dot{Q}_{gen}} = 0.678$$

2.4.5 Real Cycle

Compared to the theoretical case discussed previously, in actual applications there are some deviations based on irreversibilities in the cycle. The main issues stem from the losses in the internal and external heat transfer processes, the non-ideal absorption and desorption processes, the pressure drop, and the non-condensable gases. Figure 2.26 presents an overview of the changes (dotted lines) in the absorption cycle in an enthalpy concentration diagram as a result of the aforementioned irreversibilities.

2.5 Adsorption Cooling Cycle

Adsorption involves the distribution of molecules between two phases, one of which is a solid and the other either a liquid or a gas. Adsorption is a well known and applied technology in water treatment, in the purification of liquids, and in gas cleaning processes.

However, it was not until the 1990s that researchers started to investigate the potential use of adsorption as a refrigeration cycle (Zhang and Wang 1997; Suzuki 1993; Cho and Kim 1992).

The adsorption cycle consists of two main phases: desorption and adsorption. Initially, the system is at a low pressure and temperature; the adsorbent contained in the adsorber is saturated with refrigerant. In order to regenerate it, the desorption phase is initiated. The adsorbent is heated by an external heat source, driving the refrigerant out of the adsorbent and increasing system's pressure. The desorbed refrigerant condenses in the condenser, producing heat. The next phase is adsorption. The adsorber is cooled back to an ambient temperature and is connected to the evaporator, which causes the refrigerant's adsorption. Figure 2.27 presents the basic layout of a two-bed adsorption cycle, as described previously. Figure 2.28 presents the corresponding Clapeyron diagram for a typical adsorption refrigeration cycle. Depending on the use of the adsorption system, either heat produced from the highly exothermic adsorption process or cooling energy resulting from the endothermic evaporation process in the evaporator can be utilized.

Adsorption cooling has advantages over absorption, which can also be powered by low grade heat sources, in the simplicity of the equipment used, since there is no requirement

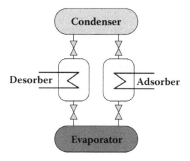

FIGURE 2.27
Adsorption process layout.

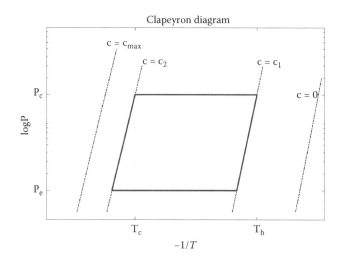

FIGURE 2.28
Clapeyron diagram for the adsorption cycle.

for the use of a rectifier. Furthermore, adsorption allows the use of heat sources with very low temperatures, while absorption systems require heat sources with at least 70°C (Wang et al. 2009b). Compared to conventional vapor compression, adsorption systems enable the utilization of waste heat or solar energy, and have lower operational costs, no moving parts, and no vibrations (Wang and Oliveira 2006; Jiangzhou et al. 2005; Wade et al. 1992).

The main drawback of adsorption technology is the limited adsorption capacity of the adsorbents, resulting in a low COP and low specific cooling power (SCP) (Richardson et al. 2007).

In terms of categorization, adsorption can be divided into to two types based on the nature of surface forces: physical adsorption and chemical adsorption.

Physical adsorption is caused by relatively weak van der Waals forces that hold the adsorbate at the surface. Multiple layers may be formed since physical adsorption is not selective. For most adsorbents, the adsorption heat released during adsorption of the refrigerant is similar to the condensation heat of the refrigerant (Wang and Oliveira 2006). Due to the small heat of adsorption, this type of adsorption is stable only at temperatures below 150°C (Inglezakis and Poulopoulos 2006). In terms of kinetics, physical adsorption is fast and does not require any energy input to be initiated, whereas chemical adsorption may need a certain level of energy in order to be activated (Murzin 2013). As a result of the weak forces applied, physical adsorption is a reversible process, unlike chemical adsorption (Richardson et al. 2007).

In chemical adsorption much greater forces are applied than in physical adsorption. The adsorbent reacts chemically with the adsorbate producing new types of molecules. A certain level of electron exchange is involved (Ruthven 2008). One of the most important differences between the two types of adsorption is that chemical adsorption is confined to a monolayer, unlike physical adsorption (Rouquerol et al. 2014).

2.5.1 Energy and Exergy Analysis of the Adsorption Cycle

Figure 2.29 presents the quantification of the exergy and energy parameters of the adsorption refrigeration cycle.

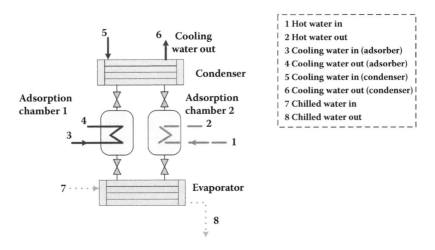

FIGURE 2.29
Schematic of a conventional two-bed adsorption refrigeration system.

A more detailed analysis of the cycle, including the adsorption and desorption processes, will be thoroughly discussed in Chapter 6. At this point, the cycle will be investigated from the point of view of the heat that is exchanged with the water streams. Hence, the cooling power produced in the evaporator is simply equal to:

$$\dot{Q}_{evap} = \dot{m}_7 c_{p,7} (T_7 - T_8) \tag{2.70}$$

The heat input of the cycle is equal to the heat delivered in the chamber that is on desorption mode, and thus the following relation can be extracted:

$$\dot{Q}_{des} = \dot{m}_1 c_{p,1} (T_1 - T_2) \tag{2.71}$$

The COP for an adsorption refrigeration cycle is equal to:

$$COP = \frac{\dot{Q}_{evap}}{\dot{Q}_{des}} \tag{2.72}$$

The Carnot COP for an adsorption cycle is calculated by the following expression (Sharonov and Aristov 2008; Meunier et al. 1998; San and Hsu 2009):

$$COP_{carnot} = \frac{T_{evap}\left(T_{des} - T_{cond}\right)}{T_{des}\left(T_{cond} - T_{evap}\right)} \tag{2.73}$$

Hence, the second law efficiency can easily be determined as follows:

$$\eta = \frac{COP_{abs,refr}}{COP_{Carnot,abs,refr}} \tag{2.74}$$

2.6 Desiccant Cooling Cycle

Given that buildings with low energy demand are often equipped with ventilation systems, desiccant cooling can be a promising option for building air conditioning applications, especially since it can be powered by low grade heat sources, like solar thermal energy and waste heat. In fact, solid sorption systems are already available on the market for such applications.

The most widely used desiccants are lithium chloride (liquid desiccant), silica gel, and zeolites (solid desiccants). Desiccant materials will be discussed more thoroughly in their respective chapter since their properties have a massive impact on the overall performance of a desiccant cooling system.

Desiccant systems base their operation on the use of a rotary dehumidifier, where the dehumidification of air takes place. The dry air is then cooled down in consecutive heat exchangers before entering the cooled room. A low grade heat source is required to regenerate the desiccant material. A simple cycle is shown in Figure 2.30. Air enters the desiccant wheel (point 1), where dehumidification and heating of the stream takes place by a return stream (point 8). The dehumidified air is then led through a rotary regenerator where it is cooled down. After the regenerator (point 3), the dry air is further cooled down in an evaporative cooler before (point 4) it enters the room. At the same time, air is removed from the room to be regenerated (point 5). The return air is initially cooled down by an evaporative cooler before it is led to the regenerator to heat up from the hot dry air of point 2. Downdraft the regenerator, the return air is further heated by an external source before entering the desiccant wheel (point 8). In the desiccant wheel, the moisture adsorbed by the desiccant to dry the entering air is then desorbed to the return air, which eventually is ejected into the environment. The process is also shown in the psychometric chart in Figure 2.31. The system presented is operated in an open cycle in ventilation mode.

Apart from the ventilation mode, desiccant cooling systems can be operated in closed loops through recirculation of the air. Figure 2.32 presents a desiccant cooling system in recirculation mode, while the corresponding psychometric chart is presented in Figure 2.33. In this configuration the air recirculates in a closed loop, while regeneration is achieved by an external air supply. In this module, the fact that the desiccant wheel operates with outside

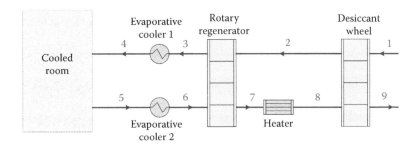

FIGURE 2.30
An experimental desiccant system. (Adapted from Kanoğlu, Mehmet et al., *Applied Thermal Engineering*, 24 (5): 919–932, 2004.)

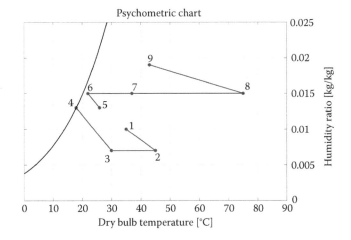

FIGURE 2.31
Psychometric chart for the open desiccant cooling system (ventilation mode) of Figure 2.30.

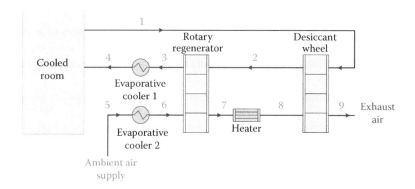

FIGURE 2.32
Desiccant cooling system in recirculation mode.

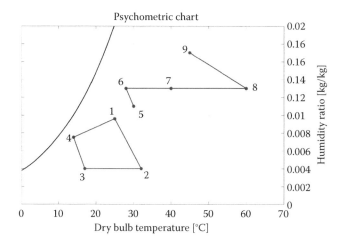

FIGURE 2.33
Psychometric chart for the desiccant cooling system of Figure 2.32.

air results in higher temperatures than in the ventilation case. On the other hand, this side effect is counterbalanced by the fact that the continuous recirculation of cold air reduces energy consumption, resulting in COP values comparable with the ventilation module.

2.6.1 Energy Considerations for Desiccant Cooling

The desiccant cooling system will be investigated thoroughly in a following chapter. This section presents an analysis of the main energy and exergy aspects of such systems. The indexing in the following equations is based on Figure 2.30. Given the fact that desiccant cooling systems are thermally driven, the COP is defined in a similar way to the respective performance indicator for absorption:

$$COP = \frac{\dot{Q}_{cr}}{\dot{Q}_{reg}} = \frac{h_5 - h_4}{h_8 - h_7} \tag{2.75}$$

To further evaluate the performance of the system, the effectiveness of its components is used. The effectiveness of the regenerator, based on the definition of the effectiveness for a heat exchanger, is defined as follows (Kanoğlu et al. 2004):

$$e_{regen} = \frac{T_2 - T_3}{T_2 - T_6} \tag{2.76}$$

In the case of the desiccant wheel, Van den Bulck et al. (1988) proposed a definition of effectiveness using the humidity ratio of the streams:

$$e_{dw} = \frac{\omega_1 - \omega_2}{\omega_1 - \omega_{2,id}} \tag{2.77}$$

where $\omega_{2,id}$ refers to the humidity ratio of exit stream 2 in an ideal case. Finally, the effectiveness of the two evaporative coolers is defined as follows:

$$e_{ev.c,1} = \frac{T_3 - T_4}{T_3 - T_{wb,3}} \tag{2.78}$$

$$e_{ev.c,2} = \frac{T_5 - T_6}{T_5 - T_{wb,5}} \tag{2.79}$$

where T_{wb} refers to the wet bulb temperature of the moist air.

Concerning the COP for the reversible case of an open desiccant cooling system, Lavan et al. (1982) proposed the following ratio:

$$COP_{rev} = \left(1 - \frac{\overline{T}_{cond}}{\overline{T}_{hs}} \right) \left(\frac{\overline{T}_{evap}}{\overline{T}_{cond} - \overline{T}_{evap}} \right) \tag{2.80}$$

where the equivalent temperatures $\overline{T}$ for each case are defined as follows:

$$\overline{T}_{cond} = \frac{h_9 - h_1 - \left(\omega_6 - \omega_3 \right) * h_w}{s_9 - s_1 - \left(\omega_6 - \omega_3 \right) * s_w} \tag{2.81}$$

where h_w and s_w refer to the enthalpy and entropy of liquid water, respectively.

$$\overline{T}_{hs} = \frac{h_7 - h_8}{s_7 - s_8} \tag{2.82}$$

$$\overline{T}_{evap} = \frac{h_4 - h_5 - (\omega_5 - \omega_4) * h_w}{s_4 - s_5 - (\omega_5 - \omega_4) * s_w} \tag{2.83}$$

Finally, given the aforementioned reversible COP definition, the second law efficiency of the open desiccant cooling system can be easily defined as:

$$\eta = \frac{COP}{COP_{rev}} \tag{2.84}$$

2.7 Organic Rankine Cycles

The organic Rankine cycle (ORC) is a modification of a conventional Rankine cycle. The main difference is that an organic fluid is used instead of water steam. In the last decade, the ORC has gathered scientific interest as a promising technology for utilizing the power from low grade heat sources such as solar, geothermal energy, and waste heat. For lower temperature heat sources, the conventional water steam Rankine cycle is not technically feasible, or leads to poor efficiencies. On the other hand, the ORC poses several advantages including: the potential for low temperature heat recovery due to the lower boiling point of the working fluids used; the operation of the expander under less severe conditions; and the use of dry organic fluids, which reduce the need for superheating (Quoilin et al. 2010; Pei et al. 2010; Tchanche et al. 2011).

Figures 2.34 and 2.35 present a typical layout and the corresponding T-s and logP-h diagrams for the cycle, respectively. As shown, the following basic processes take place in an ORC:

- 1–2: The high pressure PH working fluid is heated in the evaporator. Ideally, the process is isobaric. However, in actual applications a certain pressure drop occurs depending on stream conditions and the geometry of the evaporator. Furthermore, depending on the temperature levels of the heat source powering the ORC and the working fluid, a certain level of superheating may be applied.

- 2–3: The fluid is expanded in the expander up to the low pressure PL to produce, via the generator, the electrical power output of the cycle.

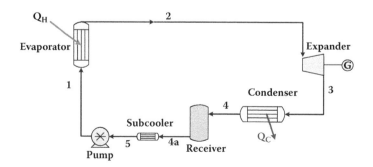

FIGURE 2.34
Standard ORC layout.

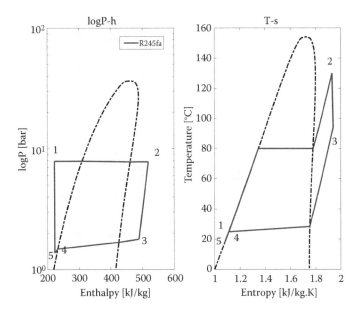

FIGURE 2.35
LogP-h and T-s diagrams for a conventional ORC system.

- 3–4: The initially superheated vapor is cooled and condensed up to saturated liquid condition in the condenser. As in all heat exchangers, there is usually a significant pressure drop during this process.
- 4–4a: A receiver tank is applied in most cases in order to ensure that the entrance of the subcooler is saturated—even in a transient state during which the exit of the condenser is not saturated.
- 4–5: In order to protect the downdraft pump from cavitation issues, a subcooler is normally implemented, cooling down the working fluid in the range of 3–10 K.
- 5–1: The pump of the cycle serves two functions, compression of the working fluid up to the high pressure PH, and its circulation throughout the cycle.

Apart from the typical subcritical ORC, research has been conducted into the transcritical ORC, in which the working fluid is compressed to pressures higher than their critical and, in the evaporator, the fluid, instead of gradually evaporating, switches to a supercritical state. Although the transcritical ORC results in higher pressures in the cycle, it improves the energetic and exergetic efficiency of the cycle and thus is worth investigating in small-scale applications. This will be discussed more thoroughly in the respective chapter (Schuster et al. 2010; Roumpedakis et al. 2015). Figure 2.36 presents one such cycle and a respective subcritical cycle using R227ea as the working fluid.

2.7.1 Energy and Exergy Considerations for the ORC

The analysis presented below is based on the ORC layout presented in Figure 2.34.

Condenser:
$$\dot{Q}_{cond} = \dot{m}(h_3 - h_4) \tag{2.85}$$

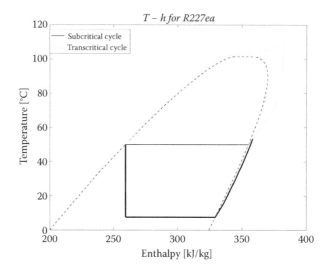

FIGURE 2.36
Comparison of a transcritical and subcritical ORC for the R227ea working fluid.

Receiver: $$\dot{Q}_{rec} = \dot{m}(h_4 - h_{4a})$$ (2.86)

Subcooler: $$\dot{Q}_{sub} = \dot{m}(h_{4a} - h_5)$$ (2.87)

Pump: $$\dot{W}_{pump} = \frac{\dot{V}_5(P_1 - P_5)}{\eta_{o,pump}}$$ (2.88)

Evaporator: $$\dot{Q}_{evap} = \dot{m}(h_2 - h_1)$$ (2.89)

Expander: $$\dot{Q}_{exp} = \dot{m}\left(h_2 - h_3\right)$$ (2.90)

Hence, the net electrical power production of the ORC is equal to:

$$\dot{W}_{el,net} = \dot{W}_{el,exp} - \dot{W}_{el,pump} = \frac{\dot{Q}_{exp}}{\eta_{mech,exp} \cdot \eta_{gen,exp}} - \frac{\dot{W}_{pump}}{\eta_{mech,pump} \cdot \eta_{gen,pump}}$$ (2.91)

And the net electrical efficiency of the ORC is:

$$\eta_{el,ORC} = \frac{\dot{W}_{el,net}}{\dot{Q}_{evap}}$$ (2.92)

On the other hand, the exergy efficiency of the ORC is defined as follows:

$$\eta_{ex,ORC} = \frac{\dot{W}_{el,net}}{\dot{m}\left[\left(h_{hs,i} - h_a\right) - T_a\left(s_{hs,i} - s_a\right)\right]}$$

(2.93)

2.7.1.1 Tutorial on the ORC

Assume the ORC that is presented in Figure 2.37. The cycle is operating with the novel fluid R1234ze and a mass flow rate of 0.5133 kg/s. Real-time application is considered, thus there is pressure drop of 0.2 bar in the condenser and evaporator, 0.1 bar in the sub-cooler, and 0.01 bar in the receiver and filter. The pump's isentropic efficiency is equal to 67.2%, and the isentropic efficiency of the expander is 61.9%. Evaporation pressure is 27.45 bar, and the lowest pressure in the cycle is equal to 8.13 bar. The maximum temperature of the cycle is equal to 100°C. The level of subcooling is 5 K. The mechanical efficiency is 0.9, and the generator's efficiency is 0.85. For simplicity, assume that the exit on the condenser, the filter, and the receiver are saturated. Determine the following:

iv. The pressure, temperature, and enthalpy of all points, as shown in Figure 2.37 and draw the T-s and the logP-h diagram of the cycle

v. The electrical power production in the expander

vi. The net electrical power output of the cycle (neglect pump fluid losses)

vii. The overall electrical efficiency of the cycle

Solution:

iv. Based on the pressure drop in the evaporator, the pressure at point 2 can be determined:

$$P_2 = P_1 - \Delta P_{evap} = 27.45 - 0.2 = 27.25 \; bar$$

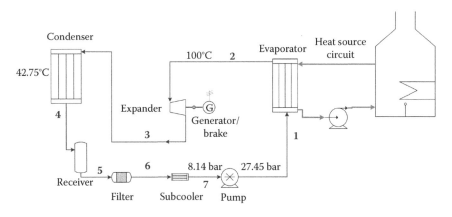

FIGURE 2.37
ORC layout investigated in the tutorial.

The lowest pressure in the cycle is at the inlet for the pump, thus at point 7. Given that and the updraft pressure drops, the pressure in every point of the cycle can be determined:

$$P_6 = P_7 + \Delta P_{sub} = 8.13 + 0.1 = 8.23 \ bar$$
$$P_5 = P_6 + \Delta P_{filter} = 8.23 + 0.01 = 8.24 \ bar$$
$$P_4 = P_5 + \Delta P_{rec} = 8.24 + 0.01 = 8.25 \ bar$$
$$P_3 = P_4 + \Delta P_{cond} = 8.25 + 0.2 = 8.45 \ bar$$

The temperature at point 2 is given to be 100°C, thus the enthalpy can be found from a fluids database to be equal to 438.6 kJ/kg. Making use of the isentropic efficiency of the expander, the enthalpy at point 3, can be calculated:

$$\eta_{exp} = \frac{h_2 - h_3}{h_2 - h_{3,is}} \Rightarrow h_3 = 425.4 \ kJ/kg$$

For the given enthalpy and pressure at point 4, the temperature is found to be 56.37°C. Given the fact that the outlet of the condenser is saturated liquid, the temperature and the enthalpy of point 4 can be easily found to be 42.75°C and 259.0 kJ/kg, respectively. In the same way, the temperature and the enthalpy of point 5 are found to be 42.66°C and 258.9 kJ/kg, while for point 6 the corresponding values are 42.60°C and 258.8 kJ/kg.

Subcooling is given to be 5 K, thus the temperature at point 7 is 37.60°C. For this temperature and the pressure already found, the corresponding enthalpy is determined to be 251.0 kJ/kg.

Finally, for the determination of point 1, the isentropic efficiency of the pump is used:

$$\eta_{pump} = \frac{h_{1,is} - h_7}{h_1 - h_7} \Rightarrow h_1 = 253.5 \ kJ/kg$$

And the corresponding temperature is 38.98°C.

An overview of the main thermodynamic data for each point, as calculated above, is presented in Table 2.2. Moreover, Figure 2.38 presents the cycle diagrams for the determined points.

TABLE 2.2

Overview of Common Refrigerants and Their Basic Properties

Point	Pressure (bar)	Temperature (°C)	Enthalpy (kJ/kg)
1	27.45	38.98	253.5
2	27.25	100	438.6
3	8.45	56.37	425.4
4	8.25	42.75	259.0
5	8.24	42.66	258.9
6	8.23	42.60	258.8
7	8.13	37.60	251.0

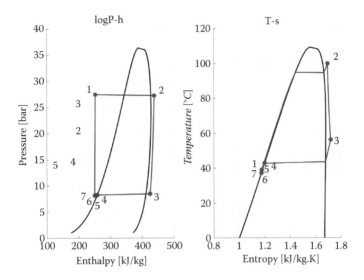

FIGURE 2.38
The logP-h and the T-s diagrams for the cycle in the tutorial.

v. The electrical power production in the expander is equal to:

$$\dot{W}_{el,exp} = \eta_{mech} \cdot \eta_{gen} \cdot \dot{m}_{R1234ze} \cdot (h_2 - h_3) = 5.186 \; kW$$

vi. The net electrical power of the cycle is equal to:

$$\dot{W}_{el,net} = \dot{W}_{el,exp} - \frac{\dot{m}_{R1234ze} \cdot (h_1 - h_7)}{\eta_{mech} \cdot \eta_{gen}} = 3.474 \; kW$$

vii. Finally, the overall electrical efficiency of the investigated cycle is equal to:

$$\eta_{el,ORC} = \frac{\dot{W}_{el,net}}{\dot{Q}_{evap}} = \frac{\dot{W}_{el,net}}{\dot{m}_{R1234ze} \cdot (h_2 - h_1)} = 3.65\%$$

2.8 Supercritical CO$_2$ Cycle

In terms of power production, the two most widely applied thermodynamic cycles are the Rankine cycle and the gas Brayton cycle. A standard gas Brayton cycle suffers the drawback of requiring a large work input for the compression of the working fluid. On the other hand, the conventional water-steam Rankine cycle requires significant super-heating to avoid high moisture contents at the outlet of the turbine, resulting in high exergy losses.

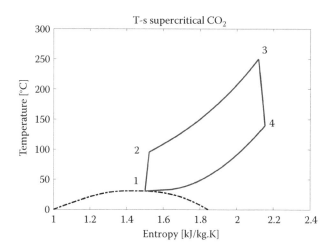

FIGURE 2.39
T-s diagram of a supercritical CO_2 cycle.

In order to avoid the aforementioned drawbacks of the two cycles, a novel cycle has been introduced, working entirely above the critical point, thus being called a supercritical cycle. Given the fact that, at all points, the working fluid is in a supercritical state, working fluids with relatively low critical temperatures are only applicable in this cycle.

Among the available working fluids, carbon dioxide (CO_2) is the most competitive for several reasons including its low critical temperature of 30.98°C, low cost, availability, and the fact that it is non-toxic (Wu 2004; Dinis Gaspar and Silva 2015).

As shown in Figure 2.39, a typical supercritical cycle consists of four main processes:

- 1–2: Carbon dioxide, which is in a slightly supercritical state, is non-isentropically compressed up to the high pressure of the cycle.
- 2–3: Heat is added to the cycle at constant pressure.
- 3–4: The working fluid is then expanded to produce work.
- 4–1: Heat is then rejected to the environment, the carbon dioxide returns to its initial state, and then the cycle restarts.

Nomenclature

A	Surface	$[m^2]$
C	Heat capacity	[W/K]
c_p	Specific heat capacity	[J/kg.K]
COP	Coefficient of performance	–
COP_{carnot}	Carnot's coefficient of performance	–
$\dot{Ex}$	Exergy flow	[W]

h	Enthalpy	[J/kg]
$\dot{m}$	Mass flow rate	[kg/s]
P	Pressure	[bar]
ΔP	Pressure drop	[bar]
$\dot{Q}$	Heat flux	[W]
s	Entropy	[J/kg.K]
$\Delta\dot{S}$	Rate of entropy change	[W/K]
T	Temperature	[K]
$\bar{T}$	Equivalent temperature	[K]
T_C	Low temperature in Carnot's cycle	[K]
T_H	High temperature in Carnot's cycle	[K]
T_0	Reference temperature	[K]
$\dot{W}$	Power	[W]
w	Concentration of refrigerant in solution	[kg/kg]
$\dot{V}$	Volumetric flow rate	[m³/s]
v	Specific volume	[m³/kg]

Greek Symbols

ε	Effectiveness	–
η	Efficiency	–
$\eta_{o,pump}$	Overall pump efficiency	–
π	Pressure ratio	–
ρ	Density	[kg/m³]
ω	Humidity ratio	[kg$_{water}$/ kg$_{dry\ air}$]

Subscripts

a	Ambient
abs	Absorber
abs,HP	Absorption chiller on heat pump mode
abs,refr	Absorption chiller on refrigeration mode
C	Cooling
c	Cold stream
comp	Compressor
cond	Condenser
cr	Cooled room
des	Desorber
dw	Desiccant wheel

el	Electrical
evap	Evaporator
ev.c	Evaporative cooler
ex	Exergy
exp	Expander
filter	Filter in ORC system
gen	Generator
geom.	Geometrical
H	Heat rejection in condenser
h	Hot stream
hex	Heat exchanger
hs	Heat source
i	Inlet
id	Ideal
irrev	Irreversible
is	Isentropic
loss	Loss
mech	Mechanical
net	Net (electrical)
o	Outlet
ORC	Organic Rankine cycle
pump	Pump
rec	Receiver
reg	Regeneration heat supplied to desiccant unit by external source
rev	Reversible
ss	Strong solution
sub	Subcooler
throttl	Throttling device
w	Water
wb	Wet bulb
ws	Weak solution
vcc	Vapor compression cycle
vol	Volumetric

References

Air Conditioning Refrigeration Institute. 1998. *Refrigeration and Air Conditioning*. 3rd ed. Upper Saddle River: Prentice-Hall.

Bejan, Adrian, and Allan Kraus. 2003. *Heat Transfer Handbook*. Hoboken, NJ: Wiley.

Boyce, Meherwan P. 2012. "Gas Turbine Engineering Handbook." Waltham, MA: Butterworth-Heinemann.

Cho, Soon-Haeng, and Jong-Nam Kim. 1992. "Modeling of a Silica Gel/Water Adsorption-Cooling System." *Energy* 17 (9): 829–839.

Dinis Gaspar, Pedro, and Pedro Dinho da Silva. 2015. *Handbook of Research on Advances and Applications in Refrigeration Systems and Technologies*. Hershey, PA: Engineering Science Reference.

Gordon, Jeffrey M., and Kim Choon Ng. 2000. *Cool Thermodynamics: The Engineering and Physics of Predictive, Diagnostic and Optimization Methods for Cooling Systems*. Cambridge: Cambridge International Science Publishing.

Granryd, Eric, högskolan Kungliga Tekniska, energiteknik Institutionen för, K. T. H. Industrial Engineering, Management, K. T. H. Royal Institute of Technology, and Technology Department of Energy. 2009. *Refrigerating Engineering*. Stockholm: Royal Institute of Technology, KTH, Department of Energy Technology, Division of Applied Thermodynamics and Refrigeration.

Hung, T. C., T. Y. Shai, and S. K. Wang. 1997. "A Review of Organic Rankine Cycles (ORCs) for the Recovery of Low-Grade Waste Heat." *Energy* 22 (7): 661–667.

Inglezakis, Vassilis J., and Stavros G. Poulopoulos. 2006. *Adsorption, Ion Exchange and Catalysis: Design of Operations and Environmental Applications*. Amsterdam: Elsevier.

Javanshir, Alireza, and Nenad Sarunac. 2017. "Thermodynamic Analysis of a Simple Organic Rankine Cycle." *Energy* 118: 85–96.

Ji, Jie, Cui He, Zheng Xiao, Chunqiong Miao, Liqi Luo, Haisheng Chen, Xinjing Zhang, Huan Guo, Yaodong Wang, and Tony Roskilly. 2017. "Simulation Study of an ORC System Driven by the Waste Heat Recovered from a Trigeneration System." *Energy Procedia* 105: 5040–5047.

Jiangzhou, S., R. Z. Wang, Y. Z. Lu, Y. X. Xu, and J. Y. Wu. 2005. "Experimental Study on Locomotive Driver Cabin Adsorption Air Conditioning Prototype machine." *Energy Conversion and Management* 46 (9–10): 1655–1665.

Kakaç S., and Hongtan Liu. 2002. *Heat Exchangers: Selection, Rating, and Thermal Design*. 2nd ed. Boca Raton: CRC Press.

Kanoğlu, Mehmet, Melda Özdinç Çarpınlıoğlu, and Murtaza Yıldırım. 2004. "Energy and Exergy Analyses of an Experimental Open-Cycle Desiccant Cooling System." *Applied Thermal Engineering* 24 (5): 919–932.

Katsanos, C. O., D. T. Hountalas, and E. G. Pariotis. 2012. "Thermodynamic Analysis of a Rankine Cycle Applied on a Diesel Truck Engine Using Steam and Organic Medium." *Energy Conversion and Management* 60: 68–76.

Koronaki, I. P., E. G. Papoutsis, and V. D. Papaefthimiou. 2016. "Thermodynamic Modeling and Exergy Analysis of a Solar Adsorption Cooling System with Cooling Tower in Mediterranean Conditions." *Applied Thermal Engineering* 99: 1027–1038.

Kuppan, T. 2000. *Heat Exchanger Design Handbook*. New York: Dekker.

Lakew, Amlaku Abie, Olav Bolland, and Yves Ladam. 2011. "Theoretical Thermodynamic Analysis of Rankine Power Cycle with Thermal Driven Pump." *Applied Energy* 88 (9): 3005–3011.

Lavan, Z., J. B. Monnier, and W. M. Worek. 1982. "Second Law Analysis of Desiccant Cooling Systems." *Journal of Solar Energy Engineering* 104 (3): 229–236.

Refprop database.

Marion, Michaël, Ionut Voicu, and Anne-Lise Tiffonnet. 2012. "Study and Optimization of a Solar Subcritical Organic Rankine Cycle." *Renewable Energy* 48: 100–109.

Meunier, F., P. Neveu, and J. Castaing-Lasvignottes. 1998. "Equivalent Carnot Cycles for Sorption Refrigeration: Cycles de Carnot Equivalents pour la Production de Froid par Sorption." *International Journal of Refrigeration* 21 (6): 472–489.

Murzin, Dmitry. 2013. *Engineering Catalysis*. Berlin: Walter de Gruyter GmbH.

Pei, Gang, Jing Li, and Jie Ji. 2010. "Analysis of Low Temperature Solar Thermal Electric Generation Using Regenerative Organic Rankine Cycle." *Applied Thermal Engineering* 30 (8–9): 998–1004.

Quoilin, Sylvain. 2011. "Sustainable Energy Conversion Through the Use of Organic Rankine Cycles for Waste Heat Recovery and Solar Applications." PhD. diss., University of Liège, Belgium.

Quoilin, Sylvain, Vincent Lemort, and Jean Lebrun. 2010. "Experimental Study and Modeling of an Organic Rankine Cycle Using Scroll Expander." *Applied Energy* 87 (4): 1260–1268.

Richardson, J. F., J. H. Harker, and J. R. Backhurst. 2007. *Coulson and Richardson's Chemical Engineering*, Vol. 2 *Particle Technology and Separation Processes*. 5th ed. Elsevier.

Roumpedakis, Tryfon, Konstantinos Braimakis, and Sotirios Karellas. 2015. "Investigation and Optimization of the Operation and Design of a Small Scale Experimental Trigeneration System Powered by a Supercritical Orc." ASME ORC 2015, 3rd International Seminar on ORC Power Systems, Brussels, 12 October.

Rouquerol, Françoise, Jean Rouquerol, Kenneth S. W. Sing, Guillaume Maurin, and Philip Llewellyn. 2014. "Introduction." In *Adsorption by Powders and Porous Solids,* 2nd ed., 1–24. Oxford: Academic Press.

Ruthven, Douglas M. 2008. "Fundamentals of Adsorption Equilibrium and Kinetics in Microporous Solids." In *Adsorption and Diffusion,* edited by Hellmut G. Karge and Jens Weitkamp, 1–43. Berlin, Heidelberg: Springer Berlin Heidelberg.

San, Jung-Yang, and Hui-Chi Hsu. 2009. "Performance of a Multi-Bed Adsorption Heat Pump Using SWS-1L Composite Adsorbent and Water as the Working Pair." *Applied Thermal Engineering* 29 (8–9): 1606–1613.

Schuster, A., S. Karellas, and R. Aumann. 2010. "Efficiency Optimization Potential in Supercritical Organic Rankine Cycles." *Energy* 35 (2): 1033–1039.

Sharonov, V. E., and Yu I. Aristov. 2008. "Chemical and Adsorption Heat Pumps: Comments on the Second Law Efficiency." *Chemical Engineering Journal* 136 (2–3): 419–424.

Soares, Claire. 2002. *Process Engineering Equipment Handbook*. New York: McGraw-Hill.

Sun, Wenqiang, Xiaoyu Yue, and Yanhui Wang. 2017. "Exergy Efficiency Analysis of ORC (Organic Rankine Cycle) and ORC-Based Combined Cycles Driven by Low-Temperature Waste Heat." *Energy Conversion and Management* 135: 63–73.

Suzuki, Motoyuki. 1993. "Application of Adsorption Cooling Systems to Automobiles." *Heat Recovery Systems and CHP* 13 (4): 335–340.

Tchanche, Bertrand F., Gr Lambrinos, A. Frangoudakis, and G. Papadakis. 2011. "Low-Grade Heat Conversion into Power Using Organic Rankine Cycles—A Review of Various Applications." *Renewable and Sustainable Energy Reviews* 15 (8): 3963–3979.

Thu, Kyaw, Young-Deuk Kim, Aung Myat, Won Gee Chun, and Kim Choon Ng. 2013. "Entropy Generation Analysis of an Adsorption Cooling Cycle." *International Journal of Heat and Mass Transfer* 60: 143–155.

Van den Bulck, E., S. A. Klein, and J. W. Mitchell. 1988. "Second Law Analysis of Solid Desiccant Rotary Dehumidifiers." *Journal of Solar Energy Engineering* 110 (1): 2–9.

Wade, L., E. Ryba, C. Weston, and J. Alvarez. 1992. "Test Performance of a 2 W, 137 K Sorption Refrigerator." *Cryogenics* 32 (2): 122–126.

Wang, L. W., R. Z. Wang, and R. G. Oliveira. 2009. "A Review on Adsorption Working Pairs for Refrigeration." *Renewable and Sustainable Energy Reviews* 13 (3): 518–534.

Wang, R. Z., and R. G. Oliveira. 2006. "Adsorption Refrigeration—An Efficient Way to Make Good Use of Waste Heat and Solar Energy." *Progress in Energy and Combustion Science* 32 (4): 424–458.

Wu, Chih. 2004. *Thermodynamic Cycles: Computer-Aided Design and Optimization, Chemical Industries*. Vol. 99. *Chemical Industries Series*. New York: M. Dekker.

Zhang, Li Zhi, and Ling Wang. 1997. "Performance Estimation of an Adsorption Cooling System for Automobile Waste Heat Recovery." *Applied Thermal Engineering* 17 (12): 1127–1139.

3

Solar Thermal Collectors

Solar collectors are the most central components of every solar energy installation. They capture and convert solar irradiation into either electric energy for solar PV applications or useful heat for solar thermal applications. In PV applications, solar energy is absorbed by PV modules to create a flow of electrons by means of the PV effect. The electricity generated can then satisfy an electrical load, be stored into an electrical storage system for standalone applications, or be fed directly into the electrical grid. Details on the operating principles, components, and technical characteristics of solar PV systems will be provided in Chapter 4. On the other hand, solar thermal applications make use of solar thermal collectors to harness solar irradiation and convert it into heat. The heat is, in turn, transferred to a heat transfer fluid (HTF)—usually water, air, or oil—that is flowing through the solar collector. The heat carried by the HTF can be used to either (1) satisfy a heating/cooling load or (2) charge a thermal energy storage (TES) system from which heat can later be discharged when solar irradiation is not available (cloudy or foggy conditions and night hours). Finally, hybrid PV-thermal solar collectors that combine PV modules with a heat extraction unit for the cogeneration of solar electricity and heat have been developed recently, motivated by the fact that solar energy captured by PV modules is not only converted to electrical energy but also to high amounts of waste heat with a detrimental effect on the PV efficiency.

Solar thermal collectors can be classified into two main categories: non-concentrating and concentrating. In non-concentrating collectors, solar irradiation is intercepted by their collecting surface area, A_c, which also serves as the solar absorbing surface, A_{abs}. The geometric concentration ratio, C_{geo}, is defined as the ratio of the surface area that intercepts solar irradiation to the area where solar energy gets absorbed:

$$C_{geo} = \frac{A_c}{A_{abs}} = \frac{\text{Surface area of collector}}{\text{Area of absorption}} \tag{3.1}$$

The conclusion is drawn that non-concentrating collectors have a geometric concentration ratio of unity. Conversely, concentrating collectors make use of concave reflective optical surfaces to collect solar radiation and redirect it toward a much smaller absorbing area ($C_{geo} > 1$). Thus, in concentrating collectors, solar radiation is focused and arrives at the solar absorbing surface, A_{abs}, at a considerably higher solar flux intensity vis-à-vis non-concentrating collectors. To achieve highly concentrated solar fluxes during the whole day, solar concentrators usually track the sun's position in the sky. The capability of solar concentrators to collect and focus solar energy can be expressed in terms of their mean flux concentration ratio $\tilde{C}$ that is defined as the ratio of the mean solar flux intensity over a targeted area A at the focal plane to the direct normal solar irradiance, I_{DNI}

$$\tilde{C} = \frac{\overline{q}_{solar}}{I_{DNI}} = \frac{Q_{solar}}{I_{DNI} \cdot A} \tag{3.2}$$

where $\bar{q}_{solar}$ is the mean solar flux intensity over area A. Non-concentrating collectors have a concentration ratio $\tilde{C}$ of unity, while the concentration ratio of concentrating collectors depends greatly on the system's type, the optical design and surface quality of the solar concentrators, and the sun tracking mechanism. High concentration ratios imply the conversion of sunlight into high-temperature process heat, which can be utilized for driving thermodynamic cycles for power generation or for process heat applications.

3.1 Non-Concentrating Solar Collectors

Non-concentrating collectors are the dominant type of collectors used in low-temperature applications. These collectors usually do not follow the sun's position in the sky and are, therefore, permanently fixed in position. This category includes three collector types:

- Flat-plate collectors
- Evacuated tube collectors
- Hybrid PV-thermal collectors

3.1.1 Flat-Plate Collectors

Flat-plate collectors (FPCs) represent the most widely used low-temperature solar thermal collector technology. A typical solar FPC configuration is shown schematically in Figure 3.1. It consists of one or multiple sheets of glazing covers, absorber plates for absorbing the solar irradiation, tubes or passages for the circulation of the HTF, and insulation

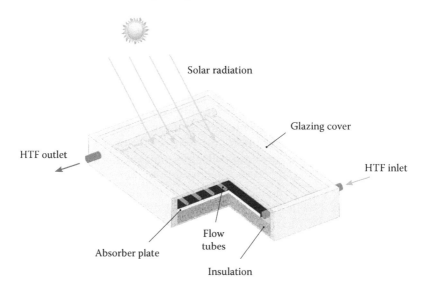

FIGURE 3.1
3D schematic view of a typical flat-plate collector.

layers to reduce heat conduction losses from the bottom and the sides of the solar collector. Solar radiation enters the FPC through the glazing covers and is then efficiently absorbed by the absorber plates. The absorbed heat is transferred to the HTF before being carried away to satisfy a thermal load or to be stored for later use. The HTF is swept out through a manifold to which all the tubes/passages are connected.

FPCs do not track the sun's position in the sky, therefore they need to be permanently fixed in position with the appropriate orientation. To achieve the best possible annual performance, collectors should be directed toward the equator—pointing south in the northern hemisphere and north in the southern hemisphere—while their tilt angle β should be equal to the latitude of the installation site. Angle variations within a range of ±10–15° apply, depending on the specific application (S. Kalogirou 2003). A schematic representation of an appropriate FPC orientation is shown in Figure 3.2.

Glazing covers are used to reduce heat convection and radiation losses. Convection losses from the absorber plate are reduced through trapping a stagnant air layer between the absorber plates and the glazing. In this way, the development of a natural convective heat exchange between the hot absorption plates and the surroundings is prevented. The insertion of a transparent honeycomb structure into the gap between the glazing covers and the absorber plates has been shown to further suppress convection losses from the collector to the surroundings (Francia 1962; Hellstrom et al. 2003). As far as the reduction of radiation losses is concerned, glazing covers—made of glass or other transparent materials with high transmissivity to short-wave radiation and low transmissivity to long-wave thermal radiation—prevent a considerable portion of the thermal radiation emitted by the absorber plates from escaping to the surroundings. Figure 3.3 shows a schematic representation of a possible path followed by an incident ray of light after entering the solar collector.

Low-iron glass has been widely used as glazing material because of its favorable optical properties over the wavelength range of interest (Rubin 1985). In particular, it exhibits a transmittance of 0.85–0.9 at normal incidence for solar radiation, while its transmittance to long-wave thermal radiation is practically negligible. The transmittance of commercially available glass materials is highly dependent on the angle of incidence, θ, of the incoming radiation and acquires a maximum value of 0.85–0.87 for $\theta = 0°$, i.e. when the direction of the ray incident on the glazing surface coincides with one of the lines perpendicular to the surface at the point of incidence (Parsons 1995). Other radiation transmitting materials used

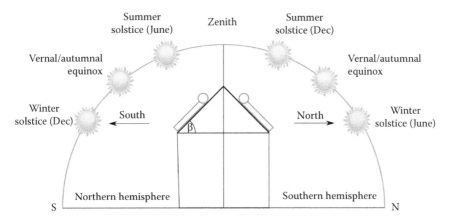

FIGURE 3.2
Orientation of a flat-plate collector in the northern and southern hemispheres.

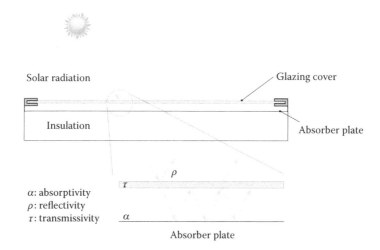

FIGURE 3.3
Possible ray paths of incident solar radiation in a flat-plate collector. The multiple reflections between the glazing cover and the absorption plate are illustrated.

to glaze flat-plate solar collectors are plastic films (Whillier 1963). The significant advantage of these materials vis-à-vis glass are their lower specific mass and higher structural flexibility. However, the high transmittance exhibited by the most commonly used plastic materials is not limited to short-wave radiation, but extends itself also toward higher wavelength bands, allowing considerable portions of the thermal radiation emitted by the absorber plate to escape to the surroundings with an adverse effect on solar collector efficiency. Additionally, their use imposes limitations on the maximum operating temperature of the collector because plastic films are not able to withstand high temperatures for long periods of time without suffering from dimensional changes and degradation to their optical properties, which creates a significant decrease in the collector's optical and overall efficiency.

Antireflective coatings are commonly applied to glazing covers to further increase their transmittance of short-wave solar radiation. The considerable impact of antireflection treatment on glazing covers is demonstrated by measuring a 6.5% increase in the useful energy output of the solar FPC at an operating temperature of 50°C (Hellstrom et al. 2003). Additionally, to prevent the deterioration of the collector's optical efficiency during operation and to keep the transmittance of the glazing covers close to its nominal value, the removal of dust, sand, and dirt deposited on the glazing is considered essential—especially in regions with dry weather and continuous lack of precipitation.

Although the majority of FPCs make use of one or multiple sheets of glazing covers to achieve higher thermal efficiency, unglazed FTCs represent a low-cost alternative for applications requiring temperatures up to ~20–25°C, such as heating swimming pools and air heating for commercial, industrial, process, and agricultural applications.

Absorber plates should be able to efficiently absorb incoming solar irradiation and retain the largest part of the thermal energy absorbed before transferring it to the HTF. This is accomplished by manufacturing absorber plates from materials with optical and thermal radiative properties that allow for the minimization of thermal losses from the plates to the surroundings and other components of the solar collector. Typically, solar absorber plates are made of copper, aluminium, or stainless steel, and are coated with selective surfaces exhibiting desired optical and radiative properties (Tripanagnostopoulos, Souliotis, and Nousia 2000; Wazwaz et al. 2002; Orel, Gunde, and Hutchins 2005; El-Sebaii and Al-Snani 2010). Selective surfaces typically consist of two thin layers on top of each other. The upper layer is typically highly absorbent of short-wave solar radiation but has a high transmittance of wavelengths corresponding to thermal radiation. The lower layer is characterized by high reflectance and low emittance of long-wave thermal radiation. Thus, the lower layer contributes to reducing radiation losses from the plate to the surroundings by retaining the largest part of the thermal radiation in the solar collector and the upper layer, allowing for the absorption and conversion of incident solar irradiation to useful thermal energy (Liu et al. 2007).

Effective heat transfer from the solar absorber plates to the HTF is of paramount importance in achieving high performance and efficiency. Efficient absorption of the heat by the HTF also prevents system overheating and its subsequent adverse effects on material stability. Typically, flow tubes are integral with or firmly bonded to the solar absorber plate to enhance heat transfer to the fluid, as shown in Figure 3.4a, Figure 3.4b, and Figure 3.4c for water-based solar collectors,

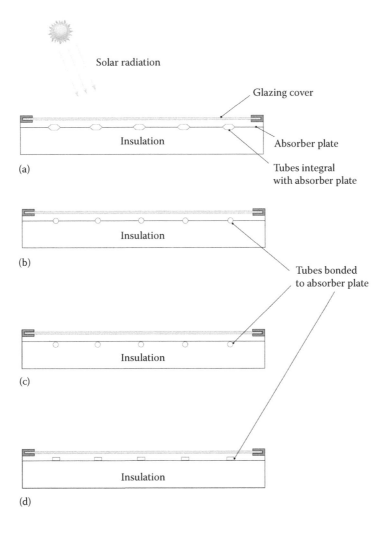

FIGURE 3.4
Schematic representation of typical water-based solar collector designs with: (a) flow tubes integral with, (b) cylindrical tubes firmly bonded to the upper surface of, (c) cylindrical tubes fastened to the lower surface of, and (d) extruded rectangular tubes bonded to the absorber plate. (Adapted from Kalogirou, S. A., *Progress in Energy and Combustion Science*, 30 (3): 231–295, 2004.)

respectively. Fluid flow tubes are most commonly made of copper due to its high corrosion resistance. Figure 3.4d shows an alternative design using rectangular-shaped tubes bonded to the top surface of the absorber plate in order to increase the heat transfer area (Kreider 1982).

Several design configurations have been demonstrated for air-based solar collectors (Klein, Beckman, and Duffie 1977), as shown in Figure 3.5. An example of a conventional air-based FPC with the air flow passing below the absorber plates is provided in Figure 3.5a. However, heat transfer with this collector design is limited by the low convective heat transfer coefficients between the absorber plates and air as well as by the fact that heat transfer only proceeds at the absorber plate-air interface, leading to significant temperature gradients throughout the air volume and perpendicular to the air flow direction. It becomes evident that higher absorption efficiencies can be obtained by effectively

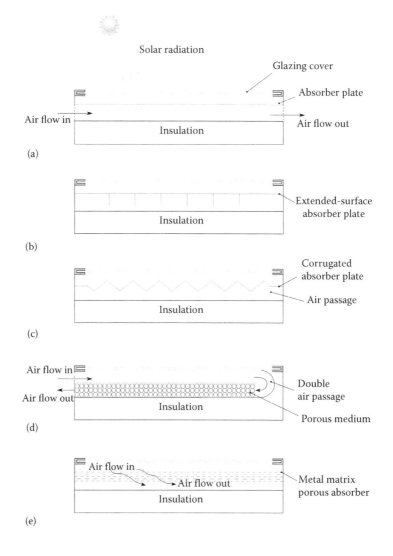

FIGURE 3.5
Cross-sectional view of typical air-based solar collector designs illustrating: (a) an FPC with an air passage below the absorber plate, (b) an FPC with a finned absorber plate, (c) an FPC with a corrugated solar absorber, (d) a double-passage solar collector, and (e) a solar collector with a metal matrix porous absorber.

increasing the heat transfer area between the absorber plates and the HTF. This is achieved by using extended-surface (Ackermann, Ong, and Lau 1995) or corrugated (Liu et al. 2007) absorber plates, as shown in Figures 3.5b and 3.5c, which also induce an increase in the residence time of the air flow in the solar collector. Even better heat transfer rates can be achieved by employing a double-passage collector (Wijeysundera, Ah, and Tjioe 1982; Ho, Yeh, and Wang 2005), in which the cold air flow initially passes over and exchanges heat with a flat-plate absorber and subsequently moves through the space below the plate which might contain a porous medium in order to enhance heat transfer to the HTF, as depicted in Figure 3.5d. The use of metal matrix porous absorbers with a large specific surface area in place of flat plates, as shown in Figure 3.5e, has also the potential to enhance heat transfer and achieve higher air temperatures as it allows the incoming radiation to spread over the whole matrix volume and become absorbed at high heat transfer rates by the air flow passing through it (Kreider 1982).

The useful thermal power output gained by any flat-plate solar absorber can be expressed by:

$$Q_u = \dot{m}_{HTF} \cdot c_{p,\,HTF} \cdot (T_o - T_i) \tag{3.3}$$

where $\dot{m}_{HTF}$ and $c_{p,HTF}$ are the mass flow rate and specific heat capacity of the HTF, respectively, and $T_o - T_i$ is the temperature difference of the HTF between the outlet and inlet of the solar collector. The useful thermal power output at steady state conditions can be also obtained by formulating the energy balance equation of the collector:

$$Q_u = A_c \left[S - U_L \cdot (T_{pm} - T_a) \right] \tag{3.4}$$

where A_c is the collector area, U_L is the collector's overall heat loss coefficient, T_{pm} is the mean absorber plate temperature, T_a is the ambient temperature, and S is the total absorbed solar irradiance:

$$S = I_{DNI} \cdot \cos\theta \cdot (\tau\alpha)_D + I_d \cdot (\tau\alpha)_d \cdot \left(\frac{1 + \cos\beta}{2} \right) + I \cdot \rho_g \cdot (\tau\alpha)_g \cdot \left(\frac{1 - \cos\beta}{2} \right) \tag{3.5}$$

where I_{DNI}, I_d, and $I = I_{DNI} \cdot \cos\theta_s + I_d$ are the direct normal, diffuse horizontal, and global horizontal solar irradiances, respectively, θ_s is the solar zenith angle, β is the tilt angle of the solar collector from the horizontal, ρ_g is the reflectivity of the ground, τ is the transmissivity of the glazing cover, α is the absorptivity of the solar absorber, and $(\tau\alpha)$ is the tau-alpha product, which represents the fraction of incoming solar irradiance that is absorbed on the plate after multiple ray reflections between the plate and the glazing. Thus, the three terms of Eq. (3.5) represent the fractions of solar radiation absorbed on the collector plate due to direct, diffuse, and ground-reflected radiation. Since the tau-alpha product depends on the angle of incidence θ, total absorbed solar irradiance can be expressed based on the total solar irradiance, $I_t = I_{DNI} \cdot \cos\theta + I_d \cdot (1+\cos\beta)/2 + I \cdot \rho_g \cdot (1-\cos\beta)/2$, and an effective tau-alpha product, $(\tau\alpha)_{eff}$:

$$S = I_t \cdot (\tau\alpha)_{eff} \tag{3.6}$$

By replacing the mean absorber plate temperature T_{pm} of Eq. (3.4) with the fluid temperature at the inlet of the solar collector, the useful thermal power output Q_u can be reformulated according to the Hottel-Whillier-Bliss equation (Hottel and Whillier 1958):

$$Q_u = A_c F_R \left[S - U_L \cdot (T_i - T_a) \right] \tag{3.7}$$

where F_R is the collector heat removal factor defined as the ratio of useful thermal power output Q_u over the theoretical heat gain that could be achieved if the entire collector is maintained at the inlet temperature of the HTF, T_i. It is dependent on the solar collector's characteristics, the HTF type, and its flow rate through the collector (Duffie and Beckman 2013). The overall heat loss coefficient, U_L, is a function of the collector's design as well as the collector's inlet and ambient temperatures. The following equation defines the thermal efficiency of the collector as the ratio of the useful thermal power output to incident solar radiation on the solar collector:

$$\eta_{th} = \frac{Q_u}{I_t \cdot A_c} = F_R \cdot (\tau\alpha)_{eff} - F_R \cdot U_L \cdot \frac{(T_i - T_a)}{I_t} \tag{3.8}$$

Note that the efficiency is largely driven by the temperature difference $(T_i - T_a)$. By experimentally determining the useful thermal power output Q_u and measuring T_i, T_a, and I_t, the instantaneous collector efficiency can be plotted against $\frac{(T_i - T_a)}{I_t}$. Fitting a linear regression model to the experimental data, the slope and intercept of the straight line obtained are equal to $F_R \cdot (\tau\alpha)_{eff}$ and $F_R \cdot U_L$, respectively. Despite the fact that the tau-alpha product, the heat removal factor F_R, and the overall heat loss coefficient U_L of a solar collector are non-constant variables, the linear regression technique can be used to determine the long-term performance of the collector.

Overall, the main disadvantage of conventional flat-plate solar collectors is their low efficiency collector temperatures above ~80°C because their thermal losses to the surroundings increase considerably with temperature. Therefore, research has focused on the development of designs that would reduce thermal losses to the ambient, and thus widen the application range of this technology. Pulling a moderate vacuum between the glazing cover and the absorber plates would contribute to suppressing convection losses but the pressure-gradient force:

$$F = \Delta P \cdot A_{glazing} = (p_{amb} - p_{collector}) \cdot A_{glazing} \tag{3.9}$$

exerted on the glazing makes this solution technologically demanding due to material stability issues. Another technological solution aiming to enhance the collector's thermal efficiency includes the use of advanced manufacturing techniques like ultrasonic welding to increase the quality of bonding between the flow tubes and the absorber plate and improve heat conduction to the tubes and HTF.

3.1.2 Evacuated Tube Collectors

Tubular designs can inherently withstand higher pressure gradient forces vis-à-vis FPCs, and are therefore used to realize the concept of evacuated solar collectors. Evacuated tube collectors (ETCs) are classified into direct ETCs and indirect ETCs.

A typical indirect ETC configuration is shown schematically in Figure 3.6. It consists of a fin-shaped absorber plate for collecting the solar irradiation that is placed inside an evacuated glass tube. A heat pipe containing the HTF is bonded to the upper surface of the absorber plate. The highest part of the heat pipe protrudes above the evacuated tube and is mounted into a heat exchanging manifold. With this arrangement, solar radiation enters the ETC through the glass tube, is efficiently absorbed by the absorber fin, and the heat is transferred to the pipe and the HTF. Unlike FPCs, where thermal energy is absorbed by the HTF in its sensible form, indirect ETCs use phase change fluids that undergo an evaporation-condensation cycle during operation. Upon absorption of solar energy, the HTF—usually methanol—vaporizes and rises to the top of the heat pipe where it transfers its heat to the fluid flowing through the manifold and then condenses. The condensed fluid returns back to the bottom of the heat pipe to repeat the cycle. Evacuated tube solar collector systems consist of an array of evacuated tubes that are all mounted to the same heat exchanging manifold.

On the other hand, direct ETCs do not make use of intermediate heat transfer stages to deliver the heating load. Typical direct evacuated tube configurations are shown schematically in Figure 3.7. Figure 3.7a depicts an evacuated tube with a design similar to the indirect evacuated tubes discussed earlier. In this arrangement, solar radiation is absorbed by a flat absorber plate enclosed in a vacuum-sealed tube, and the heat is subsequently transferred to the HTF—usually water—flowing through a U-tube firmly bonded to the plate. An alternative direct ETC configuration is shown in Figure 3.7b. It consists of two glass tubes fused together, between which the vacuum is pulled. A selective coating is applied on the inner surface of the inner glass tube to absorb the incoming solar irradiation. A concentric tube is placed inside the glass tubes and feeds the collector with water. After flowing out of the feeder tube, the water flows back to the outlet side in contact with the inner glass tube and absorbs heat. Because high temperatures are achievable with ETCs, system overheating that may lead to water evaporation should be avoided to prevent damage to the evacuated glass tubes and the overall solar hot water system.

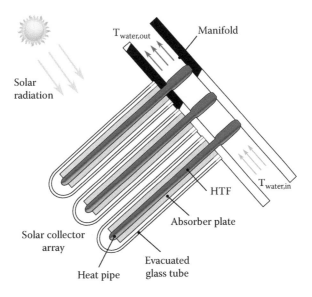

FIGURE 3.6
Cross-sectional view of an indirect evacuated tube solar collector array.

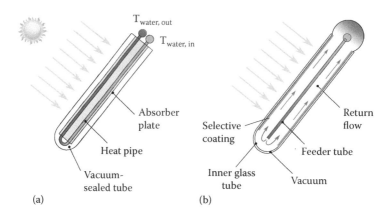

FIGURE 3.7
Different configurations of direct evacuated tube solar collectors.

Apart from their higher thermal efficiency at high collector temperatures, which typically corresponds to temperature differences between the collector and the surroundings $(T_c - T_a)$, above 50°C (International Energy Agency 2012), evacuated solar collectors also exhibit a superior performance vis-à-vis FPCs under low irradiation, low temperature, and high wind conditions because of the excellent insulation properties of the vacuum, which suppresses convection losses and prevents the inner glass tube or the heat pipe from being cooled by the ambient. On the other hand, FPCs acquire greater efficiency values at a moderate ambient temperature and $T_c - T_a$ of up to ~50°C. With regard to operation and maintenance costs, the modularity of evacuated tubes is an advantage compared to FPC designs. However, even in areas with moderate solar insolation levels, some low-cost FPCs can be more cost-efficient than ETCs. Another important disadvantage inherent in the geometrical configuration of ETCs is their discontinuous absorber area since the space between single evacuated tubes remains unexploited. Typically, space coverage (defined as the ratio of solar absorber area over ground area) acquires values in the range of 0.6–0.8.

3.1.3 Hybrid PV-Thermal Collectors (PVT)

During the operation of conventional PV cells under high solar irradiance, the generation of electricity is accompanied by the production of thermal energy and the development of relatively high temperatures throughout the PV modules. However, this has a direct effect on the PV cell efficiency since efficiency drastically decreases with increasing temperatures, according to (Skoplaki and Palyvos 2009):

$$\eta_{PV}(T) = \eta_{PV,ref} \cdot \left(1 - \zeta \cdot (T - T_{ref})\right) \tag{3.10}$$

where η_{pv} is the PV cell efficiency at temperature T (°C), $\eta_{PV.ref}$ is the cell efficiency as obtained at standard conditions (solar irradiance: $I = 1000$ W/m^2, $T_{ref} = 25$°C), and ζ (1/°C) is the solar PV cell temperature coefficient. Thermal energy generated by conventional PV cells is removed in the form of waste heat and is not utilized further for the production of useful energy, thus leading to low exergy conversion efficiency.

Enhancing the overall exergy conversion efficiency of PV cells is the main motivation behind the development of hybrid PVT solar collectors (Zondag et al. 1999). Hybrid PVT collectors consist of PV modules coupled to a heat exchanging unit that absorbs the heat produced and then transfers it to the HTF. The useful heat generated during the process is meant to satisfy thermal loads, thus achieving cogeneration of solar electricity and heat at high overall system efficiencies. Low-temperature heat extracted from the PV cells is mostly applied for domestic water heating, while high-temperature heat can be used for electricity generation via organic Rankine cycles and thermoelectric generators, or for the production of solar-assisted heating and cooling. A comprehensive review of PVT collector designs and their applications is provided by Chow (2010) and Riffat and Cuce (2011).

The useful thermal power output Q_u of a PVT collector at steady state conditions can be obtained by the modified Hottel-Whillier-Bliss equation (Florschuetz 1979):

$$Q_u^{PVT} = A_c F_R \left[S - U_L \cdot (T_i - T_a) - Q_{el} \right] \tag{3.11}$$

where Q_{el} is the electrical power output by the PV cells. The thermal efficiency of a hybrid PVT collector is then obtained by:

$$\eta_{th} = \frac{Q_u^{PVT}}{I_t \cdot A_c} \tag{3.12}$$

Several designs of hybrid PVT collectors have been demonstrated. They are categorized based on the type of HTF used (Zondag et al. 2003; Ibrahim et al. 2011). Three typical configurations of water-based PVT collectors are schematically shown in Figure 3.8. Figure 3.8a depicts a sheet-and-tube design that approximates an FPC, with the difference that PV modules are placed on top of the absorber plate. In this arrangement, the heat produced is absorbed by a water stream that is flowing in fluid tubes that are firmly bonded to the absorber plate. In the hybrid PVT collector designs shown in Figures 3.8b and 3.8c, heat is absorbed by a water stream flowing through a channel above or below the PV module, respectively. The concept depicted in Figure 3.8b takes advantage of the excellent match of the water absorption spectrum to the solar radiation spectrum. In particular, water exhibits a high absorptance at high wavelengths and thus contributes to reducing themal radiative losses from the absorber plate to the surroundings while it is fully transparent to short-wave radiation that is utilized by PV cells for electricity generation. However, construction issues may arise for large collector sizes due to the pressure exerted by the water layer on the PV module and the overlying glass cover. This problem can be circumvented by forcing the water stream flow through a channel below an opaque or transparent PV module, as depicted in Figure 3.8c. Apart from these typical configurations, several other design concepts have been proposed. Studies have mainly focused on the experimental and numerical investigation of the effect of different PV cell materials (S. Kalogirou and Tripanagnostopoulos 2005) and packing factors (Fujisawa and Tani 2001), fluid flow rates and inlet temperatures (Ji et al. 2007), weather conditions (Dubey and Tiwari 2010), and thermal absorber materials (Sandnes and Rekstad 2002) and dimensions (Huang et al. 1999; Ji et al. 2006) on the performance of water-based PVT collectors.

Air-based hybrid PVT collectors exhibit considerably lower manufacturing costs vis-à-vis water-based collectors. However, as highlighted also in the case of conventional FPCs,

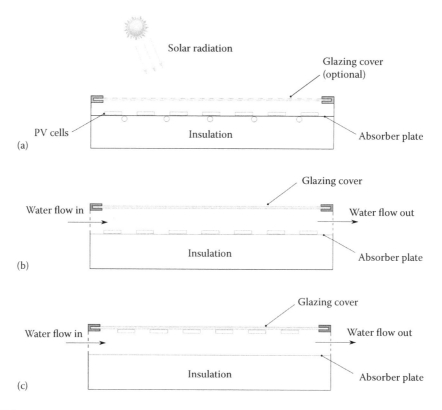

FIGURE 3.8
Common types of hybrid water-based PVT collector designs: (a) a sheet-and-tube PVT collector, (b) a collector with the PV cells bonded to the top surface of the absorber plate and a water channel above them, and (c) a PVT collector with the PV cells mounted on the lower surface of the glazing cover and a water flow below them.

heat transfer from the absorber/PV module to the air flowing inside the collector is poor. Therefore, the use of this type of collector is rather limited. Five typical configurations of air-based PVT collectors are depicted in Figure 3.9. The first design consists of PV cells firmly bonded to an absorber plate. Air passes through the air space formed between the glass cover and the absorber plate and recovers part of the thermal energy produced. However, the low metal-to-air heat transfer coefficient and the limited heat transfer area lead to a relatively low collector performance. The second design follows a similar concept except that air travels through a channel underneath the PV module, as shown in Figure 3.9b. As with conventional air-based thermal collectors, common techniques for increasing the heat transfer area include the integration of fins to the absorber plate (Figure 3.9c) and the use of corrugated or V-shaped absorber plates. A comparison of single- to double-passage PVT collectors reveals a considerably higher performance for the double-passage configuration (Figure 3.9d) due to the reduction of the glass cover temperature, which directly affects the convection and radiation heat losses of the system (Hegazy 2000; Sopian et al. 1996). Hendrie (1982) proposed a design that involves a two-stage heating of an air jet entering the bottom of the collector through a 0.25 cm-diameter hole. The air stream impinges on a tilted, perforated absorber plate positioned underneath the PV modules, rises through the absorber plate holes, and then comes into contact with the lower surface of the PV cells, as depicted in Figure 3.9e.

   Despite the constantly decreasing costs of PV modules, the levelized cost of electricity for solar PV systems is still higher than fossil-fuel powered electricity. Concentrating solar

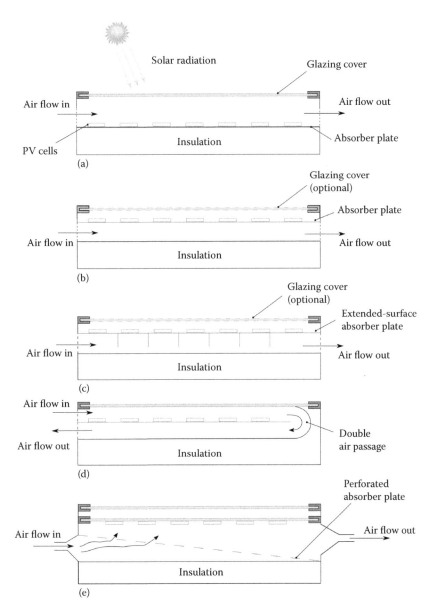

FIGURE 3.9
Typical configurations of hybrid air-based PVT collector designs: (a) PVT with an air passage above the PV cells, (b) PVT with the air passage below the PV modules, (c) PVT with a finned absorber plate, (d) double-air-passage collector, and (e) air-jet PVT design.

radiation is the key to effectively decreasing the investment costs of a PV system, as the area of PV cells required to produce the same power output decreases proportionally to the geometric concentration ratio:

$$C_{geo, PV} = \frac{A_c}{A_{PV\,cells}} = \frac{\text{Surface area of collector}}{\text{PV cell area}} \tag{3.13}$$

However, PV cell temperatures increase with higher concentration. Therefore, interest in concentrating PVs—and in the development of concentrating PVT collectors—has risen exponentially. Garg and Adhikari (1999) investigated an air-based PVT collector coupled to a non-imaging compound parabolic concentrator (CPC) and obtained higher efficiency compared to conventional PVT systems, while Rosell et al. (2005) studied the performance of a linear Fresnel concentrator positioned in tandem with a water-based PVT collector and measured efficiencies above 60%. More details about the design and geometrical configuration of concentrating collectors will be provided in Section 3.2.

3.2 Concentrating Solar Collectors

Because of the dilute nature of solar radiation, non-concentrating thermal collectors are not capable of converting sunlight to high-temperature heat, thus they are restricted to low-temperature applications. Heat delivery at higher temperatures can be achieved by decreasing the heat losses from the absorber to the surroundings. This can be attained by effectively concentrating sunlight incident on a larger area onto a smaller absorber area from which the heat losses will occur. Solar concentrator optics are typically classified into two main types:

- Non-imaging concentrators
- Imaging concentrators

Imaging concentrators are reflective optical devices that use incident solar irradiation to form an image of the light source at the focal plane of the concentrator. On the other hand the main function is to optimize the optical radiative transfer from a light source—the sun or the sun's image—to the solar absorber. Non-imaging concentrators are integrated to either non-concentrating or concentrating collectors.

3.2.1 Non-Imaging Concentrating Collectors

In contrast to imaging concentrating optics, non-imaging concentrators make use of both diffuse and direct solar radiation. The most commonly used form of non-imaging optics is the Compound Parabolic Concentrator (CPC) (Welford and Winston 1989). CPCs can be either two- or three-dimensional, corresponding to cylindrical trough and cone shapes, respectively, as illustrated in Figure 3.10.

The 2D configuration consists of two parabolic reflector segments with different focal points facing each other. It is designed to accept incoming solar radiation and only reflect rays entering the concentrator within a specific range of incidence angles to the absorber. Defining as acceptance halfangle (θ_{acc}), the angle between the axis of symmetry of the CPC and the line connecting the focus of the one parabolic mirror segment to the opposite edge of the inlet aperture, all rays entering the CPC with incidence angles less than θ_{acc} will experience multiple internal reflections in the CPC before finding their way to the absorber. Every ray with an incidence angle greater than θ_{acc} will be rejected by the CPC and reflected back out through the CPC inlet. Depending on its geometrical configuration (Mills and Giutronich 1978; O'Gallagher et al. 1982), the absorber can be positioned either at the CPC outlet aperture or the bottom region of the CPC, as depicted in Figure 3.11. Knowing that the geometric concentration ratio is defined as the ratio

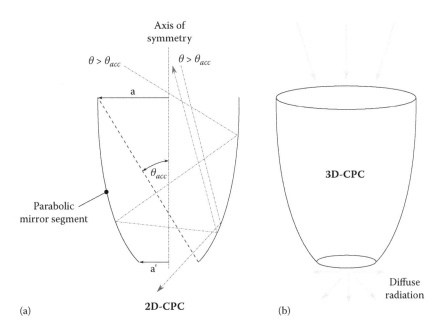

FIGURE 3.10
Schematic illustration of (a) a 2D- and (b) a 3D-compound parabolic concentrator. Possible paths of the incident solar radiation for $\theta < \theta_{acc}$ and $\theta > \theta_{acc}$, as well as the diffuse nature of the outgoing radiation at the CPC outlet aperture, are also indicated.

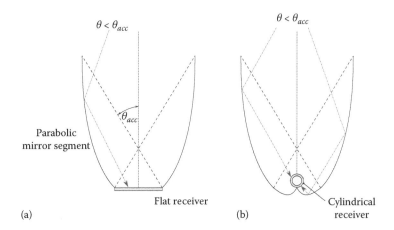

FIGURE 3.11
Concentration of solar radiation on a (a) flat-plate and (b) tubular absorber using a 2D-compound parabolic concentrator.

between the solar collector and solar absorber surface areas (Eq. 3.1), the maximum geometric concentration ratio for a 2D-CPC with a flat absorber mounted at its outlet aperture is given (Lovegrove and Pye 2012):

$$C_{geo,2D,max} = \frac{A_c}{A_{abs}} = \frac{a}{a'} = \frac{1}{\sin\theta_{acc}} \quad (3.14)$$

For a 3D-CPC, geometric limitations on acceptance of the light rays apply in one additional direction. The maximum geometric concentration ratio reads (Lovegrove and Pye 2012):

$$C_{geo,\,3D,\,max} = \frac{A_c}{A_{abs}} = \left(\frac{a}{a'}\right)^2 = \frac{1}{\sin^2\theta_{acc}} \tag{3.15}$$

Typically, concentration ratios of CPCs are in the single digits. As the upper parts of a CPC do not have a large contribution to augmenting the radiation flux reaching the solar absorber, it is common practice to truncate them in order to reduce material costs. Reducing the height of a CPC to two-thirds of its original value causes only a 10% reduction of the concentration ratio (Welford and Winston 1989).

The fundamental advantage of non-imaging optics is that concentration of sunlight is possible without the need to actively track the sun's position in the sky. Most commonly, stationary mounting of two-dimensional CPC collectors is done by aligning their long axis along the east-west direction and tilting their inlet aperture toward the equator by an angle equal to the site latitude. Alignment of a CPC trough with its long axis along the east-west direction offers the advantage that the concentrator faces the sun continuously. However, the delivery of solar energy to the absorber is limited by the sun's motion in the sky as only rays with incidence angles less than θ_{acc} will be converted to useful energy. Using a CPC with a higher acceptable angle would increase the range of hours over which sunshine collection is possible. However, according to Eq. (3.14), this will be achieved at the expense of a lower concentration ratio. The minimum acceptance angle for CPC troughs mounted along the east-west direction is 47°, corresponding to the change in the azimuth angle of the sun between the summer and winter solstices. Alternatively, a CPC with acceptance angles $2 \cdot \theta_{acc} < 47°$ can be used when the tilt angle of the concentrator is seasonally adjusted to compensate for the variations in the solar angle. The need for seasonal adjustments of the tilt angle and limitations regarding the range of operational hours of the concentrator over a day are eliminated when aligning it with its long axis along the north-south direction. However, this type of orientation requires continuous tracking of the sun's motion over the day.

Despite their low concentration ratios, non-imaging systems increase the performance of solar thermal collectors at relatively low costs. Apart from CPCs, various other geometries have also been developed (Welford and Winston 1989). The properties of prism-coupled CPCs (PCCPC), compound circular arc (CCAC), compound elliptical (CEC), compound hyperbolic (CHC), trumpet-shaped, and dielectric-filled concentrators are reviewed by Madala and Boehm (2016) and Tian et al. (2018).

3.2.2 Imaging Concentrating Collectors

Imaging concentrators are reflective optical devices that collect dilute low-flux solar radiation and focus the rays of light toward a solar absorber receiver positioned at the reflector's focal plane, where an image of the light source is formed. Assuming the sun to be a circular disk which subtends an angle of $2 \cdot \theta_{sun} = 0.533°$ (0.0093 radians) at the earth, as illustrated in Figure 3.12, reflection of sunlight by the reflecting mirrors would form a circular image of the sun at the concentrator's focal plane. To collect solar energy at maximum efficiency and acquire highly concentrated solar fluxes over the whole day, the optical axis of imaging concentrators must be continuously aligned to the incident solar irradiation. Therefore,

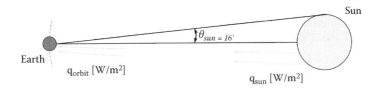

FIGURE 3.12
Solar radiation emitted by the circular solar disk reaching the earth's surface.

tracking the sun's motion in the sky is indispensable for this technology. Depending on their sun tracking mechanism, imaging concentrators can be distinguished into:

- Single-axis tracking concentrators
- Dual-axis tracking concentrators

Single-axis tracking systems follow the sun only in one direction, either east-to-west or north-to-south. However, the inaccuracies introduced by the continuously changing orientation of the sun have to be compensated for through the accurate adjustment of the concentrators. On the other hand, dual-axis tracking enables concentrating mirrors to follow the sun's motion with high accuracy, leading to even higher concentration ratios and HTF temperatures but at the expense of higher mechanical complexity and costs as well as lower reliability.

As the sun's motion is continuously followed the concentration ratio for imaging concentrators is not limited by the position of the sun but rather by the size of the sun's disk. For a perfectly specular reflector and accurate sun tracking mechanism, the maximum geometric concentration ratio for a 2D concentrator is (Lovegrove and Pye 2012):

$$C_{geo,2D,max} = \frac{1}{sin\,\theta_{sun}} \approx 215 \tag{3.16}$$

For a 3D imaging concentrator, geometric limitations on acceptance of light rays apply in one additional direction. The maximum geometric concentration ratio reads (Lovegrove and Pye 2012):

$$C_{geo,3D,max} = \frac{1}{sin^2\theta_{sun}} \approx 46257 \tag{3.17}$$

In practice, typical concentration ratios achieved by imaging concentrators are considerably lower. Losses in concentration occur due to shape irregularities and poor optical quality of the concentrating mirrors, inaccuracy of the sun tracking system, and shading effects introduced by the solar receiver or the mirror frame.

The energy absorption efficiency of a perfectly insulated (no conduction and convection losses) concentrating solar thermal absorber receiver is defined as:

$$\eta_{abs} = \frac{\alpha_{eff} \cdot Q_{solar} - Q_{rerad.losses}}{Q_{solar}} = \frac{\alpha_{eff} \cdot Q_{solar} - A_{receiver} \cdot \varepsilon_{eff} \cdot \sigma \cdot T^4}{A_{receiver} \cdot q_{solar}} \tag{3.18}$$

where Q_{solar} is the solar power intercepted by the receiver area $A_{receiver}$, $Q_{rerad.losses}$ are the thermal radiation losses from the solar absorber receiver to the surroundings, α_{eff} and ε_{eff} are the effective absorptance and emittance of the receiver, T is the average absorber temperature, and $\sigma = 5.67 \cdot 10^{-8}$ W·m^{-2}·K^{-4} is the Stefan Boltzmann constant. The difference between the two terms in the numerator of Eq. (3.18) yields the net power absorbed by the solar receiver, which can be used to drive a thermal or thermochemical process. Based on Eq. (3.2) and assuming a blackbody receiver ($\alpha_{eff} = \varepsilon_{eff} = 1$), the energy absorption efficiency can be expressed in terms of the mean flux concentration ratio $\tilde{C}$:

$$\eta_{abs} = 1 - \frac{\sigma T^4}{I_{DNI} \cdot \tilde{C}} \tag{3.19}$$

According to the second law of classical thermodynamics, the energy conversion efficiency of the system is limited by the Carnot efficiency. According to the Carnot's theorem, the thermal efficiency of any heat engine operating between two isothermal reservoirs at temperatures T_H and T_L cannot exceed the limiting value of:

$$\eta_{Carnot} = 1 - \frac{T_L}{T_H} \tag{3.20}$$

where T_L and T_H are the lower and upper temperatures of the equivalent Carnot heat engine, respectively. From a thermodynamics point of view, operation of the system at the highest possible temperature T_H should lead to the highest efficiency. However, higher temperatures have an adverse effect on the absorption efficiency of the solar concentrating system since thermal losses increase considerably with temperature. Thus, an optimal temperature T_{opt} can be defined for a perfectly insulated concentrating solar absorber receiver at which the ideal exergy efficiency of the system reaches its maximum value (Fletcher and Moen 1977):

$$\eta_{ex,ideal} = \eta_{abs} \cdot \eta_{Carnot} = \left(1 - \frac{\sigma T^4}{I_{DNI} \cdot \tilde{C}}\right) \cdot \left(1 - \frac{T_L}{T_H}\right) \tag{3.21}$$

The optimal operating temperature T_{opt} is obtained by setting:

$$\frac{\partial \eta_{ex,ideal}}{\partial T} = 0 \tag{3.22}$$

Figure 3.13 illustrates the variation of the absorption, Carnot, and the ideal exergy efficiencies for a blackbody receiver as a function of its temperature for various mean flux concentration ratios. In practice, when also considering conduction and convection heat losses from the receiver to the surroundings as well as system irreversibilities, lower optimal operating temperatures are obtained.

State-of-the-art imaging concentrators that are commercially available for large-scale collection of solar energy for thermal applications include:

- Parabolic trough collectors
- Linear Fresnel reflectors
- Central tower receivers
- Paraboloidal dish reflectors

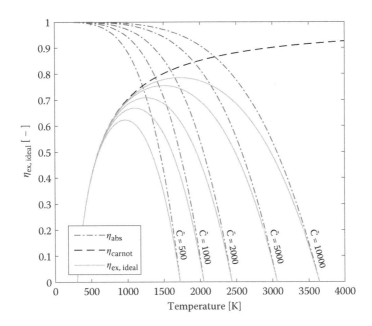

FIGURE 3.13
Variation of the absorption (η_{abs}), Carnot (η_{Carnot}), and ideal exergy ($\eta_{ex,ideal}$) efficiencies for a blackbody solar receiver as a function of the operating temperature for various mean flux concentration ratios $\tilde{C}$.

3.2.2.1 Parabolic Trough Collectors

Parabolic trough collectors (PTCs) represent the most advanced technology among high concentration solar energy systems (Mendelsohn, Lowder, and Canavan 2012). Parabolic/paraboloidal geometric configurations are central to most imaging solar concentrating systems, as mirrors of these shapes are capable of focusing sun rays parallel to their symmetry axis into a focal point, as illustrated in Figure 3.14. Important design parameters for the determination of the shape and size of such reflectors are

- The focal length f, defined as the distance between the focal point and the vertex of the parabola
- The rim angle Φ_{rim}, defined as the angle between the symmetry axis of the parabola and the mirror rim
- The aperture width w, defined as the distance between the two rims of the parabola

Specifying two of these three key parameters is sufficient to define the shape and size of a parabolic trough collector completely. The rim angle Φ_{rim} is related to the other two parameters, according to (Lovegrove and Pye 2012):

$$\frac{w}{f} = \frac{4}{\tan \Phi_{rim}} + \sqrt{\frac{16}{\tan^2 \Phi_{rim}} + 16} \tag{3.23}$$

The rim angle has an important effect on the geometric concentration ratio of a PTC. For a given aperture width, it follows that a small rim angle Φ_{rim} results to a narrow mirror which is not capable of collecting high amounts of solar radiation. On the other hand, a

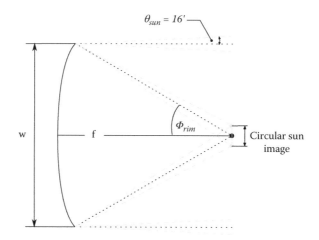

FIGURE 3.14
Concentration of solar radiation with a perfect parabolic mirror. The reflection of non-parallel rays and the creation of a circular sun image at the concentrator's focal plane are indicated.

large rim angle would facilitate the collection of an increased amount of solar energy, but the light rays captured and reflected at the region close to the mirror rims would have to travel a longer distance to reach the focal point. In this case, any geometrical imperfections of the mirrors would result in a wider spread of the reflected beams and lower concentration ratios. Decisions on the exact shape and size of parabolic mirrors depend on the requirements set on the concentration ratio as well as on economic considerations since further increase of the rim angle Φ_{rim} at an already wide angle has a negligible effect on the performance of the system (S. A. Kalogirou et al. 1994).

A typical PTC is shown schematically in Figure 3.15. It makes use of a set of 2D parabolic-shaped mirrors—typically made of silver-coated glass—that follow the sun in one only direction (single-axis tracking concentrators) and concentrate sunlight at their focal plane at mean flux concentration ratios $\tilde{C}$ in the range of 30–100. With this arrangement, incident beams on the reflectors are redirected to a usually tubular solar absorber receiver placed along their focal line. Concentrated solar radiation is efficiently absorbed by the tubular receiver—typically made of stainless steel—before being transferred in the form of thermal energy to the HTF flowing through the receiver tube. Commonly, selective coatings are applied on the surface of tubular receivers to increase their absorption efficiency by providing a high absorptance to short-wave solar radiation and low emittance for long-wave thermal radiation. Similar to non-concentrating solar collectors, the tubular receiver is usually placed inside a glass tube to reduce heat convection and radiation losses to the surroundings whose presence, however, introduces reflection losses of the incident solar radiation. The glass envelope is typically made of borosilicate glass and an antireflective coating is applied on its inner and outer surfaces to achieve higher transmittance for the visible part of the light spectrum and thus lower reflection losses. Evacuation of the space between the glass tube and the receiver at pressures of around 10^{-4} mbar (Price et al. 2002) is commonly applied to further decrease thermal losses and protect the selective coating of the receiver from degradation. HTF temperatures of up to ~400°C are typically obtained with PTCs.

The whole assembly of reflective surfaces and tubular receiver is mounted on a metal support structure that is typically made of hot-laminated steel. The design of this structure has a major impact not only on the mechanical strength but also, most importantly, on the optical

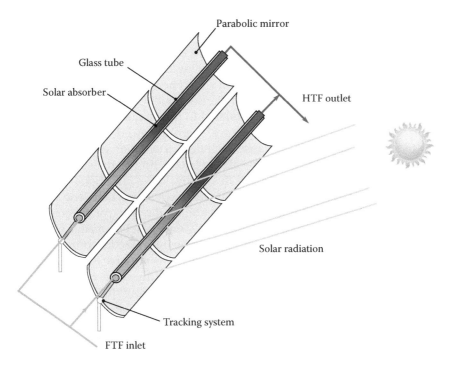

Parabolic mirror

Glass tube

Solar absorber

HTF outlet

Solar radiation

Tracking system

FTF inlet

FIGURE 3.15
Schematic configuration of a parabolic trough collector array with tubular solar receivers.

performance and cost of the solar collector field. Bending and torsional deflections of the reflecting mirrors induced by self-weight and wind loads acting on their surfaces should be kept to a minimum to reduce any deviations from the ideal parabolic shape of the mirrors and ensure concentration of the sunlight onto the solar receiver at high concentration ratios. Low torsional effects can be achieved by increasing the stiffness of the support structure or/ and by shortening the length of the solar collector assembly. However, higher structural stiffness implies an increase in the material costs with adverse effects on the economic viability of the parabolic trough system. An optimization procedure involving ray-tracing simulations and a coupled structural - cost estimation model model was developed by Weinrebe, Abul-Ella, and Schiel (2011) to investigate the interdependence of PTC field design parameters and the impact of the collector stiffness on its cost and optical efficiency. In an attempt to satisfy the competing design criteria of solar PTCs, various structural concepts using steel or fibre-glass frameworks with central torque boxes, torque tubes and double V-trusses, or strutted frames have been proposed and manufactured (S. Kalogirou et al. 1994; Lupfert et al. 2000; Lupfert et al. 2001; Kötter et al. 2012).

PTCs can be aligned with their long axis along either the north-south or the east-west direction. North-south orientated solar collectors track the sun from east to west and thus a lower diurnal variation in the collector performance can be achieved. On the contrary, solar collectors with an east-west orientation exhibit high optical losses during the hours after sunrise and before sunset due to the large incidence angles at which sun rays strike the collectors' surface. However, collectors with an east-west orientation can obtain incidence angles of $\theta = 0°$ at noon and thus reach a higher peak performance over the day compared to north-south-orientated solar collectors, for which optical efficiency losses are highest at noon. Additionally, a north-south collector field is more sensitive to seasonal variations in the sun angle vis-à-vis an east-west field, leading to more uneven annual energy yield profiles. North-south orientated collectors exhibit a higher energy yield during the summer, while east-west parabolic troughs collect higher amounts of energy during the winter because they are capable of following the seasonal movements of the sun in the sky. Overall, slightly higher amounts of solar energy can be captured on an annual basis with a north-south alignment of the PTC field but at the expense of more tracking adjustments. Optical losses of a PTC field related to the incidence angle of the rays on the collector surface—referred to as "cosine losses"—are reviewed for the various modes of tracking by (S. A. Kalogirou 2004). Because each field orientation exhibits different advantages and disadvantages, PTCs should be aligned according to the specific application in order to best match demand variations. Besides cosine losses, additional losses in the optical efficiency are induced due to shading effects when more than one collector row is installed. In particular, especially for small sun angles, the energy collected by a collector row can be significantly reduced due to the shadow on it cast by an adjacent collector row. Therefore, distance between collector rows should be optimized in order to reduce shading effects while keeping an eye on the adverse effects an excessive increase in the distance between rows might have on the economic viability of the PTC field due to increased land use and longer HTF piping. Typically, the optimum row distance, d_{row}, is estimated at approximately three times the aperture width w of the solar collector. Finally, optical losses introduced by sand or dirt depositions and their negative effect on the reflectivity of the parabolic mirrors should be kept to a minimum to maximize solar energy collection.

3.2.2.2 Linear Fresnel Reflectors

Notwithstanding the fact that parabolic mirrors are from a geometric point of view the best candidates for concentrating solar collectors, linear Fresnel technology makes use of flat or slightly concave reflectors. The linear Fresnel reflector (LFR) system consists of several rows of mirror segments, each capable of rotating along a single axis to track the sun's position in the sky and focus solar radiation onto a solar receiver that is fixed at an elevated linear tower aligned parallel with the rotational axis of the mirrors, as illustrated in Figure 3.16. Incident sunlight is absorbed on the receiver surface and transferred to a HTF before being converted into useful energy. The most common types of receiver used in linear Fresnel systems are evacuated (Mills 1995) and non-evacuated (Supernova 2011) absorber tubes, as well as inverted cavity receivers (Reynolds et al. 2004), in which the solar absorbing surface is enclosed in an insulated cavity with an aperture through which the concentrated sunlight enters. Usually 2D CPCs or other types of non-imaging secondary reflectors are used in tandem with the primary concentrating system to further augment the intensity of the solar radiation incident on the absorber receiver. The linear Fresnel collector fields can deliver useful thermal energy at temperatures up to ~450–500°C.

Fresnel systems might not make use of large, full-surface parabolic mirrors but the reflector rows are aligned in a way that imitates the shape of PTCs. All mirror rows rotate by the same number of degrees during a day. However, as a result of the fixed position of the solar receiver, each row is tilted at a different angle with respect to the sun rays so that the incident sunlight can be efficiently focused onto the solar receiver. Therefore, the amount of solar radiation intercepted by every mirror row is reduced by the factor

$$\cos\theta = \frac{I_c}{I_{DNI}} \tag{3.24}$$

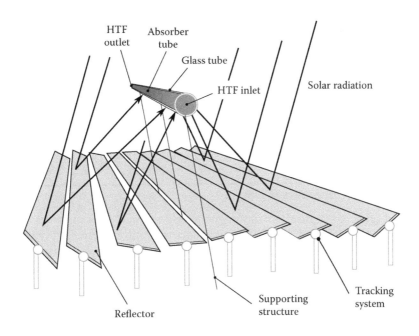

FIGURE 3.16
Schematic layout of a linear Fresnel collector array focusing incident solar radiation onto an elevated, fixed-position tubular receiver.

where θ is the angle between the sun rays and the surface normal to the collectors, I_{DNI} is the direct normal irradiance, and I_c is the radiation flux intercepted by a mirror row. Further optical losses in the system are introduced by shading, during which a mirror row prevents a portion of sun rays from reaching the row located behind it. Contrary to PTC systems, in LFR designs there is also a second process besides shading that reduces the amount of energy reaching the solar receiver. This process is called blocking and occurs when part of the radiation redirected from a reflector row does not reach the receiver but is intercepted by the rear side of the front collector row. The amount of shading and blocking is directly related to the relative receiver-mirror position as well as to the height of the linear tower height and the sun angle. Furthermore, due to the fixed elevated position of the solar receiver, the sun rays redirected from the solar collectors have to travel a longer distance vis-à-vis the one in PTC systems, and are therefore subject to atmospheric attenuation. Optical losses due to atmospheric attenuation are dependent on the specific atmospheric and weather conditions of the site as well as on the distance between the individual reflector row and the tower receiver. Although all these features induce important losses in concentration and power, the manufacturing costs of Fresnel collectors are considerably lower vis-à-vis the bulky PTCs due to their smaller size and less curved shape with a longer focal length. Additionally, Fresnel collectors eliminate the need for heavy and stiff supporting structures since they are mounted close to the ground, thus reducing not only investment costs but also the operating and maintenance costs of the solar plant as no special equipment is required to wash the collector mirrors in order to prevent degradation of their optical quality by sand or dirt depositions.

In an attempt to increase the efficiency of linear Fresnel systems, an alternative configuration involving reflector rows with alternating orientations and more than one receiver tower has been proposed. Systems adopting this configuration are referred to as compact linear Fresnel reflector systems (CLFR). The presence of more than one absorber receiver enables a design with non-uniform orientation of the mirror rows since reflected radiation can be redirected toward either of the two tower receivers located in the vicinity of the individual reflector. Thus, an alternating mirror orientation is selected in CLFR systems as the best solution in terms of minimizing the shading and blocking effects of adjacent reflectors (Mills 2013). Drastic reduction in the amounts of shading and blocking allows for construction of receiver towers with lower heights and costs vis-à-vis the original LFR systems. This mirror arrangement further enables a considerable decrease in the spacing between individual reflector rows and, therefore, more efficient land use.

3.2.2.3 Central Tower Receivers (CTR)

Considerably higher concentration ratios and working fluid temperatures—compared to those of parabolic trough and linear Fresnel collectors—can be achieved with central solar tower systems. Solar tower technology makes use of an array of individual, pole-mounted, flat, or slightly concave mirrors—referred as heliostats—with dual-axis tracking capability to redirect incident direct-beam solar radiation and concentrate it onto a centrally located, elevated absorber receiver or a secondary reflector, as illustrated in Figure 3.17. Every heliostat consists of several mirror modules that are typically made of low-iron glass and exhibit a surface reflectivity of ~0.9. The receiver at the top of the tower effectively intercepts and absorbs the incoming concentrated solar radiation. Absorbed heat is subsequently transferred in the form of thermal energy to a HTF flowing through the receiver and carried away to satisfy a load or to be stored for later use. Commonly, HTF temperatures of up to 1500°C are obtained and the heat captured by solar tower receivers is used for driving power cycles or for the provision of high-temperature process heat.

Contrary to PTC and linear Fresnel systems, where sunlight is focused along the focal line of the collectors, in solar tower systems sun rays are focused onto a single focal point

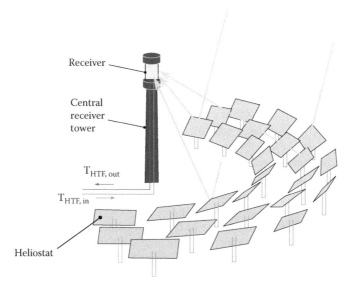

FIGURE 3.17
Schematic illustration of a central tower receiver system using an equator-facing, sun-tracking heliostat field to focus sunlight onto a central receiver mounted at the top of the tower.

leading to mean flux concentration ratios $\tilde{C}$ in the range of 500–5000. In practice, since sun rays are not completely parallel as they originate from a circular sun disk, and due to geometrical and optical imperfections of the mirrors, reflection of the sunlight by the heliostats forms a circular image of the sun at the focal plane of the heliostats, as shown in Figure 3.14. Since collection and conversion of solar energy takes place at a fixed point, parasitic thermal losses from the HTF piping to the surroundings can be reduced significantly vis-à-vis line-focusing concentrating systems. The fixed position of the solar receiver implies that the two-axis tracking heliostats are not pointing directly to the sun, i.e. the angle θ formed between the sun rays and the surface normal to the heliostat is typically not equal to zero. Specifically, the tracking mechanism turns each individual heliostat so that its surface normal bisects the angle formed between the sun rays and the line path from the heliostat to the receiver. Cosine losses are dependent on the relative position of the individual heliostat to the solar receiver and on the sun's position in the sky. Cosine losses represent the most important loss factor of a heliostat field. During morning hours, heliostats east of a north-facing receiver tower exhibit high losses while mirrors west of the field have a higher optical efficiency. Opposite trends can be observed during the hours before sunset. The effects of shading, blocking, and atmospheric attenuation on the optical efficiency of a heliostat field are similar to a linear Fresnel system. Especially at small sun angles, individual heliostats cast a shadow on mirrors located behind them and prevent them from collecting all the incident solar flux. Since shading and blocking are largely dependent on the relative receiver-heliostat position, an optimal spacing between the individual heliostats should keep these effects to a minimum while avoiding high land-use costs (Cádiz et al. 2015). In regard to the layout of the heliostat field, the distance between the furthest heliostat and the central tower should not exceed a value over which optical losses due to atmospheric attenuation increase considerably. Apart from the aforementioned losses, other factors affecting the optical efficiency of a central tower system are degradation in the reflectivity of heliostat mirrors and spillage, i.e. when a portion of the reflected radiation does not reach the receiver due to tracking inaccuracy or geometrical imperfections of the heliostats. The overall optical efficiency of the solar heliostat field is defined as:

$$\eta_{opt} = \frac{Q_{solar}}{I_{DNI} \cdot A_{heliostats}} = \eta_{cos} \cdot \eta_{shadowing} \cdot \eta_{blocking} \cdot \eta_{ref} \cdot \eta_{atten.} \cdot \eta_{spillage} \qquad (3.25)$$

where Q_{solar} is the solar power intercepted by the receiver and A is the surface area of all the heliostats comprising the solar field. It can be expressed as the product of all the optical loss factors mentioned above. Keeping optical losses at the lowest level possible is of great importance for the economic viability of a central tower system. Higher optical efficiency implies reduction of the heliostat field size and in the investment costs of a central tower system, as the solar field size typically represents ~40–50% of capital costs (Kolb et al. 2007).

There are various configurations of the heliostat field and the central tower depending on the solar receiver type. External receivers typically consist of panels of vertical tubes welded side-by-side in order to approximate a cylindrical shape. Due to the quasi-cylindrical shape of the receiver, the heliostat field in this configuration surrounds the central tower, and the solar radiation reflected by the mirrors is absorbed on the surface of the vertical tubes that carry the HTF. Designs with the heliostat field surrounding the central tower result in a lower distance between the tower and the furthest heliostat as well as in a shorter and lower-cost tower. However, as the high-temperature heat-absorbing surface of the receiver is exposed to the surroundings, high convective and thermal radiative losses are inherent to this design, leading to lower energy absorption efficiencies η_{abs}. Thus, the surface area of the receiver tubes has to be kept as small as possible and its lower limit is determined by the desired HTF temperature and heat removal capability.

In order to minimize the area over which thermal losses to the surroundings occur and thus achieve higher absorption efficiency, cavity-receiver configurations have been proposed. Decisions on the size of the aperture through which solar radiation enters the cavity receiver involve a trade-off between intercepting a high amount of solar radiation and keeping thermal radiation and convection losses to a minimum (Steinfeld and Schubnell 1993). Since solar irradiation is only captured through the receiver aperture, heliostat fields in cavity-receiver systems usually do not surround the central tower. Instead, the mirrors are positioned on one side of the tower and within a quasi-conical region that is defined by the normal to the aperture and the acceptance angle of the receiver. The receiver aperture typically faces toward the pole and is slightly tilted in order to efficiently capture the irradiation coming from the equator-facing heliostat field. However, surround heliostat fields are also applicable in a cavity-receiver configuration if multiple cavities are placed adjacent to each other so as to face different parts of the solar field.

Finally, an alternative configuration that involves positioning the solar receiver at ground level is the so-called "beam-down" technology. In this arrangement, a hyperbolic secondary reflector at the top of a solar tower is used to redirect sunlight collected by the heliostat field to an upward-facing receiver located at ground level (Segal and Epstein 2001). Although installation of the receiver close to the ground has important benefits for the performance of O&M tasks, beam-down central receiver systems are less developed than elevated tower receiver concepts and are characterized by increased optical losses due to the insertion of the hyperbolic reflector between the heliostat field and the elevated focal point. Furthermore, the use of a secondary reflector magnifies the sun image leading to lower concentration ratios. Thus, an array of CPCs is usually mounted at the receiver aperture to recover some of the lost magnification at the expense of additional investment costs.

In an attempt to reduce the magnification of the image produced at the receiver aperture, another beam-down optical system configuration which suggests the use of flat rather than curved optics for the secondary reflector and mounting the reactor at an elevated position closer to the reflector has been proposed (Vant-Hull 2014). In this way, the beam-down length is substantially reduced and relatively high concentration ratios can be achieved at the reactor aperture without the need for CPCs.

3.2.2.4 *Paraboloidal Dish Reflectors*

Paraboloidal dish reflectors (PDR) are point-focusing concentrating systems as well. They make use of a 3D paraboloidal-shaped mirror with dual-axis tracking capability to reflect incident solar radiation toward a receiver positioned close to the focal plane of the paraboloid, as illustrated schematically in Figure 3.18a. After being intercepted and absorbed by the solar receiver, solar radiation is transferred to a HTF in the form of thermal energy and is then converted into electricity either using a heat engine directly coupled to the receiver, or after being transported through the HTF piping to a central energy conversion unit. Alternatively, concentrated solar energy can be used in solar receiver-reactors as the source of process heat for driving highly-energy intensive thermochemical processes. The most common type of receiver for solar dish systems is the cavity-receiver technology that allows for reduction of convective and thermal radiation losses from the high-temperature receiver to the surroundings. With regard to the aperture size of cavity receivers, the same design principles as those used for central tower systems apply.

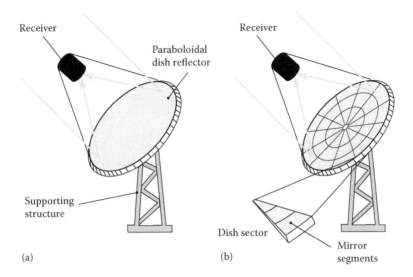

FIGURE 3.18
Schematic of a (a) full-surface and (b) multi-faceted paraboloidal dish reflector.

Contrary to CTRs, the solar receiver of dish systems is not at a fixed location but capable of following the movements of the sun tracking system as it is integrated to a supporting structure that is connected to the frame of the dish concentrator. It is this important difference that allows the optical axis of the PDRs to continuously point toward the sun, thus leading to minimal cosine losses and high optical efficiency. Dish concentrators exhibit the highest concentration ratios among all imaging concentrating collectors with typical mean flux concentration ratios $\tilde{C}$ in the range of 1000–10000. Therefore, HTF temperatures in excess of 1500°C can be achieved with dish concentrator systems.

A dish concentrator is constructed either as a large, full-surface paraboloidal mirror or consists of several parabolic-shaped mirror modules mounted side-by-side on a mirror supporting structure to approximate the prescribed paraboloidal shape, as shown in Figure 3.18b. Higher rigidity and optical quality can be achieved with full-surface concentrator designs since the entire concentrator is shaped via a forming process. This significantly restricts any geometrical imperfections that might have an adverse effect on the optical performance of the mirror but at the expense of high manufacturing costs. Segmented concentrators offer a lower-cost manufacturing solution with higher modularity. However, because the mirror segments composing the concentrator are mounted to the supporting structure and aligned individually, facet misalignments might lead to deviations from the exact paraboloidal shape and, subsequently, to losses in concentration and power (Andraka 2008). The reflective surface of dish concentrators consists typically either of silver-coated glass or of thin-glass mirrors bonded on a metallic or plastic substrate. Various concepts have been proposed for the supporting structure, most of which make use of steel frameworks, space frames, or steel trusses (Coventry and Andraka 2017). Similar to PTCs, the optical performance of a dish concentrator depends largely on the design of the support structure. Structures with high stiffness reduce the bending and torsional deflections of the paraboloidal concentrator due to the self-weight and wind loads applied on their reflective surface, and thus prevent geometrical deviations from the ideal paraboloidal shape.

3.3 Collector Applications

The technological development in the field of solar thermal collectors was largely driven by the high potential of solar energy to contribute to the drastic decrease of anthropogenic CO_2 emissions in all energy end-use sectors as highlighted in Chapter 1. Depending on the specific application, different types of collectors are used to harness solar energy and generate heat at the desired temperature level. The present section focuses on the systems developed to enable integration of solar thermal energy into various applications in the residential, commercial, and industrial sectors. Depending on the temperature level at which thermal energy is delivered, solar thermal systems are classified into:

- Low-temperature systems
- Mid- and high-temperature systems

3.3.1 Low-Temperature Solar Thermal Systems

3.3.1.1 Water Heating Systems

Solar-driven water heating systems for building applications make use of collectors that absorb incident radiation and transfer it to the HTF flowing through them to heat up water, which then can be either used directly or stored for later use. These systems can be classified into two categories depending on the driving force of the HTF circulation:

- Natural circulation systems
- Forced circulation systems

In natural circulation systems, density differences created by heating the HTF in the solar collector serve as the driving force for circulation, and, thus, the costs associated with the pumping and control equipment are avoided. In both types of systems, the HTF can be either water that is heated directly in the solar collector (direct systems), or another medium that flows through the collector tubes and then transfers the absorbed heat through a heat exchanger to a well-insulated water storage tank (indirect systems). All solar-driven water heating systems are accompanied by an auxiliary heater operated using a conventional energy source to produce heat during hours with limited or no solar insolation.

The most common configuration of natural circulation systems is the so-called thermosiphon. A typical direct thermosiphon system is illustrated in Figure 3.19a. The HTF flow rises through a typically flat-plate collector and enters the top of the storage tank, pushing low-temperature HTF from the bottom of the thermally-stratified tank toward the collector. Because the flow of the HTF is density-driven and in order to avoid reversion of the flow and mixing of the hot water stored in the tank with cold collector water during hours with no sunshine, the storage tank needs to be installed well above the solar collector. This means that most of the thermosiphon system components will be exposed to weather conditions and therefore must be protected from freezing at low ambient temperatures. Furthermore, in order to maintain the already low pressure differences in the system, low HTF mass flow rates and circulation pipes with a high diameter are used to minimize linear pressure losses.

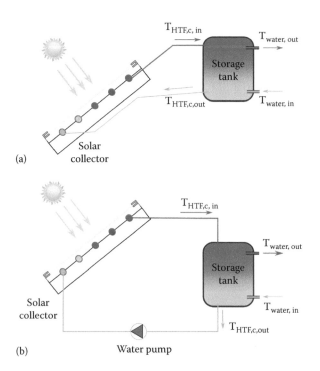

FIGURE 3.19
Schematic layout of a (a) natural-circulation thermosiphon and (b) direct, forced-circulation water heating system.

An alternative natural circulation system proposed for solar-driven domestic water heating is the integrated collector storage (ICS) system, which makes use of a 2D-CPC concentrator with a tubular solar absorber that also serves as the water storage tank. While, in principle, higher temperatures can be achieved with this configuration due to the presence of non-imaging concentration optics, ICS systems suffer from significant thermal losses from the water storage tank to the surroundings because the double role of the tank only allows for a very limited portion of it to be thermally insulated. Design guidelines and details regarding the construction and performance of ICS systems are provided by S. Kalogirou (1997).

The operating principle and configuration of forced circulation systems are similar to thermosiphon systems, with the main difference being that a pump is employed for the circulation of the HTF, as shown in Figure 3.19b. The use of a circulation pump offers higher operational flexibility and enables installation of the storage tank below the solar collector and thus inside the building, but introduces a higher level of complexity because a control system is required to optimize operation. Direct systems using water as the HTF are more common in frost-free climates and in regions with low-acidic, low-mineral water that inhibits corrosion and/or clogging of the piping system. In indirect systems, the pump circulates the HTF—typically water/ethylene glycol solutions, silicon oils, or refrigerants—from the solar collector to the storage through a closed piping loop. Thus, contrary to direct systems, an expansion tank and a pressure relief valve is integrated into the system to prevent overpressure build-up. Besides liquid media, air represents another attractive HTF for indirect water heating systems because it is freely available and has superior physical and chemical properties (non-toxic, non-corrosive, no phase changes involved). In this kind of system, solar energy captured by air-based FPCs is transferred to an air flow circulating in a closed piping loop and subsequently to the water storage tank through an air-to-liquid heat exchanger. However, as explained in Section 3.1.1, heat transfer in air-based collectors is limited by the low absorber-to-air convective heat transfer coefficient. Further disadvantages of air systems for water heating are the additional space required for air ducts and fans vis-à-vis water piping and pumps as well as the higher energy consumption losses for circulation of the HTF.

3.3.1.2 Space Heating Systems

Integration of solar thermal energy into conventional space heating systems can contribute considerably to reducing anthropogenic CO_2 emissions produced in the buildings sector, since space heating represents approximately 80% of the heat consumption in buildings in OECD countries. Therefore, solar space heating systems have recently been developed and can be classified into two main types:

- Active systems
- Passive systems

Passive systems do not involve the use of solar thermal collectors but rather use the intelligent design of building elements so that they are capable of collecting, storing, and distributing solar energy in the form of thermal energy during the winter and reject it during the summer, thus achieving thermal comfort. Proper orientation and ventilation of the building, selection of appropriate construction materials, and implementation of shading techniques are key parameters for achieving an effective solar energy control. A review of the various passive solar systems for space heating is presented in Chan, Riffat, and Zhu (2010).

Active solar-driven space heating systems are very similar to water heating systems in terms of both the collector technologies employed for harnessing solar radiation as well as the overall system configuration. Taking advantage of these similarities, solar systems for combined water and space heating have attracted considerable attention because they can provide fossil fuel energy savings in the range of 25–30% for a typical building (International Energy Agency 2012). They are most commonly designed to satisfy part of the annual space and water heating demand in order to prevent system oversizing and associated high capital costs. An auxiliary heater is employed to cover the unsatisfied heat load.

Direct air-based systems are commonly used for space heating. A typical configuration is schematically shown in Figure 3.20. They make use of air-based FTCs to capture solar radiation that is then transferred through the air flow either to the building in order to meet the thermal load or to a storage tank for later use. Packed beds arrangements of solid materials represent the most suitable and widely applied storage units for air-based solar systems since they are low-cost solutions and enable air to also be used as the HTF in the storage unit, thus simplifying the system design. Charging of the TES unit proceeds during hours of sunshine by flowing solar-heated air through the storage material. During cloudy or overcast periods, the air flow is reversed and low-temperature air enters the tank to recover the stored energy. However, an important disadvantage of this system is the high variability of the air outlet temperature during the discharging phase of the storage unit as the air temperature largely depends on the amount of thermal energy stored in the TES unit when operating the system at constant air mass flow rates. This limitation can be overcome by varying the air flow rates but low rates will result in reduction of the solar collector's performance.

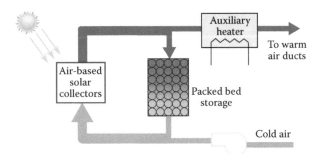

FIGURE 3.20
Schematic configuration of a direct, air-based solar system for space heating with a packed bed energy storage unit. An auxiliary heat source tops up the air temperature during hours of limited or no sunlight.

Indirect, liquid-based systems use water or water-antifreeze mixtures as HTFs, and are most commonly applied for combined water and space heating. In these systems, FPCs or ETCs with integrated CPC reflectors capture incident solar radiation. The heat absorbed is carried by the HTF to a storage unit and then transferred to the service water or the air via load heat exchangers. Liquid-based systems are more complex and expensive compared to air-based systems, but the superior thermal properties of liquid HTFs enable the use of more compact storage tanks as well as operation of the solar collector at higher efficiency.

All space heating systems are installed along with an auxiliary heat source that tops up the air temperature when solar energy alone does not cover the building heat demand. Besides conventional energy sources, liquid-based systems can also make use of water-to-air heat pumps as an auxiliary heat source. With this arrangement, heat stored in the water tank will be extracted by a HTF stream flowing through a piping network submerged in the tank. The HTF will then be compressed by an electrically driven heat pump and release its heat in the building to meet the space heating demand.

3.3.1.3 Space Cooling and Refrigeration Systems

A very promising application for solar thermal energy is the provision of comfort cooling. Interestingly, the peak cooling demand in the buildings sector coincides with the hours of maximum solar radiation during a day, thus providing the option to replace conventional electrically-powered air conditioning systems and reduce electricity demand during peak-load periods. Solar space cooling can also be provided during the night or cloudy hours of the day by utilizing thermal energy stored in a TES tank. Additionally, several types of solar cooling systems can be adapted to provide solar space and water heating during the winter. Besides space cooling, industrial refrigeration represents another attractive field for applying solar-driven cooling technologies. A comprehensive review of solar and low-energy cooling technologies for buildings is provided by Florides et al. (2002).

Solar cooling systems can be classified into three main categories:

- Mechanical cooling systems
- Thermally driven desiccant evaporative coolers
- Thermally driven absorption/adsorption coolers

Mechanical systems make use of solar energy to power a prime mover that drives the vapor compressor of a conventional vapor compression refrigeration cycle. The prime mover can be either an electric motor supplied with solar electricity generated in PV cells or a Rankine engine supplied with thermal energy captured by a solar collector. An important advantage of mechanical cooling systems is that they enable the use of conventional space cooling and refrigeration equipment. On the other hand, limiting parameters for the utilization of PV-based systems have been—until recently—the low efficiency and high costs of PV cells. For Rankine-engine-based systems, a major problem is that the rotational speed of the Rankine engine is dependent on the availability of solar energy and seldom matches the energy input required by the vapor compressor.

Thermally driven desiccant coolers are open cycle systems that are based on a combination of air dehumidification and evaporative cooling to produce cool air directly. Desiccant coolers are referred to as open cycle systems to underline the fact that the refrigerant is discarded from the system after providing the cooling effect. Desiccant coolers make use of solid (e.g. silica gel, zeolites, activated alumina) or liquid (e.g. LiBr, LiCl) desiccant materials to extract moisture from air coming from the surroundings before the air flow is driven through an evaporative cooler, where cool air is produced. Solar thermal energy is then used to heat up the desiccant material and facilitate release of its adsorbed moisture at high temperatures.

To ensure continuous adsorption of moisture by the desiccant material, a typical configuration for a system using solid desiccant materials involves a desiccant wheel which rotates through two separate compartments. In the first compartment, adsorption of moisture by part of the desiccant wheel as well as the evaporative cooling process proceed. In the second compartment, a second air stream flows through a heat exchanger to absorb solar thermal energy captured by FTCs and dehumidifies the other part of the desiccant wheel at temperatures in the range of 50–75°C before being exhausted to the surroundings.

In desiccant evaporative cooling systems using liquid desiccant materials, the liquid desiccant is sprayed into an absorption unit containing air to draw its moisture. Low-moisture air subsequently flows through an evaporative cooler, where cool air is produced and used to cover the cooling demand. On the other hand, the H_2O-diluted desiccant exits the absorption unit and is pumped through a liquid-liquid sensible heat exchanger into a regeneration unit. There it is sprayed into a stream of solar-heated regenerative air, to which the moisture is transferred, and then flows back to the absorption unit to close the cycle while the moist air stream is exhausted to the surroundings.

Finally, thermally driven adsorption and absorption coolers are closed cycle systems producing chilled water—instead of air directly—that can be supplied to any type of space conditioning and industrial refrigeration unit. Adsorption chillers consist of two compartments. The first compartment contains a low-pressure evaporator with liquid water as the refrigerant and a chamber filled with sorbent material. The sorbent material creates an extremely low-humidity condition in the compartment, thus causing the liquid water to evaporate at low temperatures. During the water phase change, heat is removed from the system. In particular, a separate water stream flowing through the evaporator is cooled down and is used to meet the cooling demand. The sorbent in the compartment is continuously cooled to enable uninterrupted adsorption of the refrigerant vapor. The second compartment contains a condenser and a second chamber filled with sorbent material. The sorbent material in this chamber is initially saturated with water vapor and is regenerated by absorbing heat from a solar-heated water stream flowing through it. The water vapor removed from the sorbent during its regeneration process releases its heat in the condenser and flows in its liquid state to the evaporator. Common sorbent materials used in solar adsorption chiller units are silica gels and zeoliths. Solar adsorption chillers are simple, very robust systems with no moving parts. They do not involve the use of toxic or corrosive materials and are capable of working at temperatures as low as 55°C. They are, however, bulky and heavy. Typically, a COP close to 0.6 is achieved at driving temperatures close to 80°C.

In solar-driven absorption chillers (Herold, Radermacher, and Klein 2016), a refrigerant-absorbent mixture contained in an absorber unit is pumped into a generator unit. Thermal energy in the form of solar-heated water is added to the generation unit and induces evaporation of the refrigerant and separation from the absorbent material, which flows back to the absorber after being cooled down in a heat exchanger. The refrigerant vapor then passes through a condenser unit where it rejects its heat and returns to its liquid state. It then circulates by means of an expansion valve into a low-pressure evaporator. There, the cooling effect is provided and the refrigerant is vaporized and returned back to the absorber, where it gets attracted by the absorbent material. Common absorbent-refrigerant pairs used in solar absorption systems are $LiBr$-H_2O and H_2O-NH_3. The H_2O-NH_3 system requires driving temperatures in the range of 95–120°C, while generators of $LiBr$-H_2O systems operate at 70–95°C. Typically, a COP in the range of 0.6–0.8 is achieved (Duffie and Beckman 2013). To increase the COP of absorption chillers to levels close to 1.1–1.2, "double effect" systems using two generator units have been introduced but require generator temperatures of 150–180°C, thus setting higher requirements on the solar collector unit.

3.3.1.4 Water Desalination Systems

Apart from the escalating energy demand of the recent years, rapid population and industrial growth created a considerable increase in the demand for fresh water. In parallel, river and lake water pollution levels are constantly rising due to disposal of industrial waste and sewage into these water resources. Thus, the production of fresh water at large quantities is restricted largely to highly energy-intensive water desalination techniques. For the production of 25 million m^3/day, conventional desalination technologies consume approximately 230 million tons of oil per year and contribute significantly to global anthropogenic CO_2 emissions (S. A. Kalogirou 2004). Replacing conventional energy sources used in water desalination plants with solar energy has the potential of generating large fossil fuel and CO_2 savings.

Solar-driven water desalination systems can be classified into two categories:

- Direct collection systems
- Indirect collection systems

Direct systems capture solar energy and produce fresh water without the use of mechanical or electrical equipment. The most representative solar collector used for direct water desalination is the so-called solar still. A schematic configuration of a double-slope, symmetrical-basin solar still is shown in Figure 3.21. It consists of a seawater-containing basin with a black bottom surface that is enclosed in an inverted V-shaped glass envelope. Solar irradiation enters through the glass cover and gets absorbed on the bottom of the basin. The temperature of the salt water rises and water vapor is produced, which subsequently rises inside the solar still and condenses on the underside of the glass cover. Distilled water takes advantage of the tilt angle of the glass cover and runs down into water collecting channels. Similar to FPCs, the glass cover has high transmittance to short-wave radiation and low emittance for long-wave thermal radiation, in order to minimize radiation losses to the surroundings. A typical

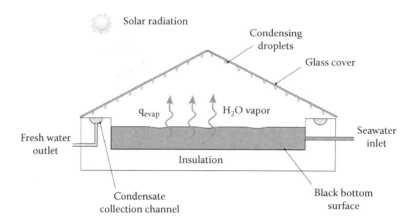

FIGURE 3.21
Cross-sectional view of a double-slope, symmetrical-basin solar still.

solar still achieves thermal efficiencies close to 35% and daily production rates of 34l/m². The low production rates of conventional solar stills are directly related to the low brine depths used as a result of the limited heat transfer area. Other important disadvantages of a conventional solar still is its significant thermal losses to the surroundings because its large surfaces are in contact with the surrounding air or the ground, and the fact that the heat of condensation released on the undersize of the cover is rejected as waste heat from the system. Design modifications with respect to the tilt angle, cover slope and shape, and the flow pattern of the seawater feed were made to improve the thermal performance of solar stills (Khalifa and Hamood 2009; Tanaka and Nakatake 2009).

Multi-effect basin stills achieve a 35% efficiency increase by making use of several compartments placed on top of one another so that the heat of condensation released in a lower compartment can be re-utilized for heating the water in the compartment above (Qiblawey and Banat 2008). Apart from the abovementioned passive designs, the thermal performance of solar stills can be increased by integrating an external source of heat like a non-concentrating (FPC, ETC, PVT) or concentrating (PTC) solar collector to the system in order to achieve higher temperatures and evaporation rates. A review of the various solar still designs is presented by Deniz (2015), whereas Sampathkumar et al. (2010) focuses on the development of a thermal model to study the performance of various types of active solar distillation systems.

Another direct solar desalination method is the solar humidification-dehumidification (HD-DHD) (Narayan et al. 2010). It was introduced to prevent direct contact between the solar collector and seawater thus preventing corrosion of the solar still components. It makes use of an air-based FTC to capture solar energy as depicted in Figure 3.22. The air stream passing through the collector is heated up and then flows through a humidifier in which seawater is sprayed. The solar-heated air stream absorbs moisture from the seawater spray and is then driven through a dehumidification unit containing a piping system for the circulation of seawater. The incoming humid air stream condenses on the external surface of the pipes, leaving fresh water at the bottom of the dehumidification unit. The heat of condensation is used to preheat the seawater before it enters the humidification unit.

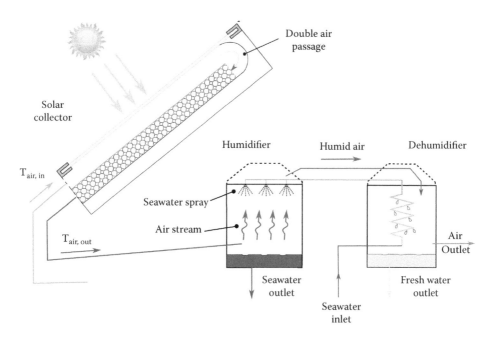

FIGURE 3.22
Schematic illustration of a solar humidification-dehumidification system using a double-passage, air-based collector.

Indirect solar desalination methods employ two separate subsystems to drive desalination: (1) a solar collector array and (2) a conventional desalination plant. They can be distinguished into two categories:

- Membrane processes
- Thermal processes

In membrane processes, electricity generated by PV cells or by a solar-driven Rankine engine can be used either for driving prime movers or for the ionization of salts contained in the seawater. On the other hand, thermally driven processes use the solar radiation as the source of the process heat required for driving water desalination. The most widely applied thermal desalination processes are multi-stage flash distillation (MSF) and multi-effect distillation (MEF) with or without vapor compression (Spiegler 1977). MSF is the predominant conventional thermal desalination process, accounting for 45% of total world desalination capacity. A solar-driven MSF system consists of multiple vessels—the so-called "stages"—each containing a heat exchanger and condensate collector positioned below as illustrated in Figure 3.23a. Every vessel is maintained at a different pressure level that corresponds to the boiling temperature of seawater of various salt concentrations. During steady state operation of the system, saltwater is pumped through the heat exchangers of the vessels. After being preheated, seawater is further heated either directly in a solar collector or indirectly in a separate heat exchanger. Hot saltwater then enters the last vessel at a temperature far above its boiling point at the pressure of the stage. Thus, a small fraction of the water evaporates, rises inside the stage, and condenses on the surface of the heat exchanger tubes. Fresh water drops into the condensate collector while the remaining unevaporated saltwater passes on to subsequent stages where the same procedure takes place at lower temperatures and pressures. Although higher temperatures are beneficial for the thermal performance of a MSF plant, a maximum temperature of 120°C is typically applied to reduce corrosion of metal surfaces. In a solar-driven process, any type of collector, whether non-concentrating (FPC, ETC, PVT) or small concentrating (PTC, LFR, CTR, PDR), can be integrated to drive the process.

MEF distillation consists of multiple vessels containing heat exchangers as well. However, in this process only the first vessel is heated using an external energy source. In a solar process, steam produced in a solar collector flows through the heat exchanger of the first vessel—referred to as "effect". Seawater is sprayed on the outer surface of the heat exchanger pipes and part of it evaporates and is passed through the heat exchanger pipes

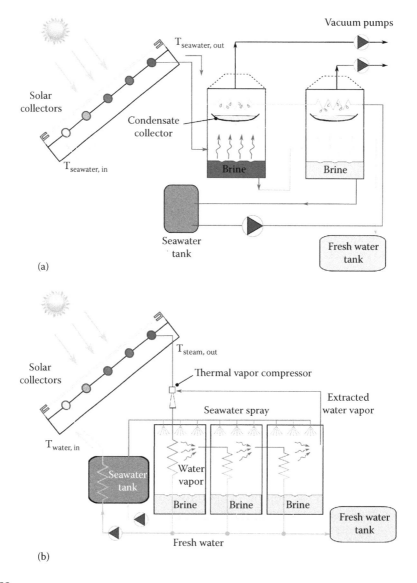

FIGURE 3.23
Schematic representation of (a) a solar-driven two-stage flash distillation system and (b) a multi-effect desalination system with thermal vapor compressor.

of the next effect as shown in Figure 3.23b. In the next effect, seawater evaporates following the procedure described above and water vapor inside the heat exchanger pipes condenses to distilled water. To further improve the thermal efficiency of the process, often part of the water vapor produced in the final effect is extracted and compressed by a thermal vapor compressor so that it can be re-employed as an additional heat source in the first effect.

3.3.2 Medium- and High-Temperature Solar Thermal Systems

3.3.2.1 Industrial Process Heat Systems

A large number of industrial units consume significant amounts of process heat. Heat production is currently based on the combustion of conventional fossil fuels, which reveals the large potential of solar thermal energy systems toward achieving considerable CO_2 savings in the industrial sector. Because the temperature at which industrial process heat is delivered varies considerably depending on the specific process, different types of solar collectors are applied in solar-driven heat production. For low- to medium-temperature applications, the solar thermal technologies used are similar to those employed in the buildings sector, such as FPCs and ETCs with or without CPC reflectors. For provision of heat at medium- to high-temperature levels, low-concentrating solar technologies are applied. Potential applications of solar thermal technologies for the provision of process heat include:

- Heating of liquid and gaseous media used at different stages of industrial processes
- Generation of steam
- Direct coupling of a process unit to a solar collecting system

The operation principle and configuration of solar systems used for heating liquids and gases is very similar to the systems used for domestic water and space heating that were described in Sections 3.3.1.1 and 3.3.1.2. The heat transfer medium in the solar collector is usually selected so as to match the fluid used in the industrial process and thus decrease system complexity. Depending on the sanitation requirements of the individual process, two main system configurations are applicable for both solar industrial water and air systems: open- and closed-circuit systems. Closed-circuit systems recycle air/water from the process side and supply it back to the solar and their configuration is almost identical to the domestic water and space heating in Section 3.3.1. On the other hand, open-circuit systems are applied in cases where no contamination of air/water is allowed in the process. In these systems, the solar-heated fluid is discarded after covering the heating demand, and fresh fluid is provided to the solar collector.

Steam generation processes are carried out at comparatively higher temperatures. Non-concentrating collectors are not adequate for delivering process heat at this temperature level because their efficiency decreases considerably with temperature. Therefore, concentrating technologies with low concentration ratios like PTCs are most commonly used for the solar-driven production of steam. Solar steam generation systems can be distinguished into three categories (S. Kalogirou, Lloyd, and Ward 1997):

- The steam-flash steam generation concept
- The direct steam generation concept
- The unfired-boiler steam generation concept

Water is circulated through the solar collectors in the first two processes. In the steam-flash process, water is pressurized before passing through the solar collector to avoid evaporation in the piping system. After leaving the solar collector, it flows through an expansion valve into a flash vessel where steam is produced. On the other hand, in the direct steam generation concept, steam is produced directly inside the solar receiver tube because no expansion valve is employed. The high operating pressures and temperatures developed in a steam-flash steam generation system result in higher thermal losses from the collector and in higher parasitic power consumption by the water pump vis-à-vis the direct steam generation concept. The latter offers higher collector performance as well as lower complexity and costs, but the development of a two-phase flow inside the piping system leads to pressure variations and exertion of high stresses on the collector components. The disadvantages of both systems can be overcome by the unfired-boiler steam generation concept, which is currently the predominant process for solar steam generation. Instead of water, another HTF is circulated in a closed loop system through the solar collector and the heat is transferred via heat exchange to a water stream for the production of steam. However, HTFs are inferior to water with respect to safety issues, chemical stability, costs, and thermal properties. Lower heat capacity suggests higher flow rates and pump energy consumption, while lower heat transfer coefficients between the solar receiver and the HTF require operation of the collector at higher temperatures and lead to lower efficiency.

3.3.2.2 Solar Thermal Power Generation Systems

Power generation represents one of the predominant solar energy applications. Besides PV cells, thermal collectors are widely used to collect solar energy to drive heat engines. Thermal power generation systems require high operating temperatures which are not obtainable with non-concentrating collectors. The exclusive use of imaging concentrators, however, implies that these applications are limited to regions with high direct normal irradiance.

Power generation systems based on PTCs or LFRs usually produce steam and supply it to a heat engine for electricity generation. The operating principle and system configuration of industrial process heat systems outlined in Section 3.3.2.1 is directly applicable to PTC-based power generation systems. The unfired-boiler steam generation concept with synthetic oils as the HTF is the most commonly used configuration. With today's technology, the temperature at which parabolic-trough-based power generation systems can deliver useful thermal energy is limited to ~398°C. At higher temperatures the synthetic oil quickly degrades and, therefore, research activities in the field have shifted their focus to overcoming this limitation. Although TES units are not indispensable to plant operation, several PTC power generating plants in operation use molten salt thermal storage systems to increase their annual energy yield and improve plant dispatchability. Plants without TES units typically rely on natural gas boilers to ensure continuous operation. Similar to PTC-based systems, LFR systems use either water or another HTF for the generation of steam. In Murcia, Spain, the linear Fresnel power plant Puerto Erado 1, which has an an installed capacity of 1.4 MW, uses water as the HTF to power a steam turbine. The technical feasibility of using molten salt was also demonstrated.

CTRs can also be utilized for solar-driven power generation. Solar energy collected by the heliostat field is redirected to a cylindrical or cavity receiver mounted at the top of the tower at the desired flux density, and the heat is carried by a HTF either to a storage tank for later use or to the power conversion unit. Several thermodynamic cycles can be driven by CTR systems. For solar power plants driving a Rankine cycle, steam is generated directly or via heat exchange with an intermediate HTF—molten salts, liquid metals—passing through the solar receiver. The majority of CTR plants in operation use an intermediate HTF for the production of steam because this configuration enables integration of a storage unit into the plant, thus improving its overall performance. Since relatively low temperatures are required for driving a Rankine cycle, a cylindrical receiver with a heliostat field surrounding the central tower would suffice for this application. When driving a Brayton or a combined Brayton-Rankine cycle, pressurized air at 5–30 bar is used as the heat transfer medium and the relatively high temperatures required at the gas turbine inlet—typically in the range of 1,000–1,600 K—are obtained by using cavity receivers. Volumetric receivers that absorb concentrated sunlight in the volume of highly-porous material have been proposed to increase the thermal performance of the system. Finally, PDRs are used to drive Rankine, Brayton, or sodium-heat engines. Also, the direct coupling of a Stirling engine to the dish receiver has gained considerable attention.

3.3.2.3 Material Processing and Thermochemical Fuel Production Systems

The mineral processing and extractive metallurgical industries are major consumers of fossil-fuel-derived electricity and of high-temperature process heat and are, consequently, an important contributor to the total GHG emissions in the industrial sector (Rankin 2012). The carbon footprint of these processes can be reduced considerably by using environmentally cleaner energy sources, like concentrated solar radiation, to provide high-temperature process heat. Processes with great potential include: (a) the processing of industrial minerals, (b) the extraction of metals from their ores, and (c) the recycling of scrap metals and metallurgical waste materials. Integration of solar energy into these processes is still at a research and development phase and, due to the high temperatures required, only high-concentrating imaging collectors, like CTRs and PDRs, are considered for potential commercial applications.

Limestone ($CaCO_3$) is one of the most important industrial minerals and its thermal decomposition to lime (CaO) represents the most energy-intensive step in cement manufacturing. The technical feasibility of a solar-driven lime production process has been experimentally demonstrated in a 10 kW_{th} solar multitube rotary kiln leading to peak lime production rates of 4 kg/h (Meier et al. 2005). The potential for integration of solar energy into the processing of industrial minerals was demonstrated also for the glass manufacture. In particular, experimental investigations have shown that the heat treatment of pure silica, ternary simple-soda-lime-silica (SLS) glass batch, and industrial SLS pellets at temperatures close to 1,400°C using concentrated sunlight as the source of the high-temperature process heat resulted in complete conversion of the raw materials to amorphous glasses (Ahmad, Hand, and Wieckert 2014).

Research in the field of solar-driven extraction of metals has largely focused on the production of Zn via the thermal dissociation or carbothermal reduction of ZnO. Both processes have been demonstrated at a pilot-plant scale of 100 kW_{th} (Villasmil et al. 2014) and 300 kW_{th} (Wieckert et al. 2007), respectively, leading to Zn production rates of ~0.2 kg/h and ~50 kg/h. The solar carbothermal reduction of Al_2O_3 was experimentally investigated under vacuum conditions in order to reduce the onset temperature of Al vapor formation, but aluminium contents of only 4–19 wt.% were due to the formation of by-products during the process (Kruesi et al. 2011). Silicon purities of 66.1–79.2 wt.% were obtained during the experimental investigation of solar-driven carbothermal reduction of SiO_2 (Loutzenhiser, Tuerk, and Steinfeld 2010). The solar co-production of Fe and synthesis gas by the combined reduction of Fe_3O_4 and reforming of CH_4 has been investigated using a fluidized bed reactor at temperatures between 1,073 K and 1,273 K, yielding Fe purities up to 68 wt.% (Steinfeld, Kuhn, and Karni 1993).

Research in the recycling of scrap metals and metallurgical processes involves the remelting of aluminium scrap using a solar rotary kiln with a capacity of 1–2 kg Al at temperatures close to 800°C (Neises-von Puttkamer et al. 2016), as well as the recovery of Zn and/or Pb via the carbothermal reduction of electric arc furnace dust (Schaffner et al. 2003) and Waelz oxide (Tzouganatos et al. 2013) in a 10 kW_{th} solar reactor.

Transportation is also a highly energy-intensive sector that currently accounts for ~26% of global energy-related CO_2 emissions (Chapman 2007), and energy demand in this sector is projected to further increase by nearly 40% by 2030 (Turton 2006). Reducing the CO_2 footprint of this sector necessitates the identification of sustainable fuel production technologies driven by renewable energy sources. Emissions can be substantially reduced by using solar energy as the source of high-temperature process heat. Therefore, research in the field of solar-driven thermochemical processes is focuses on identifying potential routes for the production of storable and transportable fuels. Solar-driven thermochemical production of fuels is envisaged to proceed along two parallel routes (Meier and Steinfeld 2010): (1) the production of H_2/CO via the decarbonization of fossil fuels via solar cracking, reforming, or gasification processes, and (2) the production of H_2/CO via H_2O/CO_2- splitting thermochemical cycles.

An extensive review on the solar-driven steam gasification of different types of carbonaceous feedstock using three reactor concepts and at a pressure of ~1 bar is presented by Piatkowski et al. (2011). Recently, a pressurized 3 kW_{th} solar reactor concept was proposed and used to perform the steam gasification of a charcoal-water slurry, leading to a solar-to-fuel energy conversion efficiency of 20% and an upgrade of the calorific content of the feedstock by 35% (Müller et al. 2017). The technical feasibility of the combined solar reduction of ZnO and CH_4 reforming for the simultaneous production of Zn metal and syngas was investigated using a 5 kWt_h vortex-flow solar reactor, leading to chemical conversions in the range of 83–100% and reactor thermal efficiencies up to 28% (Kräupl and Steinfeld 2001). Experimental investigation of the thermal cracking of CH_4 into H_2 and C in a 10 kW_{th} multi-tubular cavity-type and at temperatures in the range of 1550°C–1800°C resulted in a 75% H_2 yield (Rodat, Abanades, and Flamant 2009).

Of particular interest are H_2O/CO_2-splitting thermochemical cycles (Smestad and Steinfeld 2012), the most popular of which proceed in two steps. The first step of the cycle involves the high-temperature endothermic reduction of a metal oxide to the corresponding metal or partially reduced oxide using concentrated solar radiation. In the second step, the metal or partially reduced metal oxide is oxidized with H_2O and/or CO_2 in a non-solar exothermic process for the production of H_2 and/or CO while the corresponding metal oxide is recycled back to the solar reactor thus closing the thermochemical cycle. Various metal oxide redox pairs have been investigated for their suitability for performing H_2O/CO_2-splitting thermochemical cycles, with Fe_3O_4/FeO (Steinfeld 2012) and ZnO/Zn (Villasmil et al. 2014) being the most popular until the focus shifted to the cerium oxides redox pair CeO_2/$CeO_{2-\delta}$ (Chuch et al. 2010) and perovskite-type materials in the form $La_{1-x}Sr_xMnO_{3-\delta}$ due to their faster redox kinetics and favorable reduction extents (Scheffe, Weibel, and Steinfeld 2013).

Nomenclature

Symbols

A	Surface area	[m^2]
c_p	Specific heat capacity	[kJ/kgK]
C_{geo}	Geometric concentration ratio	[–]
$\tilde{C}$	Mean flux concentration ratio	[–]
d_{row}	Distance between individual rows of a PTC field	[m]
f	Focal length	[m]
F	Force	[N]
I_{DNI}	Direct normal solar irradiance	[W/m^2]
$\dot{m}$	Mass flow rate	[kg/s]
p	Pressure	[N/m^2]
q	Heat flux	[W/m^2]
Q	Power	[W]
$Q_{rerad,losses}$	Thermal radiation losses from the solar absorber receiver	[W]
Q_{solar}	Solar power intercepted by a receiver	[W]
S	Solar irradiance absorbed on the collector surface	[W/m^2]
T	Temperature	[°C,K]
T_{opt}	Optimal operating temperature of a solar receiver	[K]
U_L	Collector overall heat loss coefficient	[W/m^{2}K]
w	Aperture width	[m]

Greek symbols

a	Absorptivity	[–]
α_{eff}	Effective absorptance of the solar receiver	[–]
β	Tilt angle	[°]
ε_{eff}	Effective emittance of the solar receiver	[–]
ζ	Temperature coefficient of PV cell	[1/°C]
η	Efficiency	[–]
$\eta_{ex,ideal}$	Ideal exergy efficiency of solar concentrating system	[–]
θ	Angle of incidence	[°]
θ_{acc}	Acceptance half-angle of CPC	[°]
θ_s	Solar zenith angle	[°]
ρ	Reflectivity	[–]
τ	Transmissivity	[–]
$(\tau\alpha)$	Tau-alpha product	[–]
Φ_{rim}	Rim angle	[°]

Subscripts

a	Ambient
abs	Absorber
atten	Atmoshperic attenuation
c	Collector
D	Direct
d	Diffuse
eff	Effective
el	Electrical
g	Ground
i	Inlet
o	Outlet
opt	Optical
ref	Reference
t	Total
u	Useful

Acronyms

DNI	*Direct normal irradiance*
CLFR	*Compact linear Fresnel reflectors*
CPC	*Compound parabolic concentrator*
CTR	*Central tower receivers*
ETC	*Evacuated tube collectors*
FPC	*Flat-plate collectors*
HD-DHD	*Humidification-dehumidification*
HTF	*Heat transfer fluid*
LFR	*Linear Fresnel reflectors*
MEF	*Multi-effect distillation*
MSF	*Multi-stage flash distillation*
O&M	*Operation and maintenance*
PDR	*Paraboloidal dish reflectors*
PTC	*Parabolic trough collectors*
PV	*Photovoltaic collectors*
PVT	*Photovoltaic-thermal collectors*
TES	*Thermal energy storage*

References

Ackermann, J. A., L.-E. Ong, and S. C. Lau. 1995. "Conjugate Heat Transfer in Solar Collector Panels with Internal Longitudinal Corrugated Fins—Part I: Overall Results." *Forschung Im Ingenieurwesen* 61 (4): 84–92.

Ahmad, S. Q. S., R. J. Hand, and C. Wieckert. 2014. "Use of Concentrated Radiation for Solar Powered Glass Melting Experiments." *Solar Energy* 109: 174–182.

Andraka, C. E. 2008. "Cost/Performance Tradeoffs for Reflectors Used in Solar Concentrating Dish Systems." In *ASME 2nd International Conference on Energy Sustainability (ES2008)*, 10–14. Jacksonville, FL.

Cádiz, P., M. Frasquet, M. Silva, F. Martínez, and J. Carballo. 2015. "Shadowing and Blocking Effect Optimization for a Variable Geometry Heliostat Field." *Energy Procedia* 69: 60–69.

Chan, H.-Y., S. B. Riffat, and J. Zhu. 2010. "Review of Passive Solar Heating and Cooling Technologies." *Renewable and Sustainable Energy Reviews* 14 (2): 781–789.

Chapman, Lee. 2007. "Transport and Climate Change: A Review." *Journal of Transport Geography* 15 (5): 354–367.

Chow, T. T. 2010. "A Review on Photovoltaic/Thermal Hybrid Solar Technology." *Applied Energy* 87 (2): 365–379.

Chueh, W. C., C. Falter, M. Abbott, D. Scipio, P. Furler, S. M. Haile, and A. Steinfeld. 2010. "High-Flux Solar-Driven Thermochemical Dissociation of CO_2 and H_2O Using Nonstoichiometric Ceria." *Science* 330 (6012): 1797–1801.

Coventry, J., and C. Andraka. 2017. "Dish Systems for CSP." *Solar Energy* 152: 140–170.

Deniz, Emrah. 2015. "Solar-Powered Desalination." In *Desalination Updates*. InTech.

Dubey, S., and G. N. Tiwari. 2010. "Energy and Exergy Analysis of Hybrid Photovoltaic/Thermal Solar Water Heater Considering with and without Withdrawal from Tank." *Journal of Renewable and Sustainable Energy* 2 (4): 043106.

Duffie, J. A., and W. A. Beckman. 2013. *Solar Engineering of Thermal Processes*. 2nd ed. Hoboken, New Jersey: Wiley.

El-Sebaii, A. A., and H. Al-Snani. 2010. "Effect of Selective Coating on Thermal Performance of Flat Plate Solar Air Heaters." *Energy* 35 (4): 1820–1828.

Fletcher, E. A., and R. L. Moen. 1977. "Hydrogen and Oxygen from Water." *Science* 197 (4308): 1050–1056.

Florides, G. A., S. A. Tassou, S. A. Kalogirou, and L. C. Wrobel. 2002. "Review of Solar and Low Energy Cooling Technologies for Buildings." *Renewable and Sustainable Energy Reviews* 6 (6): 557–572.

Florschuetz, L. W. 1979. "Extension of the Hottel-Whillier Model to the Analysis of Combined Photovoltaic/Thermal Flat Plate Collectors." *Solar Energy* 22 (4): 361–366.

Francia, G. 1962. *A New Collector of Solar Radiant Energy-Theory and Experimental Verification-Calculation of the Efficiencies*. SAE Technical Paper 620323.

Fujisawa, T., and T. Tani. 2001. "Optimum Design for Residential Photovoltaic-Thermal Binary Utilization System by Minimizing Auxiliary Energy." *Electrical Engineering in Japan* 137 (1): 28–35.

Garg, H. P., and R. S. Adhikari. 1999. "Performance Analysis of a Hybrid Photovoltaic/Thermal (PV/T) Collector with Integrated CPC Troughs." *International Journal of Energy Research* 23 (15): 1295–1304.

Hegazy, A. A. 2000. "Comparative Study of the Performances of Four Photovoltaic/Thermal Solar Air Collectors." *Energy Conversion and Management* 41 (8): 861–881.

Hellstrom, B., M. Adsten, P. Nostell, B. Karlsson, and E. Wackelgard. 2003. "The Impact of Optical and Thermal Properties on the Performance of Flat Plate Solar Collectors." *Renewable Energy* 28 (3): 331–344.

Hendrie, S. D. 1982. *Photovoltaic/thermal Collector Development Program. Final Report*. Lexington, MA: Massachusetts Inst. of Tech., Lincoln Lab.

Herold, K. E., R. Radermacher, and S. A. Klein. 2016. *Absorption Chillers and Heat Pumps*. Florida: CRC Press.

Ho, C. D., H. M. Yeh, and R. C. Wang. 2005. "Heat-Transfer Enhancement in Double-Pass Flat-Plate Solar Air Heaters with Recycle." *Energy* 30 (15): 2796–2817.

Hottel, H. C., and A. Whillier. 1958. "Evaluation of Flat-Plate Solar-Collector Performance." In *Trans. of Conference on the Use of Solar Energy*, II: 74–104. University of Arizona.

Huang, B. J., T. H. Lin, W. C. Hung, and F. S. Sun. 1999. "Solar Photo-Voltaic/Thermal Co-Generation Collector." In *Proceedings of the ISES Bi-Annual Conference*. Jerusalem, Israel.

Ibrahim, A., M. Y. Othman, M. H. Ruslan, S. Mat, and K. Sopian. 2011. "Recent Advances in Flat Plate Photovoltaic/thermal (PV/T) Solar Collectors." *Renewable and Sustainable Energy Reviews* 15 (1): 352–365.

International Energy Agency. 2012. *Technology Roadmap Solar Heating and Cooling.* Paris, France: International Energy Agency (IEA).

Ji, J., J. Han, T. T. Chow, C. Han, J. Lu, and W. He. 2006. "Effect of Flow Channel Dimensions on the Performance of a Box-Frame Photovoltaic/thermal Collector." *Proceedings of the Institution of Mechanical Engineers, Part A: Journal of Power and Energy* 220 (7): 681–688.

Ji, J., J. P. Lu, T. T. Chow, W. He, and G. Pei. 2007. "A Sensitivity Study of a Hybrid Photovoltaic/Thermal Water-Heating System with Natural Circulation." *Applied Energy* 84 (2): 222–237.

Kalogirou, S. 1997. "Design, Construction, Performance Evaluation and Economic Analysis of an Integrated Collector Storage System." *Renewable Energy* 12 (2): 179–192.

Kalogirou, S. 2003. "The Potential of Solar Industrial Process Heat Applications." *Applied Energy* 76 (4): 337–361.

Kalogirou, S. A. 2004. "Solar Thermal Collectors and Applications." *Progress in Energy and Combustion Science* 30 (3): 231–295.

Kalogirou, S. A., S. Lloyd, J. Ward, and P. Eleftheriou. 1994. "Design and Performance Characteristics of a Parabolic-Trough Solar-Collector System." *Applied Energy* 47 (4): 341–354.

Kalogirou, S., P. Eleftheriou, S. Lloyd, and J. Ward. 1994. "Low Cost High Accuracy Parabolic Troughs Construction and Evaluation." *Renewable Energy* 5 (1–4): 384–386.

Kalogirou, S., S. Lloyd, and J. Ward. 1997. "Modelling, Optimisation and Performance Evaluation of a Parabolic Trough Solar Collector Steam Generation System." *Solar Energy* 60 (1): 49–59.

Kalogirou, S., and Y. Tripanagnostopoulos. 2005. "Performance of a Hybrid PV/T Thermosyphon System." In *World Renewable Energy Congress (WREC 2005),* 1162–1167. Aberdeen, Scotland.

Khalifa, A. and Hamood, A. 2009. Performance correlations for basin type solar stills. *Desalination,* 249: 24–28.

Klein, S. A., W. A. Beckman, and J. A. Duffie. 1977. "A Design Procedure for Solar Air Heating Systems." *Solar Energy* 19 (5): 509–512.

Kolb, G. J., R. Davenport, D. Gorman, R. Lumia, R. Thomas, and M. Donnelly. 2007. "Heliostat Cost Reduction." In *ASME 2007 Energy Sustainability Conference,* 1077–1084. Long Beach, California, USA.

Kötter, J., S. Decker, R. Detzler, J. Schäfer, M. Schmitz, and U. Herrmann. 2012. "Cost Reduction of Solar Fields with Heliotrough Collector." In *Proceedings of the SolarPACES Conference.* Morocco.

Kräupl, S., and A. Steinfeld. 2001. "Experimental Investigation of a Vortex-Flow Solar Chemical Reactor for the Combined ZnO-Reduction and CH_4-Reforming." *J. Solar Energy Eng* 123: 237–243.

Kreider, J. F. 1982. *Solar Heating Design Process: Active and Passive Systems.* New York, NY: McGraw-Hill Book Company.

Kruesi, M., M. E. Galvez, M. Halmann, and A. Steinfeld. 2011. "Solar Aluminum Production by Vacuum Carbothermal Reduction of alumina—Thermodynamic and Experimental Analyses." *Metallurgical and Materials Transactions B* 42 (1): 254–260.

Liu, T., W. Lin, W. Gao, C. Luo, M. Li, Q. Zheng, and C. Xia. 2007. "A Parametric Study on the Thermal Performance of a Solar Air Collector with a v-Groove Absorber." *International Journal of Green Energy* 4 (6): 601–622.

Loutzenhiser, Peter G., Ozan Tuerk, and Aldo Steinfeld. 2010. "Production of Si by Vacuum Carbothermal Reduction of SiO_2 Using Concentrated Solar Energy." *JOM: Journal of the Minerals, Metals and Materials Society* 62 (9): 49–54.

Lovegrove, K., and J. Pye. 2012. "Fundamental Principles of Concentrating Solar Power (CSP) Systems." In *Concentrating Solar Power Technology: Principles, Developments and Applications.* Philadelphia, PA: Woodhead Publishing.

Lupfert, E., M. Geyer, W. Schiel, A. Esteban, R. Osuna, E. Zarza, and P. Nava. 2001. "Eurotrough Design Issues and Prototype Testing at PSA." In *Proceedings of ASME Int. Solar Energy Conference. Forum 2001, Solar Energy: The Power to Choose*, 389–394. Washington, Design Issues and Prototype Testing at PSA." *Solar Engineering*, 387–392.

Lupfert, E., M. Geyer, W. Schiel, E. Zarza, R. O. Gonzalez-Anguilar, and P. Nava. 2000. "Eurotrough: A New Parabolic Trough Collector with Advanced Light Weight Structure." In Proceedings of Solar Thermal 2000 International Conference. Sydney, Australia.

Madala, S., and R. F. Boehm. 2016. "A Review of Nonimaging Solar Concentrators for Stationary and Passive Tracking Applications." *Renewable and Sustainable Energy Reviews* 71: 309–322.

Meier, A., E. Bonaldi, G. M. Cella, and W. Lipinski. 2005. "Multitube Rotary Kiln for the Industrial Solar Production of Lime." *Journal of Solar Energy Engineering* 127 (3): 386–395.

Meier, A., and A. Steinfeld. 2010. "Solar Thermochemical Production of Fuels." *Advances in Science and Technology*, 74 (1): 303–312.

Mendelsohn, M., T. Lowder, and B. Canavan. 2012. *Utility-Scale Concentrating Solar Power and Photovoltaics Projects: A Technology and Market Overview*. NREL/TP-6A20-51137. Colorado, USA: National Renewable Energy Laboratory.

Mills, D. R. 1995. "Proposed Solar Cogeneration Powerplant for 2000 Olympics." In *Proc. Solar '95 Renewable Energy: The Future Is Now*, 465–473.

Mills, D. R. 2013. "Solar Thermal Electricity." In *Solar Energy: The State of the Art*, 1st ed. Abingdon, Oxfordshire: Earthscan.

Mills, D. R., and J. E. Giutronich. 1978. "Asymmetrical Non-Imaging Cylindrical Solar Concentrators." *Solar Energy* 20 (1): 45–55.

Müller, F., P. Poživil, P. J. van Eyk, A. Villarrazo, P. Haueter, C. Wieckert, G. J. Nathan, and A. Steinfeld. 2017. "A Pressurized High-Flux Solar Reactor for the Efficient Thermochemical Gasification of Carbonaceous Feedstock." *Fuel* 193: 432–443.

Narayan, G. P., M. H. Sharqawy, E. K. Summers, J. H. Lienhard, S. M. Zubair, and M. A. Antar. 2010. "The Potential of Solar-Driven Humidification–Dehumidification Desalination for Small-Scale Decentralized Water Production." *Renewable and Sustainable Energy Reviews* 14 (4): 1187–1201.

Neises-von Puttkamer, M., M. Roeb, S. Tescari, L. de Oliveira, S. Breuer, and C. Sattler. 2016. "Solar Aluminum Recycling in a Directly Heated Rotary Kiln." In *REWAS 2016 Towards Material Resource Sustainability*. 235–240. Berlin, Germany: Springer.

O'Gallagher, J. J., K. Snail, R. Winston, C. Peek, and J. D. Garrison. 1982. "A New Evacuated CPC Collector Tube." *Solar Energy* 29 (6): 575–577.

Orel, Z. C., M. K. Gunde, and M. G. Hutchins. 2005. "Spectrally Selective Solar Absorbers in Different Non-Black Colours." *Solar Energy Materials and Solar Cells* 85 (1): 41–50.

Parsons, R. A. 1995. "Chapter 30." In *ASHRAE Handbook: Heating, Ventilating, and Air-Conditioning Applications*. Atlanta: ASHRAE.

Piatkowski, N., C. Wieckert, A. W. Weimer, and A. Steinfeld. 2011. "Solar-driven gasification of carbonaceous feedstock—A review." *Energy & Environmental Science* 4 (1): 73–82.

Price, H., E. Lupfert, D. Kearney, E. Zarza, G. Cohen, R. Gee, and R. Mahoney. 2002. "Advances in Parabolic Trough Solar Power Technology." *Journal of Solar Energy Engineering* 124 (2): 109–125.

Qiblawey, H. and Banat, F. 2008. Solar thermal desalination technologies. *Desalination*, 220: 633–644.

Rankin, John. 2012. "Energy Use in Metal Production." In . Australia: Swinburne University of Technology.

Reynolds, D. J., M. J. Jance, M. Behnia, and G. L. Morrison. 2004. "An Experimental and Computational Study of the Heat Loss Characteristics of a Trapezoidal Cavity Absorber." *Solar Energy* 76 (1): 229–234.

Riffat, S. B., and E. Cuce. 2011. "A Review on Hybrid Photovoltaic/thermal Collectors and Systems." *International Journal of Low-Carbon Technologies* 6 (3): 212–241.

Rodat, S., S. Abanades, and G. Flamant. 2009. "High-Temperature Solar Methane Dissociation in a Multitubular Cavity-Type Reactor in the Temperature Range 1823 - 2073 K." *Energy & Fuels* 23 (5): 2666–2674.

Rosell, J. I., X. Vallverdu, M. A. Lechon, and M. Ibanez. 2005. "Design and Simulation of a Low Concentrating Photovoltaic/Thermal System." *Energy Conversion and Management* 46 (18): 3034–3046.

Rubin, M. 1985. "Optical Properties of Soda Lime Silica Glasses." *Solar Energy Materials* 12 (4): 275–288.

Sampathkumar, K., T. V. Arjunan, P. Pitchandi, and P. Senthilkumar. 2010. "Active Solar Distillation—A Detailed Review." *Renewable and Sustainable Energy Reviews* 14 (6): 1503–1526.

Sandnes, B., and J. Rekstad. 2002. "A Photovoltaic/thermal (PV/T) Collector with a Polymer Absorber Plate. Experimental Study and Analytical Model." *Solar Energy* 72 (1): 63–73.

Schaffner, B., A. Meier, D. Wuillemin, W. Hoffelner, and A. Steinfeld. 2003. "Recycling of Hazardous Solid Waste Material Using High-Temperature Solar Process Heat. 2. Reactor Design and Experimentation." *Environmental Science & Technology* 37 (1): 165–170.

Scheffe, J. R., D. Weibel, and A. Steinfeld. 2013. "Lanthanum–Strontium–Manganese Perovskites as Redox Materials for Solar Thermochemical Splitting of H_2O and CO_2." *Energy & Fuels* 27 (8): 4250–4257.

Segal, A., and M. Epstein. 2001. "The Optics of the Solar Tower Reflector." *Solar Energy* 69: 229–241.

Skoplaki, E., and J. A. Palyvos. 2009. "On the Temperature Dependence of Photovoltaic Module Electrical Performance: A Review of Efficiency/power Correlations." *Solar Energy* 83 (5): 614–624.

Smestad, Greg P., and Aldo Steinfeld. 2012. "Photochemical and Thermochemical Production of Solar Fuels from H_2O and CO_2 Using Metal Oxide Catalysts." *Industrial & Engineering Chemistry Research* 51 (37): 11828–11840.

Sopian, K., K. S. Yigit, H. T. Liu, S. Kakac, and T. N. Veziroglu. 1996. "Performance Analysis of Photovoltaic Thermal Air Heaters." *Energy Conversion and Management* 37 (11): 1657–1670.

Spiegler, K. S. 1977. *Salt-Water Purification*. 2nd ed. New York, NY: Plenum Press.

Steinfeld, A. 2012. "Thermochemical Production of Syngas Using Concentrated Solar Energy." In *Annual Review of Heat Transfer*. Vol. 15, 255–275. Redding, CT: Begell House Inc.

Steinfeld, A., P. Kuhn, and J. Karni. 1993. "High-Temperature Solar Thermochemistry: Production of Iron and Synthesis Gas by Fe_3O_4-Reduction with Methane." *Energy* 18 (3): 239–249.

Steinfeld, A., and M. Schubnell. 1993. "Optimum Aperture Size and Operating Temperature of a Solar Cavity-Receiver." *Solar Energy* 50 (1): 19–25.

Supernova. 2011. "Fresnel Specialist Novatec Biosol Turns to Superheated Steam to Boost Efficiency by 50%." *Renewable Energy Magazine*. Accessed Nov. 29, 2017. https://www.renewableenergy magazine.com/solar_thermal_electric/fresnel-specialist-novatec-biosol-turns-to-superheated.

Tanaka, H. and Nakatake, Y. 2009. Increase in distillate productivity by inclining the flat plate external reflector of a tilted-wick solar still in winter. *Solar Energy*, 83: 785–789.

Tian, M., Y. Su, H. Zheng, G. Pei, G. Li, and S. Riffat. 2018. "A Review on the Recent Research Progress in the Compound Parabolic Concentrator (CPC) for Solar Energy Applications." *Renewable and Sustainable Energy Reviews* 82: 1272–1296.

Tripanagnostopoulos, Y., M. Souliotis, and T. H. Nousia. 2000. "Solar Collectors with Colored Absorbers." *Solar Energy* 68 (4): 343–356.

Turton, Hal. 2006. "Sustainable Global Automobile Transport in the 21st Century: An Integrated Scenario Analysis." *Technological Forecasting and Social Change* 73 (6): 607–629.

Tzouganatos, N., R. Matter, C. Wieckert, J. Antrekowitsch, M. Gamroth, and A. Steinfeld. 2013. "Thermal Recycling of Waelz Oxide Using Concentrated Solar Energy." *Jom* 65 (12): 1733–1743.

Vant-Hull, L. 2014. "Issues with Beam-Down Concepts." *Energy Procedia* 49: 257–264.

Villasmil, W., M. Brkic, D. Wuillemin, A. Meier, and A. Steinfeld. 2014. "Pilot Scale Demonstration of a 100-kW$_{th}$ Solar Thermochemical Plant for the Thermal Dissociation of ZnO." *Journal of Solar Energy Engineering* 136 (1): 1–11.

Wazwaz, A., J. Salmi, H. Hallak, and R. Bes. 2002. "Solar Thermal Performance of a Nickel-Pigmented Aluminium Oxide Selective Absorber." *Renewable Energy* 27 (2): 277–292.

Weinrebe, G., Z. Abul-Ella, and W. Schiel. 2011. "On the Influence of Parabolic Trough Collector Stiffness and Length on Performance." In *Proceedings of the SolarPACES Conference*, 15–18. Granada, Spain.

Welford, W. T., and R. Winston. 1989. *High Collection Nonimaging Optics Academic*. San Diego, CA: Academic Press.

Whillier, A. 1963. "Plastic Covers for Solar Collectors." *Solar Energy* 7 (3): 148–151.

Wieckert, C., U. Frommherz, S. Kräupl, E. Guillot, G. Olalde, M. Epstein, S. Santén, T. Osinga, and A. Steinfeld. 2007. "A 300kW Solar Chemical Pilot Plant for the Carbothermic Production of Zinc." *Journal of Solar Energy Engineering* 129 (2): 190–196.

Wijeysundera, N. E., L. L. Ah, and L. E. Tjioe. 1982. "Thermal Performance Study of Two-Pass Solar Air Heaters." *Solar Energy* 28 (5): 363–370.

Zondag, H. A., D. W. De Vries, W. G. J. Van Helden, R. J. C. Van Zolingen, and A. A. Van Steenhoven. 2003. "The Yield of Different Combined PV-Thermal Collector Designs." *Solar Energy* 74 (3): 253–269.

Zondag, H. A., D. W. de Vries, A. A. van Steenhoven, W. G. J. van Helden, and R. J. C. van Zolingen. 1999. "Thermal and Electrical Yield of a Combi-Panel." In *Proceedings of the ISES Bi-Annual Conference*. Jerusalem, Israel.

4

Photovoltaic-Driven Heat Pumps

4.1 Photovoltaic Systems

Photovoltaic conversion consists of the direct conversion of photons from the sun into electricity in a solid-state semiconductor device called a PV cell, which is the core of a PV system. PV modules are formed by grouping PV cells into large groups. Interconnecting several PV modules in a parallel series configuration results in a PV array (Singh 2013). Given the fact that PV cell output is normally direct current (DC) while most power consuming devices operate with alternating current (AC), a typical PV system requires an inverter. Moreover, a battery is typically implemented in the system to store DC voltages during charging mode and supply DC electrical energy during discharge mode (Ullah et al. 2013). The measuring unit for such systems is peak kilowatts (kW_p), which refers to the expected power output of the system on a clear day when the sun is directly overhead (Parida et al. 2011). Commercial systems range from a few kW_p for domestic applications up to several GW_p.

The main advantages of the technology are its very low maintenance costs, simplicity of design, high power density—the highest among renewable technologies—and its stand-alone operation, facilitating its use in a wide variety of applications. On the other hand, the main challenges of PVs are their relatively poor efficiency and the reduction of their production costs (Joshi et al. 2009; Siecker et al. 2017). Moreover, the solar radiation absorbed by the PV cells that is not converted into electrical power results in an increase of the cell temperature. High PV cell temperature has a complex effect on the cell's performance. The short circuit current increases by 0.06–0.1%, while the fill factor, the power output, and the open circuit voltage decrease by 0.1–0.2%, 0.4–0.5%, and 2–2.3 mV/°C, respectively, resulting in the cell's efficiency decline (Sargunanathan et al. 2016). According to Alzaabi et al. (2014), when the panel's temperature reaches a temperature of 50–60°C, a reduction of 3–4% in power output will occur. For this reason, sufficient cooling needs to be realized to sustain PV cell efficiency at reasonable levels. The simplest and cheapest method for cooling is with either natural or forced-air circulation. On the other hand, a more expensive but more effective method of cooling is water-heat extraction (Lamnatou and Chemisana 2017).

The PV market has grown rapidly over the last few years. Since 2000, the annual growth rate of the global production of solar cells has ranged between 40–90% (Jäger-Waldau et al. 2011). At the end of 2012, the total installed capacity of solar PVs was above 100 GW. Europe had the largest share of new installations, with Germany installing 7.6 GW and Italy adding 3.6 GW (Kumar Sahu 2015). According to the IEA, since 2010, solar PV installations have exceeded in capacity the total installed capacity over the past four decades, and by 2050 it is expected that a share of 16% of total energy production will come from PV technology. The *Renewables Global Status Report* (REN21 2016) reported that in 2015 a record number of new 50 GW solar PV installations were reported, resulting in a global total solar PV capacity of 227 GW. The results also indicated that there is significant room for improvement, because this capacity corresponds to only a bit higher than 1% of global total electricity production—out of a 23.7% share of renewables. In terms of specific countries, the ones that were reported to generate the highest percentage of their electricity demand from solar PVs were Italy with a 7.8% share, Greece with 6.5%, and Germany with 6.4%.

The rapid expansion of the PV market results in a constant decrease in the capital costs of PVs. Infante Ferreira and Kim (2014) reported the average cost of a PV panel in the Netherlands to be in the range 250–400€/m² for small-scale applications, while the corresponding range for large-scale applications was 150–250€/m².

4.1.1 PV Cell Materials

As previously mentioned, PV cells are made of semiconductor materials such as mono- or poly-crystalline silicon, gallium arsenide (GaAs), copper indium gallium selenide (CIGS), and cadmium telluride (CdTe).

The first generation of PV cells was made of single junction crystal based on silicon wafers. In order to reduce material costs, the second generation was based on thinner films. The third generation of PV cells take advantage of multi-junctions and nanotechnology, for instance a-Si/nc-Si.

GaAs is a semiconductor of similar structure to silicon. Multicrystalline GaAs PV cells usually have lower efficiency than multi- and mono-Si. However, due to their high heat resistance and their lighter weight (in comparison to multi- and mono-Si) they are a better option for concentrated photovoltaic (CPV) and space applications (Deb 1998).

Among thin film cells, CdTe and CIGS have drawn the biggest attention. Yet, the main challenges that constrain their market share are the toxicity of cadmium and the shortage of indium, respectively (El Chaar et al. 2011). When looking toward an environmentally friendly solution, researchers have used organics and polymers as materials for solar cells. Their low cost, mechanical flexibility, disposability, light weight, and the fact that they have no environmental issues create potential for this technology. However, for the time being, the efficiency of these materials is quite low compared to other available technologies (Goetzberger et al. 2003; Gorter and Reinders 2012).

As already discussed, high temperatures negatively affect the performance of a PV cell. Yet, there are some types of PV cells that are able to operate efficiently at higher temperatures. CPV systems are able to operate at higher temperatures better than flat-plate collectors, with reasonable electric conversion efficiencies for temperatures in the range 100–170°C (Meneses-Rodríguez et al. 2005).

The environmental and economic drawbacks of the aforementioned solar cell materials have led some researchers to investigate the potential of a new technology called dye-sensitized solar cell (DSSC). DSSC materials, such as titanium oxide (TiO_2), have a low production costs, are widely available, are harmless to the environment, and perform more efficiently with diffused light, thus allowing for better operation at dawn, dusk, and on cloudy days (Gong et al. 2017).

Currently, there are several commercially available PV cells technologies. The most important ones are summarized in Table 4.1. Figure 4.1 presents an overview of the

TABLE 4.1

Available PV Cell Technologies and Reported Experimental Efficiencies

PV Cells	Classification	Maximum Reported Efficiency, η (%)
Silicon	mono-Si (crystalline)	25.6
	multi-Si (multicrystalline)	20.8
	Si (thin film transfer)	21.2
III-V cells	GaAs (thin film)	28.8
	GaAs (multicrystalline)	18.4
Thin film chalcogenide	CIGS (cell)	20.5
	CdTe (cell)	21.0
Photochemical	Dye sensitized	11.9
	Organic thin film	11.0
Multi-junction devices	GaInP/ GaInAs/Ge	37.9
	a-Si/nc-Si (thin film cell)	12.7

Source: Data from Green, Martin A. et al., *Progress in Photovoltaics: Research and Applications*, 23 (1): 1–9, 2015.

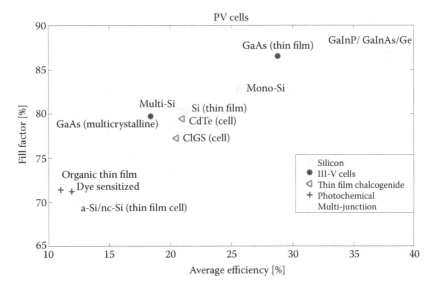

FIGURE 4.1
Fill factor and average efficiencies for the various PV cells listed in Table 4.1. (Data from Green, Martin A. et al., *Progress in Photovoltaics: Research and Applications*, 23 (1): 1–9, 2015.)

efficiencies of the same PV cell types and the corresponding values for the fill factor of each type. The efficiency values are defined under AM1.5 conditions as the maximum electric output power to incident light power (1000 W/m²) at 25°C:

$$\eta = \frac{Power\ output\ from\ the\ solar\ cell}{Incident\ light\ power} * 100\%$$

(4.1)

As Table 4.1 indicates, even though efficiencies up to 25.6% have been achieved and reported in literature for mono-Si PV cells, the respective values for commercially available PV modules are lower. Table 4.2 presents an overview of a few PV manufacturers and the reported conversion efficiencies of their PV modules.

Crystalline silicon solar cells account for 83% of the solar cell market, mainly due to their maturity, even though, as seen in Table 4.1, they are not the most efficient technology (Sarbu and Sebarchievici 2013). Figure 4.2 presents an overview of shares in the solar cell market based on data for 2010 (Tyagi et al. 2013).

TABLE 4.2

PV Manufacturers and Reported Efficiency Values for Their Products

Manufacturer	PV Module	Reported Efficiency, η (%)
Astronergy (2016)	monocrystalline (STAR series)	16.5–18.7
	multicrystalline (STAVE series)	16.0–17.1
Canadian Solar (2017)	monocrystalline (MaxPower series)	17.0–17.7
	multicrystalline (MaxPower series)	16.2–17.5
Hanwha Q CELLS (2016)	monocrystalline (Q.PEAK series)	<19.9
	multicrystalline (Q.POWER series)	<17.4
Jinko Solar (2013)	monocrystalline (Mono PERC 156MM)	20.3–21.3
	multicrystalline (Poly 156MM)	17.6–18.9
Kyocera Solar (2012)	multicrystalline (KK-series)	13.9–16.4
LONGi Solar (2016)	monocrystalline (LR6-series)	16.8–18.8
REC Solar Holdings AS (2017)	monocrystalline (MaxPower series)	16.4–18.0
	multicrystalline (TWINPEAK series)	
Sharp (Tyagi et al. 2013)	multicrystalline	14.4
	thin film	10.0
Silfab Solar Inc. (2017)	monocrystalline (SLA- SLG-series)	16.8–19.0
	monocrystalline (Bifacial)	17.2–23.1
Solaria Corporation (2017)	monocrystalline (PowerXT series)	18.7–19.3
Solartech Power Inc. (2017)	multicrystalline (F-series)	10–13.7
	multicrystalline (V-series)	16.6–17.2
	CIGS thin film (1000-series)	12.4–13.7
SolarWorld Industries GmbH (2017)	monocrystalline (SW series)	16.7–17.9
Suntech (2017)	monocrystalline (72 cell module)	17.3
	multicrystalline (72 cell module)	16.7
	monocrystalline (Hypro series)	20.9
Talesun (2017)	monocrystalline (HIPRO TP660M)	17.7–18.3
	multicrystalline (TP672P)	16.2–17.0
Trina Solar (2017)	monocrystalline (TSM-DE14A)	17.5–19.3
	multicrystalline (TSM-PD14)	16.5–17.5
Vikram Solar (2017)	multicrystalline (ELDORA series)	15.4–17.2
Yingli Solar (2017)	monocrystalline (YLM series)	16.2–18.0
	multicrystalline (YGE series)	15.3–17.1

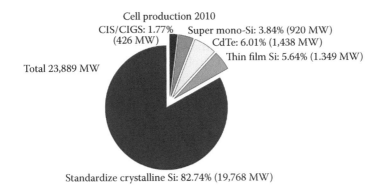

FIGURE 4.2
Solar cell market share for 2010. (Adapted from Tyagi, V. V. et al., *Renewable and Sustainable Energy Reviews*, 20 (Supplement C): 443–461, 2013.)

4.2 Solar Electric Chillers

A typical solar electric chiller—or PV-driven compression chiller—consists mainly of PV arrays, a battery, an inverter, and an electrically driven refrigeration device. A solar electric system can operate in three power configurations:

- A standalone system (Figure 4.3)
- A hybrid system, in combination with another power plant (such cases shall be discussed in a following chapter)
- A system powered solely by the grid or a grid-intertie system

In most cases, the refrigeration system is realized by a vapor compression cycle. Based on the system's capacity and the power configuration, different types of compressors are used—a presentation of the commercially available types is available in Chapter 2. In the case of a standalone system, the PV module is connected to an inverter to convert the DC electricity to AC in order to supply the motor of the compressor. Apart from the VCC, a potential cooling option powered by PVs is Peltier cooling, which takes advantage of the thermoelectric phenomenon, called Peltier effect that takes place when two materials

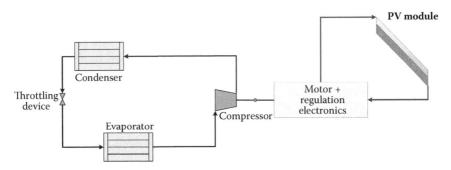

FIGURE 4.3
Schematic of a typical standalone solar electric system.

with different electric conductivities come into contact. By inducing an electric current in the closed circuit, cooling can be produced at the contact surface between two materials. However, the relatively low efficiency of the technology, in the range of 3–8%, is a major obstacle in the development of technical applications (Richter et al. 2013).

Allouhi et al. (2015) defines the overall efficiency of a solar electric refrigeration system as follows:

$$\eta_{PV,r} = \frac{Q_e}{A_{PV} \cdot I} \tag{4.2}$$

In past decades, the high initial cost of PVs and their low efficiency significantly limited the practical application of solar PV cooling. However, with the rapid expansion of the PV market and the consequent decrease in the initial cost of PVs, several theoretical and experimental investigations have recently been conducted to evaluate the feasibility of solar electric cooling. In subsequent pages, a brief review of some relevant studies found in literature will be presented.

In one of the first studies of solar electric cooling, Ayyash and Sartawi (1983) compared the initial and operating costs of a PV-assisted VCC system and a solar absorption system. The results of the simulations showed that the solar electric system could be cost competitive. Osman (1985) presented the design and operational methodology for a PV-driven cooling system, a solar air heating system, and a solar absorption cooling system for use in a solar house in Kuwait.

El Tom et al. (1991) developed a solar PV refrigerator consisting of six 40 Wp PV modules, two batteries, a charge regulator, and a refrigerator cabinet. The PV modules were connected to two circuits—an open voltage circuit of 18 V—of three modules in parallel configuration. Each battery had a capacity of 105 Ah at 12 V and served as an energy storage media for night and cloudy day requirements. The regulator consisted of a freezing and a refrigerating compartment with a 180 L capacity. The refrigerator cabinet used a 24 V DC motor compressor and operated with R12 as its working fluid. At maximum cooling, the efficiency of the refrigerator was estimated to be 0.64. At low cooling, the efficiency rose up to 0.77. Regarding the freezing compartment, it was reported to be able to provide sufficient ice to maintain a temperature of 0°C for a period of 2–3 days without active cooling.

In another study, the conversion of a conventional refrigerator to a PV-driven refrigerator was presented (Kaplanis and Papanastasiou 2006). Three commercial PV panels were used, with a peak power of 85 Wp and an open circuit voltage of 22.03 V. A 12 V battery bank—made of six Powerblock S-190 batteries—was implemented in the system, feeding the compressor in cases of inadequate power generation from the PV panels. A thermostat was implemented for variable speed control of the compressor. Control of the compressor's speed was realized by R8 (see Figure 4.4) resistance. R9 was used to preset battery protection voltage. Two conventional refrigerators were taken into consideration:

- Model FV650, with a consumption of 8.8 kWh/d during summer in the Greek climate
- Model FV100, with a consumption of 4 kWh/d, based on manufacturer's data—authors reported that when driven by PV panels the daily power load was 2.2 kWh

The first model required 22 PV panels to be driven and was found to be uneconomical.

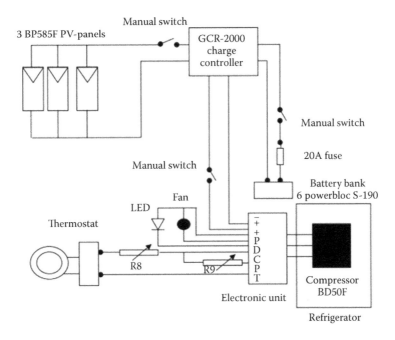

FIGURE 4.4
The retrofitted refrigeration system. (Reproduced from Kaplanis, Socrates, and Nikolaos Papanastasiou, *Renewable Energy*, 31 (6): 771–780, 2006.)

The retrofitting of conventional refrigeration led to a decrease in the useful volumetric capacity of the refrigerator by 30%. However, modifications also reduced heat losses, leading to a final power load of 1.7 kWh/d. Furthermore, the system's economics were improved by replacing the AC motor of the compressor with a DC variable-speed motor, which eliminated the need for an inverter.

Axaopoulos and Theodoridis (2009) experimentally evaluated the performance of a PV-driven ice-making system without a battery. The system, shown in Figure 4.5, consists of an ice storage tank filled with water; four hermetic compressors, each connected with a vertical plate-surface evaporator and an air-cooled condenser; a controller; and a 440 Wp PV array. Multiple compressors were preferred over a single compressor because they reduced static friction, have easy startup power requirements, and allow for a wide control range at the same time. The reported solar-to-compressor power efficiency was approximately 9.2%. Easy compressor startups allowed for operation even under low solar irradiance—as low as 150 W/m². The system's productivity was proven to be satisfactory, according to the authors, with a production rate of up to 17 kg of ice on a good day.

Bilgili (2011) investigated the performance of a PV-driven VCC system, located in the city Adana, Turkey, using simulations. A sample building with a 30 m² floor area was considered, and its cooling loads were calculated on the twenty-third of each month (May–September) based on meteorological data collected for the investigated region. The highest cooling load was estimated at 8.115 kW on August 23 at 5:00 p.m. The corresponding daily total and hourly mean cooling loads on that day were 119.5 kWh/d and 4.98 kW, respectively. Based on this data, the hourly calculations of the system's performance were conducted for different evaporating temperatures, also determining the minimum PV area to cover the compressor's power demands. Results indicated that for an evaporation

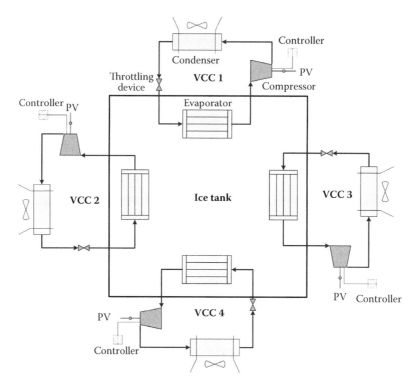

FIGURE 4.5
Schematic of a PV-driven, battery-free ice maker. (Adapted from Axaopoulos, Petros J., and Michael P. Theodoridis, *Solar Energy*, 83 (8): 1360–1369, 2009.)

temperature of 0°C, the cooling COP on July 23 was in the range of 3.04–4.07. The corresponding compressor power consumption was 0.85–2.4 kW and the heat rejection at the condenser was between 4.1–9.8 kW. Furthermore, decreasing the evaporation temperature required a larger PV surface to be achieved. For an evaporation temperature of 10°C, 18.7 m² of PV panels were required. The respective value for an evaporation temperature of −10°C was 38.7 m².

Hartmann et al. (2011) compared the performance of a solar electric system (a PV-driven mechanical compression chiller) and a solar thermal system (an adsorption chiller powered by flat-plate collectors) in terms of primary energy savings and their costs. Both systems were used to cover the heating and cooling loads of a typical building in two different European climates, Freiburg (Germany) and Madrid (Spain). The simulations were conducted with TRNSYS software for several collector and storage sizes. The basic information regarding the heating and cooling loads of the buildings in the two studied

TABLE 4.3

Technical Data for the Loads of the Two Buildings

	Freiburg	Madrid
Heating load (kWh/y)	19,337	7,288
Cooling load (kWh/a)	10,818	16,478

Source: Hartmann, N. et al., *Renewable Energy*, 36 (5): 1329–1338, 2011.

locations are presented in Table 4.3. The basic technical data for the two compared systems is listed in Table 4.4.

A conventional compression chiller powered by the grid was used as a reference for the cost and energy savings calculations, with a nominal capacity able to cover the peak thermal load of the building. Where there is an energy surplus from the solar electric system, the exceeding power is primarily used for the system's internal power consumption and secondarily supplied to the grid. On the other hand, when the PV system cannot provide sufficient energy, power from the grid is supplied to meet the chiller's demands. In the case of a solar thermal system, a storage tank is used for the storage of excess of thermal power and is connected to a backup heater for days with smaller solar coverage. According to the results of the simulations, both systems were more expensive than the conventional compression chiller. In particular, the annual cost for the conventional system was €8,859 for Freiburg and €8,140 for Madrid. On the other hand, the annual cost for the solar thermal system was 128% and 134% higher for the two cities, respectively. The solar electric system appeared to be a more competitive choice, being only 5% more expensive than the conventional chiller. In terms of the collector area, in order to achieve the same energy savings, the PV field area had to be six times smaller than the surface of the flat-plate collectors.

A similar comparison between an absorption chiller, powered by ETCs, and a PV-driven vapor compression system was conducted by Fumo et al. (2013). A conventional air-cooled chiller powered by the grid was used as the reference system. A parametric and economic study was conducted to evaluate the performance of the two solar systems. Based on the results of the simulations, 12 m² of ETCs were required to produce 3.52 kW (one ton of refrigeration), while the respective area of PV modules was 7 m². In terms of projected savings, over a project life of 25 years, it was found that the solar electric system resulted in significantly higher savings in comparison to the solar thermal system. For instance, for a PV module of 7 m², the savings per ton of refrigeration were approximately $3,500, while for an ETC area of 12 m², the savings would be negligible.

An inverter heat pump powered simultaneously by the grid and solar panels was investigated by Aguilar et al. (2014). The nominal cooling capacity of the heat pump was 3.52 kW. While on heating mode, the nominal capacity was 3.81 kW. Three 235 Wp PV panels, tilted at 30°, were directly connected to the heat pump. The heat pump, which operates with R410A as its working fluid, requires a power input of 0.86 kW on cooling mode and almost 1 kW on heating mode. Experiments were conducted for the period of July–October 2012, and the efficiency of the system was evaluated. According to the results for the investigated months, the energy efficiency ratio (EER) of the system ranged between 6.59–13.83% for cooling mode from 8 a.m.–8 p.m. each day. The corresponding range for the solar fraction was between 41.27–61.22%. When the time range of the investigations was 9 a.m. to 5 p.m. each day, the energy efficiency ratio increased to an average of 13.65,

TABLE 4.4

Technical Data for the Solar Electric and the Solar Thermal Systems

	Solar Electric	Solar Thermal
Efficiency of solar circuit (%)	15	78.9
	(PV module efficiency)	(Flat-plate collector efficiency)
COP	3	0.68
	(VCC)	(adsorption chiller)
Specific cost (€/kW)	€310	€800

while the average solar fraction was 64%. On the other hand, when the heat pump was on heating mode, the average COP was approximately 9 with an average solar fraction of 50%.

A lifecycle assessment for several solar thermal and solar electric cooling systems was conducted by Beccali et al. (2014). Six different configurations were considered in the comparative study. The reference system (system 1) was a conventional VCC system (with a nominal EER of 2.5) connected to the grid for cooling loads. For PV-assisted systems, three configurations were evaluated:

- System 2: The conventional chiller is simultaneously driven by PV panels and the grid.
- System 3: The conventional chiller is solely driven by PV panels.
- System 4: Partial-load standalone PV driving of the conventional chiller.

Two options were considered for a summer backup heat driven system:

- System 5: A backup natural-gas-fired burner to feed the absorption chiller generator.
- System 6: A conventional compression chiller to enhance the cooling production.

For heating purposes, all configurations used a natural-gas-fired burner. The systems were simulated with TRNSYS software for application in three different locations: Palermo (Italy), Zurich (Switzerland), and Rio de Janeiro (Brazil).

Results of the simulation showed that systems 2 and 3 were the most efficient in terms of the primary energy savings for all three locations. For instance, the primary energy savings for Rio de Janeiro with system 2 were 98.5% (in comparison to reference system 1), while the most efficient solar thermal system was system 6, with a primary energy savings of 32.2%. Similar results were obtained for Palermo, however, with lower values—system 3's primary energy savings were 63%. On the other hand, for Zurich, the solar thermal options were more efficient than the PV panels, with a maximum primary energy saving of 30.9% for system 6. Regarding the energy payback time (EPT), the best performing systems for Zurich and Palermo were systems 2, 5, and 6. The PV driven system resulted in an EPT of 1.9 years for Palermo, while solar thermal systems 5 and 6 resulted in EPTs of approximately 5 years. In Zurich, all three systems (2, 5, and 6) had similar EPTs, with values of 3.2, 4.4, and 4.9, respectively. For Rio de Janeiro, system 2 was the most competitive with an EPT of three years, while the EPTs of the solar thermal systems were not competitive, with values of 35 and 12 years. Systems 3 and 4, were found to have quite high EPTs for all locations, ranging between 24–38 years for system 3 and 62–100 years for system 4.

Eicker et al. (2014) conducted a comparative study between solar electric and solar thermal cooling systems based on primary energy savings for a 310 m² floor area building. Three different climatic conditions were investigated: Palermo (Italy), Stuttgart (Germany), and Madrid (Spain). The energy savings were calculated with a 30–50 kWc conventional VCC chiller connected to the grid, equipped with a 1.5 m³ chilled water storage tank. The solar electric system consisted of a PV module driving a conventional chiller. Two solar thermal systems were evaluated, driven either by flat-plate collectors or compound paraphilic collectors, both equipped with one 5 m³ hot water and 1 m³ chilled water storage tanks and an absorption chiller with a cooling capacity of 25 kW. Results indicated that, for Palermo, average net collector efficiencies were 31% and 23% for the CPCs and the FPC, respectively. The corresponding annual COPs were 3.19 for the PV-driven system, 0.79 for the CPC-driven system, and 0.77 for the FPC system. The primary energy savings were 48%, 37%, and 32%, respectively.

El-Bahloul et al. (2015) proposed the installation of a solar PV-driven VCC chiller in the city of New Borg Al-Arab, Egypt. Thermal storage is realized with the use of phase change materials (PCM). The PV panel used was a 130 Wp single multicrystalline solar module with a nominal efficiency of 13%. A commercial 50 L compressor cooler was implemented in the system, operating with R134a as its working fluid.

Two sets of experiments were conducted based on the use or non-use of the PCMs at zero- and full-load conditions for the period of June–September 2014. The evaporator temperature was set at 5°C and -10°C for zero- and full-load operation, respectively. Over a test period of four working days, the maximum compressor power requirement was 68.5 W, while the overall achieved COP was 2.28 during the PCM full-load free operation. On the other hand, for the PCM full-load operation, the achieved overall COP was 1.32, over a test period of six working days.

Li et al. (2015) experimentally tested the performance of a solar PV air conditioner for heating and cooling applications in Shanghai, China. Four different working modes were evaluated: (I)/(II) cooling during daytime/nighttime in summer, and (III)/(IV) heating during daytime/nighttime in winter. The achieved solar fraction during daytime in summer was around 80%, while the respective solar fraction for heating in daytime was found to be 5%. On the other hand, during summer, the cooling power output throughout the day was higher than the respective value for the heating output in winter, when in the evening only a part of the heating load could be covered. The solar cooling COP was found to be around 0.32, while the respective solar heating COP was 0.37, which was concluded to be less competitive than the choice of a solar collector. The inverter efficiency for the investigated system was found to be 0.70–0.80, while the PV's module efficiency was relatively low (average efficiency: 12.4%).

Esposito et al. (2015) conducted another comparative study of solar thermal and solar electric cooling for application in a hospital located in Florence, Italy. The refrigeration peak load of the hospital was calculated to be 1 MW. A conventional backup chiller was considered for the case of low cooling production from the solar cooling system. The solar thermal system consists of a double-effect H_2O-LiBr absorption chiller powered by parabolic trough collectors (PTC), while a thermal storage tank is also considered. A schematic of the two compared systems is presented in Figure 4.6. The solar electric cooling system consists mainly of the multi-Si PV panels, tilted at 36° with a south orientation, driving

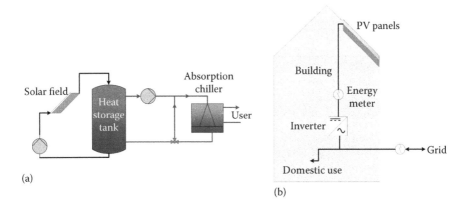

FIGURE 4.6
Schematics of the (a) solar thermal and (b) solar electric cooling systems. (Adapted from Esposito, F. et al., *Energy Procedia*, 81: 1160–1170, 2015.)

a conventional compression chiller that is also connected to the grid. The PV panels are installed in strings to work with multiple inverters to reduce shading effects. Main specifications for the two systems are listed in Tables 4.5 and 4.6.

Three design strategies were considered:

- Solar fraction maximization
- Restrictions of the available hospital space for the solar field are taken into account
- Maximization of revenues based on the incentives provided by the country's government

According to the results, it was found that in terms of energy performance the solar thermal option was more competitive because it was able to provide a maximum of 540 MWh on an annual basis, when the solar field was equal to 403 m². The PV-driven system required significantly larger areas to provide the same energy output, thus making this option less competitive from an energy point of view. The economic analysis, on the other hand, showed that solar electric cooling was more competitive, providing in all cases higher net present values (NPV)—in the case of revenue maximization, the solar thermal NPV was €36,000, while the respective value for the solar electric system was €111,000.

Torres-Toledo et al. (2016) designed and experimentally investigated the performance of a PV-driven ice maker. A controller has been implemented in the system to adapt the

TABLE 4.5

Specifications of the PV Modules and Inverters

Parameter	Value
Maximum power of PV module (Wp)	235
Efficiency (%)	14.6
Open circuit voltage (V)	36.9
Inverter peak power (kW)	20

Source: Data from Esposito, F. et al., *Energy Procedia*, 81: 1160–1170, 2015.

TABLE 4.6

Specifications of the Solar Thermal Cooling System

Parameter	Value
Parabolic Trough Collectors	
Aperture surface (m²)	9.16
Working fluid	THERMINOL VP-1
Working fluid temperature in/out (°C)	166/216
Absorption Chiller	
Cooling capacity (kW)	233
Nominal COP (-)	1.41
Generator driving temperature (°C)	180
Cooling water temperature in/out (°C)	30/37
Chilled water temperature in/out (°C)	14/7

Source: Data from Esposito, F. et al., *Energy Procedia*, 81: 1160–1170, 2015.

compressor's operation based on the availability of solar energy. A 600 Wp PV array was considered, tilted at 35°, and two batteries have also been implemented with a total capacity of 65 Ah at 24 V. The proposed system was installed in Sidi Bouzid, Tunisia. Design ice capacity of the solar ice maker was set at 12 kg of ice per day. The thermal storage is realized by 25 2 L plastic cans with a total capacity of 50 kg of ice. One-year simulations were initially carried out showing that the solar ice maker was able to produce the desired daily amount of ice for 89% of the days in a typical year. Furthermore, experiments were conducted on a test rig under two weather profiles. The results of the experiments are listed in Table 4.7.

A comparative technical and economical evaluation of a converted DC chiller and a conventional AC chiller, both powered by solar PV, was conducted by Opoku et al. (2016). The DC chiller was realized from the conversion of a domestic AC refrigerator by replacing the AC compressor with a variable-speed DC compressor. According to the results, both chillers, operating at a compressor speed of 3000 rpm, were able to achieve evaporator temperatures in the range of –10–2°C. The economic assessment showed that the DC chiller achieved a significant total cost reduction—approximately 18% less than the AC chiller.

Huang et al. (2016) designed an air conditioning system driven solely by PV panels. For the needs of the investigation, six air conditioners with different PV surfaces were experimentally evaluated. The main challenge of the proposed scheme concerned the sizing of the PV panels to provide sufficient energy to drive the air conditioning system, even on days with low solar irradiation. To ensure a steady power supply in the compressor, a battery was implemented. Furthermore, as shown in Figure 4.7, a capacitor was also installed to suppress the surge power at the compressor startup, while a controller was used to control the battery charge/discharge as well as for data logging purposes. The air

TABLE 4.7

Experimental Results for Two Weather Profiles

	July 18	September 10
Average ambient temperature (°C)	32.6	25.2
Global horizontal irradiance (kWh/m²)	7.5	1.9
PV performance ratio (%)	77.8	85.5
Energy conversion efficiency (%)	5.4	9.0
Ice output (kWh)	6.15	5.84
COP (-)	1.76	1.93

Source: Torres-Toledo, Victor et al., *Solar Energy*, 139 (Supplement C): 433–443, 2016.

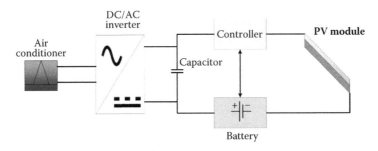

FIGURE 4.7
Schematic of the standalone PV-driven air conditioner. (Adapted from Huang, Bin-Juine, *Renewable Energy*, 88 (Supplement C): 95–101, 2016.)

conditioning system was used to cover the cooling loads of a 28 m² floor area low energy house. The total cooling load of the house was estimated to be 2.2 kW during the summer. The instantaneous operation probability (OPB) and the runtime fraction (RF) of the air conditioning system were measured. According to the results, for a solar irradiation greater than 600 W/m², the OPB was greater than 0.98 when the ratio of maximum PV power to load power exceeded 1.71. On the other hand, the RF was around 1.0 for daily solar radiation greater than 13 MJ/m², with a ratio of maximum PV power to load power greater than 3.

Li et al. (2018) investigated a grid-connected central air conditioning system powered simultaneously by multi-Si PV panels and by the grid (when the PV module cannot provide all the required power input), installed in an 14,220 m² office building located in Zhuhai, China. Potential excess energy generation from the PV module was supplied to the grid. In this study, the operational data for 2015 of the aforementioned system was presented. The main parameters of the system, a schematic of which is also shown in Figure 4.8, are listed in Table 4.8.

The performance of the system was evaluated under different weather conditions (sunny, cloudy, and rainy days) based on typical performance indicators, including the

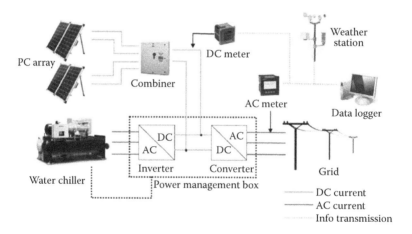

FIGURE 4.8

Schematic of the solar PV air conditioning system. (Reproduced from Li, Y. et al., *Renewable Energy*, 126: 1113–1125, 2018.)

TABLE 4.8

Overview of the Main Parameters for the Solar PV Air Conditioning System

Parameter	Value
Maximum power of PV module (Wp)	250
Efficiency STC (%)	15.3
Open circuit voltage (V)	38.4
Chiller rated power (kW)	362
Chiller rated COP (-)	6.8
Cooling tower rated power (kW)	(2 x) 18.5

Source: Data from Li, Y. et al., *Renewable Energy*, 126: 1113–1125, 2018.

TABLE 4.9

Results of the Solar PV Air Conditioning System Performance

Parameter	Sunny Day	Cloudy Day	Rainy Day
Ambient temperature min/max (°C)	28/33	28/33	25/30
Chilled water inlet/outlet temperature (°C)	28.2/11.2	28.1/11.3	27.6/10.5
Solar fraction (%)	70.4	51.1	17.1
NSF (%)	82.4	54.6	19.5
SER (%)	14.5	6.3	12.5
COP	6.47	6.48	6.21

Source: Data from Li, Y. et al., *Renewable Energy*, 126: 1113–1125, 2018.

solar fraction, the net solar fraction (NSF), and the surplus energy ratio (SER). The results of the investigations are summarized in Table 4.9, proving that the system could operate efficiently even on rainy days.

Liu et al. (2017) investigated the performance of a quasi-grid-connected PV-powered DC air conditioning system. The system's main components included PV panels, a controller, a battery, and a corresponding management module, a main power circuit, a relay circuit, a power factor correction (PFC) circuit, the DC motor, and the air conditioning circuit. The quasi-grid system is realized by converting the AC electrical power from the grid into DC and combining it with the DC power delivered by the PV panels. The calculated EER value of the proposed system was up to 18.28, which as stated by the authors was 4.6 times higher than the respective value of a conventional air conditioner. When winter heating was also considered, the payback period of the proposed system was estimated to be around seven years, a value that, according to the authors, can be further decreased when the aforementioned system is used in applications with higher electricity tariffs.

4.3 Photovoltaic-Thermal Systems

In an attempt to enhance the relatively low efficiency of PV systems and allow for cogeneration of electricity and heat, the coupling of PVs with solar thermal components was proposed in the late 1970s. The systems employing this concept were called photovoltaic-thermal (PVT) systems. As already mentioned, proper cooling of the PV cells must be employed in order not to suffer from efficiency drops. This waste heat can be exploited for several applications, including space heating, industrial process heating and/or pre-heating and crop drying, offering the advantage of employing a cogeneration concept in smaller space with only a relatively low added cost (Brahim and Jemni 2017). There are several PVT classifications, the most important of which are listed below (Lamnatou and Chemisana 2017):

- Based on the working fluid of the thermal subsystem, e.g. PVT-air, PVT-water etc.
- Based on the type of circulation (natural/forced) of the working fluid
- Based on component configurations, especially regarding the PV subsystem (absorber design, number of passes, etc.)

One field of emerging scientific interest concerns the combination of concentrated PV and a PVT systems, a novel hybrid system called the concentrated photovoltaic-thermal (CPVT) system (Sharaf and Orhan 2015).

The first investigated PVT collector was introduced by Wolf (1976) and used for a 167 m² single family residence. A full year simulation was conducted using meteorological data for Boston, Massachusetts. The annual thermal load of the house was estimated at 33,385 kWh, while 4,600 kWh more were required for hot water. The PV area was assumed to be 50 m². Results indicated that significant energy savings were obtained, showing the potential of the proposed technology.

A similar concept was modeled and investigated by Kern Jr and Russell (1978) for applications in several regions of the United States. Results indicated that from an economic point of view, the hybrid system had a high initial cost at the time of the survey. Furthermore, it was concluded that the proposed systems were more competitive in northern regions, where high heat loads were recorded.

Hendrie (1979) developed a model and experimentally studied an air- and a liquid-type PVT collector. According to the experimental data obtained, a maximum electrical efficiency of 6.8% was achieved. Furthermore, it was concluded that the production of electricity, as expected, had a negative effect on the systems thermal efficiencies. More specifically, the thermal efficiency of the air collectors decreased from 0.4 to 0.329 when electrical power was simultaneously produced. The respective reduction for the liquid collector was less severe, with an initial value 0.452 under no power generation to 0.404.

Suzuki and Kitamura (1980) retrofitted two liquid FPCs by attaching silicon solar cells. An overview of the measuring system for the investigated hybrid collectors is presented in Figure 4.9. The average efficiency of the implemented solar cells was 12% at AM 1.15 and an ambient temperature of 28°C. However, the PV arrays efficiencies were lower as realized by the experiments due to reflection and absorption losses, with measured values of 8.84% and 9.20% for the two setups. By using water as the coolant of the solar cells, the thermal efficiencies obtained by the two systems were 0.72 and 0.77, respectively.

Sharan et al. (1986) investigated the performance of CPVT system based on a linear Fresnel reflector (LFR). The system was realized by attaching PV cells to the sides of a rectangular channel receiver using an adhesive. The HTF of the solar thermal system was

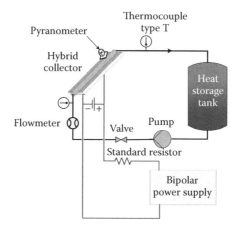

FIGURE 4.9
Schematic of the measuring system for the PVT system. (Adapted from Suzuki, Akio, and Susumu Kitamura, *Japanese Society of Applied Physics*, 19 (S2): 79, 1980.)

circulating through the channel. A parametric study was conducted to evaluate the influence of several parameters, including the concentration ratio, the mass flow rate of the HTF, the absorber size, and the LFR geometry on the system's performance. Results of the simulations showed that, for a concentration ratio of 6.4 and HTF mass flow of 20 g/s, an electrical output of 60 W and a thermal output of 0.6 kW occurred. The corresponding HTF temperature at the outlet of the receiver was 32°C, while the cell temperature was 34.5°C.

Two years later, Hamdy et al. (1988) developed a theoretical model for a CPVT setup. The model used silicon PV cells under concentrated illumination and applied it in CPV and CPVT systems. Simulations were conducted with TRNSYS software. Parabolic collectors with an aperture area of 6 m² were used. The HTF used in the receiver was a mineral oli (caloria). Based on the results of the simulations, an hourly electric efficiency in the range of 20–25% was obtained, at an average cell temperature of 60°C. The daily thermal output was calculated to be 51.4 MJ, while the corresponding electrical output was 12.4 MJ.

Ricaud and Roubeau (1994) reported the performance and economical assessment of a hybrid solar module called "Capthel," used for cogeneration of power and heat. According to the results presented by the authors, the system was able to obtain a thermal efficiency of 0.66 for a solar irradiance of 1000 W/m². The economic feasibility of such an investment for application on a private house and a commercial building was investigated. The analysis showed that a return of investment of 24 years for the domestic application and 21 years for the commercial building was obtained, highlighting the potential of the proposed system.

Garg and Adhikari (1999) conducted a theoretical investigation on the performance of a hybrid PVT air-heating collector coupled with a compound parabolic concentrator. A parametric study was conducted to investigate the influence of parameters such as the collector length, the solar cells area, and the air mass flow rate on the system's efficiency and thermal and electrical outputs. A maximum thermal efficiency of almost 0.59 was reported, for a collector area of 2 m², and an air mass flow rate of 100 kg m^{-2} h^{-1}. The optimum thermal output of 1.5 kW/m² was obtained for an air mass flow rate higher than 400 kg m^{-2} h^{-1}. The corresponding electrical power output was 80 W/m².

Kalogirou (2001) developed a model to simulate the performance of a hybrid PVT solar system using TRNSYS software. For the purposes of the simulations, meteorological data for Nicosia, Cyprus, was used. The investigated system enhanced the annual average efficiency of the PV system from 2.8% to 7.7%. The corresponding overall efficiency when the thermal output it also taken into account was measured to be almost 32%. The lifecycle analysis presented for the aforementioned system indicated that the payback period was equal to 4.6 years.

Vokas et al. (2006) investigated a PVT system for residential heating and cooling applications. The thermal output from the solar collectors was used to drive an absorption chiller. Based on an F-chart analysis for a domestic application in Athens, Greece, it was estimated that the average coverage of heating and cooling from a 30 m² PVT system would be in the range of 47.8% and 25.0%, respectively.

Joshi and Tiwari (2007) conducted a first and second law analysis on a hybrid PVT air collector, showing that the exploitation of the waste heat from the PVs enhanced exergy efficiency by 2–3%. Moreover, the decrease of the cell temperature resulted in an electrical exergetic efficiency of 12% and an overall second law efficiency of almost 15%.

Mittelman et al. (2007) investigated the energetic and economic feasibility of a triple-junction cell for a CPVT system producing cooling and power. The system consisted mainly of the 2,660 m² CPVT circuit, a 1 MW single-effect H_2O-LiBr absorption chiller, and a backup natural-gas-fired heater (see Figure 4.10). For the analysis, a global solar radiation

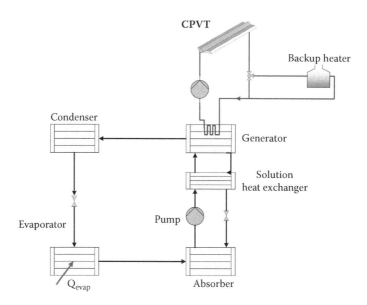

FIGURE 4.10
Schematic of the hybrid CPVT absorption system. (Adapted from Mittelman, Gur et al., *Energy Conversion and Management*, 48 (9): 2481–2490, 2007.)

of 900 W/m² and an electrical efficiency of 37% at the nominal working point were considered. According to the results of the simulations, when the coolant outlet temperature increased from 50°C to 150°C, the electrical efficiency decreased by 3%—at 50°C the electrical efficiency was 23%. The corresponding thermal efficiency for the aforementioned range was approximately 60%. The PV cell temperature was reported to be approximately 10–30 K higher than the coolant outlet temperature, resulting in a rated electric power of 518 kW$_e$ and cooling power of 1.0 MW$_c$.

For an economic analysis, a reference conventional VCC chiller with a COP of 6.36 was considered. A discount rate of 5%, a solar collector lifetime of 20 years, and a chiller's lifetime of 16 years were assumed. The proposed CPVT system was found to be more competitive than a solar cooling system driven by FPCs, with an installation cost of 3.5–4 $/Wp, depending on the conventional energy price. In comparison to a conventional chiller, the CPVT system was more preferable only when the installation costs were less than 1.5 $/Wp.

Xu et al. (2011) designed and experimentally investigated a CPVT system using truncated CPCs, coupled with a heat pump and a water storage tank, as shown in Figure 4.11. The interconnection of the two subsystems was realized by operating the receiver of the CPVT system as the evaporator of the heat pump. The CPVT unit consisted of six modules, each with an aperture area of 1.58 m². The heat pump cycle used R134a as the working fluid. Experiments were conducted on the aforementioned setup showing that for an outlet water temperature in the range of 30–70°C, the corresponding heating COP and the EER were between 6.9–3.1 and 5.1–2.5, respectively. The heat removal from the PV cells resulted in an increase of the average electrical efficiency by 4.6%, resulting in an efficiency for the CPVT system of 17.5%.

Zhao et al. (2011) optimized the design and analyzed the resulting performance of a PVT system, using both non-concentrated and concentrated solar radiation. The two main system components were the PV module and a direct absorption collector (DAC) unit.

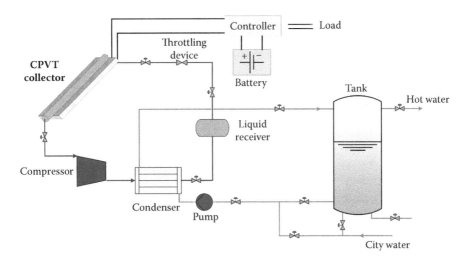

FIGURE 4.11
Schematic of the CPVT system coupled with a heat pump. (Adapted from Xu, Guoying et al., *Applied Thermal Engineering*, 31 (17): 3689–3695, 2011.)

The results of the simulation indicated that the two subsystems can independently exploit the infrared and visible parts of solar radiation. The DAC unit absorbed 89% of the infrared radiation, while 84% of the visible light was transmitted to the PV cells. Furthermore, the experiment showed that increasing the solar irradiance has a significant effect on the system's performance. When solar irradiance was increased from 800 W/m² to 8000 W/m², the HTF temperature exiting the collector was as high as 196°C, corresponding to a thermal efficiency of 0.40 and lifting the electrical efficiency of the PVT system from 12% to 22%.

Li et al. (2011) evaluated the performance of a CPVT system using 2 m² of trough collectors. For the PV module, four types of cells were experimentally evaluated: a single crystalline silicon solar cell array, a polycrystalline silicon cell array, a super cell array, and a GaAs cell array. According to the results, the GaAs cell array achieved the highest electrical efficiency. Furthermore, a similar system with a collection area of 10 m² was experimentally tested using a GaAs cell array and a concentrating silicon cell array. The GaAs cell array performed better, resulting in an instantaneous electrical efficiency of almost 9.9% in comparison to 7.5% achieved by the silicon cell array. The corresponding thermal efficiencies were 49.8% for the GaAs cell array and 42.4% for the silicon cell array.

Teo et al. (2012) experimentally investigated a PVT system implementing a parallel array of ducts designed for cooling purposes. Two sets of experiments were conducted: with and without the active cooling. The results indicated a positive effect on the active cooling in the performance of the system, increasing the solar cells efficiency from 8–9% to 12–14%.

Al-Alili et al. (2012) proposed a hybrid PVT system to produce power and heat to drive a VCC and a solid desiccant cycle, respectively. The main components of the system were the solar collector, a thermal storage tank, a battery, a backup heater, the desiccant wheel cycle, the VCC unit with a cooling capacity of 17.5 kW, and a heat recovery wheel. The system's performance was simulated with TRNSYS software. A parametric analysis was conducted by varying the CPVT collector area, the storage tank volume, and the number of the batteries. The overall COP for the optimum CPVT area was found to reach up to 0.68. The respective values for an absorption chiller powered by ETCs and a PV module driving a conventional chiller were 0.34 and 0.29.

Usama Siddiqui et al. (2012) conducted a parametric analysis studying the effect of water inlet velocity and the inlet temperature on the performance of a PVT collector. For an inlet velocity in the range of 0.01–0.1 m/s, the average PV cell temperature decreased from 41.1°C to 30.6°C, resulting in an increase in electrical efficiency of about 1.2%. On the other hand, increasing the inlet temperature of the HTF from 4°C to 45°C increased the PV cell temperature from 14.5 °C to 50.1 °C, resulting in a decrease in electrical efficiency of almost 4%.

Calise et al. (2013) developed a dynamic model to investigate the behavior of a CPVT system used for cooling, heating, and electricity applications in a building located in Naples, Italy. A 996 m² CPVT system and a 325 kW double-effect absorption chiller were considered for the simulations that were conducted with TRNSYS over the period of a year. The model's results indicated that the system was able to produce 733.3 MWh/y of thermal energy and 302.8 MWh/y of electricity, with an average thermal efficiency of 32% and an electrical efficiency of 13.3%. As a reference for the economic analysis, an air-to-water electric-driven heat pump was considered. Simulations revealed that the primary energy savings achieved by the CPVT system were 84.4% and the payback period was approximately 15.2 years.

In another study, Buonomano et al. (2013) presented a case study of a 1200 m² building. Two building locations were investigated in Italy: Milan and Naples. To drive a single-effect absorption chiller, 130 m² of ETCs and CPVT collectors were used. The model was validated against data derived from TRNSYS. The maximum PES was achieved in Naples with a reported value of 159%, compared to a reference case of a water-water electric chiller and a traditional gas-fired heater for cooling and heating, respectively. The respective maximum PES in Milan was 98%.

Lin et al. (2014) investigated the combination of PVT technology with PCMs for application in Sydney, Australia. The PVT system was integrated into the ceiling ventilation system so as to collect thermal energy and store it in the PCMs. The system was able to operate in four different modes in winter: daytime heating mode, daytime PCM charging, daytime PCM charging and space heating, and nighttime PCM discharging. The respective modes in summer were nighttime direct cooling, nighttime PCM charging, nighttime PCM charging and cooling, and daytime PCM discharging. A 68 m² net zero energy modular house was used for the simulations. According to the simulations, during winter the average thermal efficiency was 12.5% and the respective electrical efficiency was 8.31%, resulting in a power output of 1.35 kW. During summer operation, the corresponding thermal and electrical efficiencies were 13.6% and 8.26%, while the power output increased to 1.98 kW.

Sanaye and Sarrafi (2015) conducted a multi-objective optimization for a trigeneration (power, heating, and cooling) system. The main system components were 2 m² PV panels, CPVT collectors, 2 m² of ETCs, and a single-effect absorption chiller. An overview of the proposed system is presented in Figure 4.12.

The system was proposed to cover the heating and cooling loads of a case study 150 m² building in Tehran, Iran. The cooling and heating loads of the building were estimated to be 8 kW and 3.7 kWh, respectively. For the economic investigation, a reference system consisting of a grid-connected heat pump for the space heating/cooling loads and a gas-fired water heater for the domestic hot water was considered. According to the results of the simulations, for optimum standalone operation of the CPVT-based trigeneration system, a total of nine CPVT collectors, five PV panels, a 1.97 m³ water storage tank, and a 34 kWh battery were required. The corresponding exergy efficiency of the system was 9.1%, while the relative annual benefit (RNAB) was equal to approximately $6,280/y.

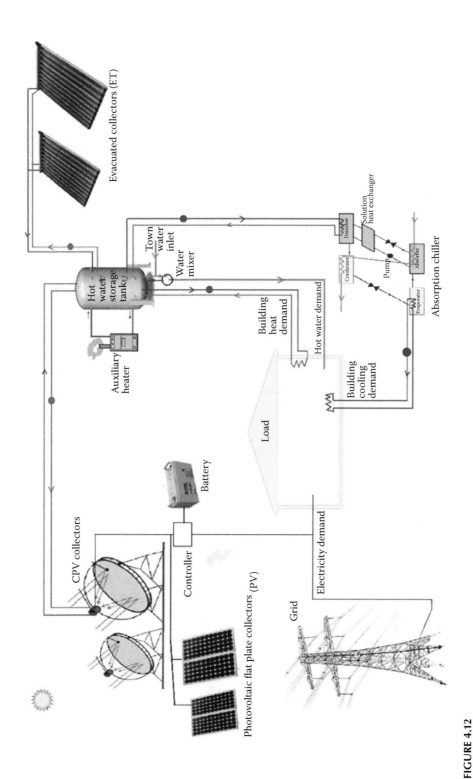

FIGURE 4.12
Schematic of the novel trigeneration CPVT-based system. (Reproduced from Sanaye, Sepehr, and Ahmadreza Sarrafi, *Renewable Energy*, 80 (Supplement C): 699–712, 2015.)

Calise et al. (2016) developed a dynamic model and presented a thermo-economic analysis for a polygeneration system consisting of PVT collectors driving a water-to-water electric heat pump and a zeolite-water adsorption chiller. The rated electrical efficiency of the PV panels was 16%. The heat pump's nominal heating capacity was equal to 8 kW, while the respective cooling capacity was 7 kW. The system was designed to provide power, space heating/cooling, and domestic hot water for a small residential building. During winter operation, hot water from the PVT collectors primarily drove the evaporator of the heat pump, while, during summer operation, the heat from the PVT collectors was led to the adsorption chiller. Throughout the year, the excess of solar energy was converted into hot water for domestic use. For the economic analysis, a reference system consisting of a natural gas boiler for the production of domestic hot water and a reversible air-to-air heat pump for space heating/cooling (with a heating COP of 3) was considered. The systems were evaluated for a case study building measuring 100 m², located in Naples, Italy. The results showed that the total energy efficiency of the PVT system was equal to 0.49, with a heating COP for the heat pump over 4, and a cooling COP for the adsorption chiller of 0.55. From an economic point of view, the novel system was not competitive unless a subsidy of 50% or more of the investment costs was provided.

Bianchini et al. (2017) investigated the potential of a PVT system located in Forli, Italy, based on data collected from the system's remote monitoring. The results indicated that PV cooling resulted in a 1–3% increase in the electric yield of the system. At an average outlet temperature of 40°C, the system was able to produce 835 kWh/m² of electricity and 1600 kWh/m² of heat. On a yearly basis, the system was able to produce approximately 1360 kWh/y of electricity, while the respective thermal production ranged between 267–443 kWh/m², depending on the average inlet temperature of the cooling fluid. The investigated system was also economically compared with separate PV and flat-plate solar collectors. According to the results, the PVT system was able to be competitive when its installation costs were in the range of €3,700–4,700/kWp.

Cai et al. (2017) investigated the performance of a novel hybrid solar PVT ground-source heat pump system using simulations. The proposed system can be divided into three main subsystems: the PVT system, the heat pump, and the terminal system. The PVT circuit consists of the PVT collector, the heat storage tank, the circulating pump, and the cooling tower. The terminal system consists of floor radiation pipe-coils for heating in winter and a fan-coil for space cooling in summer. The aforementioned system was installed at the Qingyun community demonstration project in Dalian, China. The building, with a total area of 1,288 m², was designed as a workplace operating between 8 a.m. and 5 p.m. Peak power generation was 45.4 kW. The selected heat pump had a nominal heating capacity of 79.4 kW and a cooling capacity of 71.8 kW. The experiments were conducted in 2015. During the heating period, the system's power generation measured equal to 20.08 MWh. On March 12, 2015, the PVT configuration reduced the PV module temperature by 26%, down to 28°C, resulting in a 2% increase in PV efficiency. The highest electrical efficiency of the PVT module was reported to be 15%.

Nomenclature

A_{pv}	PV panels surface	[m²]
COP	Coefficient of performance	–
EER	Energy efficiency ratio	–
EPT	Energy payback time	[years]
I	Solar radiation	[W/m²]
Q	Heat	[J]
$\dot{Q}$	Heat flux	[W]
RF	Runtime fraction	–
SER	Surplus energy ratio	–

Greek Symbols

θ	Temperature	[°C]
η	Second law efficiency	–

Subscripts

evap	Evaporation
PV,r	Photovoltaic-driven refrigeration system

Abbreviations

AC	Alternating current
a-Si	Amorphous silicon
CdTe	Cadmium telluride
CIGS	Copper indium gallium selenide
CPC	Compound paraphilic collectors
CPV	Concentrated photovoltaic
CPVT	Concentrated photovoltaic-thermal
DSSC	Dye-sensitized solar cell
ETC	Evacuated tube collectors
FPC	Flat-plate collectors
GaAs	Gallium arsenide
GaInAs	Gallium indium phosphide
GaInP	Gallium indium arsenide
Ge	Germanium
DAC	Direct absorption collector

DC	Direct current
LFR	Linear Fresnel reflector
nc-Si	Nanocrystalline silicon
NPV	Net present value
NSF	Net solar fraction
OPB	Operation probability
PCM	Phase change materials
PFC	Power factor correction
PTC	Parabolic trough collectors
PV	Photovoltaic
PVT	Photovoltaic-thermal
STC	Standard test conditions
VCC	Vapor compression cycle

References

Aguilar, Francisco J., Pedro V. Quiles, and Simón Aledo. 2014. "Operation and Energy Efficiency of a Hybrid Air Conditioner Simultaneously Connected to the Grid and to Photovoltaic Panels." *Energy Procedia* 48 (Supplement C): 768–777.

Al-Alili, A., Y. Hwang, R. Radermacher, and I. Kubo. 2012. "A High Efficiency Solar Air Conditioner Using Concentrating Photovoltaic/Thermal Collectors." *Applied Energy* 93 (Supplement C): 138–147.

Allouhi, A., T. Kousksou, A. Jamil, P. Bruel, Y. Mourad, and Y. Zeraouli. 2015. "Solar Driven Cooling Systems: An Updated Review." *Renewable and Sustainable Energy Reviews* 44 (Supplement C): 159–181.

Alzaabi, A. A., N. K. Badawiyeh, H. O. Hantoush, and A. K. Hamid. 2014. "Electrical/Thermal Performance of Hybrid PV/T System in Sharjah, UAE." *International Journal of Smart Grid and Clean Energy* 3 (4): 385–389.

Astronergy. "Products: PV Modules- Crysalline Series." Accessed November 2017. http://www.astronergy.com/products.php.

Axaopoulos, Petros J., and Michael P. Theodoridis. 2009. "Design and Experimental Performance of a PV Ice-Maker Without Battery." *Solar Energy* 83 (8): 1360–1369.

Ayyash, S., and M. Sartawi. 1983. "Economic Comparison of Solar Absorption and Photovoltaic-Assisted Vapour Compression Cooling Systems." *International Journal of Energy Research* 7 (3): 279–288.

Beccali, Marco, Maurizio Cellura, Pietro Finocchiaro, Francesco Guarino, Sonia Longo, and Bettina Nocke. 2014. "Life Cycle Performance Assessment of Small Solar Thermal Cooling Systems and Conventional Plants Assisted with Photovoltaics." *Solar Energy* 104 (Supplement C): 93–102.

Bianchini, Augusto, Alessandro Guzzini, Marco Pellegrini, and Cesare Saccani. 2017. "Photovoltaic/Thermal (PV/T) Solar System: Experimental Measurements, Performance Analysis and Economic Assessment." *Renewable Energy* 111 (Supplement C): 543–555.

Bilgili, Mehmet. 2011. "Hourly Simulation and Performance of Solar Electric-Vapor Compression Refrigeration System." *Solar Energy* 85 (11): 2720–2731.

Brahim, Taoufik, and Abdelmajid Jemni. 2017. "Economical Assessment and Applications of Photovoltaic/Thermal Hybrid Solar Technology: A Review." *Solar Energy* 153 (Supplement C): 540–561.

Buonomano, A., F. Calise, and A. Palombo. 2013. "Solar Heating and Cooling Systems by CPVT and ET Solar Collectors: A Novel Transient Simulation Model." *Applied Energy* 103 (Supplement C): 588–606.

Cai, Junjie, Zhenhua Quan, Tianyao Li, Longshu Hou, Yaohua Zhao, and Mengliang Yao. 2017. "Performance Study of a Novel Hybrid Solar PV/T Ground-Source Heat Pump System." *Procedia Engineering* 205 (Supplement C): 1642–1649.

Calise, Francesco, Massimo Dentice d'Accadia, Rafal Damian Figaj, and Laura Vanoli. 2016. "A Novel Solar-Assisted Heat Pump Driven by Photovoltaic/Thermal Collectors: Dynamic Simulation and Thermoeconomic Optimization." *Energy* 95 (Supplement C): 346–366.

Calise, Francesco, Massimo Dentice d'Accadia, Adolfo Palombo, and Laura Vanoli. 2013. "Dynamic Simulation of a Novel High-Temperature Solar Trigeneration System Based on Concentrating Photovoltaic/Thermal Collectors." *Energy* 61 (Supplement C): 72–86.

Canadian Solar. "Solar Panels." Accessed November 2017. https://www.canadiansolar.com/solar-panels/dymond.html.

Deb, Satyen K. 1998. "Recent Developments in High Efficiency Photovoltaic Cells." *Renewable Energy* 15 (1–4): 467–472.

Eicker, Ursula, Antonio Colmenar-Santos, Lya Teran, Mariela Cotrado, and David Borge-Diez. 2014. "Economic Evaluation of Solar Thermal and Photovoltaic Cooling Systems through Simulation in Different Climatic Conditions: An Analysis in Three Different Cities in Europe." *Energy and Buildings* 70 (Supplement C): 207–223.

El-Bahloul, Asmaa Ahmed M., Ahmed Hamza H. Ali, and Shinichi Ookawara. 2015. "Performance and Sizing of Solar Driven dc Motor Vapor Compression Refrigerator with Thermal Storage in Hot Arid Remote Areas." *Energy Procedia* 70 (Supplement C): 634–643.

El Chaar, L., L. A. lamont, and N. El Zein. 2011. "Review of Photovoltaic Technologies." *Renewable and Sustainable Energy Reviews* 15 (5): 2165–2175.

El Tom, O. M. M., S. A. Omer, A. Z. Taha, and A. A. M. Sayigh. 1991. "Performance of a Photovoltaic Solar Refrigerator in Tropical Climate Conditions." *Renewable Energy* 1 (2): 199–205.

Esposito, F., A. Dolci, G. Ferrara, L. Ferrari, and E. A. Carnevale. 2015. "A Case Study Based Comparison Between Solar Thermal and Solar Electric Cooling." *Energy Procedia* 81: 1160–1170.

Fumo, N., V. Bortone, and J. C. Zambrano. 2013. "Comparative Analysis of Solar Thermal Cooling and Solar Photovoltaic Cooling Systems." *Journal of Solar Energy Engineering* 135 (2): 346–355.

Garg, H. P., and R. S. Adhikari. 1999. "Performance Analysis of a Hybrid Photovoltaic/Thermal (PV/T) Collector with Integrated CPC Troughs." *International Journal of Energy Research* 23 (15): 1295–1304.

Goetzberger, Adolf, Christopher Hebling, and Hans-Werner Schock. 2003. "Photovoltaic Materials, History, Status and Outlook." *Materials Science and Engineering: R: Reports* 40 (1): 1–46.

Gong, Jiawei, K. Sumathy, Qiquan Qiao, and Zhengping Zhou. 2017. "Review on Dye-Sensitized Solar Cells (DSSCs): Advanced Techniques and Research Trends." *Renewable and Sustainable Energy Reviews* 68 (Part 1): 234–246.

Gorter, T., and A. H. M. E. Reinders. 2012. "A Comparison of 15 Polymers for Application in Photovoltaic Modules in PV-Powered Boats." *Applied Energy* 92 (Supplement C): 286–297.

Green, Martin A., Keith Emery, Yoshihiro Hishikawa, Wilhelm Warta, and Ewan D. Dunlop. 2015. "Solar Cell Efficiency Tables (Version 45)." *Progress in Photovoltaics: Research and Applications* 23 (1): 1–9.

Hamdy, M. A., F. Luttmann, and D. Osborn. 1988. "Model of a Spectrally Selective Decoupled Photovoltaic/Thermal Concentrating System." *Applied Energy* 30 (3): 209–225.

Hanwha Q CELLS. "Q CELLS Solar Panels." Accessed November 2017. https://www.q-cells.com/en/index/products/solar-panels.

Hartmann, N., C. Glueck, and F. P. Schmidt. 2011. "Solar Cooling for Small Office Buildings: Comparison of Solar Thermal and Photovoltaic Options for Two Different European Climates." *Renewable Energy* 36 (5): 1329–1338.

Hendrie, Susan D. *Evaluation of Combined Photovoltaic/Thermal Collectors.* (Massachusetts Inst. of Tech., Lexington, USA, Lincoln Lab, 1979).

Huang, Bin-Juine, Tung-Fu Hou, Po-Chien Hsu, Tse-Han Lin, Yan-Tze Chen, Chi-Wen Chen, Kang Li, and K. Y. Lee. 2016. "Design of Direct solar PV Driven Air Conditioner." *Renewable Energy* 88 (Supplement C): 95–101.

Infante Ferreira, Carlos, and Dong-Seon Kim. 2014. "Techno-Economic Review of Solar Cooling Technologies Based on Location-Specific Data." *International Journal of Refrigeration* 39: 23–37.

International Energy Agency. *Technology Roadmap*. Solar Photovoltaic Energy. (2014).

Jäger-Waldau, Arnulf, Márta Szabó, Fabio Monforti-Ferrario, Hans Bloem, Thomas Huld, and R Lacal Arantegui. "Renewable Energy Snapshots 2011." *JRC Report*. Available at http://re.jrc.ec .europa. eu/refsys/pdf/RE_Snapshots_2011. pdf.

Jinko Solar. "Solar Cells." Accessed November 2017. https://www.jinkosolar.com/product_257 .html?lan=en.

Joshi, Anand S., Ibrahim Dincer, and Bale V. Reddy. 2009. "Performance Analysis of Photovoltaic Systems: A Review." *Renewable and Sustainable Energy Reviews* 13 (8): 1884–1897.

Joshi, Anand S., and Arvind Tiwari. 2007. "Energy and Exergy Efficiencies of a Hybrid Photovoltaic– Thermal (PV/T) Air Collector." *Renewable Energy* 32 (13): 2223–2241.

Kalogirou, Soteris A. 2001. "Use of TRNSYS for Modelling and Simulation of a Hybrid PV–Thermal Solar System for Cyprus." *Renewable Energy* 23 (2): 247–260.

Kaplanis, Socrates, and Nikolaos Papanastasiou. 2006. "The Study and Performance of a Modified Conventional Refrigerator to Serve as a PV Powered One." *Renewable Energy* 31 (6): 771–780.

Kern Jr, E. C., and M. C. Russell. *Combined Photovoltaic and Thermal Hybrid Collector Systems*. (Massachusetts Inst. of Tech., Lexington, USA. Lincoln Lab, 1978).

Kumar Sahu, Bikash. 2015. "A Study on Global Solar PV Energy Developments and Policies with Special Focus on the Top Ten Solar PV Power Producing Countries." *Renewable and Sustainable Energy Reviews* 43 (Supplement C): 621–634.

KYOCERA Fineceramics GmbH. YOCERA Fineceramics GmbH. Esslingen, Germany. https://www .irishellas.com/files/Kyocera-Quality-Brochure_May-2012_EN.pdf.

Lamnatou, Chr, and D. Chemisana. 2017. "Photovoltaic/Thermal (PVT) Systems: A Review with Emphasis on Environmental Issues." *Renewable Energy* 105 (Supplement C): 270–287.

Li, M., G. L. Li, X. Ji, F. Yin, and L. Xu. 2011. "The Performance Analysis of the Trough Concentrating Solar Photovoltaic/Thermal System." *Energy Conversion and Management* 52 (6): 2378–2383.

Li, Y., G. Zhang, G. Z. Lv, A. N. Zhang, and R. Z. Wang. 2015. "Performance Study of a Solar Photovoltaic Air Conditioner in the Hot Summer and Cold Winter Zone." *Solar Energy* 117 (Supplement C): 167–179.

Li, Y., B. Y. Zhao, Z. G. Zhao, R. A. Taylor, and R. Z. Wang. 2018. "Performance Study of a Grid-Connected Photovoltaic Powered Central Air Conditioner in the South China Climate." *Renewable Energy* 126: 1113–1125.

Lin, Wenye, Zhenjun Ma, M. Imroz Sohel, and Paul Cooper. 2014. "Development and Evaluation of a Ceiling Ventilation System Enhanced by Solar Photovoltaic Thermal Collectors and Phase Change Materials." *Energy Conversion and Management* 88 (Supplement C): 218–230.

Liu, Zhongbao, Ao Li, Qinghua Wang, Yuanying Chi, and Lingfei Zhang. 2017. "Performance Study of a Quasi Grid-Connected Photovoltaic Powered DC Air Conditioner in a Hot Summer Zone." *Applied Thermal Engineering* 121 (Supplement C): 1102–1110.

LONGi Solar. "Products-Module-Conventional." Accessed November 2017. http://en.longi-solar .com/Home/Products/module/id/2_.html.

Meneses-Rodríguez, David, Paul P. Horley, Jesús González-Hernández, Yuri V. Vorobiev, and Peter N. Gorley. 2005. "Photovoltaic Solar Cells Performance at Elevated Temperatures." *Solar Energy* 78 (2): 243–250.

Mittelman, Gur, Abraham Kribus, and Abraham Dayan. 2007. "Solar Cooling with Concentrating Photovoltaic/thermal (CPVT) Systems." *Energy Conversion and Management* 48 (9): 2481–2490.

Opoku, R., S. Anane, I. A. Edwin, M. S. Adaramola, and R. Seidu. 2016. "Comparative Techno-Economic Assessment of a Converted DC Refrigerator and a Conventional AC Refrigerator Both Powered by Solar PV." *International Journal of Refrigeration* 72 (Supplement C): 1–11.

Osman, M. G. 1985. "Performance Analysis of a Solar Air-Conditioned Villa in the Arabian Gulf." *Energy Conversion and Management* 25 (3): 283–293.

Parida, Bhubaneswari, S. Iniyan, and Ranko Goic. 2011. "A Review of Solar Photovoltaic Technologies." *Renewable and Sustainable Energy Reviews* 15 (3): 1625–1636. REN21. 2016. Renewables 2016 Global Status Report. Paris.

REC Solar Holdings AS. "Products & Solutions." Accessed November 2017. http://www.recgroup.com/en/products-solutions.

Ricaud, A., and P. Roubeau. "'Capthel,' a 66% Efficient Hybrid Solar Module and the 'Ecothel' Co-Generation Solar System." Photovoltaic Energy Conversion, 1994. Conference Record of the 24th IEEE Photovoltaic Specialists Conference 1994, 1994 IEEE First World Conference.

Richter, Christoph, D. Lincot, and Christian A. Gueymard. 2013. *Solar Energy*. New York, NY: Springer, 2013.

Sanaye, Sepehr, and Ahmadreza Sarrafi. 2015. "Optimization of Combined Cooling, Heating and Power Generation by a Solar System." *Renewable Energy* 80 (Supplement C): 699–712.

Sarbu, Ioan, and Calin Sebarchievici. 2013. "Review of Solar Refrigeration and Cooling Systems."*Energy and Buildings* 67 (Supplement C): 286–297.

Sargunanathan, S., A. Elango, and S. Tharves Mohideen. 2016. "Performance Enhancement of Solar Photovoltaic Cells Using Effective Cooling Methods: A Review." *Renewable and Sustainable Energy Reviews* 64 (Supplement C): 382–393.

Sharaf, Omar Z., and Mehmet F. Orhan. 2015. "Concentrated Photovoltaic Thermal (CPVT) Solar Collector Systems: Part II—Implemented Systems, Performance Assessment, and Future Directions." *Renewable and Sustainable Energy Reviews* 50 (Supplement C): 1566–1633.

Sharan, S. N., S. S. Mathur, and T. C. Kandpal. 1986. "Analysis of an Actively Cooled Photovoltaic-Thermal Solar Concentrator Receiver System Using a Fin-Type Absorber." *Solar & Wind Technology* 3 (4): 281–285.

Siecker, J., K. Kusakana, and B. P. Numbi. 2017. "A Review of Solar Photovoltaic Systems Cooling Technologies." *Renewable and Sustainable Energy Reviews* 79 (Supplement C): 192–203.

Silfab Solar Inc. "Products." Accessed November 2017. http://www.silfab.ca/products/.

Singh, G. K. 2013. "Solar Power Generation by PV (Photovoltaic) Technology: A Review." *Energy* 53 (Supplement C): 1–13.

The Solaria Corporation. "Solaria PowerXT | Rooftop Solutions." Accessed November 2017. http://www.solaria.com/residential-commercial.

SOLARTECH POWER INC. "Products." Accessed November 2017.

SolarWorld Industries GmbH. "Sunmodule Plus SW290/300 MONO." Accessed November 2017. https://www.solarworld.de/fileadmin/swi_downloads/produkte/sunmodule/datenblaetter/en/swi_db_sunmodule_plus_290-300_mono_en.pdf.

Suntech. "Product Center." Accessed November 2017. http://www.suntech-power.com/menu/gxzj.html.

Suzuki, Akio, and Susumu Kitamura. 1980. "Combined Photovoltaic and Thermal Hybrid Collector." *Japanese Society of Applied Physics* 19 (S2): 79.

Talesun. "Products." Accessed November 2017. http://www.talesun-eu.com/products/.

Teo, H. G., P. S. Lee, and M. N. A. Hawlader. 2012. "An Active Cooling System for Photovoltaic Modules." *Applied Energy* 90 (1): 309–315.

Torres-Toledo, Victor, Klaus Meissner, Philip Täschner, Santiago Martınez-Ballester, and Joachim Müller. 2016. "Design and Performance of a Small-Scale Solar Ice-Maker Based on a DC-Freezer and an Adaptive Control Unit." *Solar Energy* 139 (Supplement C): 433–443.

Trina Solar. "Products and Solutions." Accessed Nov. 2017. http://www.trinasolar.com:81/us/product.

Tyagi, V. V., Nurul A. A. Rahim, N. A. Rahim, and Jeyraj A. L. Selvaraj. 2013. "Progress in Solar PV Technology: Research and Achievement." *Renewable and Sustainable Energy Reviews* 20 (Supplement C): 443–461.

Ullah, K. R., R. Saidur, H. W. Ping, R. K. Akikur, and N. H. Shuvo. 2013. "A Review of Solar Thermal Refrigeration and Cooling Methods." *Renewable and Sustainable Energy Reviews* 24: 499–513.

Usama Siddiqui, M., A. F. M. Arif, Leah Kelley, and Steven Dubowsky. 2012. "Three-Dimensional Thermal Modeling of a Photovoltaic Module Under Varying Conditions." *Solar Energy* 86 (9): 2620–2631.

Vikram Solar Limited. "Products-PV Modules." Accessed November 2017. https://www.vikramsolar .com/product-category/pv-modules/#.

Vokas, G., N. Christandonis, and F. Skittides. 2006. "Hybrid Photovoltaic—Thermal Systems for Domestic Heating and Cooling—A Theoretical Approach." *Solar Energy* 80 (5): 607–615.

Wolf, Martin. 1976. "Performance Analyses of Combined Heating and Photovoltaic Power Systems for Residences." *Energy Conversion* 16 (1): 79–90.

Xu, Guoying, Xiaosong Zhang, and Shiming Deng. 2011. "Experimental Study on the Operating Characteristics of a Novel Low-Concentrating Solar Photovoltaic/Thermal Integrated Heat Pump Water Heating System." *Applied Thermal Engineering* 31 (17): 3689–3695.

Yingli Solar. "Our Products." Accessed November 2017. http://www.yinglisolar.com/en/products /solar-modules/.

Zhao, Jiafei, Yongchen Song, Wei-Haur Lam, Weiguo Liu, Yu Liu, Yi Zhang, and DaYong Wang. 2011. "Solar Radiation Transfer and Performance Analysis of an Optimum Photovoltaic/Thermal System." *Energy Conversion and Management* 52 (2): 1343–1353.

5

Absorption Cooling Heat Pumps

Absorption cooling was extensively popular in the first years of refrigeration applications, mainly for ice production purposes (Kalogirou 2014). Absorption cooling's low COP has been the main obstacle in the expansion of the absorption refrigeration market. However, the fact that absorption chillers have a very low energy consumption enables them to be used in remote areas, and hence they are considered the most widely distributed cooling system worldwide. The most commonly used working pairs for absorption chillers are water-lithium bromide (H_2O-LiBr) and ammonia-water (NH_3-H_2O). For applications above 5°C, H_2O-LiBr systems are preferred for reasons that will be discussed in Section 5.1.1. Absorption chillers are commercially available in a wide range of capacities, from several kWs to, more commonly, hundreds of kWs. For single-effect absorption machines, the COP is in the range of 0.7–0.8. Double-effect systems may raise the COP to 1.2. Most commercial solar cooling applications are equipped with absorption chillers because there is significant experience with this technology in comparison to other potential thermally driven options.

5.1 Absorption Applications and Performance Data

5.1.1 Working Pairs

Several working pairs have been suggested and investigated in literature. Marcriss (1978) carried out a survey discussing the potential coupling of several refrigerant compounds with adsorbents based on select criteria, including the positive deviation from Raoult's law in order to enhance the rated COP of the system and the prevention of crystallization issues under certain conditions. As mentioned earlier, the most widely used working pairs for absorption chillers are H_2O-LiBr and NH_3-H_2O.

In H_2O-LiBr systems, because water serves as the refrigerant of the cycle, there is a temperature limitation due to water freezing. This means that H_2O-LiBr absorption chillers must operate with a temperature minimum of 5°C to avoid ice formation and permanent damage to the system. Hence, for close-to- and sub-zero temperatures in the evaporator, NH_3-H_2O is preferred, with such systems being able to generate temperatures as low as −60°C. The respective temperatures for desorption range between 70°C and 120°C. H_2O-LiBr systems, in which water is the refrigerant, benefit from water's high specific evaporation enthalpy and the low volatility of LiBr. On the other hand, such systems suffer from crystallization issues, as LiBr is not totally soluble in water, and when LiBr concentrates beyond a certain limit, it crystallizes and causes problems in the normal operation of certain components. This can occur due to an air leakage in the system, electric power failures, or a lower temperature in the condenser. To reduce crystallization issues, H_2O-LiBr absorption chillers are cooled by wet cooling towers (Eicker 2014). Another means of prevention is the implementation of crystallization sensors by the manufacturers of absorption chillers in order to shut down and clean the machine in case of a crystallization incident (Dinçer 2003).

As a working pair, NH_3-H_2O has advantages due to its high affinity and high stability (Kurem and Horuz 2001). The fact that the refrigerant (ammonia) is lighter than air allows for more compact designs in comparison to H_2O-LiBr systems which have a high specific volume of water vapor. NH_3-H_2O systems demand special attention due to the toxicity and volatility of ammonia. High volatility makes the use of a rectifier necessary in such systems; otherwise, water would accumulate in the evaporator, decreasing the efficient operation of the system. Another limitation of NH_3-H_2O systems is the corrosivity of ammonia, prohibiting the use of copper, aluminum, zinc, and respective alloys. Furthermore, because of the ammonia vapor pressure curve, such systems normally operate under high pressure, raising safety concerns relevant to their operation. H_2O-LiBr absorption chillers, at the same time, operate under partial vacuum and with a non-toxic and non-flammable working pair, making them a safer choice. The high pressure of NH_3-H_2O systems results in increased capital costs in comparison to H_2O-LiBr absorption chillers. On the other hand, H_2O-LiBr can be corrosive to certain metals and is also expensive as a pair. In order to reduce the corrosion effects of LiBr, some additives may be added.

Regarding the temperature range of NH_3-H_2O absorption chillers, evaporation can take place at temperatures as low as −60°C, given the fact that the freezing point of NH_3 is as low as −77°C, allowing for use in both air conditioning and industrial refrigeration applications. The corresponding desorption temperatures range between 70 and 120°C. The typical COP values for single-stage NH_3-H_2O absorption chillers are 0.4–0.6, while H_2O-LiBr values are in the range of 0.5–0.7. The main field of application for these absorption chillers is air conditioning equipped with fan coils or cooled ceilings.

Apart from the two aforementioned commercial working pairs, several investigations have been carried out for potential working pair replacements in order to enhance system performance. Balamuru et al. (2000) simulated the performance of a ternary mixture consisting of NH_3 (refrigerant) and H_2O-NaOH (absorbent) in an absorption refrigeration cycle. The influence of the addition of NaOH was investigated. It was determined that increasing the mass concentration of NaOH from 0% to 30% had a positive effect on the COP, moving the optimum to lower generator temperatures. The maximum COP reported was approximately 0.58, at a generator temperature of 60°C. The addition of NaOH was also found to have a positive effect on the cooling capacity of the system, with a capacity in the range of 4–10 kW, for a generator temperature of 60–160°C, respectively.

Using simulations, Pilatowsky et al. (2001, 2004) tested the performance of a monomethylamine-water absorption chiller powered by evacuated tube collectors coupled with a conventional backup heater. The system was simulated for potential application in milk cooling in the rural regions of Mexico. The results of the modeling process indicated that the investigated system could chill water down to 5–10°C, at low generation temperatures in the range 60–80°C, with a COP of 0.15–0.70 (condensation temperature = 25°C). Collector efficiency ranged between 58–82%, resulting in a solar COP of 0.22–0.32.

De Lucas et al. (2004) investigated the replacement of a binary H_2O-LiBr with a ternary mixture, H_2O-(LiBr:CHO$_2$K=2:1) by mass ratio, in an absorption refrigeration system. The simulations were carried out for an absorption chiller with a 1580 kW nominal capacity at a condenser temperature of 46.1°C and an evaporator temperature of 5.5°C. The results showed that the binary mixture COP was 0.75 (generator temperature = 101.7°C, generator heat input = 2084.67 kW, absorber temperature = 40°C). The respective COP value for the ternary mixture was 0.85 (generator temperature = 66°C, generator heat input = 1850.21 kW, absorber temperature = 15°C).

Aphornratana (2005) carried out an extensive review for investigated absorption cooling schemes. According to Aphornratana, R22 and R21 have been pointed out due to their favorable solubility with a variety of organic solvents. Among these solvents, dimethyl ether of tetraethylene glycol (DMETEG) and dimethylformamide (DMF) have been remarked as optimal choices. As already mentioned, the crystallization issues of the conventional absorption working pairs are one of the main limitations of the temperature levels of the cycle. Herold et al. (1991) proposed the use of an aqueous ternary hydroxide working fluid, with sodium, potassium, and caesium hydroxide as the components of the mixture (in a 40/36/24 proportion). The properties of the mixture allowed for an enhancement in the temperature lift of the cycle, and thus more suitable conditions for water heating applications were achieved. Dan and Murthy (1989) compared the performance of refrigerants R21 and R22 paired with DMETEG or DMF and concluded that the working pair R21-DMETEG was the most efficient for a single-effect absorption cycle.

Environmental concerns regarding ozone depletion have resulted in the investigation of alternative working pairs—mainly with environmentally friendly refrigerants—including R123a-DMETEG, R123a-DMF, and R123a-trifluoroethanol.

5.1.2 Absorption Units

Apart from the conventional schemes already discussed in Chapter 2, several more advanced absorption cycles have been introduced and investigated. One interesting modification is the von Platen-Munters system, which introduces a third fluid in order to remove the pump of a conventional absorption system, thus resulting in a module that is free of moving parts. Most commonly, the three-fluid system consists of the conventional NH_3-H_2O working pair along with hydrogen, which is a neutral gas and is used for pressure elevations, hence it is called a carrier gas. The amount of ammonia, hydrogen, and water is such that, at an appropriate pressure, ammonia can condense at room temperature. Ammonia vapor mixed with hydrogen circulates through a heat exchanger (HEX1 in Figure 5.1) toward the absorber, where ammonia is absorbed and hydrogen rises toward the evaporator through a secondary stream of HEX1. The NH_3-H_2O solution then flows toward the generator to complete the conventional absorption cycle, as discussed in Chapter 2.

The GAX (generator-absorber exchange absorption cycle) is another well-known modification of the conventional single-effect absorption cycle. Based on this module, the high temperature solution exiting the generator is forced to pass through the generator heat exchanger, where it transfers part of the heat to the cold refrigerant, and then is led to the GAX-HEX. In the GAX-HEX, a two-phase heat transfer takes place on both streams—generation in the cold stream and absorption in the hot stream. Apart from GAX-HEX, there is another heat exchanger, the absorber heat exchanger (AHX), that is used for cooling purposes by the cold solution in the cycle, as shown in Figure 5.2. In a situation with no GAX-HEX, the cycle is then named AHX, working at a lower temperature level and lower COP.

The broader working temperature range of the GAX cycle requires the use of more stable solutions for the range of its application, hence in most cases a NH_3-H_2O solution is used as the working fluid. Due to the heat recovery step in the GAX, this configuration enhances the COP to levels above 0.8. However, it requires driving steam temperatures as high as 160°C, hence it is not currently available for commercial solar cooling applications.

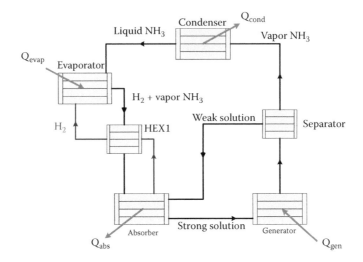

FIGURE 5.1
Overview of a von Platen-Munters absorption system.

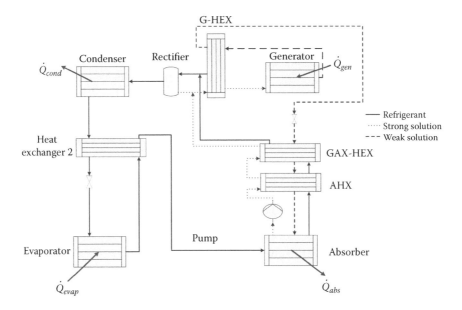

FIGURE 5.2
Schematic of the GAX absorption cycle. (Adapted from Jawahar, C. P., and R. Saravanan, *Renewable and Sustainable Energy Reviews*, 14 (8): 2372–2382, 2010; Velázquez, N., and R. Best, *Applied Thermal Engineering*, 22 (10): 1089–1103, 2002.)

Kim (2007) presented an overview of the COP of several modifications of an absorption cycle with respect to the driving temperature, as shown in Figure 5.3

Zhu et al. (2008) conducted experiments on a GAX absorption system using NH_3-NaSCN as the working pair. The experimental absorption system was powered by a 5 kW electrical hot water heater. The experimental results showed that this system could operate more efficiently at lower generator and evaporator temperatures in comparison to conventional

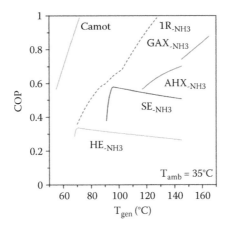

FIGURE 5.3
COP performance of several NH₃-H₂O absorption cycles (SE: single effect, HE: half-effect). (Reproduced from Kim, D. S. et al., "Solar Absorption Cooling", PhD Dissertation, Delft, 2007.)

working pairs. The maximum COP obtained by the system was 0.481 for a driving temperature of 100°C (corresponding to a generator outlet temperature 68.8°C), an evaporator temperature of 5°C, and a cooling water temperature of 23.2°C. The corresponding cooling output was 0.824 kW for cooling. The maximum thermal output on heating mode was 4 kW.

5.1.3 Theoretical Investigations on Absorption Units

Given the fact that absorption cooling is a rather old technology, several relevant investigations have been conducted in terms of simulations and experiments. Gordon and Ng (1995) developed a general thermodynamic model for absorption chillers in steady state conditions, useful for diagnostic reasons, and validated the accuracy of the simulations against the manufacturer's data from the period of research.

Bruno et al. (1999) investigated the economic feasibility of integrating an NH_3-H_2O absorption chiller instead of a conventional vapor compression machine into the existing energy systems of process plants. A case study for the integration of one such absorption chiller in a Bayer petrochemical plant in Tarragona, Spain, was compared to the results of implementing a conventional chiller. The results of the analysis showed that the absorption chiller would be more attractive than the conventional machine only if the cogeneration plant scaled up. At the investigated capacity, the mechanical compression chiller was more attractive.

Ezzine et al. (2004) carried out an energetic and exergetic analysis of a double-effect double generator NH_3-H_2O absorption chiller. The simulations were conducted in FORTRAN77. Several investigations were conducted to evaluate the influence of the evaporator pressure and the chiller driving temperature. The optimal COP reported was equal to 0.78, for an evaporator pressure of 3.5 bar. In terms of the second law analysis, the absorber-generators were identified as the components responsible for the highest exergy destruction.

Şencan et al. (2005) carried out an exergetic analysis of a H_2O-LiBr absorption cycle used for cooling and heating applications. A model was developed in Fortran language to investigate the behavior of the system under varying heat sources and chilled water inlet temperatures. Based on the results of the simulations, it was concluded that an increase in the heat source temperature would influence the COP of the system positively, however it would decrease the exergetic efficiency, both in cooling and heating modes. In heating mode, increasing the chilled water inlet temperature would increase both the second law efficiency as well as the COP. On the other hand, in cooling mode, an increase in the chilled water inlet temperature would degrade the exergetic efficiency of the system.

Kaynakli and Yamankaradeniz (2007) carried out a first and second law analysis of a single-effect 10 kW H_2O-LiBr absorption cooling system. Several investigations were carried out to identify the effect of the evaporator, condenser, and generator temperatures on the entropy generation for each single component, and the total system entropy generation, the COP, and the solution distribution ratio. It was found that with increases in generator temperature, the solution distribution ratio decreased while the COP increased. Regarding entropy generation, it was concluded that an increase in generator and/or evaporator temperature and/or a decrease in condenser temperature would minimize exergy losses and thus enhance the system's performance.

Kaynakli and Kilic (2007) used simulations to investigate the performance of a H_2O-LiBr absorption refrigeration system. The developed model was validated against data found in literature. After the validation, the model was used to investigate the influence of operating temperatures and the system's heat exchangers on its performance. Based on the results of the calculations, it was found that a generation temperature of 85°C was the optimum for the second law efficiency. For higher generation temperatures, the COP was able to further increase but the second law efficiency was found to decrease. Increasing the absorption and condensation temperatures was also found to damage the COP, with temperatures above 45°C dramatically decreasing the COP. On the other hand, the second law efficiency increased until 40°C, and at higher temperatures it also decreased. Furthermore, it was found that the use of a solution heat exchanger could increase the COP from 0.57 to 0.82.

Figueredo et al. (2008) reported the modeling of a double-stage H_2O-LiBr absorption chiller with a nominal capacity of 200 kW. The machine was able to operate during the summer in double-stage mode with a driving temperature of 170°C supplied by a natural-gas-fired heater, in single-stage mode at a driving temperature of 90°C supplied by solar collectors, or in combined mode with both temperatures. During winter, the chiller could operate in double-lift mode for heating driven by the heat from the natural gas heater. By varying the loads and optimizing the distribution of the total area between the system's components, the maximum COP obtained was 1.23.

Kaushik and Arora (2009) carried out an energetic and exergetic analysis of a single-effect and a series flow double-effect H_2O-LiBr system. The results indicated that the maximum COP of the single-effect cycle was in the range 0.6–0.75, with an optimum generator temperature of 91°C. The maximum COP of the double effect ranged between 1.00–1.28, with a generator temperature of 150 °C. Concerning the exergetic efficiency, which in both cases had a maximum value around 20%, the optimum for the single effect was achieved at a 60°C generator temperature, while the optimum for the double effect was achieved at 95 °C.

Gebreslassie et al. (2010) conducted an exergy analysis for the half-, single-, double-, and triple-effect H_2O-LiBr absorption cycle. The results of the simulations, carried out with EES software, showed, as expected, that the triple effect increased both the COP and the exergetic efficiency in comparison to the other investigated cycles. Furthermore, the effect of the heat source temperature was analyzed, showing that an increase in its value caused the COP to decrease at a slow rate while also severely damaging exergetic efficiency.

Gomri (2010) used simulations to investigate and compare the first and second law performance of single-, double-, and triple-effect H_2O-LiBr absorption cycles with a cooling capacity of 300 kW. Based on the results of the simulations, it was concluded that the triple-effect absorption chiller had the maximum COP with a value of approximately 1.85, which, however, required a far higher generator temperature (above 125°C). The respective optimum COP value for the double effect was 1.3 (at generator temperatures in the range of 95–125°C). The optimum COP for the single effect was approximately 0.8, with the main advantage that the generator temperature was as low as 60°C, thus allowing the exploitation of extremely low-grade heat sources. Various evaporation temperatures were investigated for the three cycles. In all three cases, it was found that the maximum COP was achieved at the highest evaporation temperature (Figure 5.4, T_{cond} = 33°C). In terms of the exergetic efficiency, the three cycles had comparable results, with maximum efficiencies in the range of 23–25%. However, in this case, the effect of the evaporation temperature was reversed, with the lowest evaporation temperature resulting in the highest efficiency (Figure 5.5, T_{cond} = 33°C).

Somers et al. (2011) modeled the performance of a single- and a double-effect H_2O-LiBr absorption chiller with Aspen Plus software and verified it against data from corresponding literature. The results of the model for the single-effect system showed a COP of 0.738 at a cooling capacity of 10.77 kW (with a generation temperature of 89.9°C, a condenser temperature of 40.2°C, and an evaporator temperature of 1.3°C). The results for the double effect showed a COP of 1.387 and a cooling output of 354.4 kW. The corresponding deviation of the model's predictions from the literature was up to 2.6 % for the single-effect case and up to 4.7 % for the double effect.

Bendaikha and Larbi (2012) assessed the potential implementation of a proton exchange membrane fuel cell (PEMFC) in an air conditioning system consisting of a H_2O-LiBr absorption chiller powered by a geothermal source. The simulations were conducted using meteorological data for Saïda, Algeria. The results of the simulations showed that the efficiency of the hybrid system was around 70% on heating mode and 30–45% in cooling mode. However, in an overall feasibility study, the authors concluded that the system offers a competitive solution for an off-grid site.

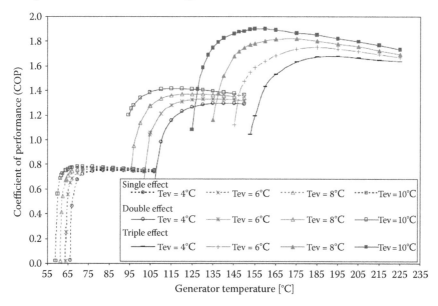

FIGURE 5.4
Behavior of the COP for the single-, double- and triple-effect absorption chiller for several generator and evaporation temperatures. (Reproduced from Gomri, Rabah, *Energy Conversion and Management*, 51 (8): 1629–1636, 2010.)

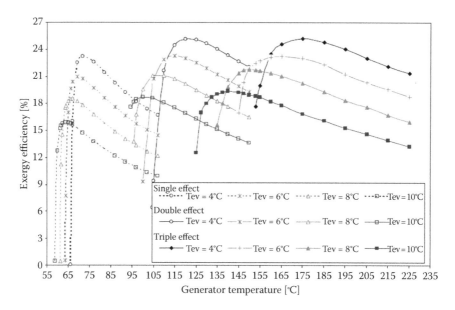

FIGURE 5.5
Behavior of the exergetic efficiency for the single-, double- and triple-effect absorption chiller for several generator and evaporation temperatures. (Reproduced from Gomri, Rabah, *Energy Conversion and Management*, 51 (8): 1629–1636, 2010.)

Labus et al. (2013) evaluated the accuracy of four steady state models for the modeling of small-scale absorption chillers by comparing the model's predictions with the experimental data for the secondary water streams collected from a 12 kW absorption chiller test bench. The four evaluated models were the following: the adapted Gordon-Ng model (GNA), the adapted characteristic equation model, the multivariate polynomial regression model (MPR), and the artificial neural network model (ANN). The GNA model was found to be the less accurate model, based on R^2 and coefficient of variation (CV) statistical indicators. According to the findings, the most accurate method was ANN, with an R^2 in the range of 0.998 and a CV <2%.

Aman et al. (2014) modeled and analyzed the energetic and exergetic performance of a 10 kW NH_3-H_2O absorption chiller used for residential air conditioning purposes. Based on the investigated scale and temperature range of the generator, the use of flat-plate collectors is suggested by the authors to provide the required heat input in the generator. The first and second law performance of the absorption chiller was evaluated for a generator temperature of 80°C, a condenser temperature of 30°C, and an evaporator temperature of 2°C. The results reported a COP of 0.6 at a heat input of 16.77 kW, with a second law efficiency of 32.01%. The component with the highest exergy losses was the absorber, with a contribution of 62% of the total exergy losses.

Gong and Boulama (2015) carried out a second law analysis on a H_2O-LiBr absorption chiller to identify the importance of the difference between condensation and absorption temperature. Based on the results of the simulations, it was found that increasing the temperature difference between the absorption and condensation decreased both first and second law efficiency. On the other hand, when the two temperatures were equal, an optimum was found for the common temperature in order to minimize the exergy losses.

Mansouri et al. (2015) modeled the steady state performance of a commercial gas-fired NH_3-H_2O absorption chiller with Aspen Plus software, and validated the results of the simulations

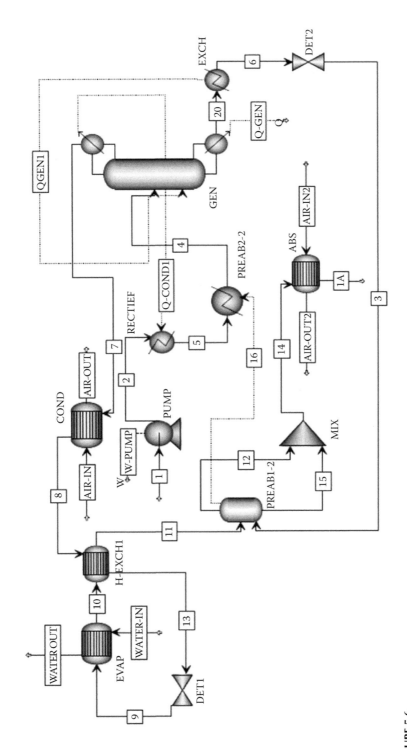

FIGURE 5.6
Overview of the ASPEN model of the single-effect absorption machine. (Reproduced from Mansouri, Rami et al., *Energy*, 93 (Part 2): 2374–2383, 2015.)

against data provided by the manufacturer. Figure 5.6 presents an overview of the ASPEN model layout. The heat transfer performance of the system's heat exchangers was measured at a cooling air temperature of 35°C. Furthermore, the temperatures in the cycle were validated against experimental data reported in the literature for two cooling air temperatures.

Sochard et al. (2017) developed a model to simulate the performance of a solar absorption chiller. The model was evaluated and validated against data from literature for a GAX absorption cycle with NH_3-H_2O as the working pair. The results were in decent agreement with the literature, showing a maximum COP of 0.6 and a cooling output of approximately 8 kW.

5.1.4 Experimental Investigations on Absorption Units

Asdrubali and Grignaffini (2005) used experiments to investigate the performance of a single-effect H_2O-LiBr chiller manufactured by Yazaki. The nominal capacity of the chiller was 17 kW and the nominal driving temperature was equal to 88°C. The nominal chilled water output temperature was 6°C. The experiments reported that the maximum efficiency of the chiller was measured at driving temperatures in the range of 70°C, with a COP around 0.42 and a corresponding exergetic efficiency of 15%.

Arivazhagan et al. (2006) experimentally investigated the performance of a double-stage half-effect absorption cooling system. The working pair used was R134a-DMAC (N,N-dimethylacetamide). Data from the experiments was collected by varying the condensing temperature (between 20–25°C), the absorber temperature (between 20–35°C), and the generator's temperature (between 50–75°C). The results showed that the 1 kW absorption chiller was able to produce an evaporator temperature of –1°C when driven by a heat source temperature as low as 55°C and deliver chilled water at 8°C. The optimum generator temperature range was identified to be between 65–70°C, with a corresponding COP of 0.36 at a generator temperature equal to 70°C.

Erickson (2007) presented the performance results of a steam-fired demonstration absorption chiller/heat pump over an eight-month period. The cooling capacity of the chiller was 351.7 kW (100 tons), while hot water production was around 940 kW, powered by 586 kW of 5.5 bar steam, which was produced by natural-gas-fired boilers. The nominal COP of the chiller was reported to be around 0.6. Prior to this setup, chilling was supplied by an ammonia vapor compression chiller. Hot water was directly supplied by the natural-gas-fired boilers. Hence, the savings in both natural gas and electricity from the demonstration project were estimated to exceed $110,000 on an annual basis, resulting, according to the author, in a payback period of approximately 1.8 years.

Zetzsche et al. (2009) developed a prototype 10 kW NH_3-H_2O absorption chiller at the Institute of Thermodynamics and Thermal Engineering (ITW) in Stuttgart, Germany. At a nominal driving temperature of 100°C, a cooling water temperature of 27–32°C, and a chilled water temperature of 14°C, the reported COP of the chiller was equal to 0.72.

Another application of an absorption chiller is in the Bangkok Suwarnabhumi International Airport cogeneration plant, in Thailand. The gas turbine of the cycle is able to produce electricity up to 55 MW in the two generators of the plant. The exhaust gases are then led through a heat recovery steam generator to produce high-pressure steam that drives a 13.6 MW generator. After exiting the turbine of the steam cycle, the expanded steam is used as the hot source for the eight installed H_2O-LiBr absorption chillers that can provide a cooling capacity of 59 MW. The nominal temperature of the chilled water exiting the chillers is 5–7 °C, which is able to cover the space cooling needs of the passenger buildings, the hotels, and other buildings within the airport (Jaruwongwittaya and Chen 2010). Figure 5.7 presents an overview of the aforementioned cogeneration plant.

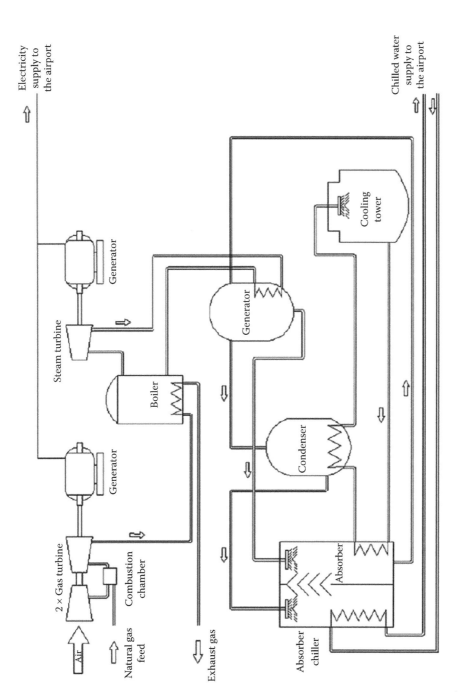

FIGURE 5.7

Schematic of the Suwarnabhumi International Airport cogeneration plant. (Reproduced from Jaruwongwittaya, Tawatchai, and Guangming Chen, *Renewable and Sustainable Energy Reviews*, 14 (5): 1437–1444, 2010.)

AlQdah (2011) evaluated the performance of a NH_3-H_2O absorption air conditioning system driven by exhaust heat. The system was designed for automobile applications. The results showed that for an exhaust temperature in the range of 200–320°C, the corresponding cooling output was between 4.5–6.2 kW, which is within the range of cooling loads for such an application. The maximum measured COP was approximately 1.1 for a condensation temperature of 40°C and an evaporation temperature of 14°C (with a generation temperature of 80°C).

Le Lostec et al. (2013) developed a model to investigate the performance of a NH_3-H_2O absorption chiller. The model was validated by previously reported experimental data from Le Lostec et al. (2012). Based on the experimental results, the tested absorption chiller had a maximum COP of 0.6, with a maximum cooling output of 10 kW (with a generator temperature 86°C, a condenser temperature of 34°C, an absorber temperature of 31°C, and an evaporator temperature of 19°C). Two main sets of simulations were carried out to evaluate the effect of the evaporator and desorber temperature on the chiller's performance. Results of the simulations indicated that the COP decreased by 25% when the evaporator temperature decreased by 10 K. On the other hand, an increase in the desorber temperature of 10 K resulted in a 4% drop in the COP.

Lamine and Said (2014) studied the performance of an industrial H_2O-LiBr absorption chiller—model 16JB manufactured by Carrier—installed in the production unit of Henkel detergent, in Algeria. The experimental data was analyzed using Fortran programming language for varying generator and evaporator temperatures. The results showed an obtained maximum COP in the range of 0.75 for a generator temperature of 85–100°C, an evaporator temperature of 5°C, a condenser temperature of 30°C, and an absorber temperature of 38°C.

Prasartkaew (2014) modified and experimentally tested the performance of a commercial H_2O-LiBr absorption chiller with a nominal capacity of 7 kW. The chiller's working pair solution was adjusted to cover the cooling load of a test room, which was approximately 4.5 kW. The heat input for the generator was supplied by a 20 kW electrical boiler. The experiments were conducted over three days with different hot water temperatures, 81.3, 84.4, and 90°C, respectively. The corresponding measured values for the heat input were 8 kW for a temperature difference of 6 K, 8.8 kW for a temperature difference of 6.7 K, and 10 kW for a temperature difference of 10 K. The respective cooling outputs were 4.2, 5.1, and 5.3 kW, while the corresponding average COP for the three measurements was 0.52, 0.58, and 0.53, respectively. The nominal COP of the chiller before the retrofitting was 0.6.

Beausoleil-Morrison et al. (2015) conducted experiments on a small-scale (35 kW) commercial single-effect H_2O-LiBr absorption chiller. The absorption chiller was powered by three in-series configuration heaters with a total capacity of 65 kW, supplying the generator with water at temperatures within a range of 70–93°C. The cooling load was simulated by a flat-plate heat exchanger in series with an electric heater with a capacity of 19 kW. The chilled water inlet temperature was reported to be between 15–19°C. Thirty-six steady state experiments were conducted to measure the chillers performance at several working points. The cooling capacity was found to vary between 6.9–40.5 kW, while the chiller COP was in the range 0.56–0.83. Based on the experimental data, a quasi-steady-state model was developed and validated, showing a satisfactory accuracy.

Zamora et al. (2015) developed and reported the part-load performance of an ammonia-lithium nitrate absorption chiller. The experimental procedure included variations of the thermal load keeping the hot and cooling water temperatures constant. At full load operation, the prototype chiller was able to produce a cooling output of 10.1 kW (generator temperature of 90°C, a condenser temperature of 35°C, evaporator temperature of 15.5°C) with a COP of 0.61. By gradually decreasing the load of the chiller from 100% to 30%, the COP was decreased to 0.53.

TABLE 5.1

Technical Design and Experimental Data from the Three Prototype Chillers

Parameter	1st Chiller	2nd Chiller	3rd Chiller
Capacity (kW)	5	5	100
Evaporator working range (°C)	−10/18	9/18	−10/18
Condenser working range (°C)	15/45	23/33	18/45
Generator working range (°C)	60/130	60/95	70/180
Charge of NH_3-H_2 solution (kg_{NH3} per kW)	0.48	0.36	0.48
Experimental max reported COP (−)	0.71	0.67	0.58
Corresponding second law efficiency (%)	13.7	20.9	22.3

Source: Helm, M. et al., *International Journal of Refrigeration*, 32 (4): 596–606, 2009.

Xu et al. (2015) developed and evaluated the performance of a 50 kW variable effect H_2O-LiBr absorption chiller. The nominal generation temperature of the prototype chiller is 125 °C, the condensation temperature is 40 °C, the absorption temperature 35°C, and the evaporator temperature is equal to 5°C. Several sets of steady state experiments were conducted to measure the chiller's performance. The COP of the chiller was measured in the range of 0.69–1.08. At the maximum COP operating point, the respective cooling capacity was 51.9 kW, at a heat source temperature of 120°C, a chilled water inlet of 13.5°C, and a cooling water inlet of 27.6°C.

Franchini et al. (2015) designed and developed a prototype 5 kW H_2O-LiBr absorption chiller that was installed at the Energy System and Turbomachinery Laboratory at Bergamo University. Results from preliminary tests of the setup revealed a COP of 0.358 at an effective cooling capacity of 3.25 kW, a heat source temperature of 88.2°C, a condensation temperature of 34.9°C, and an evaporator temperature of 8.5°C, resulting in a chilled water inlet/outlet temperature of 19.8/16.6°C.

Boudéhenn et al. (2016) developed and experimentally evaluated the performance of three prototype absorption chillers with cooling capacities 5, 5, and 100 kW respectively. The three NH_3-H_2O prototype chillers were developed between 2010 and 2015. The performance of the three chillers was tested and the summary of the main technical data and results is presented in Table 5.1.

5.1.5 Market Status

After 1945, following Carrier Corporation's pioneering H_2O-LiBr absorption technology, absorption technology enjoyed significant success in the U.S. refrigeration market for many years. However, by 1975, the low cost of electricity led to the decline of this market (Herold 1995). Recently, the rising costs of conventional electricity production, along with increasing environmental concerns, have again driven attention toward absorption refrigeration systems. Absorption systems have an advantage over the conventional ones in terms of their high reliability, their more efficient exploitation of low-grade heat sources, their lower noise levels, and their simpler implementation.

Absorption chillers can be used for a wide range of applications, including:

- Petroleum and chemical industry (gas liquefaction, distillation processes)
- Food industry (food product freezing, storage and drying processes)
- Breweries (for storage and refrigeration of the product)

- HVAC applications (both industrial and domestic)
- In cogeneration units (also to provide cooling)
- In geothermal/solar applications since they are ideal for lower-grade heat sources as previously mentioned

TABLE 5.2

Overview of the Absorption Chiller Market

Manufacturer (Country)	Model	Working Pair	Capacity (kW)	COP (–)
AGO (Germany)	Congelo	NH_3-H_2O	50–150	0.51
Baelz	Bee/Bumblebee	H_2O-LiBr	50/160	0.8
Broad Air Conditioning Co. (China)	BCT	H_2O-LiBr	16–500	1.1
Carrier Corporation (USA)	16LJ	H_2O-LiBr	90–4,000	n/a
	16TJ	H_2O-LiBr	350–2,500	n/a
	16NK	H_2O-LiBr	352–4,652	n/a
Century Corporation (South Korea)	AR-D	H_2O-LiBr	98–193	n/a
ClimateWell AB (Sweden)	ClimateWell Solar Chiller	$LiCl$-H_2O	7	0.7
Colibri B.V./Stork B.V. (Netherlands)	ARP	NH_3-H_2O	100–>2,500	Up to 0.8
EAW (Germany)	Wecagral SE	H_2O-LiBr	15–200	0.71–0.75
En-Save (Germany)	En-Save Cold	NH_3-H_2O	30–100	n/a
Heinen & Hopman (Netherlands)	SWM60-SWM1200	H_2O-LiBr	150–5,000	Up to 0.8
Helioclim (France)	(pilot plant)	NH_3-H_2O	10	n/a
Jiangsu Huineng (China)	RXZ	H_2O-LiBr	10–175	0.70
Krloskar Pneumatic Company (India)	KVAC-SA/DA	H_2O-LiBr	211–2,400	n/a
LG A/C (South Korea)	WCDH	H_2O-LiBr	350–5,275	1.51
	WCMH	H_2O-LiBr	98.4–3,587	0.80
Meibes System-Technik (Germany)	n/a	H_2O-LiBr	5	n/a
Phoenix (Germany)	n/a	H_2O-LiBr	10	0.74
Pink (Austria)	PC19	NH_3-H_2O	19	0.63
Robur (Italy)	GA ACF	NH_3-H_2O	17.7	Up to 0.9
Rotartica (Spain)	Solar 045	H_2O-LiBr	11	0.67
Sakura (Japan)	SHL	H_2O-LiBr	10.5–176	0.71–0.80
Slarice GmbH (Germany)	n/a	NH_3-H_2O	25–several MWs	n/a
Thermax (India)	LT 1/2/3/5	H_2O-LiBr	35–171	0.78
World Energy Europe (Ireland)	S050-S1500 (single effect)	n/a	176–5,274	Up to 0.81
	SWHH100-1500 (double effect)	n/a	352–5,274	Up to 1.48
Yazaki (Japan)	WFC SC	H_2O-LiBr	17.6–175.8	0.70
York (USA)	YIA	H_2O-LiBr	422–4,840	

Sources: Boudéhenn, François et al., *Energy Procedia*, 91: 707–716, 2016; Wang, R. Z. et al., *International Journal of Refrigeration*, 32 (4): 638–660, 2009; Jakob, Uli, "Solar Cooling Technologies", in *Renewable Heating and Cooling*, 119–136, Woodhead Publishing, 2016; Henning, Hans-Martin, *Applied Thermal Engineering*, 27 (10): 1734–1749, 2007.

Absorption chillers are already quite widespread, with several manufacturers across the world, as shown in Table 5.2. Medium- and larger-scale commercial absorption chillers have been available for many years, while smaller-scale chillers have become a field of intense focus in recent years.

As is evident from Table 5.2, commercial absorption chillers are mostly available at cooling capacities higher than 10 kWs. The main challenge at the moment is to expand the market for smaller, domestic-scale applications, and thus develop cost-efficient directly fired absorption chillers. Furthermore, another research focus is the optimization of efficiency by increasing the driving temperature or introducing hybrid schemes. The market is even smaller for indirect-fired chillers, which have proven to be most appropriate for solar cooling applications. Most commercial absorption chillers operate with driving temperatures in the range of 90–100°C in the case of single-effect chillers, resulting in COPs (normally) up to 0.8, as shown in Table 5.2. Theoretically, further increases in the driving temperature would increase the COP of a single-effect cycle even more. However, operating at higher temperatures may lead to significant operational issues, mainly because solar collectors that are able to operate at such a temperature range are very costly. Furthermore, passing this temperature limit may increase the possibility of boiling phenomena, which, in order to prevent, requires further pressurization of the system, further increasing costs. For small scale applications, there are currently no reported double-effect (indirectly fired) commercial systems, mainly as a result of high costs. Furthermore, to achieve COPs in the range of 1.4, double-effect chillers require higher driving temperatures, as high as 140–160°C. Triple-effect chillers that can increase the COP up to 1.7. have been developed at lab scale, however, the fact that they require driving temperatures of 200°C and have a high system complexity has ruled out their commercial competitiveness for the time being (Grossman 2002).

Regarding heat rejection in the absorption chiller, water-cooled systems are considered to have high initial as well as high operational costs due to the use of a cooling tower. At the same time, the use of a cooling tower carries the risk of bacteria contamination, which can cause severe malfunctions in the system's operation. More specifically, the initial cost of a cooling tower can be as high as €200–250/kWc (Schweigler et al. 2007), while the total specific cost of an absorption chiller ranges between €400–1,000/kWc (Kim et al. 2007). Furthermore, the huge loss of water in such systems makes the use of cooling towers for smaller applications in most solar cooling applications less attractive. Helm et al. (2009) carried out extensive research and reported the temperature range and dependency between absorption chillers, solar collectors, and the different options for a heat rejection system. The conclusions of this survey are presented in Table 5.3.

TABLE 5.3

Operational Data for the Different Options for a Solar Collector-Absorption Chiller-Heat Rejection System

	Absorption Chiller			Flat-Plate Collector	
Cooling System	Chilled Water T (°C)	Cooling Water T (°C)	Hot Water T (°C)	Driving T (°C)	Efficiency (%)
Wet cooling tower	15–18	27–35	80–75	80–85	58
Dry air cooler	15–18	40–45	105–100	105–110	46
Dry air cooler/ Latent heat storage	15–18	32–40	90–85	90–95	53

Source: Reproduced from Helm, M. et al., *International Journal of Refrigeration*, 32 (4): 596–606, 2009.

In conclusion, the use of conventional cooling technologies versus the use of absorption chillers is an economic trade-off. The operational costs of an absorption chiller are lower than a vapor compression chiller, but the initial cost is higher. Hence, a specific feasibility investigation based on the cost of fuel and energy savings has to be carried out to estimate the competitiveness of one such selection.

5.2 Solar Cooling with Absorption Chillers

Commercial solar absorption cooling dates back to the 1970s, when Arkla Industries Inc. introduced 10 kW and 75 kW cooling capacity units. The nominal driving temperature was approximately 90°C for a cooling water temperature of 29°C and a chilled water temperature of 7°C (M. Sayigh and Khoshaim 1981; Ali et al. 2008; Sayigh and Bahadori 1979; Janzen et al. 1981). This machine was implemented in more than 100 demonstration projects in the United States and the cost of Solaire-36 was approximately $3,000 in 1977 (Maidment and Paurine 2012).

Until the 1990s, the Japanese company Yazaki was the main manufacturer of low- and medium-scale absorption chillers designed for solar cooling projects. However, in recent years, the market for low-scale solar cooling projects has started to expand and new companies, mainly in Europe, are beginning to develop similar prototypes. In terms of medium-scale chillers, the WFC-10 single-effect H_2O-LiBr absorption chiller manufactured by Yazaki was the most widely used commercial product.

Another early application of solar absorption cooling was the cooling of a winery in southern France (this case study will be discussed more thoroughly in the following pages). The 52 kW cooling capacity unit (model Yazaki WFC-15) was installed in 1991, and was powered by 130 m² of vacuum tube collectors. With a driving temperature of 80°C, the reported COP was approximately 0.58 (Eicker 2009).

5.2.1 Theoretical Investigations on Solar Absorption Cooling

Blinn et al. (1979) developed a transient model for a H_2O-LiBr absorption air conditioning system to simulate the performance of a solar-powered system in Miami, Florida, and Charleston, South Carolina, in the United States. The system performance was found to degrade significantly as a result of the chiller's transients. These negative effects would diminish when the room's thermostat dead-band was increased. The system was powered by flat-plate solar collectors, and it was found that the maximum solar fraction was obtained at the lowest source temperature. Furthermore, a comparison of the auxiliary strategies was carried out, leading to the conclusion that vapor compression achieved the highest solar fraction compared to parallel or in series auxiliary heat.

Hawlader et al. (1993) investigated the performance of a solar-powered open cycle absorption cooling system with H_2O-LiBr and H_2O-LiCl as the tested working pairs. Figure 5.8 presents an overview of the experimental setup. A parametric study was conducted to identify the critical parameters for the operation of the collector/regenerator of the system. Correlations for mass and heat transfer were developed based on the experimental data. Based on the correlations, a simulation program was then developed. The program found that the solution flow rate, the solar irradiance, the ambient humidity, and the inlet concentration of the solution strongly influenced the water evaporation rate. Generation

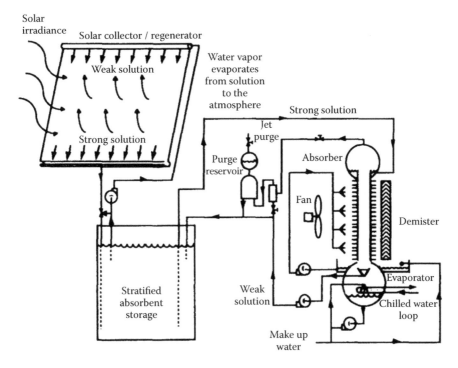

FIGURE 5.8
Schematic of the solar-powered open cycle absorption cooling test rig. (Reproduced from Hawlader, M. N. A. et al., *Solar Energy*, 50 (1): 59–73, 1993.)

efficiencies were measured to be in the range of 38–67%, while respective cooling capacities were between 31–72 kW.

Sorour and Ghoneim (1994) carried out a feasibility study in terms of energetic and economic performance for solar and heating applications in several regions in Egypt. A simulation model for the thermo-economic analysis of a solar and heating application was developed with reference to the system shown in Figure 5.9. According to the results of the simulations, the optimum solar collector surface varied significantly within the region,

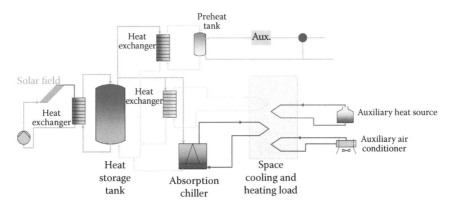

FIGURE 5.9
Schematic of the solar-powered system simulated by Sorour and Ghoneim. (Adapted from Sorour, M. M., and A. A. Ghoneim, *Renewable Energy*, 5 (1): 489–491, 1994.)

ranging from 12 to 24 m² for evacuated tube collectors and from 21 to 55 m² for flat-plate collectors. At the same time, the cost per unit of energy for the solar and heating application was in the range of 68–83% of the corresponding conventional fuel cost based on the prices at the time of the report.

Oh et al. (1994) investigated the performance of a gas-fired, double-effect, air-cooled H_2O-LiBr absorption heat pump with a nominal capacity of 7 kW. Cycle simulation was carried out to evaluate the performance of the heat pump in cooling mode. The results of the simulation showed that there is an optimum in the solution distribution ratio to maximize cooling production, with a COP of approximately 1.27. Furthermore, it was found that an increase in the inlet temperature of cooling air above 37.5°C decreased the COP, but raised the possibility of corrosion due to the high temperature of the first generator.

Ghaddar et al. (1997) developed and simulated the performance of a solar absorption system in Beirut, Lebanon. According to the economic analysis that was carried out, based on the cost of conventional cooling production at the time, a solar cooling system could be competitive only in the case that it was used also for domestic water heating reasons. Furthermore, it was estimated that in order for the system to be powered 100% by solar power, a minimum collector surface of 23.3 m² was required per refrigeration ton. The optimum water storage capacity would be in the range 1000–1500 l.

Chen and Hihara (1999) proposed the implementation of a compressor into the conventional absorption cycle to add a further amount of heat apart from the solar collectors. The overall added heat to the generator would be controlled by adjusting the mass flow through the installed compressor. To evaluate the proposed configuration, numerical simulations were conducted showing that the COP was in the range of 0.85, with a far steadier behavior than the conventional solar-powered absorption cycle.

Florides et al. (2002) modeled and investigated the performance of a solar-assisted H_2O-LiBr absorption cooling system in Nicosia, Cyprus, with simulations using TRNSYS software. The results of the system sizing showed that the optimal system consisted of a 15 m² compound parabolic collector tilted 30° and a 600 l hot water storage tank. For the simulations, an absorption chiller based on an Arkla model WF-36 with a nominal capacity of 18 kW was considered. The investigated building was a typical Cypriot house of 196 m². Based on the annual cooling and hot water production for a typical Cypriot house, 49% of the required energy could be provided by solar energy.

Elsafty and Al-Daini (2002) compared the economic feasibility of a solar-assisted vapor absorption air conditioning system and a conventional vapor compression system for the needs of a five floor student hospital in Alexandria, Egypt. The peak and hourly loads for the hospital were calculated for the month of August. By comparing the present values of the schemes, it was found that the total cost for a vapor compression system was 11% lower than the respective cost of a single-effect absorption system. A double-effect vapor absorption system would be the most economical choice, with a total cost 30% lower than the conventional vapor compression system.

Rivera and Rivera (2003) modeled a solar driven absorption chiller with ammonia-lithium nitrate mixture as the working pair. Solar energy is harvested by a compound parabolic collector (CPC) which also serves as the generator-absorber of the system. Figure 5.10 presents and overview of the system modeled. For the simulations, meteorological data was used from a station installed at the Energy Research Centre of the National University of Mexico in Temixco, Morelos, Mexico. Depending on weather conditions, the theoretical efficiency of the CPC was in the range of 0.33–0.78. The COP of the system measured in the range of 0.15–0.40, at a generation temperature of 120°C and a condensation temperature of 40–44°C, resulting in an ice-production capacity of up to 11.8 kg.

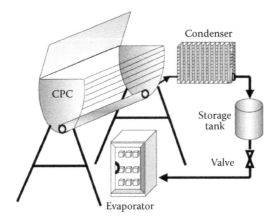

FIGURE 5.10
Schematic of the solar absorption cooling system using a CPC. (Reproduced from Rivera, C. O., and W. Rivera, *Solar Energy Materials and Solar Cells*, 76 (3): 417–427, 2003.)

Assilzadeh et al. (2005) developed a model using TRNSYS software and carried out simulations for a solar cooling system located in Malaysia. The considered system was powered by evacuated tube collectors. For the absorption air conditioning unit, a single-effect H_2O-LiBr unit based on the Arkla WF-36 was considered. The results showed that the optimum system based on the climatic conditions of Malaysia for a 3.5 kW (1 ton) cooling capacity required 35 m² solar collectors, tilted at 20°. This configuration achieved a solar fraction of approximately 60%.

Balghouthi et al. (2005) carried out a feasibility study for a solar cooling system based on absorption in Tunisia. The system was optimized for a 10 kW H_2O-LiBr chiller. Based on the results of the simulations that were carried out with TRNSYS software, a 30 m² flat-plate collector area tilted at 35° was required. Regarding the storage tank for the hot water, a 1 m³ storage tank with a maximum temperature of 120°C was proposed.

Using simulations, Mazloumi et al. (2008) investigated the performance of a solar-powered single-effect H_2O-LiBr absorption cooling system located in Ahwaz, Iran. The system was designed to cover the peak cooling load of a typical house, which is equal to 17.5 kW for the month July. The results of the simulations showed that a minimum of 57.6 m² parabolic trough collectors were required. The optimum collector efficiency for the design day of July was estimated to be 69%. The chiller's COP was also simulated for the design days of July and September, and ranged between 0.67–0.76.

Ortega et al. (2008) developed a 1D model to describe the two-phase behavior of a compound parabolic concentrator used as a generator in a NH_3-H_2 absorption chiller. The measurements for the non-tracking CPC were taken for a site in Temixco, Morelos, Mexico, during a typical day in 1996. The maximum reported cooling output was around 4.8 kW at a temperature of −10°C. The corresponding solar and overall efficiencies were 46.3% and 21.2%, respectively. The solar fraction was 100% and the chiller's COP was equal to 0.458.

Eicker and Pietruschka (2009) carried out an economic analysis using TRNSYS software of a solar-driven 15 kW H_2O-LiBr absorption system for use in office buildings. The results of the analysis showed that the cooling cost for long operation hours in Madrid was in the range of €180–200/MWh. The cooling cost increased to €270/MWh for shorter cooling periods.

Kim and Infante Ferreira (2009) investigated the performance of a low temperature driven half-effect H_2O-LiBr absorption chiller. The 10 kW air-cooled chiller was powered by flat-plate solar collectors and was used for air conditioning in an extremely hot climate. The results of the simulations showed that for a driving temperature of 90°C and an ambient temperature of 35°C, the direct and the indirect air-cooled chillers that were investigated were delivering chilled water at temperatures of 5.7°C and 7.8°C, respectively. The corresponding COP and the cooling output for the two cases was 0.38 and 12.8 kW for the direct and 0.36 and 9 kW for the indirect air-cooled chiller, respectively. The effect of a further increase in the ambient temperature was investigated, showing a significant negative impact on the COP and the cooling output. In the direct air-cooled chiller—in which the most severe decrease occurred—the COP decreased to 0.31 with a cooling output of 4.8 kW.

Using simulations, Vargas et al. (2009) designed and investigated a solar-powered cooling and heating setup using a solar collector, a gas burner, a hot water storage tank, a hot water heat exchanger, and an absorption chiller. The simplified model that was developed consisted of fundamental and empirical correlations, and was used to simulate the system's transient and steady state behavior under different working conditions. In order to maximize the exergy input rate, an optimal set of three heat capacity rates that characterize the system were identified.

Using TRNSYS software, Al-Alili et al. (2010) designed and simulated the performance of a solar air conditioning system for Abu Dhabi, United Arab Emirates. The model consisted of 34 m² of evacuated tube collectors powering a 10 kW absorption system. The results of the optimization showed that the minimum auxiliary heater consumption was equal to 1845 kWh on annual basis. The minimum total cost for the investigated system with a solar fraction of 0.5 was equal to $72,203.

Koroneos et al. (2010) analyzed the potential installation of a 70 kW commercial H_2O-LiBr absorption chiller powered by flat-plate solar collectors in a medical center in Igoumenitsa, Greece. The results of the simulations showed a capital cost for such a system to be approximately €138,000, which is a lot higher than the €30,360 required for a conventional system. Given the estimated initial cost, the payback period was estimated to be approximately 24 years.

Onan et al. (2010) developed a model with Matlab software to simulate the hourly performance of a solar-powered 105 kW absorption system used in villa applications. Meteorological data was used for Mardin, Turkey, during the period of May 15–September 15. The ambient temperature during the period of the analysis was in the range of 13.2–40.3°C. The highest losses were reported in the solar collectors, with a first law efficiency of 26.2–61.9% and an average second law efficiency of 6%.

Tsoutsos et al. (2010) investigated the design of a solar cooling system for a hospital. The system's performance was evaluated for three different cities in Greece—Sitia, Athens, Thessaloniki—and was compared with the results of a case in Basel, Switzerland. The solar cooling system consisted of 500 m² of flat-plate collectors with selective surfaces, a 70 kW H_2O-LiBr absorption chiller, an auxiliary 50 kW vapor compression chiller, and an 87 kW auxiliary preheater. Four alternative scenarios in terms of different solar heating and cooling fractions were compared. In the optimal scenario, the solar fraction was 74.23% for cooling and 70.78% for heating. The total cost for this scenario was estimated to be €173,992 with a payback period of 11.5 years if no funding subsidies were procured. In a case of 40% funding, the payback period decreased to 6.9 years.

Sarabia Escriva et al. (2011) carried out simulations to investigate the performance of a single-effect H_2O-LiBr absorption chiller directly connected to 8 m² of vacuum solar collectors, tilted 15°, for air conditioning applications in several locations in Spain. The results showed that maximum hourly cooling power was in the range of 5 kW for most locations, while the reported efficiency of the coupled system was in the range of approximately 30%.

Ozgoren et al. (2012) investigated the performance of a 3.5 kW solar cooling system located in Adana, Turkey, on an hourly basis. For the investigation, a NH_3-H_2O absorption unit was considered, while for the solar field, evacuated tube collectors were used. The evaluation of the system was for a day with a maximum temperature (July 29). The cooling COP varied significantly throughout the day, ranging from 0.243–0.454. The heating COP varied between 1.243–1.454. Maximum collector efficiency was achieved at 13.00 with a value of 78.4% for a solar irradiation intensity of 719 W/m². The results of the simulations led to the conclusion that such a system is suitable for domestic/office cooling applications in the investigated region from a performance point of view.

Beccali et al. (2012) carried out a life cycle analysis with TRNSYS software for a 12 kW NH_3-H_2O solar absorption cooling and heating system. Two configurations were investigated—an auxiliary gas burner to produce an extra heat input and an auxiliary electric chiller—for two different locations: Palermo, Italy, and Zürich, Switzerland. According to the results of the simulations, the estimated life cycle was approximately 25 years. Primary energy savings were in the range of 28–33% for cooling and 34% for heating in the case of Zürich. In the case of Palermo, the respective values were 42–46% for cooling and 84% during heating mode.

Caciula et al. (2013) developed a model to simulate the performance of a NH_3-H_2O absorption chiller powered by compound parabolic collectors. The results of the simulations showed a maximum COP around 0.73 and an exergetic efficiency of 18%, for a generator temperature of 74°C and an evaporator temperature of 6°C. In the case of an evaporator temperature equal to –3°C, the COP shifted toward a generator temperature of 90°C with a value of 0.66, while the exergetic efficiency was around 19.5%. For both cases mentioned above, the condensation temperature was kept constant at 30°C.

Vasilescu and Infante Ferreira (2014) investigated the performance of a solar-powered double-effect parallel-flow absorption system for subzero evaporator temperatures with simulations. The working pair was a NH_3-$LiNO_3$ solution. The system was designed to cover the cooling loads of a pork slaughterhouse located in Naples, Italy. The cooling load of the building was in the range 100–600 kW. The chilled fluid inlet temperature was set at –20°C. Parabolic trough collectors were considered in order to supply the required high driving temperature at the generator. The effect of the distribution ratio of the strong solution was investigated in terms of the cooling output and the COP. The optimal value for the distribution ratio was 0.65, resulting in a cooling output of approximately 600 kW, with a COP around 0.87.

Ssembatya et al. (2014) studied the performance of a solar cooling system located in the United Arab Emirates/Gulf region with simulations. The solar cooling system is powered by 128 m² of vacuum tube solar collectors. A 35.2 kW single-effect vapor absorption chiller is used for the production of the cooling effect. Three storage tanks, 1 m³ each, have also been considered; one integrated with a heat exchanger for stratified charging hot water, one for backup storage of hot water, and one for chilled water storage. The simulations were carried out with TRNSYS. The results showed that the system operates at maximum efficiency throughout the year except in the summer period, during which low solar fraction cooling results in less efficient performance. The measured COP for a typical day in April varied from 0.4 to 0.7.

Lazzarin (2014) energetically and economically compared the performance of solar-powered single- and double-effect absorption chillers, solar adsorption systems, and conventional compression chillers driven by PV modules. For the solar thermally driven cooling systems, the choices of flat-plate, evacuated tube, and parabolic trough collectors were considered. Comparing the required specific collector area for 1 kWh cooling production showed that evacuated tube collectors driving double-effect absorption chillers were the most competitive choice, with 0.24 m² per kWh. The PV systems also showed favorable results despite their low efficiency, mainly due to high EER, which was equal to 4 for water-cooled chillers and 3 for air-cooled chillers, with 0.27 m² per kWh in the water-cooled case and 0.36 m² per kWh in the air-cooled compression chiller case. Regarding capital costs, a PV module driving a water-cooled chiller was the second most competitive choice for a daily production of 10 kWh, with a cost of €2,067. The flat-plate collectors driving a single-effect absorption chiller had a cost of €1,917. The most expensive choice was the evacuated tube collectors driving a double-effect absorption chiller at a cost of €2,244. In the case of the air-cooled chiller, it was reported that the PV module was 25% cheaper than the most economical thermally driven air-cooled chiller—€2,656 compared to €3,300.

Using simulations, Ketfi et al. (2015) investigated the influence of several parameters, including the efficiency of the heat exchangers and generator temperature in the performance of a commercial 70 kW single-effect H_2O-LiBr absorption chiller powered by solar energy. The simulations were conducted in a Matlab programming environment. Based on the results of the simulations, it was observed that the COP increased with increases in the generator and evaporator temperatures, while the COP decreased with increases in the absorber and condenser temperatures. The maximum achieved COP was 0.78 for a generation temperature of 92°C, an absorber/condenser temperature of 30°C, and an evaporator temperature of 7°C. At full load, the required heat input was in the range of 90 kW, which can be supplied by 225.5 m² of Cube France flat-plate collectors or 175.1 m² of evacuated tube collectors according to the calculations of the model.

Esposito et al. (2015) carried out a case study to investigate the potential of solar thermal and solar electric cooling for a large-scale hospital application. The peak load for refrigeration was considered to be equal to 1 MW. For the case study, it was considered that for both the solar thermal and the solar electric option, a backup electric chiller would be installed to cover any loads that could not be satisfied by the solar cooling system. Based on the results of the performance and economic analysis, it was found that the solar thermal system, when used only for the space conditioning needs, was not competitive. However, if the solar thermal system was also used for heating purposes, as it commonly is, the CO_2 emission savings and the money savings turn the investment into a profitable one. On the other hand, the solar electric option turns a profit after approximately 13 years, yet it requires a larger space to produce the same amount of energy.

Shirazi et al. (2016) presented a parametric study focusing on the feasibility of single-, double- and triple-effect absorption chillers powered by solar energy for heating and cooling applications. The systems investigated, as shown in Figure 5.11, consisted of a solar collector circuit, a storage tank, a backup heater, an absorption unit, coils for cooling and heating, and a control system and equipment. The influence of several parameters including the collector area, the DNI fraction, and the insulation of the storage tank was investigated. It was found that beyond a certain level, increasing the collector area to enhance the solar fraction was not cost efficient. Furthermore, such systems were found to be infeasible both energetically and economically for low DNI fraction (<50%) regions.

Utham et al. (2016) carried out transient simulations with TRNSYS software to determine the optimal solar-powered absorption air conditioning system for a case in Gujarat, India. For the simulation, an office building of 1000 m² was considered. The absorption chiller used in the model was based on the commercial product WFC-SC30, manufactured by Yazaki, with a nominal capacity of 105.5 kW (30 tons). The results showed that a total of 35 m² of evacuated tube collectors tilted at 24.2°, along with a storage tank of 0.5 m³, were the optimal combination to provide the required cooling loads for the building.

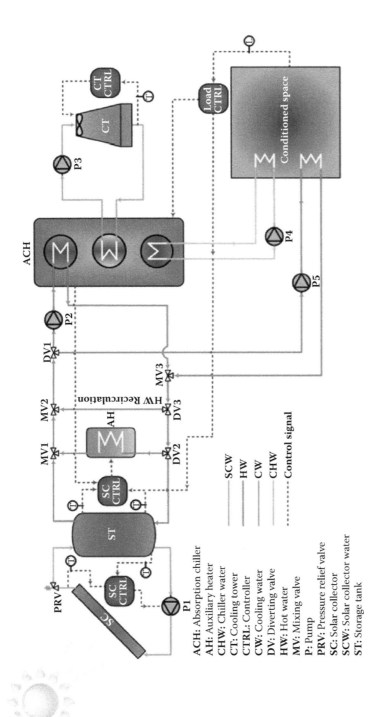

FIGURE 5.11

Overview of the solar cooling and heating system investigated by Shirazi et al. (From Shirazi, Ali et al., *Energy Conversion and Management*, 114: 258–277, 2016.)

Abdullah et al. (2016) carried out simulations with TRNSYS 17 software to optimize the performance of an absorption chiller powered solely by solar heat to cover the cooling and heating needs of a typical Australian house. For the purposes of the simulations, three different locations in Australia were considered: Brisbane, Adelaide, and Melbourne. The results showed that the total cost for a 20-year life cycle in the optimized model was $53,387AU for Brisbane, $51,639AU for Adelaide, and $32,816AU for Melbourne. In all cases, it was found that the optimized system was far less cost efficient than the option of an inverted reverse-cycle air conditioner, with a cost increase of at least 28%. However, as the authors state, the reduced electrical consumption (more than 50%), the reduction in carbon dioxide emissions, and the 75% less critical peak kW power create the potential to avoid high investments in new electrical infrastructure and the consequent increase in electricity prices.

Lubis et al. (2016) developed a model to investigate the potential application of a novel commercial single-double-effect absorption chiller powered by solar collectors in Asian tropical climates. The experimental data used to validate the model was based on a solar-powered cooling system operating with the aforementioned chiller. The system, located in Indonesia, consists of 181.04 m² of evacuated tube collectors. A 1 m³ hot water storage tank is also implemented in the system. The nominal capacity of the single-effect chiller is 239 kW and it can operate either on single-effect mode or in single-double-effect mode, utilizing two heat sources. The second heat source is a natural-gas-fired heater and has a separate generator and condenser to utilize the heat input. The absorber and the evaporator are common for both streams from the different heat sources, as shown in Figure 5.12. Based on results from the simulations, a COP of 1.93 was reported on single-double-effect mode, with a hot water temperature—from the solar collectors—of 90°C and cooling water temperature of 34°C. The results indicated, according to the authors, that the investigated system offers a competitive alternative for use in tropical regions of Asia.

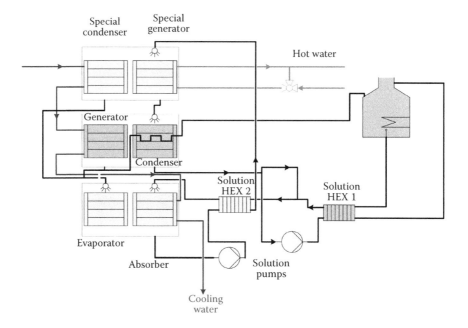

FIGURE 5.12
Schematic of the single-double-effect absorption chiller. (Adapted from Lubis, Arnas et al., *Renewable Energy*, 99: 825–835, 2016.)

Porumb et al. (2017) developed a mathematical model to simulate the performance of a single-stage solar H_2O-LiBr absorption chiller. The results of the model were validated using data from literature and showed a maximum deviation in the range of 5.5%. Several working conditions were evaluated and the optimal temperature working range was determined, achieving in the best case scenario a COP of 0.81 at a hot water temperature of 100°C, a cooling water inlet of 27°C, and a chilled water inlet temperature of 18°C.

In terms of system design, there are not yet general rules for the sizing of solar cooling system components. More specifically, while in solar thermal systems the typical storage volumes per m² of the solar collector's surface range between 50 and 100 l, the respective range for solar cooling projects may range from volumes smaller than 30 l per m² to cases with volumes as high as 100 l per m².

5.2.2 Dynamic Simulations of Solar Absorption Cooling Systems

Absorption is a dynamic phenomenon, thus when one such cycle needs to be investigated in terms of time calculations rather than time average input data, a dynamic simulation is required. Several studies have been carried out, providing models for the dynamic simulation of absorption chillers.

Kim and Park (2007) developed a lumped parameter dynamic model to simulate the performance of a commercial 10.5 kW single-effect NH_3-H_2O absorption chiller. The results of the simulation revealed the existence of an optimum volume for the generator and an optimum mass and concentration of the strong solution for the maximization of the cooling capacity of the system.

Kohlenbach and Ziegler (2008) developed a dynamic model to evaluate the performance of an absorption chiller. The model was time discretized and was based on internal energy and mass balances. The dynamic effect was implemented with the addition of thermal and mass storage terms, along with time delay functions to realize the inertia of the system. The results of the model were validated based on a single-effect H_2O-LiBr absorption chiller operating with generator temperatures between 75–85°C, showing adequate accuracy.

Zambrano et al. (2008) developed a dynamic model and validated its results based on experimental data from a 35 kW solar absorption cooling plant in Seville, Spain. The solar field consists of 151 m² of flat-plate collectors, while a 2.5 m³ hot water storage tank is implemented along with a 60 kW auxiliary gas-fired heater. The results of the experiment used for the validation are similar to Syed et al. (2005). Comparison of the measured data and the data from the model proved that the dynamic model is satisfactorily accurate, with a relative error less than 3%.

Matsushima et al. (2010) developed a dynamic model to simulate the transient behavior of triple-effect absorption chillers. The model is a combination of object-oriented formulation and parallel processing to allow for the investigation of more complex configurations than the triple-effect cycle. The results of the model were validated with actual performance data and showed adequate accuracy, with a maximum deviation of 10% from the experimental data. The model was used for the design of a prototype with a COP of 1.6.

Calise (2012) developed a dynamic model for an innovative solar heating and cooling system. The system consisted of a double-effect H_2-LiBr absorption chiller powered by parabolic trough collectors and a biomass-fired heater as a backup. A case study was presented in which the aforementioned system was used for heating and cooling a university hall. The maximum cooling load of the building was estimated to be 250 kW. The simulations were carried out with TRNSYS software. The results showed that the primary energy savings of the system were above 80%, proving the economic competence of such a system. Furthermore, the study was extended by changing the location of the hall and evaluating scenarios for several cities in the Mediterranean, proving the potential of solar cooling and heating applications, especially in hot climates.

Zinet et al. (2012) developed a dynamic model for a 15 kW single-effect H_2O-LiBr absorption chiller. The heat exchangers of the system are solved based on the NTU method, except for the evaporator-absorber (a custom falling film evaporator-absorber is used and is solved with more analytical heat and mass transfer modeling).

Evola et al. (2013) modeled the dynamic behavior of single-stage solar-powered H_2O-LiBr absorption chiller. The results of the model were validated against a commercial 4.5 kW absorption chiller manufactured by Rotartica. Validation was realized by comparison of the secondary water circuits (temperature and mass flows). Experimental measurements were conducted on two hot days (August 14 and September 8). The highest thermal COP reported was in the range of 0.74. The maximum relative error in terms of the COP and thermal loads of the components did not exceed 5%, hence proving that the model was decently accurate. Validation was followed by a set of simulations to evaluate the influence of the hot and the cold storage tank volumes. The results of the simulations showed that increasing the hot storage tank from 0.2 m^3 to 1 m^3 would increase the average daily thermal COP from 0.705 to 0.728, while the cooling production would decrease by 0.5 kWh/day, from 39.2 to 38.7 kWh/day. On the other hand, increasing the storage for the chilled water from 0.2 m^3 to 1 m^3 would have a beneficial impact on both the cooling production (increasing it from 38.6 to 39.6 kWh/day) and the average thermal COP (increasing it from 0.708 to 0.722).

Ochoa et al. (2016) developed a dynamic model based on a finite difference method in a Matlab environment to simulate the transient performance of a single-effect H_2O-LiBr absorption chiller. The results of the model were validated based on data from experiments conducted at the test rig at the Mechanical Engineering Department of the Federal University of Pernambuco–UFPE, in Brazil. The test rig consisted of a 30 kW micro turbine as the heat source, a 0.5 m^3 hot water buffer, a 1.2 m^3 chilled water storage tank, a 90 kW cooling tower, and a 35 kW absorption chiller. An overview of the apparatus is presented in Figure 5.13. The deviation between the simulation results and the experimental data regarding the temperatures of the secondary water circuits was in the range of 0.3–5% (relative error). The reported COP was 0.61 for a driving temperature of 95°C.

5.2.3 Performance Data from Experimental Setups

Nakahara et al. (1977) developed a solar absorption system based on a 32.2 m^2 array of flat-plate collectors with selective surfaces, a variable tilt angle, and a 7 kW single-effect H_2O-LiBr absorption chiller. The absorption chiller, manufactured by Yazaki, had a working temperature input/output of 75/65°C for the generator, 18/25°C for the condenser, and 12.5/7.5°C for the evaporator. The setup also included a 2.5 m^3 hot storage tank and a backup 14 kW gas-fired heater. Based on the experiments carried out, it was found that increasing the generator's temperature would result in an increase in the cooling capacity from approximately 3.2 kW to 6.9 kW. On the other hand, the optimum temperature for the COP would be lower, around 75°C, with a value of 0.8.

Van Hattem and Actis Dato (1981) designed, modeled, and evaluated the actual performance of a small-scale solar absorption cooling system in northern Italy. The absorption machine was a single-stage H_2O-LiBr chiller with a cooling capacity of 4 kW, powered by an array of 36 m^2 flat-plate collectors with selective surfaces. The theoretical results were

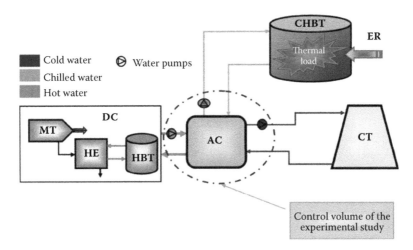

FIGURE 5.13
Overview of the experimental apparatus at the Federal University of Pernambuco–UFPE. (Reproduced from Ochoa, A. A. V. et al., *Energy Conversion and Management*, 108: 30–42, 2016.)

found to be in sufficient agreement with the measured data, estimating an overall system efficiency of 11% on an annual average basis.

Bong et al. (1987) investigated and reported the performance of a solar-powered air conditioning system consisting of a 7 kW H_2O-LiBr absorption chiller and 32 m² of heat pipe collectors, installed in Singapore. An auxiliary 9 kW heater was implemented in the system, along with a 2 m³ hot water storage tank and a 2 m³ chilled water tank with two separate compartments. The performance of the system was measured on a cloudy, an average, and a sunny day. The solar fraction on the average day was 52.1 %, while on the sunny day it was as high as 87.9 %. The average COP of the chiller was 0.669, while the average solar COP was 9.9 %.

Al-Karaghouli et al. (1991) evaluated the performance of a solar cooling system installed in the Solar Energy Research Center, in Iraq, during the summer season. Two 60 ton (211 kW) H_2O-LiBr chillers were installed along with two 40 ton (140.6 kW) heat pump chillers and auxiliary equipment, including two auxiliary boilers, five cooling towers, and four air-handling units, as shown in Figure 5.14. The chillers were powered by 1,577 evacuated tube solar collectors. Based on the analysis of the results, it was found that solar collection efficiency was 49% on a daily average basis. The chiller's COP was 0.618 and the solar fraction was 60.4% of the total load.

Bell et al. (1996) designed a H_2O-LiBr absorption chiller at Coventry University, in the United Kingdom. The experimental absorption chiller was realized by an evaporator/absorber cell and a generator/condenser cell. The results of the experiment showed that there is an optimum for the generator temperature (in each set of operating conditions). Specifically, for an evaporation temperature of 5°C and an absorption temperature of 32°C, the optimum generator temperature was found to be 68°C, yielding a COP of approximately 0.72.

Hammad and Zurigat (1998) reported the behavior of a 1.5 ton absorption refrigeration unit powered by 14 m² flat-plate solar collectors. The investigated system, shown in Figure 5.15, consists of four shell-and-tube heat exchangers—generator, absorber, condenser, and solution HEX—and a fin-and-tube evaporator. The system was located in the city of Amman, Jordan. Cooling of the condenser and absorber was realized by the city's main

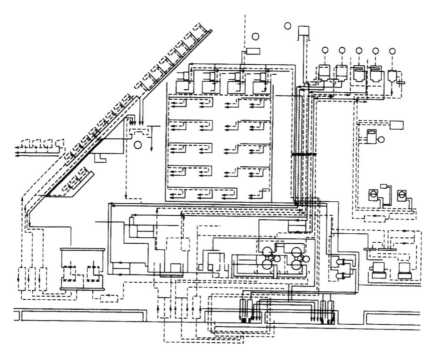

FIGURE 5.14
Schematic of the solar cooling system in the Solar Energy Research Center, Iraq. (Reproduced from Al-Karaghouli, A. et al., *Energy Conversion and Management*, 32 (5): 409–417, 1991.)

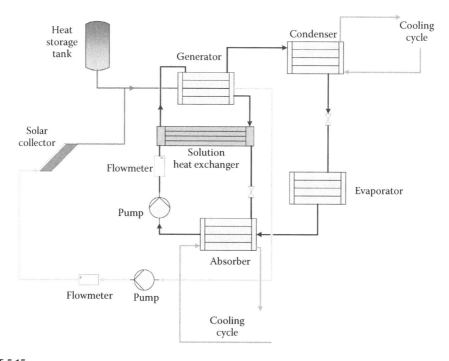

FIGURE 5.15
Schematic of the system experimentally investigated by Hammad and Zurigat. (Adapted from Hammad, M., and Y. Zurigat, *Solar Energy*, 62 (2): 79–84, 1998.)

water grid. The solar irradiance, measured in April and May, ranged between 900 and 1100 W/m², resulting in hot water temperatures in the solar collector between 75–97°C. The corresponding maximum generator temperature was equal to 80°C, resulting in a COP value of 0.85.

Best and Ortega (1999) presented the results of the Sonntlan Mexicali Solar Cooling Project, in Mexico, for the period of 1983–1986. The projects supplied six houses with solar cooling, including passive elements for cooling load reduction (from 18 kW in the case of a conventional house, to 7 kW). In total, 288 flat-plate collectors, 1.1 m² each, were installed on the roofs of the six houses (316 m² of solar collectors in total), along with two heat storage tanks of a total volume of 30 m³. The required cooling load was covered by a 90 kW Arkla WFB 300 Solaire H_2O-LiBr chiller. The temperature range in the chiller was between 70–90°C for hot water, 25–28°C for cooling water, and 7–11°C for chilled water exiting the absorption chiller. After certain modifications to the system—including bypassing the heat exchanger between the collectors and the hot water storage tanks and the decreasing the length of the pipeline for the hot water and the chilled water—the yearly solar fraction increased from 59% to 75%, while the average COP varied between 0.53–0.73.

De Francisco et al. (2002) experimentally evaluated the performance of a solar-powered 2 kW NH_3-H_2O absorption chiller prototype. A parabolic cylindrical collector supplied the chiller with thermal oil at a temperature of 140°C, resulting in generator temperatures around 120°C. The highest achieved COP was in the range of 0.5 for a generator temperature of 95°C and an evaporator temperature of 6°C. The performance degraded rapidly at higher generator temperatures, with a COP of approximately 0.35 for a generator temperature of 120°C.

The absence of many commercially available small-scale chillers at the beginning of the 2000s lead Storkenmaier et al. (2003) to develop a 10 kW single-effect H_2O-LiBr chiller within the framework of the German "Solarthermie 2000plus." The chiller was designed to be driven by a moderate temperature of approximately 85°C, resulting in a nominal chilled water temperature of 15°C. The corresponding COP of the chiller was equal to 0.74 for a cooling water temperature of 27°C. The chiller was powered by a 40 m² solar collector field.

Syed et al. (2005) evaluated the performance of a 35 kW solar-powered single-effect H_2O-LiBr chiller based on experiments carried out in summer 2003. The solar cooling system was used to cover the cooling loads of a typical house in Madrid, Spain, and was powered by 49 .9 m² flat-plate solar collectors. The chiller operated at partial load with relatively poor efficiency, with a maximum cooling capacity of 7.5 kW and a period average COP of 0.34. The corresponding generation temperature was 57–67°C, while the absorption temperature was in the range of 32–36°C. The main reason for this performance was the size of the solar system, which was originally designed to power a 10 kW chiller.

Richter and Safarik (2005) reported on two small-scale solar-powered NH_3-H_2O absorption cooling plants in Germany with a common COP of approximately 0.54. The first system was a 15 kW air conditioning system driven by a source temperature of 95°C and a nominal evaporating temperature of 3°C. The second plant had a nominal capacity of 20 kW with a nominal evaporating temperature of -6°C and a driving temperature of 100°C.

Lokurlu and Müller (2005) reported on the performance of a solar-powered double-effect H_2O-LiBr chiller installed at a hotel in Turkey. Parabolic trough collectors with a 180 m² aperture area were used to drive the generator of the absorber. The nominal generation temperature was 144°C, and the nominal capacity of the chiller was equal to 110 kW.

Pongtornkulpanich et al. (2008) designed and reported on the performance of a 10 ton (35.17 kW) solar-assisted single-effect H_2O-LiBr absorption air conditioning system at the

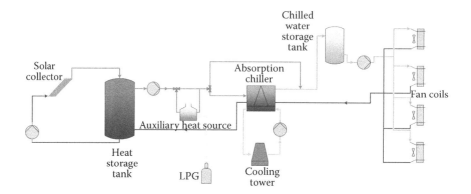

FIGURE 5.16
Schematic of the solar cooling system installed at the School of Renewable Energy Technology, Phitsanulok, Thailand. (Adapted from Pongtornkulpanich, A. et al., *Renewable Energy*, 33 (5): 943–949, 2008.)

School of Renewable Energy Technology in Phitsanulok, Thailand. The setup has been in operation since 2005 and covers the air conditioning needs of the main testing buildings. Seventy-two m² of evacuated tube solar collectors supply the required heat at a temperature of 70–95°C. A 0.4 m³ hot water storage tank was also installed along with a backup LPG-fired heater. The interconnection of the components is shown in Figure 5.16. Based on the performance analysis of the system, a solar fraction of 81% was estimated, while the rest, 19%, was supplied by the auxiliary heater. An economic analysis was also carried out, showing that the initial costs for the solar absorption system are rather high in comparison to a conventional vapor compression system. This is, however, compensated at some point by lower operating costs.

Ali et al. (2008) evaluated the performance of an integrated free-cooling and solar-assisted single-effect H_2O-LiBr absorption chiller. The system is located in Oberhausen, Germany and has been in operation since August 2002, providing air conditioning for a floor space of 270 m². The solar field consists of 108 m² vacuum tube collectors, supplying hot water to a 10 ton (35.17 kW) absorption chiller. As shown in Figure 5.17, a hot water storage tank of 6.8 m³ and a 1.5 m³ chilled water storage tank are also implemented. Based on the results during the cooling months of 2005 and 2006, free cooling reached up to 70% on a monthly average basis in June 2006. The monthly average solar fraction ranged between 31.1–100%, while the annual average over a five-year period was 60%. During sunny days with a clear sky, the chiller's COP was in the range of 0.37–0.81, while the corresponding collectors' field efficiency was between 0.352–0.492. Based on the results, the specific collector area was found to be equal to 4.23 m²/kW$_c$.

Rodríguez Hidalgo et al. (2008) carried out an experimental investigation of a solar absorption cooling unit at Universidad Carlos III de Madrid, in Spain. The single-effect H_2O-LiBr absorption chiller, model WFC10 manufactured by Yazaki, was powered by a 50 m² flat-plate collectors field located on campus. The experiments were carried out in the summer of 2004 and showed that the cooling capacity of the system ranged between 6–10 kW at a generator heat input of 10–15 kW. This resulted in a mean cooling period of 6.5 hours with complete solar autonomy for the average day of the investigated season. Furthermore, the feasibility of the system for air conditioning applications was evaluated by calculations on a 90 m² single detached house. It showed that a seasonal solar fraction of 56% could be achieved with the setup.

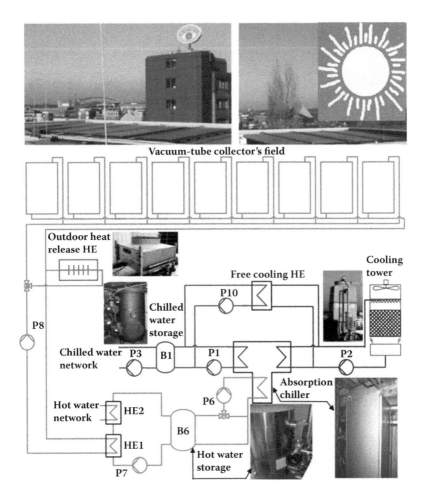

Vacuum-tube collector's field

FIGURE 5.17
Schematic of the solar absorption system located in Oberhausen, Germany. (Reproduced from Ali, Ahmed Hamza H. et al., *Solar Energy*, 82 (11): 1021–1030, 2008.)

Helm et al. (2009) presented performance results from a pilot 10 kW solar H_2O-LiBr absorption heating and cooling installation, in which the wet cooling tower has been replaced by a dry cooling tower in conjunction with latent heat storage. The latent heat storage is realized by using a phase change material ($CaCl_2$-$6H_2O$) in a temperature range of 27–29°C with a storage capacity of 120 kWh to absorb the solar gain from the 40 m² solar collector field. Operational data has been collected and reported on since fall 2007. Based on the data, on a hot day with a 28°C ambient temperature, the chiller was able to produce at least 10 kW at a driving temperature of 80–90°C and a COP of approximately 0.72.

Rosiek and Batlles (2009) analyzed the performance of the solar-powered air conditioning system installed at the Solar Energy Research Center (CIESOL), in Almeria, Spain. The main components of the solar air conditioning system are the 70 kW single-effect H_2O-LiBr absorption chiller, model WFC SC20 manufactured by Yazaki, and a 60 m² flat-plate solar collector array, which are shown in Figure 5.18. Furthermore, an auxiliary 100 kW heater has been installed along with two 5 m³ hot water storage tanks. Based on the calculations carried out, the energy demand of the building was estimated to be 8,124 kWh for heating

FIGURE 5.18
View of the flat-plate collector field at CIESOL building. (Reproduced from Rosiek, S., and F. J. Batlles, *Renewable Energy*, 34 (6): 1423–1431, 2009.)

and 13,255 kWh for cooling. The seasonal performance of the solar air conditioning system was measured for the months of July, August, and September, and it was found that the average cooling production was 40.3 kW, with a COP of 0.66, 0.62, and 0.42 for the three respective months. Furthermore, an estimation of the energy and CO_2 savings was presented. According to the results of the calculations, the total energy savings for both heating and cooling were estimated to be 17,197 kWh/year and the corresponding CO_2 savings 12,898 kg/year.

Agyenim et al. (2010) developed and tested a prototype solar refrigeration system at Cardiff University, in the United Kingdom. The main components of the system were the 12 m² vacuum tube solar collector, the 4.5 kW H_2O-LiBr absorption chiller, a 1 m³ cold water storage tank, and a 6 kW fan coil. The experiments were conducted during the summer and autumn of 2007. Based on the results, it was found that the average COP was 0.58 for a hot, sunny day with an ambient temperature of 24°C. The chilled water outlet temperature was as low as 7.4°C, pointing to the potential of the system for domestic cooling applications.

Bermejo et al. (2010) reported on the performance of a 174 kW double-effect H_2O-LiBr absorption chiller, model Broad BZH15, installed at the Engineering School of Seville, in Spain. The chiller is powered by 352 m² of Fresnel linear concentrating collectors and a direct-fired auxiliary natural gas burner. At nominal point, the water, pressurized at 13 bar, leaves the collectors at a temperature of 180°C. The high temperature generator operates at a temperature of 145°C, the condenser inlet/outlet temperature is 30/37°C, while the evaporator inlet/outlet temperature is 12/7°C, and the nominal COP is 1.34. Experiments were conducted between May and October 2008. Based on the experimental data, the average solar heat contribution in the generator was 75% of the total heat input. The corresponding COP throughout a day was in the range 0.85–1.30, with a daily average cooling power was 135 kW. The daily average collector efficiency was 35–40%, resulting in a solar COP of 0.44.

Marc et al. (2010) carried out an experimental investigation of the performance of a solar absorption setup used for cooling four classrooms of 170 m² in total located in Saint Pierre, Reunion Island, without any backup system. The chiller used was a H_2O-LiBr single-effect absorption chiller, with a nominal capacity of 30 kW at a generator temperature of 90°C, an absorber temperature of 30°C, and an evaporator temperature of 11°C. Two storage tanks of

1.5 m³ and 1 m³ were installed for the hot and cold water, respectively, providing 45 minutes of autonomy for hot and cold water production. The solar field was composed of 90 m² of double-glazed flat-plate solar collectors. The system proved, via the experiments, to be capable of covering the cooling loads of the classrooms, achieving a temperature difference of 6 K between interior and exterior temperature. The experiments were conducted between March and June 2008 and showed an average refrigeration production of 42 kWh/day, while the average theoretical COP of the system for the same period was estimated to be 0.335.

Qu et al. (2010) developed and reported on the performance of a solar absorption cooling and heating setup, installed at Carnegie Mellon University, in the United States, to cover the thermal loads of a building. A 52 m² of linear parabolic trough collector supplied heat to a 16 kW double-effect H_2O-LiBr absorption chiller. The experiments were conducted from February to September 2007. The efficiency of the solar collectors was in the range of 33–40%, while the chiller's COP was approximately 1.0–1.1 delivering a maximum cooling output of 12 kW, resulting in a solar COP around 0.33–0.4. Based on the experimental results, a model was developed to simulate the potential application of this system in a building in Pittsburgh, Pennsylvania, and it was found that the system could achieve a solar fraction of 39% for cooling and 20% for heating.

Ortiz et al. (2010) developed a model to predict the performance of a solar cooling application installed at the Mechanical Engineering Building of the University of New Mexico, in Albuquerque, and validated its results based on experimental data. A schematic of the actual setup is presented in Figure 5.19, and consists mainly of a hybrid solar collector field, with 124 m² flat-plate collectors and a 108 m² of vacuum tube solar collectors, a 70 kW H_2O-LiBr single-effect absorption chiller, model SH20 manufactured by Yazaki, and several auxiliary components. Experimental data from the setup was also reported by Mammoli et al. (2010). The COP of the absorption chiller was measured to be between 0.65–0.77, while the corresponding system's COP was in the range of 0.53–0.65.

Abdulateef et al. (2009) conducted experiments on a 5.28 kW solar-powered NH_3-H_2O absorption chiller installed at University Kebangsaan, in Bengali, Malaysia. Solar power was collected by 10 m² of evacuated tubes, providing the generator with hot water at 60–100°C. A cooling tower was implemented for heat rejection in order to maintain a cooling water temperature of 18–30°C. The experiments showed a maximum COP of 0.58, at an 85°C generator temperature, a 28°C condenser temperature, a 30°C absorber temperature, and a 16°C evaporator temperature.

A steady state analysis of the performance of a solar-powered absorption cooling system installed at the sports center of the University of Zaragoza, Spain, was conducted by Monné et al. (2011). A total 30 m² of flat-plate collectors supplied heat to a commercial 4.5 kW air-cooled single-effect H_2O-LiBr absorption chiller, model Rotartica 045. The experiments were conducted during 2007 and 2008. The average COP over the period of the experiments was around 0.5, while the corresponding solar COP was approximately 0.16. The average chilling capacity was 5.8 kW for 2007 and 4.4 kW for 2008.

Moreno-Quintarar et al. (2011) experimentally investigated the performance of a solar-powered intermittent absorption refrigeration system installed at Centro de Investigación en Energía of the Universidad Nacional Autónoma de México. The evaluated ternary mixture was NH_3-$LiNO_3$-H_2O, and its results were compared to the working pair NH_3-$LiNO_3$ for an 8 kg ice production application. The system was powered solely from a 2.54 m² compound parabolic concentrator with a cylindrical receiver serving as generator/absorber. The evaporator temperatures were as low as -11°C (with generator temperatures in the range 87–112°C), resulting in a solar COP of 0.098, which was 24% higher than the binary mixture case.

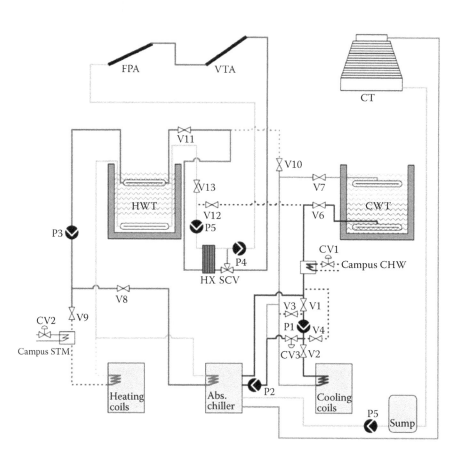

FIGURE 5.19
Schematic of the solar cooling installation at the Mechanical Engineering Building at the University of New Mexico. (Reproduced from Ortiz, M. et al., *Energy and Buildings*, 42 (4): 500–509, 2010.)

Bujedo et al. (2011) evaluated the results of several control strategies for the part load operation of a solar-powered absorption cycle. A 77.5 m² hybrid solar field (40 m² of vacuum tube collectors in parallel with 37.5 m² of flat-plate collectors) supplied the required heat to drive the generator of the 35 kW H_2O-LiBr absorption chiller, a Yazaki model WFC-10. Each collector field was connected to a 2 m³ hot storage tank, while another 1 m³ tank was implemented for the chilled water storage. The system was designated to air condition 200 m² of offices at Building One of the Fundacion Cartif in Boecillo, Spain. The rest of the building is conditioned by water-to-air electric chillers, as shown in Figure 5.20. The new control strategies proved to enhance the solar fraction between 7–12%, while the COP improved by 44–48%, reaching an average daily value of 0.57.

González-Gil et al. (2011) reported on the performance of a direct air-cooled single-effect H_2O-LiBr absorption prototype designed for solar air conditioning applications. The experimental setup, installed at Madrid, Spain, was powered by 48 m² of flat-plate collectors. According to the experimental results, conducted during summer 2010, the cooling

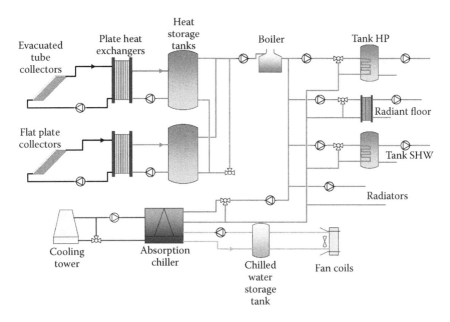

FIGURE 5.20
Schematic of the solar cooling installation at Building One of the Fundacion Cartif in Boecillo, Spain. (Adapted from Bujedo, Luis A. et al., *Solar Energy*, 85 (7): 1302–1315, 2011.)

output of the chiller was in the range of 2–3.8 kW—85% of the nominal capacity—delivering chilled water at 14–18°C. The driving temperature was in the range of 85–110°C, while the absorption temperature was always lower than 46°C, and the condensation temperature was always lower than 50°C. The corresponding COP ranged between 0.38–0.66, with solar fractions between 40–100%, based on the cooling loads of a 40 m² test room.

Achuthan et al. (2011) developed and reported on the performance of a compact solar refrigeration system based on a H_2O-LiBr absorption chiller. The authors attached micro nozzles in the spray pipes of the evaporator and the absorber, enhancing the heat transfer by 20%. The temperature levels in the set of experiments conducted were 75–93°C for the generator inlet, 25–35°C for the condenser inlet, and 19–22°C for the evaporator outlet temperature. The corresponding COP increased from 0.3 to 0.6 after the modification.

Balghouthi et al. (2012) conducted experiments on a solar absorption cooling installation at the Center for Energy Research and Technology (CRTEn), in Bordj-Cédria, Tunisia. The cooling load of the 150 m² laboratory connected with the system has a mean cooling load of 4.5 kW, with a peak load of 10.9 kW. Thirty-nine m² of linear parabolic trough collectors provided, via a heat transfer fluid, the required heat input to a 16 kW double-effect H_2O-LiBr absorption chiller. The maximum reported cooling output was 12 kW when the evaporator temperature was equal to 8°C. The COP was in the range of 0.80–0.91, resulting in a solar COP between 0.10–0.43. A drain backup night storage was implemented in the system, enhancing the average solar fraction from 54% to 77%. The operation of the solar cooling system resulted in a significant reduction in CO_2 emissions, with an annual saving of 3000 kg.

Darkwa et al. (2012) investigated both with simulations and experiments the performance of a solar-powered absorption system. A total of 220 m² of evacuated U-tube solar collectors were installed at the Center for Sustainable Energy Technologies building at the University of Nottingham, Ningbo, in China. Four storage tanks were implemented for hot water storage with a total capacity of 16 m³. The chiller in the installation was a H_2O-LiBr absorption chiller with a nominal capacity of 55 kW, at a driving temperature of 90°C, and it delivered chilled water at a temperature of 7°C. Meteorological data was collected in August 2010 over a seven-day period, while the rest data was collected on August 19, 2010. Based on the results of the experiments, it was found that collector efficiency was around 61%, in comparison to the 69% theoretical value. The absorption chiller's COP was 0.69, which agreed with the nominal COP of 0.7 provided by the manufacturer.

Lizarte et al. (2012) conducted experiments to evaluate the performance of a solar-assisted 4.5 kW air-cooled single-effect H_2O-LiBr absorption chiller designed for domestic applications. The solar apparatus consisted of a 42.2 m^2 vacuum flat-plate collector field, a 25 kW plate heat exchanger, and a 1.5 m^3 hot water storage tank. Figure 5.21 presents an overview of the solar-powered absorption cooling system, which in terms of the experiments was used to provide air conditioning for a 40 m^2 room located in Madrid, Spain. The experiments were conducted for ten days in the summer of 2009. Based on the experimental data, it was concluded that the average COP was 0.53, the solar COP was 0.06, the solar fraction during the experimental period was 47%, and the mean collector efficiency was 0.27. On days with outlet dry bulb temperatures of 38°C, the cooling output of the chiller was equal to 4.2 kW, while the evaporator outlet temperature was 16°C.

Ayadi et al. (2012) evaluated the performance of a solar cooling system designed to cover the heating and cooling demands of 172 m^2 office buildings, located in northern Italy. The 17.6 kW H_2O-LiBr single-effect absorption chiller, a Yazaki model WFC-SC5, was powered by 61.1 m^2 flat-plate collectors. A reversible heat pump was installed as backup, and a 5 m^3 hot water storage tank and a 1 m^3 cold water storage tank were also implemented in the system. Performance data was collected for 2011, and based on the findings, it was estimated that the thermal efficiency of the solar collectors was between 30–40%, and the absorption chiller's average COP was 0.55.

Yin et al. (2012, 2013) conducted experiments on a mini-scale solar absorption cooling system powered by 96 m^2 of evacuated U-tube solar collectors. The nominal capacity of the H_2O-LiBr absorption chiller was 8 kW, while a 3 m^3 tank was also installed in the setup for hot water storage. The setup was installed and evaluated at Shanghai Jiao Tong University, in China. At nominal operation, the chiller was supplied with hot water at 70–95°C with a mass flow rate of 1.1 kg/s, and was able to deliver chilled water at 9°C with a mass flow rate 0.42 kg/s. The continuous operation performance of the system was evaluated under a nine-hour period, and the results showed a cooling capacity of 4.6 kW with an average COP of 0.31. Furthermore, in the 50 m^2 test room that was used, the effect of replacing conventional fan coils with radiant cooling panels was investigated. According to the findings of the authors, the cooling output increased by 23.5% due to this change.

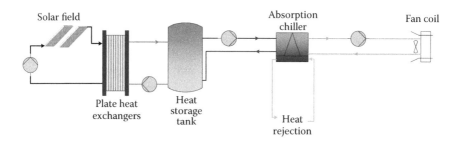

FIGURE 5.21
Schematic of the solar absorption cooling facility. (Adapted from Lizarte, R. et al., *Energy and Buildings*, 47 (Supplement C): 1–11, 2012.)

Winston et al. (2014) developed and reported on the performance of a solar-powered cooling system. The 53.3 m² of non-tracking external compound parabolic concentrators (XCPC) provide the required heat to drive a 23 kW double-effect H_2O-LiBr absorption chiller. Experimental data was collected during the summers of 2011 and 2012. The temperature working range of the solar collectors was 160–200°C, with an average daily efficiency of 36.7%. The reported average COP on a daily basis for a typical day was approximately 0.99, instantaneous COP was in the range 0.769–1.181, and the corresponding solar COP was 0.363. The respective results for the daily average COP on a cloudy day were 1.019, with instantaneous COP having quite broad variations in the range of 0.617–2.236.

Alsaqoor and AlQdah (2014) compared the performance of an NH_3-H_2O absorption chiller powered by different sources—electrical energy, LPG, and solar energy. The experiments were carried out at Tafila Technical University, in Jordan. For the solar energy scenario, flat-plate collectors were used. Based on the experiments carried out, it was found that electrical energy resulted in the most efficient operation of the absorption chiller, with a measured COP of 0.463, while the respective value for the LPG was 0.244, but for the flat-plate collectors was only 0.109.

Albers (2014) reported the performance and the control strategy of the solar absorption chiller installed at the Federal Environment Agency, in Dessau, Germany. The system was installed to replace an adsorption chiller. The electrical efficiency in comparison to the former setup increased by 35%, while water consumption reduced 70%. Regarding the thermal efficiency, the previous setup had a COP of 0.47, in comparison to 0.76 for the absorption chiller—an improvement of 62%.

Weber et al. (2014) presented the performance data for a solar-powered NH_3-H_2O absorption chiller. The solar field consisted of 132 m² of linear concentrating Fresnel collectors providing hot water or steam at 160–200°C to drive two identical 12 kW absorption chillers (model ACF60-00 LB manufactured by Robur). The chiller's nominal COP was 0.6, providing cooling temperatures between -10°C and 0°C. An ice storage system made of four 0.3 m³ units was implemented for performance testing. The cooling load was simulated by three 10 kW heating elements. The experimental data was collected between 2011 and 2012. For driving temperatures of 120°C, the cooling output was only 5 kW, with a thermal energy efficiency ratio (EER) of 0.3. For typical days, the average thermal EER was around 0.6 and the cooling output varied between 10–26 kW. The implementation of the ice storage proved to perform its work satisfactorily, with a storage capacity of 110 kWh.

Bolocan et al. (2015) developed a prototype solar absorption system with NH_3-H_2O as the absorption chiller's working pair. The prototype's system performance was evaluated by simulations carried out in an EES environment along with experimental data for the 5 kW absorption system installed in Brasov, Romania. The maximum reported COP was equal to 0.43.

5.2.4 Rethymno Village Hotel Solar Absorption System

Rethymno Village Hotel is another example of an installed commercial solar-powered single-stage absorption unit. The H_2O-LiBr absorption chiller (model SOLE Climasol XZR 30/105) has an installed capacity of 105 kW. The heat input is supplied by 500 m² of flat-plate collectors (250 flat-plate collectors of 2 m² each), and there are seven storage tanks installed totaling 7 m³ (Karagiorgas et al. 2007).

Rethymno Village Hotel is located on the island of Crete, in southern Greece. Solar energy in the Rethymno Village Hotel is used to provide space cooling during the summer and heating during the winter. The air-conditioning system is realized by fan coils that are connected to the single-stage absorption chiller. This system is the first commercial application of solar cooling that used flat-plate collectors instead of evacuated tube collectors. The system is comprised of the solar collector field, the absorption chiller, the hot water storage tanks, the piping system, and a wet cooling tower.

The flat-plate collector system (manufactured by SOLE S.A.) provides the absorption chiller with hot water at a temperature of 75–80°C via hot storage tanks. The solar collectors have a south orientation, with a tilt angle of approximately 30°, which is the optimum angle to maximize the annual energy collection. In terms of the arrangement of the solar collectors, they are placed in rows to reduce to minimum shading effects.

During spring and autumn, when required, the solar collectors are directly connected to the fan coils circulating water at approximately 50–55°C for space heating needs. During the summer period, water is led from the collectors to the storage tanks and then to the absorption chiller, which operates 7–8 hours on average each day (Karagiorgas et al. 2007). The average COP for these water supply conditions is 0.6. The air that is cooled down by the absorption chiller is supplied through the fan coils at a temperature of 10–12°C. Figure 5.22 provides an overview of the temperature profiles of the absorption chiller based on measurements carried out by Karagiorgas et al. (2007). The heat rejection from the absorption chiller is realized by an installed cooling tower. In case there is a need for additional heat, an auxiliary 290 kW_{th} boiler has been installed. If there is a need for additional cooling, an electrically driven chiller, with a cooling capacity of 170 kW_c, is used. A secondary boiler has also been installed with a thermal capacity of 175 kW_{th} to provide sanitary hot water in the case that the solar collectors cannot cover these requirements (for example, on cloudy days). The electrical requirements for the pump of the absorption chiller are estimated to be 0.5 kW_e, while the respective electrical consumption of the water circuits was 1.5 kW_e.

The system has been operational since August 2000. According to the technical assessment carried out by the Center for Renewable Energy Sources and Saving (CRES), the annual solar energy harvested is 650 MWh and the total building load is 1.5 GWh, leading to an approximate cover of 43% of the total load from solar energy. Based on the performance of the absorption chiller, the energy savings are estimated to be 40 MWh/month, which is equivalent to savings of €3,800/month. The overall project had a total investment

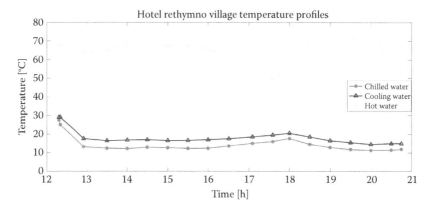

FIGURE 5.22
Temperature profiles of the Rethymno Village Hotel solar cooling system. (Adapted from Karagiorgas, M. et al., "Operation and measurement results of the solar cooling installation in Rethymnon village hotel", 28th AIVC Conference on Building Low Energy Cooling and Advanced Ventilation Technologies in the 21st Century, Crete, Greece, September 2007, 2007.)

cost of €264,000, subsidized at a percentage of 50% by the Greek Ministry of Development. The total investment cost for the solar collectors was €7,500.

5.2.5 Demokritos Research Center Solar Absorption System

Demokritos Research Center, located in Athens, Greece, is another case of a solar heating and cooling application developed by the National Technical University of Athens. The system is connected to provide heating and cooling at the laboratory and offices of the Solar and Energy Systems Lab. The floor area of the lab, which is located underground, is equal to 320 m², and its total volume is equal to 1,250 m³ (Drosou et al. 2014). Two separate air conditioning units with a capacity of 25 kW comprise the already installed conventional system. The connection of the solar collectors to the existing plant is realized by supplying the existing heat exchanger unit. The single-stage LiBr-H_2O absorption chiller, a Yazaki Aroace WFC-10, has a nominal output of 35.2 kW, chilling water at a nominal temperature of 7°C with a thermal COP of 0.7. The working range of the heat input is 70–95°C. Heat rejection from the chiller is realized by a cooling tower. Backup heating is provided through a 120 L storage tank equipped with a 50 kW electrical resistance. Figure 5.23 presents a simplified schematic of the solar cooling installation.

The solar field consists of 80 flat-plate collectors (model Foco Ikarus A3), with selective surfaces (2 m² each) and an average collector efficiency of 55%. Two storage tanks of 5 m³ and 2 m³, respectively, have been also installed for the hot water supplied by the collectors. An intermediate circuit is introduced to transfer the heat from the solar collectors toward the chiller. Both the intermediate and the solar circuit have a mixture of 15/85 ethylene glycol-water mixture as their working fluid. An overview of the main technical characteristics of the solar absorption installation at the Demokritos Research Center is presented below in Table 5.4.

5.2.6 Centre for Renewable Energy Sources and Saving Solar Cooling System

Another application of solar cooling is the plant installed at the Centre for Renewable Energy Sources and Saving (CRES) in Pikermi, Athens, Greece. The total surface of the building is 427 m², while the respective volume is 1296 m³. The solar cooling installation

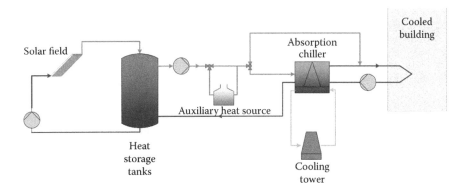

FIGURE 5.23
Simplified schematic of the solar cooling and heating installation at the Demokritos Research Center. (Adapted from Zervos, Arthouros, Demonstrating the Efficiency of Solar Space Heating and Cooling in Buildings, edited by SACE, Project Demonstration Report, available at: http://www.energycon.org/sace/HE01.pdf, 2002.)

TABLE 5.4

Basic Technical Data for the Demokritos Research Center Solar Absorption System

General Information	
Application	Space cooling and heating
Location	Athens, Greece
Total cooling area	320 m²
Start of operation	August 2002
Solar System	
Collector type	Flat plate with selective surfaces
Collector area	160 m²
Tilt angle	30°
Storage	
Hot storage	7 m³ (1 tank of 5 m³ and 1 tank of 2 m³)
Cold storage	–
Absorption Chiller	
Model	Yazaki Aroace WFC-10
Cooling Capacity	35.2 kW$_c$
Auxiliary Equipment	
Backup electrical resistance	50 kW
Performance	
Solar fraction	n/a
Chiller's COP	0.7

FIGURE 5.24
Overview of the solar collector field located on the roof of the CRES building. (Courtesy of V.N. Drosou.)

has been in operation since December 2011. The demonstration plant was developed as a part of the HIGH COMBI project. The system consists of solar collectors, an underground thermal energy storage system, a H₂O-LiBr absorption chiller, a wet cooling tower, and an 18 kW conventional heat pump. The heat pump operates with R410A as the working fluid and has a nominal electrical COP of 7. The solar collectors, shown in Figure 5.24, are of the flat-plate type with selective surfaces (model Climasol 2.67 m² manufactured by Sole S.A.), with a total surface of 149.5 m² (56 solar collectors in total). They are placed in eight rows, tilted at 30°, and their working medium is water. This data is courtesy of V.N. Drosou.

FIGURE 5.25
Image of the underground storage tank at the CRES solar cooling and heating system. (Courtesy of V.N. Drosou.)

FIGURE 5.26
View of the absorption chiller installed at the CRES solar cooling and heating system. (Courtesy of V.N. Drosou.)

The underground energy storage system has a total volume of 58 m³, with a height-to-diameter ratio (H/D) equal to 1.15. The walls are made of St-37 with a thickness of 7 mm. Figure 5.25 presents an image of the interior of the storage tank.

As can be seen in Figure 5.26, the absorption chiller (model Climasol XZR, manufactured by Sole S.A.) has a nominal capacity of 35 kW at a driving hot water temperature of 75°C (at a volumetric flow rate of 10 m³/hr), a cooling water temperature of 32°C (a volumetric flow rate of 25 m³/hr), and a chilled water outlet temperature of 7°C (a volumetric flow rate of 6 m³/hr). The corresponding COP is approximately 0.6.

In heat mode, the solar system supplies the building with 45°C hot water. The heat pump serves as backup, to cover any heat needs that the solar field is unable to cover. On the other hand, in cooling mode, as shown in Figure 5.27, the solar setup supplies the building with 7°C chilled water at nominal operation. Again the heat pump serves as an auxiliary chiller in case the absorption chiller is incapable of covering the cooling load of the building. The absorption chiller is normally driven by the heat from the solar collectors or the

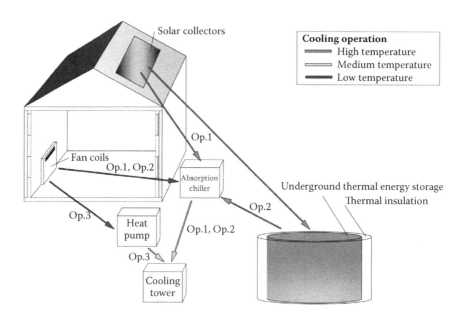

FIGURE 5.27
Overview of the CRES solar thermal installation in cooling mode. (Reproduced from Drosou, Vassiliki N. et al., *Renewable and Sustainable Energy Reviews*, 29 (Supplement C): 463–472, 2014.)

underground storage system. During the summer, any excess energy from the solar collectors is stored in the underground storage system, which may reach temperatures up to 90°C. During autumn and spring, when the loads are relatively low, excess solar energy is again stored in the underground system, raising the tank temperature to 95°C.

The designed solar fraction is 85% of the total energy demand of the building. While, based on measurements reported from Drosou et al. (2014), the actual solar fraction is around 70%. According to Drosou et al. (2016), the annual cooling demands were estimated to be 19.5 MWh/a, which refers to the period of May–September, while the respective heating loads were 12.3 MWh/a for the period of October–April. Figure 5.28 presents the average behavior of the energy demand as a fraction of the peak load on a daily, weekly, and yearly basis, showing that the use of the solar heating and cooling system is during working hours—between 8:00 a.m. and 5:00 p.m.—five days per week, throughout the year. Table 5.5 presents an overview of the basic aforementioned technical data for the CRES application.

Along with the applications discussed above, Table 5.6 displays an overview of several more existing solar absorption cooling applications in Greece.

5.2.7 ISI Pergine Business Center

ISI Pergine Business Center is a 9,815 m² two-story building located in the industrial area of Pergine, Trento, in Italy. The building is owned by the Tecnofin Trentina S.p.A.

The solar-powered absorption chiller is a 70 kW single-effect H_2O-LiBr absorption chiller that is able to provide 40–50°C hot water during the winter period and 14–17°C chilled water during the summer period. The absorption chiller is powered by a 265 m² solar field consisting of flat-plate collectors with selective surfaces, tilted 30°. The heating

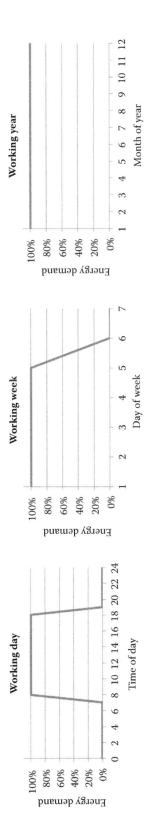

FIGURE 5.28

Overview of the energy demand curve on a daily, weekly, and annual basis for the CRES building. (Courtesy of V.N. Drosou.)

TABLE 5.5

Basic Technical Data for the CRES Solar Absorption System

General Information	
Application	Space cooling and heating
Location	Pikermi, Athens, Greece
Total cooling area	427 m²
Start of operation	December 2011
Solar System	
Collector type	Flat plate with selective surfaces
Collector area	149.5 m²
Tilt angle	30°
Storage	
Hot storage	58 m³ underground storage tank
Cold storage	–
Absorption Chiller	
Model	Climasol XZR (manufacturer: Sole S.A.)
Cooling capacity	35 kW$_c$
Auxiliary Equipment	
Backup heat pump	18 kW
Performance	
Solar fraction	70% (measured)
Chiller's COP	0.6

demand during the winter period is 230 kW, while the respective cooling demand during the summer period is 188 kW (Allouhi et al. 2015). Heat rejection is realized by a 175 kW wet cooling tower.

The achieved power production of the absorber is 108 kW (the respective power supply by the solar collectors is approximately 145 kW). The rest of the energy demand is covered by an auxiliary electric compression chiller with a capacity of 120 kW. Based on the power

TABLE 5.6

Overview of Existing Solar Absorption Cooling Systems in Greece

Owner (Location)	Type of Collector	Collector Area (m²)	Cooling Capacity (kW$_c$)	Start of Operation
American College (Athens)	Evacuated tube	615	168	1984
Rethymno Village Hotel (Rethymno)	Flat plate with selective surfaces	450	105	2000
Lentzakis S.A. (Rethymno)	Flat plate with selective surfaces	448	105	2002
Demokritos Research Center (Athens)	Flat plate with selective surfaces	160	35.2	2002
Sol Energy Hellas A.E. (Palaio Faliro)	Flat plate	78.6	35.1	2007
CRES (Athens)	Flat plate with selective surfaces	149.5	35	2011

Source: Tsoutsos, T. et al., *Energy and Buildings*, 42 (2): 265–272, 2010.

TABLE 5.7

Basic Technical Data for the ISI Pergine Business Center (SOLAIR)

General Information	
Application	Space cooling and heating
Location	Pergine, Trento, Italy
Total cooling area	9,815 m²
Start of operation	2004
Solar System	
Collector type	Flat plate with selective surfaces
Collector area	265 m²
Tilt angle	30°
Storage	
Hot storage	2 m³ storage tank
Cold storage	–
Absorption Chiller	
Model	n/a
Cooling capacity	70 kW$_c$
Auxiliary Equipment	
Backup electric compression chiller	120 kW
Connection via heat exchanger to district heating network	
Performance	
Solar fraction	65% (summer)
Chiller's COP	n/a

estimates, the solar fraction is approximately 65%, resulting in significant CO_2 emissions savings. Furthermore, the system is connected via a heat exchanger to the district heating network for extra heat supply.

The solar cooling and heating system has been in operation since 2004, and its initial cost was €540,000 (32% of the total was a subsidy from the province of Trento). The total energy savings are reported to be approximately 71,700 kWh for heating and 48,900 kWh for cooling. Table 5.7 presents an overview of the basic technical data for the ISI Pergine Business Center.

5.2.8 GICB Building Solar Cooling Application

The solar cooling application in the wine cellar of the GICB building, located in the city of Banyuls sur Mer, France, is one of the oldest solar cooling installations in Europe and the first of its kind in France, installed in 1991. Cooling is dedicated to the purpose of wine conservation and supplies three floors via three air-handling units: Level 0 is used for dispatching and Levels -1 and -2 are used for bottle storage (with a capacity of approximately 3 million bottles of wine).

The solar field consists of 130 m² of evacuated tube collectors, tilted 15° with a south-southwest orientation. Hot water leaving the solar collectors at a temperature between 60–95°C can be stored in a 1 m³ buffer tank. The chiller of the installation is a 52 kW single-effect H_2O-LiBr absorption chiller (model WFC 15, manufactured by Yazaki). The heat rejection from the absorption chiller is realized by a wet cooling tower with a capacity of

TABLE 5.8

Basic Technical Data for the GICB

General Information	
Application	Cooling for wine conservation
Location	Banyuls sur Mer, France
Total cooling area	3,500 m²
Start of operation	1991
Solar System	
Collector type	Evacuated tube collectors
Collector area	130 m²
Tilt angle	15°
Storage	
Hot storage	1 m³ storage tank
Cold storage	–
Absorption Chiller	
Model	Yazaki WFC 15
Cooling capacity	52 kW$_c$
Auxiliary Equipment	
Backup chiller	none
Performanc	
Solar fraction	n/a
Chiller's COP	0.57

180 kW. An overview of the key technical features of the GICB solar absorption cooling installation is provided in Table 5.8.

The temperature difference between input and output achieved by the system is regulated to 4°C to avoid thermal shock phenomena. The average COP of the chiller is 0.57. Based on results published by SOLAIR, the annual cooling production from May until the end of September was estimated at 17,000 kWh. The setup is equipped with a remote control device to allow off-site monitoring of the system's operation. By assuming an average cost saving of €0.05/kWh, Tecsol estimated the annual savings in terms of power consumption to be approximately €850. The total capital cost of the project was €294,500 (40% of which was a subsidy).

5.2.9 Agència de la Salut Pública

Another solar cooling and heating system was installed in 2007 in an office and laboratory building at the Public Health Agency of Barcelona, Spain.

Eighty-two m² of flat-plate collectors with selective surfaces, tilted 30°, provide the required heat to drive a 35 kW single-effect H_2O-LiBr absorption chiller (model WFC-SC10, manufactured by Yazaki). A hot water mixture (water-glycol 30%) from the solar collectors is stored in two 3 m³ storage tanks. Another 1 m³ storage tank has been implemented for the chilled water. A typical driving temperature provided by the solar collectors is around 80°C. A 98 kW wet cooling tower is also installed for the heat rejection of the absorption chiller. The two hot storage tanks can either work in series or in parallel. The parallel configuration allows for different temperature set points, allowing for separate operation during periods in which both cooling and heating is needed.

TABLE 5.9

Basic Technical Data for the Building Powered by the Solar Thermal System at the Public Health Agency of Barcelona

General Information	
Application	Space cooling, heating, and domestic hot water
Location	Barcelona, Spain
Total cooling area	2,597 m²
Start of operation	July 2007
Solar System	
Collector type	Flat plate with selective surfaces
Collector area	82 m²
Tilt angle	30°
Storage	
Hot storage	6 m³ (2 storage tanks of 3 m³)
Cold storage	1 m³
Absorption Chiller	
Model	Yazaki WFC-SC10
Cooling capacity	35 kW$_c$
Auxiliary Equipment	
Backup compression Chiller	323 kW
Backup heater	508 kW (gas fired)
Performance	
Solar fraction	20%
Chiller's COP	n/a

Due to its use, the building has extensive thermal demands. The system provides the building with space cooling, heating, and domestic water, covering approximately the 20% of the building's needs. The other 80% of the buildings needs is supplied by a 508 kW gas-fired boiler and a conventional 323 kWc compression chiller. The key technical characteristics of the solar cooling setup in Public Health Agency of Barcelona are listed in Table 5.9.

The total investment cost was approximately €310,000, while the annual maintenance costs are approximately €5,700.

5.2.10 Inditex Arteixo Offices

The solar cooling installatation at the main offices of Inditex at Arteixo, A Coruña, in Spain, is a part of the new integrated energy system in the facility, consisting of a 5 MW cogeneration plant, a 850 kW wind turbine, and a 1,626 m² (gross area) of flat-plate collectors with selective surface (Cetinkaya et al. 2011). The solar collectors are placed in the main building, where the offices are mainly located. The building has two stories with 10,000 m² each. The new system is able to cover 50% of the Arteixo facility's energy loads. Space heating, cooling, and domestic hot water was supplied, prior to the solar cooling setup, by two electric heat pumps and an electric cooler. Hot water had a supply/return temperature of 55/45°C, while the respective temperatures for cold water were 7/12°C. In the current installation, hot water can be stored in two 30 m³ storage tanks with a working temperature range of 55–80°C (Dalenbäck 2009). The absorption chiller has as its working pair H_2O-LiBr and a nominal capacity of 170 kW$_c$. An overview of the available in literature

TABLE 5.10

Basic Technical Data for the Inditex, Spain, Application

General Information	
Application	Space cooling and heating and domestic hot water
Location	Arteixo, A Coruña, Spain
Total cooling area	20,000 m²
Start of operation	2003
Solar System	
Collector type	Flat plate with selective surfaces
Collector area	1,626 m²
Tilt angle	n/a
Storage	
Hot storage	60 m³ (2 storage tanks x 30 m³)
Cold storage	–
Absorption Chiller	
Model	n/a
Cooling capacity	170 kW$_c$
Auxiliary Equipment	
Backup electric cooler	
2 x Backup electric heat pumps	
Performance	
Solar fraction	n/a
Chiller's COP	n/a

technical specifications for the solar absorption cooling installation at Inditex, Spain, is provided in Table 5.10. The total capital cost was estimated to be €900,000 and was 11% subsidized by the Galician Regional Ministry for Industry and Trade and 33% subsidized by the IDAE Spain (Institute for Energy Diversification and Saving).

5.2.11 The Technical College for Engineering in Butzbach

The Technical College of Butzbach, Germany, is an example of a cooling system driven solely by solar energy without any backup heat sources. At the site, two ventilation systems were already installed, with a 1,250 m³/hr volumetric flow rate each, which was, however, insufficient to cover all cooling loads during summer. The building is used throughout the

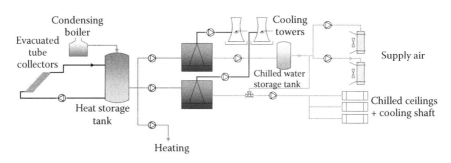

FIGURE 5.29

Schematic of the solar cooling installation at the Technical College at Butzbach, Germany.

TABLE 5.11

Basic Technical Data for the Technical College in Butzbach, Germany

General Information	
Application	Space cooling
Location	Butzbach, Germany
Total cooling area	n/a
Start of operation	n/a
Solar System	
Collector type	Evacuated tube collectors with CPC-mirror
Collector area	60 m^2
Tilt angle	30°
Storage	
Hot storage	3 m^3
Cold storage	1 m^3
Absorption Chiller	
Model	SK Sonnenklima Suninverse
Cooling capacity	2 × 10 kW$_c$
Auxiliary Equipment	
Backup boiler	28 kW (natural-gas-fired condensing boiler used for space heating only)
Performance	
Solar fraction	n/a
Chiller's COP	n/a

summer, with increased demands due to its high occupation and thermal loads because of the use of computer equipment.

The solar cooling installation was subsidized in the frame of German Solarthermie 2000plus. Cooling coils, chilled ceilings, and cooling panels were implemented along with the existing ventilation units. Cooling production is realized by two 10 kW absorption chillers powered by 60 m^2 of evacuated tube collectors with a compound parabolic concentrator mirror, tilted at 30°, working with pure water. Space heating during the winter period is provided by a 28 kW natural-gas-fired condensing boiler. Hot water from the solar collectors can be stored in a 3 m^3 storage tank, while chilled water can be stored in a 1 m^3 tank. Heat rejection from the chillers is realized by a wet cooling tower. An overview of the system is presented in Figure 5.29, while its key technical data is listed in Table 5.11.

The use of two absorption chillers allows for two different modes of operation, based on the needs of the building. There can be a single temperature level, in which case the chillers work in a parallel configuration. In the other mode, there is a low temperature level for air dehumidification and cooling provided by the first chiller, while the other chiller works on a higher temperature level providing cooling effect for the chilled ceilings and the cooling panels.

5.2.12 The Jiangmen Solar Absorption System

A large-scale solar absorption cooling installation has been developed in Jiangmen, China, as part of the ninth Five-Year Research Project (1995-2000).

The facility in Jiangmen is a 24-story building, consisting of hotels, business centers, entertainment places, and an education center. The solar system consists of a 500 m^2 modified flat-plate collectors powering a two-stage H$_2$O-LiBr absorption chiller with a nominal capacity

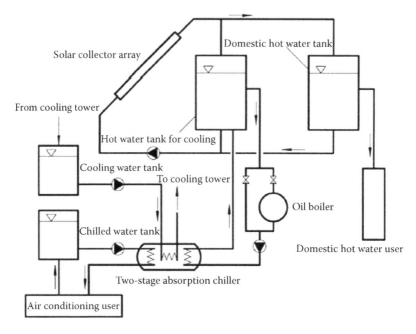

FIGURE 5.30
Overview of the solar cooling and heating application in Jiangmen. (Reproduced from Zhai, X. Q., and R. Z. Wang, *Renewable and Sustainable Energy Reviews*, 13 (6):1523–1531, 2009.)

TABLE 5.12

Basic Technical Data for the Jiangmen Application

General Information	
Application	Space cooling and heating
Location	Jiangmen, China
Total cooling area	600 m²
Start of operation	April 1999
Solar System	
Collector type	Modified flat plate
Collector area	500 m²
Tilt angle	n/a
Storage	
Hot storage	1 storage tank (capacity n/a)
Cold storage	1 storage tank (capacity n/a)
Absorption Chiller	
Model	Custom made
Cooling capacity	100 kW$_c$
Auxiliary Equipment	
Backup oil boiler	
Performance	
Solar fraction	n/a
Chiller's COP	0.373-0.458

of 100 kW$_c$ (Li et al. 1999). The nominal hot water temperature driving the absorber is 75°C, and the nominal chilled water output temperature is 9°C, with a capacity for a temperature drop between the input and output of the chilled water in the range of 12–17°C (Zhai and Wang 2009). The total area supplied with space cooling by the solar cooling system is 600 m². An auxiliary oil boiler is also installed at the site as a backup to the solar collectors. Each water circuit has a storage tank, as shown in Figure 5.30. Based on data reported by Sumathy et al. (2002), the absorption chiller was able to deliver chilled water at temperatures as low as 7°C, even for very low-grade driving temperatures (62°C). The data was collected on certain days in April and May. It was found that the chiller's capacity was in the range of 66.7–108.4 kW, with a corresponding COP of 0.373–0.458, for a hot water temperature of 60.8–72.6°C, a cooling water temperature in the range of 28.2–30.6°C, and a chilled water outlet temperature between 6.8–12.0°C. An overview of the main characteristics of Jiangmen solar cooling installation is listed in Table 5.12.

5.3 Process Model

In the following section, the model developed to simulate the operation of an investigated H$_2$O-LiBr single-effect absorption chiller is presented.

5.3.1 Basic Assumptions

In order to simplify the process, the following assumptions were made:

  i. Temperature, pressure, and concentration of LiBr are considered to be homogeneous within each component.
  ii. Pressure losses within the components are neglected, thus there are two pressure levels in the cycle: $P_{high}(=P_{cond}=P_g)$ and $P_{low}(=P_{abs}=P_e)$.
  iii. Power consumption of the pump is neglected.
  iv. No heat losses to the environment are considered.
  v. The expansion devices are adiabatic.
  vi. No changes in kinetic and potential energy are considered.
  vii. The exit from the condenser is saturated liquid.
  viii. The exit from the evaporator is saturated vapor.

5.3.1.1 Generator

The solution of each separate component consists of a system of three main balances: a mass balance, a volume balance, and an enthalpy balance (Evola et al. 2013; Xu et al. 2016). The mass balance consists of the following equations for the solution and the vapor:

$$\frac{d\,M_{sol,g}}{dt} = \dot{m}_{sol,g,in} - \dot{m}_{sol,g,out}\ \dot{m}_{vap,g,des} \tag{5.1}$$

$$\frac{d\,M_{vap,g}}{dt} = \dot{m}_{vap,des,g} - \dot{m}_{vap,g,out} \tag{5.2}$$

The solution concentrations can be determined by a mass balance for the LiBr in the generator, as follows:

$$x_{sol,g}\frac{d\,M_{sol,g}}{dt} + M_{sol,g}\frac{d\,x_{sol,g}}{dt} = \dot{m}_{sol,g,in}x_{sol,g,in} - \dot{m}_{sol,g,out}x_{sol,g,out} \tag{5.3}$$

where $\dot{m}_{vap,g,des}$ stands for the mass flow rate of the desorbed vapor, $M_{vap,g}$ stands for the total mass of vapor inside the generator, and $M_{sol,g}$ is the total mass of the solution in the generator. The volume balance for the generator is expressed by Formula (5.4):

$$V_g = \frac{M_{sol,g}}{\rho_{sol,g}} + \frac{M_{vap,g}}{\rho_{vap,g}} \tag{5.4}$$

where V_g stands for the total inner volume of the generator. The enthalpy balance in the generator can be expressed as follows:

$$\frac{d(M_{sol,g}c_{p,sol,g}\overline{T}_g)}{dt} = \dot{m}_{sol,g,in}h_{sol,g,in} - \dot{m}_{sol,g,out}h_{sol,g,out} - \dot{m}_{vap,g,des}h_{vap,g,des} + \dot{Q}_g \tag{5.5}$$

$$\dot{Q}_g = (aA)_g\,\Delta T_{lm,g} \tag{5.6}$$

where aA is the overall heat transfer coefficient of the generator (W/K), and ΔT_{lm} refers to the mean logarithmic temperature difference of the two streams in the generator. The respective enthalpy balance from the driving heat side can be simplified using the logarithmic mean temperature difference (LMTD) method, as shown below (in terms of the subscripts, it is considered that the heat transfer fluid is hot water, thus the subscript $_{hw}$):

$$T_{hw,out} = \overline{T}_g + (T_{hw,in} - \overline{T}_g)\exp\left[\frac{(aA)_g}{\dot{m}_{hw}c_{p,hw}}\right] \tag{5.7}$$

5.3.1.2 Absorber

The aforementioned balances for the generator can also be easily applied also for the case of the absorber with proper modifications to the inlet and outlet streams. For instance, the mass and volume balances are presented below:

$$\frac{d\,M_{sol,abs}}{dt} = \dot{m}_{sol,abs,in} - \dot{m}_{sol,abs,out} + \dot{m}_{vap,abs} \tag{5.8}$$

$$x_{sol,abs} \frac{d\, M_{sol,abs}}{dt} + M_{sol,abs} \frac{d\, x_{sol,abs}}{dt} = \dot{m}_{sol,abs,in} x_{sol,abs,in} - \dot{m}_{sol,abs,out} x_{aol,abs,out} \qquad (5.9)$$

$$V_{abs} = \frac{M_{sol,abs}}{\rho_{sol,abs}} + \frac{M_{vap,abs}}{\rho_{vap,abs}} \qquad (5.10)$$

where $\dot{m}_{vap,abs}$ stands for the mass flow rate of the absorbed vapor, Mvap,abs stands for the total mass of vapor inside the absorber, $M_{sol,abs}$ is the total mass of solution in the absorber, and V_{abs} is the total inner volume of the absorber. On the other hand, the enthalpy balances for the cooling water (cw) and the solution can be expressed as follows:

$$\frac{d(U_{sol,abs})}{dt} = \dot{m}_{sol,abs,in} h_{sol,abs,in} - \dot{m}_{sol,abs,out} h_{sol,abs,out} + \dot{m}_{vap,abs} h_{vap,abs} - \dot{Q}_{abs} \qquad (5.11)$$

$$T_{cw,abs,out} = \overline{T}_{abs} + (T_{cw,abs,in} - \overline{T}_{abs}) \exp\left[-\frac{(aA)_{abs}}{\dot{m}_{cw,abs} c_{p,cw,abs}} \right] \qquad (5.12)$$

$$\dot{Q}_{abs} = (aA)_{abs} \Delta T_{lm,abs} \qquad (5.13)$$

5.3.1.3 Condenser

In a similar way, the vapor and liquid refrigerant mass, volume, and enthalpy balances can be expressed for the case of the condenser:

$$\frac{d\, M_{l,cond}}{dt} = \dot{m}_{l,cond} - \dot{m}_{l,out,cond} \qquad (5.14)$$

$$\frac{d\, M_{vap,cond}}{dt} = \dot{m}_{vap,in,cond} - \dot{m}_{l,cond} \qquad (5.15)$$

$$V_{cond} = \frac{M_{l,cond}}{\rho_{l,cond}} + \frac{Mvap,cond}{\rho_{vap,cond}} \qquad (5.16)$$

$$\frac{d(U_{l,cond})}{dt} = \dot{m}_{vap,in,cond} h_{vap,in,cond} - \dot{m}_{l,out,cond} h_{l,out,cond} - \dot{Q}_{cond} \qquad (5.17)$$

$$\dot{Q}_{cond} = (aA)_{cond} \Delta T_{lm,cond} \tag{5.18}$$

As in the cases of the absorber and the generator, the respective enthalpy balance from the cooling water side can be simplified using the logarithmic mean temperature difference method, as shown below (this analysis is considered a separate cooling stream for the absorber and the condenser):

$$T_{cw,cond,out} = \bar{T}_{cond} + (T_{cw,con,ind} - \bar{T}_{cond})\exp\left[-\frac{(aA)_{cond}}{\dot{m}_{cw,cond}C_{p,cw,cond}}\right] \tag{5.19}$$

5.3.1.4 Evaporator

The case of the evaporator is very similar to that of the condenser, with just a proper modification of the inlet and outlet streams. The mass and volume balances are presented below:

$$\frac{d\,M_{l,e}}{dt} = \dot{m}_{l,in,e} - \dot{m}_{l,e} \tag{5.20}$$

$$\frac{d\,M_{vap,e}}{dt} = \dot{m}_{vap,e} - \dot{m}_{vap,out,e} \tag{5.21}$$

$$V_e = \frac{M_{l,e}}{\rho_{l,e}} + \frac{M_{vap,e}}{\rho_{vap,e}} \tag{5.22}$$

The respective enthalpy balance for the refrigerant is shown in Eq. (5.23):

$$\frac{d(U_{vap,e})}{dt} = \dot{m}_{l,in,e}h_{l,in,e} - \dot{m}_{vap,out,e}h_{vap,out,e} - \dot{Q}e \tag{5.23}$$

$$\dot{Q}_e = (aA)_e \Delta T_{lm,e} \tag{5.24}$$

The chiller water temperature at the outlet is calculated using the logarithmic mean temperature difference method, as shown below (Bergman and Incropera 2011):

$$T_{chw,out} = \bar{T}_e + (T_{chw,in} - \bar{T}_e)\exp\left[-\frac{(aA)_e}{\dot{m}_{chw}C_{p,chw}}\right] \tag{5.25}$$

5.3.1.5 Solution Heat Exchanger

For an H$_2$O-LiBr absorption chiller, the use of a solution heat exchanger is necessary, as already discussed. The solution heat exchanger is simply modeled as a conventional heat exchanger, thus only energy balances are required for each stream (Marc et al. 2015, Xu et al. 2016):

$$\frac{d(U_{h,i})}{dt} = \dot{m}_h (h_{h,i} - h_{h,out}) - \dot{Q}_{shex} \tag{5.26}$$

$$\frac{d(U_{c,i})}{dt} = \dot{m}_c (h_{c,i} - h_{c,out}) - \dot{Q}_{shex} \tag{5.27}$$

$$\dot{Q}_{shex} = (aA)_{shex} \Delta T_{lm,shex} \tag{5.28}$$

5.3.1.6 Heat Transfer Considerations

A deciding factor in the accuracy of the model is the selection of the heat transfer correlations used for the determination of the heat transfer coefficient for each separate heat exchanger in the absorption chiller. Depending on the flow, several different correlations have been proposed in literature (Bejan and Kraus 2003; Chemieingenieurwesen 2010; Bergman and Incropera 2011). Table 5.13 presents some formulas that were already evaluated for relevant systems.

5.3.1.7 System Pressures

System pressures can be determined either by using a thermodynamic library for the mixture, such as RefProp (Lemmon et al. 2010), or by applying the Clausius-Clapeyron equation (Ochoa et al. 2016) in the case of the condenser and the evaporator (between time i and time $i-1$):

$$\ln \frac{P_{cond,i}}{P_{cond,i-1}} = \frac{h_{fg}}{R} \left(\frac{1}{T_{cond,i-1}} - \frac{1}{T_{cond,i}} \right) \tag{5.35}$$

$$\ln \frac{P_{evap,i}}{P_{evap,i-1}} = \frac{h_{fg}}{R} \left(\frac{1}{T_{evap,i-1}} - \frac{1}{T_{evap,i}} \right) \tag{5.36}$$

TABLE 5.13

Heat Transfer Correlations Used in Absorption Chiller Modeling

Single-Phase Heat Transfer		Source
$Nu = c\,\mathrm{Re}^{0.8}\,\mathrm{Pr}^{d}\begin{cases} heating:\ c = 0.0243\ d = 0.4 \\ cooling:\ c = 0.0265\ d = 0.3 \end{cases}$	(5.29)	(Winterton 1997)
$Nu = \dfrac{(f/8)(\mathrm{Re}-1000)\mathrm{Pr}}{1+12.7\sqrt{f/8}\ (\mathrm{Pr}^{2/3}-1)},\ f = \left[0.079\ln(\mathrm{Re})-1.64\right]^{-2}$	(5.30)	(Gnielinski 2013)
Absorber		
$a_{abs} = \dfrac{\lambda_{sol}}{\delta}\left[0.029\left(\dfrac{4\Gamma}{\mu}\right)\mathrm{Pr}^{0.344}\right]$	(5.31)	(Ochoa et al. 2016)
Generator		
$a_{g} = 5554.3 \cdot \Gamma^{0.236}$	(5.32)	(Ochoa et al. 2016)
Boiling		
$a_{boil} = \dfrac{0.62\lambda_{sol}}{d_{tube}}\left[\dfrac{g(\rho_{l}-\rho_{v})\lambda_{l}h'_{fg}\,d_{tube}^{3}}{\nu_{v}\lambda_{v}(T_{surf}-T_{sat})}\right]$	(5.33)	(Marc et al. 2015)
Condensation		
$a_{cond} = 0.729\dfrac{\lambda_{l}}{d_{tube}}\left(\dfrac{Ra_{l}}{Ja_{l}}\right)^{1/4}$	(5.34)	(Taborek et al. 1983, Chemieingenieurwesen 2010)

5.3.1.8 Overall Masses

The mass flow rate of the strong solution can be calculated with the following expression:

$$\dot{m}_{ssol} = Cd_{g}A_{p,g}\sqrt{\frac{2\rho_{ssol}(P_{cond} - P_{e} + \rho_{ssol}gZ_{g})}{\xi}} \tag{5.37}$$

$$Z_{g} = \frac{M_{sol,g}}{\rho_{ssol}A_{p,g}} \tag{5.38}$$

where Z_{g} stands for the level of liquid inside the generator, $A_{p,g}$ is the surface of the pipe section, the pressure loss coefficient ξ is usually considered equal to 0.6, and the discharge coefficient Cd_{g} is also assumed to be equal to 0.6.

Nomenclature

A	Surface	[m²]
a	Overall heat transfer coefficient	[W m⁻²K⁻¹]
Cd$_g$	Discharge coefficient	–
c_p	Specific heat capacity	[J kg⁻¹K⁻¹]
d_{tube}	Tube diameter	[m]
f	Darcy friction factor	–
h	Enthalpy	[J kg⁻¹]
h_{fg}	Heat of vaporization	[J kg⁻¹]
Ja	Jakob number	–
M	Total mass	[kg]
$\dot{m}$	Mass flow	[kg s⁻¹]
Nu	Nusselt number	–
P	Pressure	[Pa]
Pr	Prandtl number	–
$\dot{Q}$	Heat flux	[W]
R	Gas constant	[J/kg.K]
Ra	Rayleigh number	–
Re	Reynolds number	–
T	Temperature	[K]
$\bar{T}$	Average temperature	[K]
T_{sat}	Saturation temperature	[K]
T_{surf}	Surface temperature	[K]
t	Time	[s]
U	Internal energy	[J]
V	Volume	[m³]
x	Mass fraction of refrigerant	[kg kg$_{sol}$⁻¹]
$x_{sol,g}$	Mass fraction of refrigerant left in the generator	[kg kg$_{sol}$⁻¹]
Z_g	Level of liquid inside the generator	[m]
ΔT_{lm}	Mean logarithmic temperature	[K]

Greek Symbols

δ	Thermal layer	[m]
θ	Temperature	[°C]
η	Second law efficiency	–
λ	Thermal conductivity	[W m⁻¹ K⁻¹]

μ	Dynamic viscosity	[kg m^{-1} s^{-1}]
ν	Kinematic viscosity	[m^2 s^{-1}]
ξ	Pressure loss coefficient	–
ρ	Density	[kg m^{-3}]

Subscripts

a	Ambient
abs	Absorber
ads	Adsorber
boil	Boiling
c	Cold side
chw	Chilled water
cond	Condenser
cw,abs	Cooling water supplied in absorber
cw,cond	Cooling water supplied in condenser
des	Desorbed
e	Evaporator
g	Generator
h	Hot side
hw	Hot water
in	Inlet
l	Liquid
out	Outlet
p,g	Pipe section of generator
shex	Solution heat exchanger
sol	Solution
ssol	Strong solution
vap	Vapor

Abbreviations

AHX	Absorption heat exchanger
ANN	Artificial neural network
COP	Coefficient of performance
COP$_{carnot}$	Carnot's coefficient of performance
CPC	Compound parabolic concentrator
CRES	Center for Renewable Energy Sources and Saving
CV	Coefficient of variation
DMAC	N,N-dimethylacetamide
DMETEG	Dimethyl ether of tetraethylene glycol
DMF	Dimethyl formamide

EER	Energy efficiency ratio
GAX	Generator-absorber exchange (absorption cycle)
GNA	Gordon-Ng model
HE	Half-effect cycle
HVAC	Heating, ventilation and air conditioning
LPG	Liquefied petroleum gas
MPR	Multivariate polynomial regression
PEMFC	Proton exchange membrane fuel cell
PTC	Parabolic trough collectors
SE	Single-effect cycle
XCPC	External compound parabolic concentrators

References

Abdulateef, Jasim, Mohammed Alghoul, Azami Zaharim, and Kamaruzzaman Sopian. 2009. "Experimental Investigation on Solar Absorption Refrigeration System in Malaysia." Proceedings of the 3rd Wseas Int. Conf. On Renewable Energy Sources.

Abdullah, Gazinga F., Wasim Saman, David Whaley, and Martin Belusko. 2016. "Optimization of Standalone Solar Heat Fired Absorption Chiller for Typical Australian Homes." *Energy Procedia* 91: 692–701.

Achuthan, M., A. Venkataraman, and R. Rathnasamy. 2011. "Experimental Analysis on the Performance and Characteristics of Compact Solar Refrigeration System." *Distributed Generation & Alternative Energy Journal* 26 (3): 66–80.

Agyenim, Francis, Ian Knight, and Michael Rhodes. 2010. "Design and Experimental Testing of the Performance of an Outdoor Libr/H2O Solar Thermal Absorption Cooling System with a Cold Store." *Solar Energy* 84 (5): 735–744.

Al-Alili, A., Y. Hwang, R. Radermacher, and I. Kubo. 2010. "Optimization of a Solar Powered Absorption Cycle Under Abu Dhabi's Weather Conditions." *Solar Energy* 84 (12): 2034–2040.

Al-Karaghouli, A., I. Abood, and N. I. Al-Hamdani. 1991. "The Solar Energy Research Center Building Thermal Performance Evaluation During the Summer Season." *Energy Conversion and Management* 32 (5): 409–417.

Albers, Jan. 2014. "New Absorption Chiller and Control Strategy for the Solar Assisted Cooling System at the German Federal Environment Agency." *International Journal of Refrigeration* 39: 48–56.

Ali, Ahmed Hamza H., Peter Noeres, and Clemens Pollerberg. 2008. "Performance Assessment of an Integrated Free Cooling and Solar Powered Single-Effect Lithium Bromide-Water Absorption Chiller." *Solar Energy* 82 (11): 1021–1030.

Allouhi, A., T. Kousksou, A. Jamil, P. Bruel, Y. Mourad, and Y. Zeraouli. 2015. "Solar Driven Cooling Systems: An Updated Review." *Renewable and Sustainable Energy Reviews* 44 (Supplement C): 159–181.

AlQdah, Khaled S. 2011. "Performance and Evaluation of Aqua Ammonia Auto Air Conditioner System Using Exhaust Waste Energy." *Energy Procedia* 6 (Supplement C): 467–476.

Alsaqoor, Sameh, and Khaled S AlQdah. 2014. "Performance of a Refrigeration Absorption Cycle Driven by Different Power Sources." *Smart Grid and Renewable Energy* 5 (07): 161–169.

Aman, J., D. S. K. Ting, and P. Henshaw. 2014. "Residential Solar Air Conditioning: Energy and Exergy Analyses of an Ammonia–Water Absorption Cooling System." *Applied Thermal Engineering* 62 (2): 424–432.

Aphornratana, Satha. 2005. "Research on Absorption Refrigerators and Heat Pumps." *International Energy Journal* 17 (1): 1–19.

Arivazhagan, S., R. Saravanan, and S. Renganarayanan. 2006. "Experimental Studies on HFC Based Two-Stage Half Effect Vapour Absorption Cooling System." *Applied Thermal Engineering* 26 (14): 1455–1462.

Asdrubali, F, and S Grignaffini. 2005. "Experimental Evaluation of the Performances of a H2O–Libr Absorption Refrigerator Under Different Service Conditions." *International Journal of Refrigeration* 28 (4): 489–497.

Assilzadeh, F., S. A. Kalogirou, Y. Ali, and K. Sopian. 2005. "Simulation and Optimization of A Libr Solar Absorption Cooling System with Evacuated Tube Collectors." *Renewable Energy* 30 (8): 1143–1159.

Ayadi, Osama, Alberto Mauro, Marcello Aprile, and Mario Motta. 2012. "Performance Assessment for Solar Heating and Cooling System for Office Building in Italy." *Energy Procedia* 30 (Supplement C): 490–494.

Balamuru, Vinay G., Osama M. Ibrahim, and Stanley M. Barnett. 2000. "Simulation of Ternary Ammonia–Water–Salt Absorption Refrigeration Cycles." *International Journal of Refrigeration* 23 (1): 31–42.

Balghouthi, M., M. H. Chahbani, and A. Guizani. 2012. "Investigation of a Solar Cooling Installation in Tunisia." *Applied Energy* 98 (Supplement C): 138–148.

Balghouthi, Moncef, Mohamed Hachemi Chahbani, and Amenallah Guizani. 2005. "Solar Powered Air Conditioning as a Solution to Reduce Environmental Pollution in Tunisia." *Desalination* 185 (1): 105–110.

Beausoleil-Morrison, Ian, Geoffrey Johnson, and Briana Paige Kemery. 2015. "The Experimental Characterization of a Lithium Bromide–Water Absorption Chiller and the Development of a Calibrated Model." *Solar Energy* 122: 368–381.

Beccali, Marco, Maurizio Cellura, Sonia Longo, Bettina Nocke, and Pietro Finocchiaro. 2012. "LCA of a Solar Heating and Cooling System Equipped with a Small Water–Ammonia Absorption Chiller." *Solar Energy* 86 (5): 1491–1503.

Bejan, Adrian, and Allan Kraus. 2003. *Heat Transfer Handbook*. Hoboken, NJ: Wiley.

Bell, I. A., A. J. Al-Daini, Habib Al-Ali, R. G. Abdel-Gayed, and L. Duckers. 1996. "The Design of an Evaporator/Absorber and Thermodynamic Analysis of a Vapor Absorption Chiller Driven by Solar Energy." *Renewable Energy* 9 (1): 657–660.

Bendaikha, Wahiba, and Salah Larbi. 2012. "Hybrid Fuel Cell and Geothermal Resources for Air-Conditioning Using an Absorption Chiller in Algeria." *Energy Procedia* 28: 190–197.

Bergman, T. L., and Frank P. Incropera. 2011. *Fundamentals of Heat and Mass Transfer*. 7th ed. Hoboken, NJ: Wiley.

Bermejo, Pablo, Francisco Javier Pino, and Felipe Rosa. 2010. "Solar Absorption Cooling Plant in Seville." *Solar Energy* 84 (8): 1503–1512.

Best, R., and N. Ortega. 1999. "Solar Refrigeration and Cooling." *Renewable Energy* 16 (1): 685–690.

Blinn, JC, JW Mitchell, and JA Duffie. 1979. *Modeling of Transient Performance of Residential Solar Air Conditioning Systems*. Wisconsin Univ., Madison (USA). Solar Energy Lab.

Bolocan, Sorin, Florea Chiriac, Alexandru Serban, George Dragomir, and Gabriel Nastase. 2015. "Development of a Small Capacity Solar Cooling Absorption Plant." *Energy Procedia* 74: 624–632.

Bong, T. Y., K. C. Ng, and A. O. Tay. 1987. "Performance Study of a Solar-Powered Air-Conditioning System." *Solar Energy* 39 (3): 173–182.

Boudéhenn, François, Sylvain Bonnot, Hélène Demasles, Florent Lefrançois, Maxime Perier-Muzet, and Delphine Triché. 2016. "Development and Performances Overview of Ammonia-Water Absorption Chillers with Cooling Capacities from 5 to 100 kW." *Energy Procedia* 91: 707–716.

Bruno, J. C., J. Miquel, and F. Castells. 1999. "Modeling of Ammonia Absorption Chillers Integration in Energy Systems of Process Plants." *Applied Thermal Engineering* 19 (12): 1297–1328.

Bujedo, Luis A., Juan Rodríguez, and Pedro J. Martínez. 2011. "Experimental Results of Different Control Strategies in a Solar Air-Conditioning System At Part Load." *Solar Energy* 85 (7): 1302–1315.

Caciula, B., V. Popa, and L. Costiuc. 2013. "Theoretical Study on Solar Powered Absorption Cooling System." *Termotehnica* 1: 130–134.

Calise, Francesco. 2012. "High Temperature Solar Heating And Cooling Systems for Different Mediterranean Climates: Dynamic Simulation and Economic Assessment." *Applied Thermal Engineering* 32 (Supplement C): 108–124.

Cetinkaya, Balkan, Richard Cuthbertson, Graham Ewer, Thorsten Klaas-Wissing, Wojciech Piotrowicz, and Christoph Tyssen. 2011. *Sustainable Supply Chain Management: Practical Ideas for Moving Toward Best Practice*. Berlin: Springer Science & Business Media.

Chemieingenieurwesen, V. DI-Gesellschaft Verfahrenstechnik und. 2010. "VDI Heat Atlas." Berlin: Springer.

Chen, Guangming, and Eiji Hihara. 1999. "A New Absorption Refrigeration Cycle Using Solar Energy." *Solar Energy* 66 (6): 479–482.

Dalenbäck, Jan-Olof. 2009. "Large-Scale Solar Heating and Cooling Systems in Europe." In *Proceedings of ISES World Congress 2007 (Vol. I–Vol. V): Solar Energy and Human Settlement*, edited by D. Yogi Goswami and Yuwen Zhao, 799–803. Berlin, Heidelberg: Springer Berlin Heidelberg.

Dan, Phuong Dung, and S Srinivasa Murthy. 1989. "A Comparative Thermodynamic Study of Fluorocarbon Refrigerant Based Vapour Absorption Heat Pumps." *International Journal of Energy Research* 13 (1): 1–21.

Darkwa, J., S. Fraser, and D. H. C. Chow. 2012. "Theoretical and Practical Analysis of An Integrated Solar Hot Water-Powered Absorption Cooling System." *Energy* 39 (1): 395–402.

De Francisco, A., R. Illanes, J. L. Torres, M. Castillo, M. De Blas, E. Prieto, Garcı, x, and A. a. 2002. "Development and Testing of a Prototype of Low-Power Water–Ammonia Absorption Equipment for Solar Energy Applications." *Renewable Energy* 25 (4): 537–544.

De Lucas, Antonio, Marina Donate, Carolina Molero, José Villaseñor, and Juan F. Rodríguez. 2004. "Performance Evaluation and Simulation of a New Absorbent for an Absorption Refrigeration System." *International Journal of Refrigeration* 27 (4): 324–330.

Dinçer, İbrahim. 2003. *Refrigeration Systems and Applications*. Chichester, West Sussex, England: Wiley.

Drosou, Vassiliki, Panos Kosmopoulos, and Agis Papadopoulos. 2016. "Solar Cooling System Using Concentrating Collectors for Office Buildings: A Case Study for Greece." *Renewable Energy* 97 (Supplement C): 697–708.

Drosou, Vassiliki N., Panagiotis D. Tsekouras, Th I. Oikonomou, Panos I. Kosmopoulos, and Constantine S. Karytsas. 2014. "The HIGH-COMBI Project: High Solar Fraction Heating and Cooling Systems with Combination of Innovative Components and Methods." *Renewable and Sustainable Energy Reviews* 29 (Supplement C): 463–472.

Eicker, Ursula. 2009. *Low Energy Cooling for Sustainable Buildings*. Oxford: Wiley-Blackwell.

Eicker, Ursula. 2014. *Energy Efficient Buildings with Solar and Geothermal Resources*. United Kingdom: Wiley.

Eicker, Ursula, and Dirk Pietruschka. 2009. "Design and Performance of Solar Powered Absorption Cooling Systems in Office Buildings." *Energy and Buildings* 41 (1): 81–91.

Elsafty, A., and A. J. Al-Daini. 2002. "Economical Comparison between a Solar-Powered Vapour Absorption Air-Conditioning System and a Vapour Compression System in the Middle East." *Renewable Energy* 25 (4): 569–583.

Erickson, D. C. 2007. "100 Ton Absorption Chiller/Heat Pump Demonstrates the Real Cost of Saving Energy." *ASHRAE Transactions* 113 (2): 90–94.

Esposito, F., A. Dolci, G. Ferrara, L. Ferrari, and E. A. Carnevale. 2015. "A Case Study Based Comparison between Solar Thermal and Solar Electric Cooling." *Energy Procedia* 81: 1160–1170.

Evola, G., N. Le Pierrès, F. Boudehenn, and P. Papillon. 2013. "Proposal and Validation of a Model for the Dynamic Simulation of a Solar-Assisted Single-Stage Libr/Water Absorption Chiller." *International Journal of Refrigeration* 36 (3): 1015–1028.

Ezzine, N. Ben, M. Barhoumi, Kh Mejbri, S. Chemkhi, and A. Bellagi. 2004. "Solar Cooling with the Absorption Principle: First And Second Law Analysis of an Ammonia—Water Double-Generator Absorption Chiller." *Desalination* 168 (Supplement C): 137–144.

Figueredo, Gustavo R., Mahmoud Bourouis, and Alberto Coronas. 2008. "Thermodynamic Modelling of a Two-Stage Absorption Chiller Driven at Two-Temperature Levels." *Applied Thermal Engineering* 28 (2): 211–217.

Florides, G. A., S. A. Kalogirou, S. A. Tassou, and L. C. Wrobel. 2002. "Modelling and Simulation of an Absorption Solar Cooling System For Cyprus." *Solar Energy* 72 (1): 43–51.

Franchini, Giuseppe, Ettore Notarbartolo, Luca E. Padovan, and Antonio Perdichizzi. 2015. "Modeling, Design and Construction of a Micro-scale Absorption Chiller." *Energy Procedia* 82: 577–583.

Gebreslassie, Berhane H., Marc Medrano, and Dieter Boer. 2010. "Exergy Analysis of Multi-Effect Water–Libr Absorption Systems: From Half to Triple Effect." *Renewable Energy* 35 (8): 1773–1782.

Ghaddar, N. K., M. Shihab, and F. Bdeir. 1997. "Modeling and Simulation of Solar Absorption System Performance in Beirut." *Renewable Energy* 10 (4): 539–558.

Gnielinski, V. 2013. "On Heat Transfer in Tubes." *International Journal of Heat and Mass Transfer* 63: 134–140.

Gomri, Rabah. 2010. "Investigation of the Potential of Application of Single Effect and Multiple Effect Absorption Cooling Systems." *Energy Conversion and Management* 51 (8): 1629–1636.

Gong, Sunyoung, and Kiari Goni Boulama. 2015. "Advanced Exergy Analysis of an Absorption Cooling Machine: Effects of the Difference between the Condensation and Absorption Temperatures." *International Journal of Refrigeration* 59: 224–234.

González-Gil, A., M. Izquierdo, J. D. Marcos, and E. Palacios. 2011. "Experimental Evaluation of A Direct Air-Cooled Lithium Bromide–Water Absorption Prototype For Solar Air Conditioning." *Applied Thermal Engineering* 31 (16): 3358–3368.

Gordon, J. M., and Kim Choon Ng. 1995. "A General Thermodynamic Model for Absorption Chillers: Theory and Experiment." *Heat Recovery Systems and CHP* 15 (1): 73–83.

Grossman, Gershon. 2002. "Solar-Powered Systems for Cooling, Dehumidification and Air-Conditioning." *Solar Energy* 72 (1): 53–62.

Hammad, M., and Y. Zurigat. 1998. "Performance of a Second Generation Solar Cooling Unit." *Solar Energy* 62 (2): 79–84.

Hawlader, M. N. A., K. S. Novak, and B. D. Wood. 1993. "Unglazed Collector/Regenerator Performance for a Solar Assisted Open Cycle Absorption Cooling System." *Solar Energy* 50 (1): 59–73.

Helm, M., C. Keil, S. Hiebler, H. Mehling, and C. Schweigler. 2009. "Solar Heating and Cooling System with Absorption Chiller and Low Temperature Latent Heat Storage: Energetic Performance and Operational Experience." *International Journal of Refrigeration* 32 (4): 596–606.

Henning, Hans-Martin. 2007. "Solar Assisted Air Conditioning of Buildings—An Overview." *Applied Thermal Engineering* 27 (10): 1734–1749.

Herold, Keith E. 1995. "Design Challenges in Absorption Chillers." *Mechanical Engineering* 117 (10): 80.

Herold, Keith E., Reinhard Radermacher, Lawrence Howe, and Donald C. Erickson. 1991. "Development of an Absorption Heat Pump Water Heater Using An Aqueous Ternary Hydroxide Working Fluid." *International Journal of Refrigeration* 14 (3): 156–167.

Hidalgo, M. C. Rodríguez, P. Rodríguez Aumente, M. Izquierdo Millán, A. Lecuona Neumann, and R. Salgado Mangual. 2008. "Energy and Carbon Emission Savings in Spanish Housing Air-Conditioning Using Solar Driven Absorption System." *Applied Thermal Engineering* 28 (14): 1734–1744.

Jakob, Uli. 2016. "Solar Cooling Technologies." In *Renewable Heating and Cooling*, 119–136. Woodhead Publishing.

Janzen, A. F., R. K. Swartman, and International Symposium on Solar Energy. 1981. *Solar Energy Conversion II: Selected Lectures from the 1980 International Symposium On Solar Energy Utilization Held In London (Ontario), August 1980.* Toronto: Pergamon Press.

Jaruwongwittaya, Tawatchai, and Guangming Chen. 2010. "A Review: Renewable Energy with Absorption Chillers in Thailand." *Renewable and Sustainable Energy Reviews* 14 (5): 1437–1444.

Jawahar, C. P., and R. Saravanan. 2010. "Generator Absorber GHeat Exchange Based Absorption Cycle—A Review." *Renewable and Sustainable Energy Reviews* 14 (8): 2372–2382.

Kalogirou, Soteris A. 2014. *Solar Energy Engineering: Processes and Systems.* Amsterdam: Elsevier/ Academic Press.

Karagiorgas, M., P. Kouretzi, L. Kodokalou, and Lamaris. P. 2007. "Operation and measurement results of the solar cooling installation in Rethymnon village hotel." 28th AIVC Conference on Building Low Energy Cooling and Advanced Ventilation Technologies in the 21st Century. Crete, Greece, September 2007.

Kaushik, S. C., and Akhilesh Arora. 2009. "EnergyaAnd Exergy Analysis of Single Effect and Series Flow Double Effect Water–Lithium Bromide Absorption Refrigeration Systems." *International Journal of Refrigeration* 32 (6): 1247–1258.

Kaynakli, Omer, and Muhsin Kilic. 2007. "Theoretical Study on the Effect of Operating Conditions on Performance of Absorption Refrigeration System." *Energy Conversion and Management* 48 (2): 599–607.

Kaynakli, Omer, and Recep Yamankaradeniz. 2007. "Thermodynamic Analysis of Absorption Refrigeration System based on Entropy Generation." *Current Science*: 472–479.

Ketfi, O., M. Merzouk, N. Kasbadji Merzouk, and S. El Metenani. 2015. "Performance of a Single Effect Solar Absorption Cooling System (Libr-H2O)." *Energy Procedia* 74: 130–138.

Kim, Byongjoo, and Jongil Park. 2007. "Dynamic Simulation of a Single-Effect Ammonia–Water Absorption Chiller." *International Journal of Refrigeration* 30 (3): 535–545.

Kim, D. S., and C. A. Infante Ferreira. 2009. "Air-Cooled Libr–Water Absorption Chillers for Solar Air Conditioning in Extremely Hot Weathers." *Energy Conversion and Management* 50 (4): 1018–1025.

Kim, D. S., H. Van der Ree, and C. A. Infante Ferreira. 2007. "Solar Absorption Cooling." PhD Dissertation, Delft.

Kohlenbach, P., and F. Ziegler. 2008. "A Dynamic Simulation Model for Transient Absorption Chiller Performance. Part I: The Model." *International Journal of Refrigeration* 31 (2): 217–225.

Koroneos, C., E. Nanaki, and G. Xydis. 2010. "Solar Air Conditioning Systems and their Applicability— An Exergy Approach." *Resources, Conservation and Recycling* 55 (1): 74–82.

Kurem, E., and I. Horuz. 2001. "A Comparison between Ammonia-Water and Water-Lithium Bromide Solutions in Absorption Heat Transformers." *International Communications in Heat and Mass Transfer* 28 (3): 427–438.

Labus, Jerko, Joan Carles Bruno, and Alberto Coronas. 2013. "Performance Analysis of Small Capacity Absorption Chillers by Using Different Modeling Methods." *Applied Thermal Engineering* 58 (1–2): 305–313.

Lamine, Chougui Mohamed, and Zid Said. 2014. "Energy Analysis of Single Effect Absorption Chiller (LiBr/H2O) in an Industrial Manufacturing of Detergent." *Energy Procedia* 50: 105–112.

Lazzarin, Renato M. 2014. "Solar Cooling: PV or Thermal? A Thermodynamic and Economical Analysis." *International Journal of Refrigeration* 39: 38–47.

Le Lostec, Brice, Nicolas Galanis, and Jocelyn Millette. 2012. "Experimental Study of An Ammonia-Water Absorption Chiller." *International Journal of Refrigeration* 35 (8): 2275–2286.

Le Lostec, Brice, Nicolas Galanis, and Jocelyn Millette. 2013. "Simulation of an Ammonia–Water Absorption Chiller." *Renewable Energy* 60 (Supplement C): 269–283.

Refprop database.

Li, J. H., W. B. Ma, Qing Jiang, Z. C. Huang, and W. H. XIa. 1999. "A 100kW Solar Air-Conditioning System." *Acta Energiae Solaris Sinica* 20 (3): 239–243.

Lizarte, R., M. Izquierdo, J. D. Marcos, and E. Palacios. 2012. "An Innovative Solar-Driven Directly Air-Cooled Libr–H2O Absorption Chiller Prototype for Residential Use." *Energy and Buildings* 47 (Supplement C): 1–11.

Lokurlu, A., and G. Müller. 2005. "Experiences with the Worldwide First Solar Cooling System Based on Trough Collectors Combined with Double Effect Absorption Chillers." International Conference Solar Air-Conditioning, Bad Staffelstein, Germany.

Lubis, Arnas, Jongsoo Jeong, Kiyoshi Saito, Niccolo Giannetti, Hajime Yabase, Muhammad Idrus Alhamid, and Nasruddin. 2016. "Solar-Assisted Single-Double-Effect Absorption Chiller for Use in Asian Tropical Climates." *Renewable Energy* 99: 825–835.

Martinez, A.R., 1981. "A Three And A Half Ton Solar Absorption Air Conditioner's Performance In Riyadh, Saudi Arabia." In *Solar Cooling and Dehumidifying*, 49. Pergamon Press.

Macriss, Robert A. 1978. *Analysis of Advanced Conceptual Designs for Single-Family-Size Absorption Chillers.* Department of Energy.

Maidment, G. G., and A. Paurine. 2012. "Solar Cooling and Refrigeration Systems." In *Comprehensive Renewable Energy*, 481–494. Oxford: Elsevier.

Mammoli, Andrea, Peter Vorobieff, Hans Barsun, Rick Burnett, and Daniel Fisher. 2010. "Energetic, Economic and Environmental Performance of a Solar-Thermal-Assisted HVAC System." *Energy and Buildings* 42 (9): 1524–1535.

Mansouri, Rami, Ismail Boukholda, Mahmoud Bourouis, and Ahmed Bellagi. 2015. "Modelling and Testing the Performance of a Commercial Ammonia/Water Absorption Chiller Using Aspen-Plus Platform." *Energy* 93 (Part 2): 2374–2383.

Marc, O., F. Lucas, F. Sinama, and E. Monceyron. 2010. "Experimental Investigation of a Solar Cooling Absorption System Operating Without Any Backup System Under Tropical Climate." *Energy and Buildings* 42 (6): 774–782.

Marc, Olivier, Frantz Sinama, Jean-Philippe Praene, Franck Lucas, and Jean Castaing-Lasvignottes. 2015. "Dynamic Modeling and Experimental Validation Elements of a 30 Kw Librh/H2O Single Effect Absorption Chiller for Solar Application." *Applied Thermal Engineering* 90: 980–993.

Matsushima, H., T. Fujii, T. Komatsu, and A. Nishiguchi. 2010. "Dynamic Simulation Program with Object-Oriented Formulation for Absorption Chillers (Modelling, Verification, and Application to Triple-Effect Absorption Chiller)." *International Journal of Refrigeration* 33 (2): 259–268.

Mazloumi, M., M. Naghashzadegan, and K. Javaherdeh. 2008. "Simulation of Solar Lithium Bromide–Water Absorption Cooling System with Parabolic Trough Collector." *Energy Conversion and Management* 49 (10): 2820–2832.

Monné, C., S. Alonso, F. Palacín, and J. Guallar. 2011. "Stationary Analysis of a solar LiBr–H2O Absorption Refrigeration System." *International Journal of Refrigeration* 34 (2): 518–526.

Moreno-Quintarar, G., W. Rivera, and R. Best. 2011. "Development of a Solar Intermittent Refrigeration System for Ice Production." World Renewable Energy Congress, Linköping, Sweden.

Nakahara, Nobuo, Yasuyuki Miyakawa, and Mitsunobu Yamamoto. 1977. "Experimental Study on House Cooling and Heating with Solar Energy Using Flat Plate Collector." *Solar Energy* 19 (6): 657–662.

Ochoa, A. A. V., J. C. C. Dutra, J. R. G. Henríquez, and C. A. C. dos Santos. 2016. "Dynamic Study of a Single Effect Absorption Chiller Using the Pair LiBr/H2O." *Energy Conversion and Management* 108: 30–42.

Onan, C., D. B. Ozkan, and S. Erdem. 2010. "Exergy Analysis of a Solar Assisted Absorption Cooling System on an Hourly Basis in Villa Applications." *Energy* 35 (12): 5277–5285.

Ortega, N., O. García-Valladares, R. Best, and V. H. Gómez. 2008. "Two-Phase Flow Modelling of a Solar Concentrator Applied as Ammonia Vapor Generator in an Absorption Refrigerator." *Renewable Energy* 33 (9): 2064–2076.

Ortiz, M., H. Barsun, H. He, P. Vorobieff, and A. Mammoli. 2010. "Modeling of a Solar-Assisted HVAC System with Thermal Storage." *Energy and Buildings* 42 (4): 500–509.

Ozgoren, Muammer, Mehmet Bilgili, and Osman Babayigit. 2012. "Hourly Performance Prediction of Ammonia–Water Solar Absorption Refrigeration." *Applied Thermal Engineering* 40 (Supplement C): 80–90.

Pilatowsky, I., W. Rivera, and J. R. Romero. 2004. "Performance Evaluation of a Monomethylamine—Water Solar Absorption Refrigeration System for Milk Cooling Purposes." *Applied Thermal Engineering* 24 (7): 1103–1115.

Pilatowsky, I., W. Rivera, and R. J. Romero. 2001. "Thermodynamic Analysis of Monomethylamine–Water Solutions in a Single-Stage Solar Absorption Refrigeration Cycle at Low Generator Temperatures." *Solar Energy Materials and Solar Cells* 70 (3): 287–300.

Pongtornkulpanich, A., S. Thepa, M. Amornkitbamrung, and C. Butcher. 2008. "Experience with fully operational solar-driven 10-ton LiBr/H2O single-effect absorption cooling system in Thailand." *Renewable Energy* 33 (5):943–949.

Porumb, Raluca, Bogdan Porumb, and Mugur Balan. 2017. "Numerical Investigation on Solar Absorption Chiller with LiBr-H2O Operating Conditions and Performances." *Energy Procedia* 112 (Supplement C): 108–117.

Prasartkaew, Boonrit. 2014. "Performance Test of a Small Size LiBr-H2O Absorption Chiller." *Energy Procedia* 56: 487–497.

Qu, Ming, Hongxi Yin, and David H. Archer. 2010. "A Solar Thermal Cooling and Heating System for a Building: Experimental and Model Based Performance Analysis and Design." *Solar Energy* 84 (2): 166–182.

Richter, L., and M. Safarik. 2005. "Solar Cooling with Ammonia Water Absorption Chillers." International Conference Solar Air Conditioning, Bad Staffelstein, Germany.

Rivera, C. O., and W. Rivera. 2003. "Modeling of an Intermittent Solar Absorption Refrigeration System Operating with Ammonia–Lithium Nitrate Mixture." *Solar Energy Materials and Solar Cells* 76 (3): 417–427.

Rosiek, S., and F. J. Batlles. 2009. "Integration of the Solar Thermal Energy in the Construction: Analysis of the Solar-Assisted Air-Conditioning System Installed in CIESOL Building." *Renewable Energy* 34 (6): 1423–1431.

Sarabia Escriva, Emilio J., Edwin V. Lamas Sivila, and Victor M. Soto Frances. 2011. "Air Conditioning Production by a Single Effect Absorption Cooling Machine Directly Coupled to a Solar Collector Field. Application to Spanish Climates." *Solar Energy* 85 (9): 2108–2121.

Sayigh, A. A. M., and Mehdi N. Bahadori. 1979. *Solar Energy Application in Buildings*. New York: Academic Press.

Schweigler, Christian, Stefan Hiebler, Christian Keil, Holger Köbel, Christoph Kren, and Harald Mehling. 2007. "Low-Temperature Heat Storage for Solar Heating and Cooling Applications." *ASHRAE Transactions* 113 (1). Dallas 2007.

Şencan, Arzu, Kemal A. Yakut, and Soteris A. Kalogirou. 2005. "Exergy Analysis of Lithium Bromide/Water Absorption Systems." *Renewable Energy* 30 (5): 645–657.

Shirazi, Ali, Robert A. Taylor, Stephen D. White, and Graham L. Morrison. 2016. "A Systematic Parametric Study and Feasibility Assessment of Solar-Assisted Single-Effect, Double-Effect, and Triple-Effect Absorption Chillers for Heating and Cooling Applications." *Energy Conversion and Management* 114: 258–277.

Sochard, Sabine, Lorenzo Castillo Garcia, Sylvain Serra, Yann Vitupier, and Jean-Michel Reneaume. 2017. "Modelling a Solar Absorption Chiller Using Positive Flash to Estimate the Physical State of Streams and Theoretical Plate Concept for the Generator." *Renewable Energy* 109 (Supplement C): 121–134.

SOLAIR. "GICB building in Banyuls sur Mer, France." Accessed October 2017. http://www.solair-project.eu/185.0.html.

SOLAIR. "ISI Pergine, Business Centre, Trento, Italy." Accessed Oct. 13, 2017. http://www.solair-project.eu/212.0.html.

Somers, C., A. Mortazavi, Y. Hwang, R. Radermacher, P. Rodgers, and S. Al-Hashimi. 2011. "Modeling Water/Lithium Bromide Absorption Chillers in ASPEN Plus." *Applied Energy* 88 (11): 4197–4205.

Sorour, M. M., and A. A. Ghoneim. 1994. "Feasibility Study of Solar Heating and Cooling Systems at Different Localities in Egypt." *Renewable Energy* 5 (1): 489–491.

Ssembatya, Martin, Manoj K. Pokhrel, and Rajesh Reddy. 2014. "Simulation Studies on Performance of Solar Cooling System in UAE Conditions." *Energy Procedia* 48 (Supplement C): 1007–1016.

Storkenmaier, F., M. Harm, C. Schweigler, F. Ziegler, P. Kohlenbach, and T. Sengewald. 2003. "Small-Capacity LiBr Absorption Chiller for Solar Cooling or Waste-Heat Driven Cooling." 30th International Congress of Refrigeration, Washington, USA.

Sumathy, K., Z. C. Huang, and Z. F. Li. 2002. "Solar Absorption Cooling with Low Grade Heat Source—A Strategy of Development in South China." *Solar Energy* 72 (2): 155–165.

Syed, A., M. Izquierdo, P. Rodríguez, G. Maidment, J. Missenden, A. Lecuona, and R. Tozer. 2005. "A Novel Experimental Investigation of a Solar Cooling System in Madrid." *International Journal of Refrigeration* 28 (6): 859–871.

Taborek, J., G. F. Hewitt, Naim Hamdia Afgan, Heat International Center for, and Transfer Mass. 1983. *Heat Exchangers: Theory and Practice*. New York: McGraw-Hill.

Tsoutsos, T., E. Aloumpi, Z. Gkouskos, and M. Karagiorgas. 2010. "Design of a Solar Absorption Cooling System in a Greek Hospital." *Energy and Buildings* 42 (2): 265–272.

Tsoutsos, Theocharis, Michaelis Karagiorgas, George Zidianakis, Vassiliki Drosou, Aris Aidonis, Zacharias Gouskos, and Costas Moeses. 2009. "Development of the Applications of Solar Thermal Cooling Systems in Greece and Cyprus." *Fresenius Environmental Bulletin* 18: 1367–1380.

Utham, Ganesh, Sagarkumar M Agravat, Bela Jani, and Jignasha Bhutka. 2016. "Modelling, Transient Simulation and Economic Analysis of Solar Thermal Based Air Conditioning System in Gujarat." *Smart Grid and Renewable Energy* 7 (08): 233–246.

Van Hattem, D., and P. Actis Dato. 1981. "Description and performance of an active solar cooling system, using a LiBr-$_{HO}$ Absorption Machine." *Energy and Buildings* 3 (2): 169–196.

Vargas, J. V. C., J. C. Ordonez, E. Dilay, and J. A. R. Parise. 2009. "Modeling, Simulation and Optimization of a Solar Collector Driven Water Heating and Absorption Cooling Plant." *Solar Energy* 83 (8): 1232–1244.

Vasilescu, Catalina, and Carlos Infante Ferreira. 2014. "Solar Driven Double-Effect Absorption Cycles for Sub-Zero Temperatures." *International Journal of Refrigeration* 39 (Supplement C): 86–94.

Velázquez, N., and R. Best. 2002. "Methodology for the Energy Analysis of an Air Cooled GAX Absorption Heat Pump Operated by Natural Gas and Solar Energy." *Applied Thermal Engineering* 22 (10): 1089–1103.

Wang, R. Z., T. S. Ge, C. J. Chen, Q. Ma, and Z. Q. Xiong. 2009. "Solar Sorption Cooling Systems for Residential Applications: Options and Guidelines." *International Journal of Refrigeration* 32 (4): 638–660.

Weber, Christine, Michael Berger, Florian Mehling, Alexander Heinrich, and Tomas Núñez. 2014. "Solar Cooling with Water–Ammonia Absorption Chillers and Concentrating Solar Collector—Operational Experience." *International Journal of Refrigeration* 39: 57–76.

Winston, Roland, Lun Jiang, and Bennett Widyolar. 2014. "Performance of a 23KW Solar Thermal Cooling System Employing a Double Effect Absorption Chiller and Thermodynamically Efficient Non-Tracking Concentrators." *Energy Procedia* 48: 1036–1046.

Winterton, R. H. S. 1997. *Heat Transfer*, no. 50 *Oxford Chemistry Primers*. Oxford: Oxford University Press.

Xu, Yu-jie, Shi-jie Zhang, and Yun-han Xiao. 2016. "Modeling the Dynamic Simulation and Control of a Single Effect LiBr–H2O Absorption Chiller." *Applied Thermal Engineering* 107 (Supplement C): 1183–1191.

Xu, Z. Y., R. Z. Wang, and H. B. Wang. 2015. "Experimental Evaluation of a Variable Effect LiBr–Water Absorption Chiller Designed for High-Efficient Solar Cooling System." *International Journal of Refrigeration* 59: 135–143.

Yin, Y. L., Z. P. Song, Y. Li, R. Z. Wang, and X. Q. Zhai. 2012. "Experimental Investigation of a Mini-Type Solar Absorption Cooling System under Different Cooling Modes." *Energy and Buildings* 47 (Supplement C): 131–138.

Yin, Y. L., X. Q. Zhai, and R. Z. Wang. 2013. "Experimental Investigation and Performance Analysis of a Mini-Type Solar Absorption Cooling System." *Applied Thermal Engineering* 59 (1–2): 267–277.

Zambrano, Darine, Carlos Bordons, Winston Garcia-Gabin, and Eduardo F. Camacho. 2008. "Model Development and Validation of a Solar Cooling Plant." *International Journal of Refrigeration* 31 (2): 315–327.

Zamora, Miguel, Mahmoud Bourouis, Alberto Coronas, and Manel Vallès. 2015. "Part-Load Characteristics of a New Ammonia/Lithium Nitrate Absorption Chiller." *International Journal of Refrigeration* 56: 43–51.

Zervos, Arthouros. 2002. Demonstrating the Efficiency of Solar Space Heating and Cooling in Buildings. Edited by SACE. Project Demonstration Report. Available at: http://www.energycon.org/sace/HE01.pdf

Zetzsche, M., T. Koller, T. Brendel, and H. Müller-Steinhagen. 2009. "Solar Cooling with an Ice-Storage Back-Up System." 3rd International Conference Solar Air-Conditioning, Palermo, Italy.

Zhai, X. Q., and R. Z. Wang. 2009. "A Review for Absorbtion and Adsorbtion Solar Cooling Systems in China." *Renewable and Sustainable Energy Reviews* 13 (6):1523–1531.

Zhu, Linghui, Shujun Wang, and Junjie Gu. 2008. "Performance Investigation of a Thermal-Driven Refrigeration System." *International Journal of Energy Research* 32 (10): 939–949.

Zinet, Matthieu, Romuald Rulliere, and Philippe Haberschill. 2012. "A Numerical Model for the Dynamic Simulation of a Recirculation Single-Effect Absorption Chiller." *Energy Conversion and Management* 62: 51–63.

6

Adsorption Cooling Heat Pumps

6.1 Adsorbents

There are three main types of adsorbents: physical, chemical, and composite adsorbents. In general, the desired features for an adsorbent include a large internal surface area, regeneration capability, and slow aging in order to preserve adsorptive capacity despite continuous recycling.

6.1.1 Physical Adsorbents

The most commonly applied physical adsorbents include activated carbon, zeolite, and silica gel. Furthermore, in recent years some novel physical adsorbents have been tested for cooling applications. Below, some key features of the most common physical adsorbents are described.

Activated carbon: Activated carbon is produced by materials such as wood, coal, and fossil oil. The microcrystal of the activated carbon is a six element carboatomic ring (Wang and Oliveira 2006; Cecen and Aktas 2011). The spaces between the individual microcrystallites are called pores. Most of the adsorption takes place in these micropores. Thus, for adsorption applications, the most important parameter to be controlled is the pore structure. Depending on the nature of the application, different pore volumes and pore size distributions in activated carbons are desired. Figure 6.1 presents the SEM image of one such type of activated carbon. Hence, for liquid adsorption, relatively large pores in a few nanometer size are preferable (Inagaki and Kang 2014). Activated carbon materials are divided into three categories: powder activated carbon (PAC) with a particle size of 1–150 nm, granular activated carbon (GAC) with a particle size of 0.5–4 mm, and the extruded activated carbon (EAC) with a particle size of 0.84 mm (Chen 2017). In general, activated carbons are a competitive solution given their high surface area and their low cost. Specifically, GACs and PACs have reported adsorption capacities ranging from 0.25 to 5.7 mg/g in the case of ammonia adsorption (Zheng et al. 2016).

Zeolites: The high regeneration costs of activated carbon have created the need for alternative low cost adsorbents including zeolites. Zeolite is a crystalline aluminosilicate with a three-dimensional framework structure of AlO_4 and SiO_4 that forms uniformly sized pores at a size of a molecule (Čejka 2007; Kulprathipanja and Wiley 2010). The basic structure of a typical zeolite is presented in Figure 6.2. The pores preferentially adsorb molecules that fit inside the pores and do not adsorb too-large molecules, hence acting as sieves on a molecular sieve. Zeolite crystallization takes place under hydrothermal conditions, from gels containing silica and alumina, in the presence of organic compounds as structure-directing agents (SDA) (Lu and Zhao 2004). There are many factors affecting the formation of zeolites, including source materials, solvent, SDA, gel composition, pH value, and crystallization conditions (Liu and Yu 2016). As of 2016, 232 types of zeolite materials have been identified by the

FIGURE 6.1
SEM image of activated carbon. (Reproduced from Pflitsch, Christian et al., *Carbon*, 79: 28–35, 2014.)

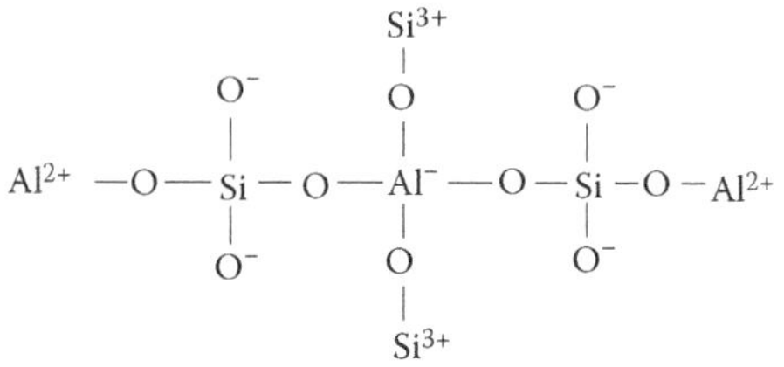

$$Al^{2+} - O - \underset{\underset{O^-}{|}}{\overset{\overset{O^-}{|}}{Si}} - O - \underset{\underset{O}{|}}{\overset{\overset{Si^{3+}}{|}}{\overset{|}{Al^-}}} - O - \underset{\underset{O^-}{|}}{\overset{\overset{O^-}{|}}{Si}} - O - Al^{2+}$$

with Si^{3+} at bottom under central Al.

FIGURE 6.2
Basic zeolite structure.

Structure Commission of the International Zeolite Association (Baerlocher and McCusker 2016). The main reasons for the steady increase in the use of zeolites include the large number of commercially available structures, their high structural ability, and safety and environmental considerations (Martínez and Corma 2013).

Apart from Si and Al, other elements can be introduced to the framework of Figure 1.3, resulting in what is known as zeotypes, including, among others, aluminophosphates (AlPOs) and silicoaluminophosphates (SAPOSs). In general, based on the synthesis technique, different properties can be achieved from zeotypes. Utchariyajit et al. (2008) investigated an alternative method to synthesize $AlPO_4$-5 and found that the morphology of zeotypes is influenced by the composition of the reaction mixture, the crystallization time and temperature, and the addition of HF acid. On the other hand, Sandoval et al. (2009) investigated the synthesis of ANA zeotypes by hydrothermal reaction of natural clinker, avoiding the use of SDAs and pure chemical SiO_2 and Al_2O_3 sources, in order to reduce production costs. Kim et al. (2014) studied the use of ferroaluminophosphate as an adsorbent and developed the adsorption isotherms of FAM-Z01 for water adsorption. Based on this study, the adsorption capacity of FAM-Z01 was found to be much higher than that of commercial silica gels.

Apart from adsorption, zeotypes can be used for catalysts, ion exchange, and other applications. However, there is currently no commercial product due to the fact that their procedure synthesis and the employed raw materials are quite expensive, which leads to high final product costs (Comyns 2009). One more restriction in their use is the high affinity for water vapor that zeolites have (Pistocchini et al. 2016).

Silica gels: Silica gel is one of the most popular adsorbents. Silica gels owe their wide use to their large capacity, low cost, market availability, and their ease of regeneration at approximately 150°C, in contrast to zeolites, which require 350°C (Yang 2003, Kim et al. 2017). This type of adsorbent is appropriate for low grade waste heat sources (Wu and Li 2009). Silica gel is a porous, granular form of silica with small particle size of 2–5 nm and a large surface-to-volume ratio (Shahata 2016). Silica gel is synthesized either from polymerization of $Si(OH)_4$ or from aggregation of colloidal silica. As a water adsorbent, silica gel is characterized by average adsorptive capacity (Tso and Chao 2012). Adsorption systems with silica gel-water working pairs have been widely investigated either in single or multiple stage systems (Mitra et al. 2015; Chua et al. 2001; Thu et al. 2011).

In order to enhance the properties of silica gel, various modifications have been investigated. Fang et al. (2014) investigated the influence of metallic ions doped on silica gel and found that the mean pore size, the total pore volume, the thermal stability, and the adsorptive capacity were enhanced. Tangkengsirisin et al. (1998) studied the influence of the addition of activated carbon in silica geland found that it enhanced the desorption rate and regeneration temperature.

Metal organic frameworks: Metal organic frameworks (MOFs) belong to a recently developed class of adsorbents that show attractive features for adsorption chillers. Indeed, since they are characterized by huge specific surface area as well as the low temperature requirements for desorbing their water content, MOFs have drawn much scientific interest in recent years (Tatlier 2017). Nevertheless, at their current stage of development, the main issues are (1) their hydrothermal cycling stability, which is really limited and results in a significant drop in their adsorption capacity, and (2) their production cost, which is still too high for practical applications (Henninger et al. 2012; Küsgens et al. 2009)

6.1.2 Chemical Adsorbents

In the search for heat and mass transfer intensification, new chemical adsorbents have been developed to enhance the adsorption system's performance and make them competitive with respect to conventional vapor compression units (Aristov et al. 2007). The large adsorption capacity and low evaporating temperature are the main advantages of chemical adsorbents. The main disadvantages of chemical adsorbents are a low thermal conductivity that results in slower reaction, and the durability of the chemical adsorbents with repeated reaction cycles (Oliveira and Wang 2007; Fujioka and Suzuki 2013). Wang, Chen et al. (2009a) investigated the adsorption of ammonia in several metal chlorides and their composites. It was found the composite adsorbents present a better performance and thus allow for higher refrigeration capacities, and they can also be desorbed at relatively low temperatures, allowing for exploitation of waste heat sources for the required heat.

Calcium chloride: Calcium chloride ($CaCl_2$) is one of the most widely used chemical adsorbents. Calcium chloride has a decent potential for use as a chemical adsorbent for methanol and ethanol vapors (Srivastava and Eames 1998). It has a very high adsorption capacity in a solid state, being able, at ambient temperature (and anhydrous state), to adsorb moisture up to 90% of its own weight and even higher percentages in the case of ammonia (N'Tsoukpoe et al. 2015). $CaCl_2$ remains solid until saturated, while after saturation it dissolves in water and thus it can be used for low temperature applications. A major drawback of calcium chloride is the agglomeration of the adsorbent, with a severe effect on the heat and mass performance of the adsorption cycle (Wang et al. 2009b). For two-stage adsorption cycles, it has been found that the combination of $CaCl_2$ and $BaCl_2$ maximize the cycle's performance with a COP of approximately 0.3 (Jiang et al. 2015; Hu et al. 2011).

Lithium chloride: Based on the thermodynamic values of ammonia on salts, lithium chloride ($LiCl_2$) can be an alternative for ammonia adsorption refrigeration (Kiplagat et al. 2010).

However, since lithium chloride faces the same problems as most salts, focus has moved toward the use of composite adsorbents of lithium chloride in silica gel, expanded graphite, and others (Maggio et al. 2009; Kiplagat et al. 2010).

Metal oxides: Apart from metal chlorides, an alternative type of chemical adsorbents is metal oxides. Among metal oxides, titanium oxide, zirconium oxide, and magnesium oxide are the most widely used for chemisorption. In the case of metal oxide adsorbents, oxygen is used as the refrigerant. As in the case of metal chlorides, a major drawback of these adsorbents is swelling and agglomeration (Wang et al. 2009b). Precipitation of hydroxides is used as the main production method for metal oxides, followed by a partial dehydration at elevated temperatures (Sing 1998).

6.1.3 Composite Adsorbents

Composite adsorbents have been developed and investigated recently in the search for improved heat transfer rates that will eventually allow for more efficient cooling systems. Composite adsorbents are made from porous media and a combination of one or more physical and chemical adsorbents, such as silica gel, expanded graphite, metal chlorides, and zeolite (Wang and Oliveira 2006).

Zhong et al. (2007) developed and investigated a composite adsorbent based on $BaCl_2$ as a working pair with ammonia and found that there was a significant improvement on the ammonia uptake, achieving a COP of 0.598 for an evaporating temperature of 15°C, a regeneration temperature of 61°C, and a condensation temperature of 35°C in a standard adsorption cycle.

Veselovskaya et al. (2010) developed a novel $BaCl_2$/vermiculite adsorbent for application in an ammonia single-stage adsorption chiller. The test rig was initially designed for an active carbon-ammonia working pair. The experimental results for the heat production showed a COP of 0.52–0.55 and a SCP in the range of 300–680 W/kg for a low grade heat source of 90°C, with a condensation temperature of 30°C and an evaporation temperature of 10°C.

In order to avoid agglomeration issues, Li et al. (2009) proposed the use of a composite adsorbent based on $CaCl_2$ for the adsorption of ammonia. Li et al. also proposed the treatment of $CaCl_2$ on a graphite solution instead of the simple mixture of the two components, and experimentally proved that the resulting composite adsorbent presented enhanced homogeneity and thus enhanced mass transfer properties.

Ye et al. (2014) tested several composite adsorbents from activated carbon fiber cloth and $CaCl_2$ in a water vapor cooling system. By varying the impregnation time, different amounts of $CaCl_2$ were measured in each case and the corresponding COP was calculated. The composite adsorbent that was developed was characterized by large sorption rates, and resulted in a COP of 0.7 for air conditioning applications (evaporation temperature of 10°C, condensation temperature of 33.75°C, and regeneration temperature of 88.4°C).

Tso and Chao (2012) developed a composite adsorbent through the impregmentation of microporous activated carbon with silica gel and $CaCl_2$. The composite adsorbent was tested in a water adsorption unit. The novel composite achieved an ideal COP of 0.7 compared to 0.37, which is the value for the COP if raw activated carbon was used. The calculations were carried out for an evaporating temperature of 5°C, a regeneration temperature of 115°C, and a condensation temperature of 27°C. The respective values for SCP were 378 W/kg for the composite adsorbent and 65 W/kg for the raw activated carbon.

Lu and Wang (2014) experimentally evaluated a single-stage solar cooling system with a (LiCl)/silica gel-methanol working pair. Based on the results, it was concluded that, in cooling mode (evaporation temperature of 15.2°C, condensation temperature of 31.4°C and regeneration temperature of 85.2°C), the investigated working pair could achieve a maximum COP of 0.41, with a SCP of 225 W/kg. On the other hand, for cold storage application (evaporation temperature of –4°C, condensation temperature of 25°C, and regeneration temperature of 88°C), the COP of the system was reduced to 0.13.

San and Hsu (2009) developed a theoretical model to predict the performance of SWS-1L composite adsorbent for a four-bed water adsorption heat pump, and compared it with silica gel as an adsorbent. The results from the theoretical analysis showed that compared to the values for silica gel, the COP improved by 51% when SWS-1L was used.

El-Sharkawy et al. (2016) studied the use of consolidated composite adsorbents based on activated carbon powder Maxsorb III and expanded graphite. Several different compositions were evaluated on an ethanol adsorption chiller. It was found that the thermal conductivity increased with an increase in the percentage of the expanded graphite and the packing density. Specifically, for a packing density of 650 kg m⁻³, the resulting thermal conductivity was 0.74 W m⁻¹ K⁻¹, a value more than 10 times higher than the respective value of Maxsorb III.

Gordeeva et al. (Gordeeva and Aristov 2011; Gordeeva et al. 2009b; Gordeeva et al. 2009a) evaluated the performance of a $LiCl/SiO_2$ composite adsorbent in a methanol adsorption chiller. The measured COP was in the range of 0.32–0.4 while the optimal SCP was estimated to be 2.5 kW/kg ($\theta_e = 10°C$, $\theta_c = 30°C$ and $\theta_{des} = 85°C$).

6.2 Adsorption Refrigerants

Depending on the adsorbent, there are several refrigerants that can be used for an adsorption cycle, with water, ammonia, methanol, and ethanol being the most widely used.

Water was the first adsorbent to be used thanks to its availability, cost, and its absence of environmental impact (Freni et al., 2016). On the other hand, the main disadvantage of water is that it cannot be used for ice making or refrigeration below 0°C applications.

Ammonia is extensively used in adsorption cooling systems, especially with metal chloride or metal chloride composite adsorbents (Anyanwu & Ogueke 2007; Ponomarenko et al. 2010; Tokarev et al. 2010; Veselovskaya et al. 2010). The main advantages of ammonia include the high enthalpy of vaporization, thermal stability, no ozone depletion, low global warming potential, and a low freezing point. The main disadvantage of ammonia is the fact that it is toxic (Dakkama et al. 2017).

Ethanol is another widely used refrigerant thanks to its low freezing point (–114°C), non-toxicity, high thermal stability, and good latent heat of evaporation (El-Sharkawy et al. 2008; Rezk et al, 2013). Brancato et al. (2015) performed theoretical calculations on the performance of ethanol adsorption chillers with carbonaceous and composite adsorbents. The SG/LiBr composite was found to be the most efficient adsorbent, resulting in a COP of 0.64 for refrigeration applications ($\theta_e = -2°C$, $\theta_c = 30°C$, and $T_{des} = 90°C$) and a COP of 0.72 for air conditioning applications ($\theta_e = 7°C$, $\theta_c = 30°C$, and $\theta_{des} = 90°C$).

Methanol is also used as refrigerant in adsorption applications. The main advantages of methanol include its high latent heat of vaporization and its non-toxicity (Frazzica et al. 2016). On the other hand, methanol is corrosive, leading to the need for corrosion-resistant materials and an increase in cost of equipment. Li et al. (2004) compared, for the same operating conditions, the performance of methanol and ethanol in a solar adsorption ice maker using activated carbon as adsorbent. From the simulations, it was concluded that the performance of methanol is approximately 3–5 times higher than ethanol.

6.3 Adsorption Working Pairs

Given the fact that the most novel working pairs based on composite adsorbents have been already presented in the previous section, the most widely used working pairs will be discussed below.

6.3.1 Zeolite-Water

Zeolite-water is one of the first working pairs to be investigated for adsorption cooling applications (Grenier et al. 1988; Meunier 1994). Solmus et al. (2010) carried out experimental work to determine the characteristics of zeolite-water. The experimental setup consisted of a long corrugated tube, which was a zeolite canister, a combination of an evaporator and condenser, a water bath to control the pressure in the canister, an oven, and a vacuum circulating pump. The analysis showed that the maximum adsorption capacity was approximately 12%. Furthermore, it was proved that the level of adsorption increased with water vapor pressure and with a decrease in zeolite temperature.

6.3.2 Silica Gel-Water

Silica gel-water is one of the most widely used working pairs for adsorption cooling applications. It is an adsorption working pair that can exploit industrial waste heat to provide cooling. For a single-stage adsorption cycle with partial vacuum pressure and an evaporation temperature of 6.7°C, condensation temperature of 29.4°C, and regeneration temperature of 80°C, the theoretical COP for refrigeration can be as high as 0.68, while the corresponding specific cooling effect was 217.3 kJ/kg, based on simulations carried out by Loh et al. (2009). Chua et al. (1999) evaluated the performance of the working pair in a two-bed adsorption chiller, and found that the maximum COP was approximately 0.46 for a cooling output of 14.1 kW (θ_e = 14.8°C, θ_c = 31.1°C, and θ_{des} = 86.3°C). Chang et al. (2007) investigated the influence of operating parameters on the performance of a silica gel-water working pair in a closed-type adsorption chiller. It was found that, under an 80°C hot water temperature, a 30°C cooling water temperature, and a 14°C chilled water temperature the obtained COP was 0.45, with a corresponding SCP of 176 W/kg.

6.3.3 Activated Carbon-Ammonia

Researchers have studied several working pairs with activated carbon as the adsorbent. Activated carbon-methanol and activated carbon-ammonia working pairs are the most extensively discussed. The main disadvantage of an activated carbon-ammonia working pair is the poor heat transfer in the solid, which leads to high cycle times and hence to low SCP values (Tamainot-Telto & Critoph, 1997).

Xu et al. (2016) carried out simulations on single- and multi-bed adsorption cycles to realize the performance of activated carbon-ammonia working pairs for both refrigeration and ice making applications and to identify the optimum operating conditions. For each combination of condensation, evaporation, and heat source temperatures, the optimum number of beds was specified and the corresponding COP was calculated. For instance, for a condensation temperature of 30°C, an evaporating temperature of 5°C, and a heat source temperature of 90°C, it was found that the optimum number of beds is one, with a corresponding COP of 0.33. Critoph and Metcalf (2004) measured the performance of an activated carbon-ammonia working pair in plate-type single-stage adsorption chiller for several activated carbon thicknesses. For a thickness of 1 mm, a COP of 0.345 was measured with a maximum SCP of 6.5 kW/kg in a two-bed adsorption chiller ($\theta_e = 15°C$, $\theta_c = 30°C$, and $\theta_{des} = 200°C$). Metcalf et al. (2012) carried out simulations to predict the efficiency of an activated carbon-ammonia pair in two- and four-bed adsorption chillers. The results of the simulations showed that the cooling COP was higher in the four-bed case, with a maximum value of approximately 0.55 when the SCP was 200 W/kg, at an evaporation temperature 3°C, a condensation temperature of 50°C, and a regeneration temperature of 200°C.

6.3.4 Calcium Chloride-Methanol

Metal chlorides, and specifically calcium chloride, are used in most cases as adsorbents for either ammonia or methanol adsorption. As a working pair, calcium chloride-ammonia's main advantage is its large adsorption capacity. Its most important challenge is agglomeration, which is a compromise to the high heat and mass transfer (Wang, Wang et al., 2009).

Lai and Li (1996) investigated the potential use of a CaCl2-methanol working pair in an adsorption chiller for either standalone operation or for implementation in a CHP. It was concluded that since the CHP operation resulted in an increased regeneration temperature, a significant increase in the COP was identified.

6.3.5 Working Pair Comparison Investigations

Apart from the standard working pairs that were already presented, several working pair comparisons have been published in literature (Freni et al. 2016; San and Lin 2008; Allouhi et al. 2015; Habib et al. 2013; Cui et al. 2005).

San and Lin (2008) compared the performance of three working pairs—activated carbon-methanol, silica gel-water, and molecular sieves-water—in a four-bed adsorption chiller. The results of the mathematical modeling, as shown in Figure 6.3, suggest that the performance of activated carbon-methanol and silica-gel-water are comparable in terms of

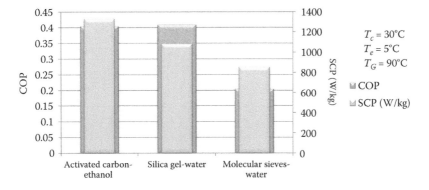

FIGURE 6.3
Working pair comparison. (Adapted from San, Jung-Yang, and Wei-Min Lin, *Applied Thermal Engineering*, 28 (8–9): 988–997, 2008.)

COP, with values of approximately 0.405. However, SCP is higher for the activated carbon-methanol working pair, with a value of 1.33 kW/kg.

Allouhi et al. (2015) carried out simulations on seven well established working pairs to determine the optimum for a solar cooling application with respect to efficiency. For the simulations, a single-stage solar adsorption cooling system was considered, and several combinations for the evaporating temperature and the condensing temperature were tested to calculate the solar COP. The adsorption capacity was only evaluated for an air conditioning case while the hot source temperature was 110°C. The results of the analysis are presented in Figure 6.4, with activated carbon fiber (ACF)-methanol proving to be the most efficient in terms of adsorptive capacity, while silica gel-water presented the highest solar COP.

On the other hand, Cui et al. (2005) investigated the performance of two environmental friendly working pairs, composite NA-water and composite NB-ethanol, and compared it to conventional working pairs zeolite 13x-water, zeolite 13x-ethanol, and activated carbon-ethanol. The refrigeration capacity was measured for the optimal desorption temperature of each working pair. As was calculated, and is also shown in Figure 6.5, the most efficient working pair proved to be NA-water with a refrigeration capacity of 522 kJ/kg at a desorption temperature of 100°C.

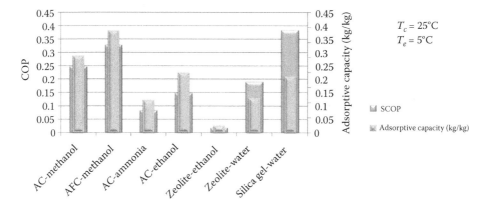

FIGURE 6.4
Solar COP and adsorptive capacity for the working pairs investigated. (Adapted from Allouhi, A. et al., *Energy*, 79: 235–247, 2015.)

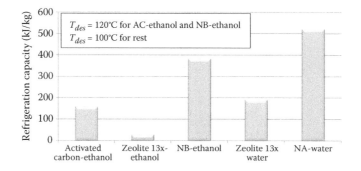

FIGURE 6.5
Refrigeration capacity comparison of environmental friendly working pairs with conventional working pairs. (Adapted from Cui, Qun et al., *Energy*, 30 (2–4): 261–271, 2005.)

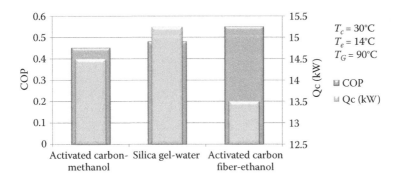

FIGURE 6.6
Working pair comparison. (Adapted from Habib, Khairul et al., *Applied Thermal Engineering*, 72 (2): 266–274, 2014.)

Habib et al. (2014) modeled the performance of three different working pairs—activated carbon-methanol, activated carbon fiber-ethanol, and silica gel-water—in a two-bed, solar-powered adsorption cooling cycle. By varying the regeneration temperature and the cycle time, the system's performance for each case was evaluated. It was found that the optimum performance for the cases of methanol and ethanol was met at a regeneration temperature of 90°C, while the optimum temperature for silica gel-water was 75°C. Figure 6.6 presents the values of the three working pairs for a regeneration temperature of 90°C. As is shown, activated carbon fiber-ethanol was the most efficient, with a COP of approximately 0.55. However, with the same operating conditions, a maximum cooling capacity of 15.3 kW was achieved with a silica gel-water working pair.

6.4 Adsorption Chiller Applications

In the following section, a review of state-of-the-art adsorption refrigeration units is presented. Flannery et al. (2017) proposed the coupling of a free-piston Stirling engine with a zeolite-water adsorption chiller, as can be seen in Figure 6.7. The Stirling engine has a power output of 1-2 kW$_e$, while the adsorption chiller has a cooling capacity of 4 kW and is powered by the waste heat from the jacket water-glycol mixture of the Stirling engine. The adsorption cycle is powered by either the waste heat of the Stirling engine cooling jacket or the waste heat from the main truck engine. Based on the experimental data collected, the average COP was approximately 0.42, while the net electrical efficiency of the system was 13% ($\theta_e = 18$°C, $\theta_c = 41$°C, and $\theta_{des} = 80$–90°C).

Chen et al. (2010) experimentally investigated the performance of a compact silica gel-water adsorption chiller. The adsorption chiller consists of two adsorption chambers and a chilled water tank. The heat exchangers—condensers and evaporators—are shell-and-tube type. Vacuum valves were avoided by applying three-way valves in order to enhance the reliability of the chiller. For average hot water, condensation, and evaporation temperatures of 82.1°C, 31.6°C, and 12.3°C respectively, the obtained cooling power was 9.6 kW with a COP of 0.49.

Sadeghlu et al. (2015) carried out simulations to evaluate the performance of a novel cascade adsorption cooling system, with composite Zeolite 13x/CaCl$_2$-water as the working pair for the upper cycle and silica gel-water for the bottom cycle. The coupling of the two

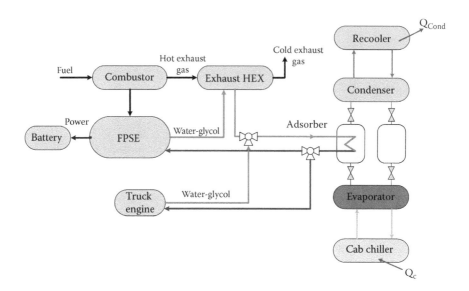

FIGURE 6.7
A schematic diagram of the experimental Stirling engine-adsorption chiller unit. (Adapted from Flannery, Barry et al., *Applied Thermal Engineering*, 112: 464–471, 2017.)

cycles consists of the use of a shell-and-tube heat exchanger, with water from the bottom cycle condensing in the tube side and the evaporation of the upper cycle water stream taking place in the shell side. The cycle operates in two stages. In the first stage, the elements SE1 and SE3 (Figure 6.8) are supplied with cooling water and operate as adsorbers, while hot water from a hot water storage tank is supplied to elements SE2 and SE4, which operate as desorbers. In the second stage, the operation of the elements is reversed so that SE2 and SE4 now operate as adsorbers, being connected with the cooling water stream. A sensitivity analysis was carried out to determine the critical parameters of the system. Activation energy—the energy barrier that has to be supplied to the system for the initiation of the desorption/adsorption process—was identified as the most crucial for both working pairs. Furthermore, an investigation into the effect of the desorption and condensation temperature on the COP was performed. According to the results of the investigation, for a desorption temperature of 75°C and a condensation temperature of 30°C, the calculated system's COP was equal to 0.275 ($T_e \approx 10°C$).

Xia et al. (2009) developed and evaluated a silica gel-water adsorption chiller. The two-chamber experimental unit includes a condenser, an evaporator, and an adsorber inside each vacuum chamber, and one methanol evaporator at the bottom of the system. Regarding evaporation, there are two water evaporators, as shown in Figure 6.9, and one methanol evaporator in order to form a heat pipe loop based on gravity to enable heat recovery in the system. Furthermore, mass recovery was applied to enhance the cooling output. For a condensing temperature of 30.5°C, an evaporating temperature of 16.5°C, and a maximum generation temperature of 84.4°C, the COP was 0.43 and the SCP was 104.6 W/kg. The cooling capacity was as high as 10.88 kW. Furthermore, it was concluded that the application of mass recovery enhanced the cooling power by 65% and the COP by more than 30%.

Gong et al. (2012) experimentally investigated the performance of a $CaCl_2$/silica gel-methanol two-bed adsorption chiller. As can be seen in the schematic of Figure 6.10, the experimental setup includes two chambers that have one adsorber, one condenser, and one evaporator each, while the third chamber is a heat pipe evaporator. Mass recovery

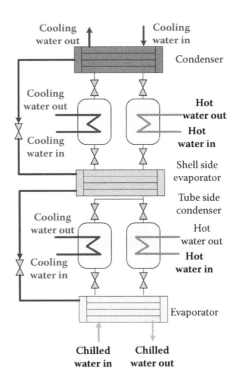

FIGURE 6.8
The combined adsorption cooling system proposed by Sadeghlu et al. (Adapted from Sadeghlu, Alireza et al., *Applied Thermal Engineering*, 87: 185–199, 2015.)

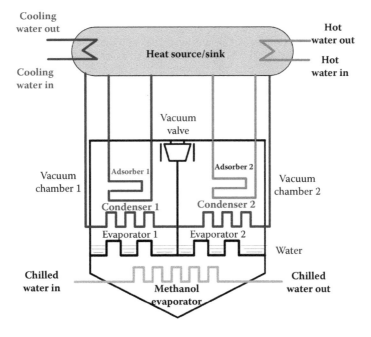

FIGURE 6.9
Schematic of the proposed silica gel-water adsorption chiller investigated by Xia et al. (Adapted from Xia, Z. Z. et al., *International Journal of Thermal Sciences*, 48 (5): 1017–1025, 2009.)

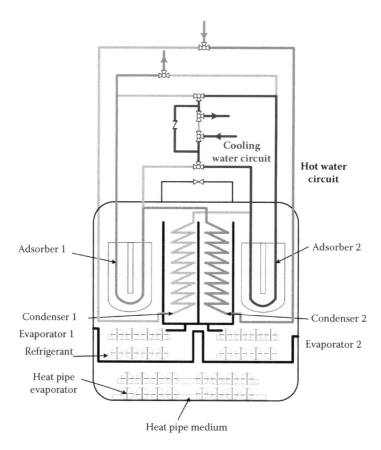

FIGURE 6.10
The experimental CaCl$_2$/silica gel-methanol adsorption chiller setup of Gong et al. (From Gong, L. X. et al., *International Journal of Refrigeration*, 35 (7): 1950–1957, 2012.)

is applied by opening valve V7 and vapor flow from desorption bed toward the adsorption bed, enabling re-adsorption and re-desorption through the pressure change on both chambers. Several experiments were carried out by varying the heating/cooling time, the mass recovery time, the hot temperature inlet, and the evaporator temperature. For a cooling time of 680 s and a mass recovery time of 60 s, the overall optimum for the COP was as high as 0.41, with a cooling capacity of 4.99 kW ($\theta_e = 15.3$°C, $\theta_c = 29.8$°C, and $\theta_{des} = 84.8$°C).

Jiang et al. (2014) measured the performance of a novel cascade cogeneration system by coupling an organic Rankine cycle with R245fa as the working medium and a CaCl$_2$/BaCl$_2$-ammonia two-stage adsorption cycle. The coupling of these two systems (Figure 6.11) is based on the use of the jacket water of the ORC condenser's and the water that heats R245fa, in both the boiler and the superheater of the ORC, to provide the required heating and cooling loads for the desorption and the adsorption phase, respectively. The two-stage adsorption chiller consists of two adsorbers, a middle temperature salt (MTS) adsorber and a low temperature salt (LTS) adsorber, a single condenser, and a single evaporator. When the MTS adsorber is supplied with cooled water and the adsorption phase starts, the LTS bed is connected with hot water at approximately 95°C to start the desorption phase. The cooling effect is produced in the single evaporator of the adsorption chiller. During the experimental investigation, the system's performance was observed for the varying

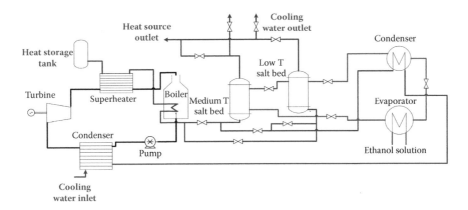

FIGURE 6.11
The cascade system proposed by Jiang et al. (From Jiang, Long et al., *Energy*, 71: 377–387, 2014.)

heat input in the ORC. For a low-grade heat source, the ORC produced 0.53 kW, while for the two-stage adsorption chiller a maximum COP of 0.201 and a SCP of 220.3 W/kg were reported, as well as a condensing temperature of 30°C, and an evaporation temperature of 0°C.

Lu and Wang (2013) compared the performance of a $CaCl_2$/activated carbon-ammonia and a silica gel/LiCl-methanol adsorption chiller. Both chillers consisted of three chambers, two of which were equipped with an adsorber, a heat exchanger, and a condenser, while the third chamber was equipped with the evaporator of the adsorption unit. The two systems were compared at nominal operation. At an evaporating temperature of –21°C, the $CaCl_2$/activated carbon chiller operated with a COP of 0.26 and a SCP value of 474 W/kg ($\theta_c =$ 29°C and $\theta_{des} = 144$°C). On the other hand, at an evaporating temperature of 15°C, the silica gel/LiCl adsorption chiller obtained a COP of 0.41 and a SCP value of 244 W/kg ($\theta_c = 31$°C and $\theta_{des} = 85$°C). Furthermore, it was found that mass recovery was more beneficial for the performance of $CaCl_2$/activated carbon chiller, reporting an improvement of 53.8% on the COP, compared to 15.4% for the silica gel/LiCl adsorption chiller.

Li et al. (2010) introduced a novel adsorption refrigeration unit based on composite $CaCl_2$/expanded graphite-ammonia with a reduced number of valves. The developed system includes two adsorbers, each with 37 adsorption unit tubes, two condensers, two evaporators, and three valves. Li et al. carried out several experiments to identify the response of the system and found that the performance was enhanced with an increasing evaporating temperature and a decreasing condensation temperature. The maximum COP that was obtained was equal to 0.27, with a SCP of 422.2 W/kg, at a condensing temperature of 25°C, an evaporation temperature of –15°C, and a heating source temperature of 140.3°C.

Hu et al. (2009) investigated the performance of a two-bed mass recovery adsorption cooling system. The proposed system, presented in Figure 6.12, offers a potential improvement in the cooling capacity. In this scheme, the desorber and adsorber are continuously supplied with hot and cold water, respectively. This increase in the heat load enhances the adsorption/desorption process, allowing for higher cooling production. The applied working pair in the cycle is zeolite/aluminum foam-water. By varying the cycle time and the adsorbent layer thickness, it was found that the system's performance increased with cycle time and thickness of the layer. The optimum values for the COP were obtained for a cycle time of 100 minutes and a layer thickness of 20 mm. In this configuration, for a condensing temperature of 40°C, an evaporating temperature of 10°C, and a maximum

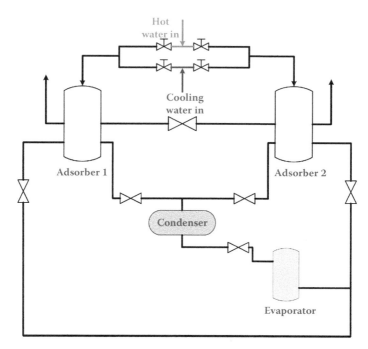

FIGURE 6.12
Mass recovery cycle. (Adapted from Hu, Peng et al., *Energy Conversion and Management*, 50 (2): 255–261, 2009.)

generation temperature of 400°C, the COP was equal to 0.56, while the corresponding SCP was approximately 190 W/kg.

Vodianitskaia et al. (2017) experimentally investigated the adsorption kinetics and the overall performance of a single-bed adsorber with silica gel-water. The adsorber of the unit is a finned tube, in which silica gel is in loose grains, and is supplied the required heating and cooling load for the desorption and the adsorption phase, respectively, through a secondary fluid, which is also water, that is flowing from a thermostatic bath. Based on the experimental data collected, it was found that an increase in particle size enhances the thermal conductivity of the grains. Regarding the system's performance, for a condensing temperature of 30°C, an evaporating temperature of 15°C, and a maximum generation temperature of 80°C the COP was equal to 0.53, while the corresponding SCP was approximately 68 W/kg. For the tested range, the maximum obtained COP was equal to 0.58.

Wang and Zhang (2009) simulated the performance of a multi-cooling tubes adsorption heat pump. Each cooling tube can be considered as a separated adsorption unit, with the adsorber on the top side, the plain tube condenser in the middle, and the evaporator at the bottom part of the cooling tube. When multiple cooling tubes are implemented, two larger adsorbers can be formed. The condenser of the multiple cooling tubes unit is then a shell-and-tube heat exchanger, and the evaporator is a heat pipe consisting of the evaporators of the cooling tubes. The working principle is that of a single-stage adsorption chiller. The working pair applied is silica gel-methanol. In the proposed configuration, for a condensing temperature of 30°C, an evaporating temperature of 15°C, and a maximum generation temperature of 85°C, the COP was equal to 0.48 at a cycle time such that the corresponding SCP was maximized at a value of 86 W/kg.

Bao et al. (2014) investigated the performance of a novel two-stage chemical adsorption cogeneration system using a combination of silica gel and calcium chloride as an adsorbent

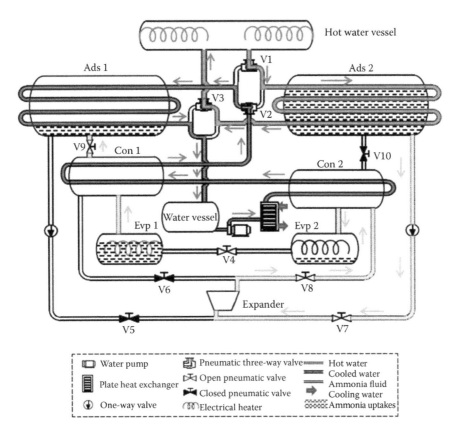

FIGURE 6.13
The novel adsorption cogeneration system proposed by Bao et al. (Reproduced from Bao, Huashan et al., *Energy*, 72: 590–598, 2014.)

and ammonia as the refrigerant. The major challenges that were identified included the capacity mismatch between the two main components and the constraints on both the desorption and expansion processes. A process diagram of the novel setup is presented in Figure 6.13. The power generation module is achieved through heating of Adsorber 1. The heat causes a change in the thermodynamic state of the salt that results in a release of high-temperature and high-pressure ammonia, which is eventually led to the expander and produces the work output. The cooling power generation, on the other hand, is achieved by supplying Adsorber 2 with cooling water from the heat sink, resulting in the adsorption of ammonia from the evaporator. The cooling effect is produced by the evaporation of ammonia in the evaporator. Based on the experiments carried out for a condensing temperature of 27–35°C, an evaporating temperature of –2.5°C, and an average generation temperature of 125°C, the COP was equal to 0.20, at a cycle time of 13 min, resulting in a SCP value equal to 364 W/kg.

Pan et al. (2014) developed and tested the performance of a composite $CaCl_2$/activated carbon-ammonia working pair in a novel adsorption refrigerator. The prototype, presented in Figure 6.14, consists of a shell-and-tube adsorber with finned tubes, a condenser, two evaporators, and four valves that control the sequence of the phases. The unit's cycle consists of four stages. In the first stage, adsorber A1 is connected with the heating fluid circuit and is on the desorption phase, while adsorber A2 is connected with the cooling circuit and adsorbs ammonia. The desorbed ammonia is then condensed in

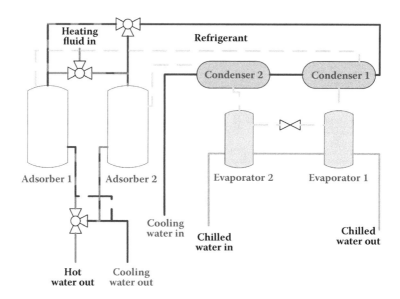

FIGURE 6.14
The prototype adsorption cooling unit proposed from Pan et al. (From Pan, Q. W. et al., *Applied Thermal Engineering*, 72 (2): 275–282, 2014.)

condenser C1 and stored in evaporator E1, while the adsorption in A2 drives the evapo-ration of ammonia in E2, which is producing cooling effect. The second stage consists of mass recovery from E1 to E2 through valve V4. In the next stage, adsorber A1 is in adsorption phase and produces cooling output in evaporator E1, while adsorber A2 is in desorption phase. Finally, in the fourth stage of the cycle, there is a mass recovery step from evaporator E2 to E1 through valve V4. The system's performance was evalu-ated under several different operating conditions. The maximum COP reported was approximately 0.23, while the corresponding SCP was approximately 230 W/kg, for a condensing temperature of 25°C, an evaporating temperature of –5°C, and a maximum generation temperature of 130°C.

Freni et al. (2007) developed a novel sorption chiller using a lightweight, finned tube heat exchanger coated with a SWS-1L composite layer. Glycol was used as the refrigerant for the measurements. The COP measured with the proposed configuration was approxi-mately 0.25, while the SCP was in the range of 150–200 W/kg for a cycle time of 10–20 minutes and for a condensing temperature of 35°C, an evaporating temperature of 7–12°C, and a maximum generation temperature of 95–10°C.

In the search for better performing adsorption chillers, Freni et al. (2012) investigated a system similar to the previous one using a new composite based on silica gel, the SWS-8L, with water as the refrigerant. The schematic of the new experimental setup is presented in Figure 6.15. The heating and cooling loads required for the desorption and adsorption phase, respectively, are supplied by a 24 kW electric boiler and a 14 kW water chiller. The temperature of the secondary streams in the evaporator and condenser are controlled by two thermocryostats. By varying the desorption, the condensation temperature, and the cycle time, and with an evaporating temperature of 15°C, the performance of the setup was measured. For a cycle time of 10 minutes, a condensation temperature of 30°C, and a desorption temperature of 90°C, the system's COP was equal to 0.3 and the SCP was as high as 389 W/kg.

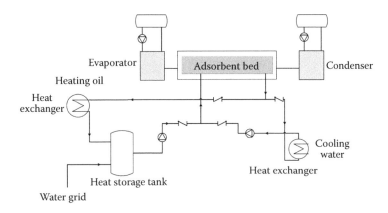

FIGURE 6.15
The compact bed that was used to investigate the performance of SWS-8L-water working pair. (From Freni, Angelo et al., *International Journal of Refrigeration*, 35 (3): 518–524, 2012.)

Tierney et al. (2017) developed a novel adsorption chiller that allows the heat exchanger, where the adsorption takes place, to directly be illuminated by concentrated irradiation. The working pair was activated carbon cloth-ethanol. Figure 6.16 presents the schematic of the experimental setup. The adsorber consisted of a fin assembly in a glass tube, forming a tube-in-tube heat exchanger. The adsorbent filled in the gaps between the fins, while the secondary fluid flowed through the inner tube, transferring the required heat and cooling load for the desorption/adsorption. In terms of the working principle, it is already described in a typical single-stage adsorption cycle. The maximum measured

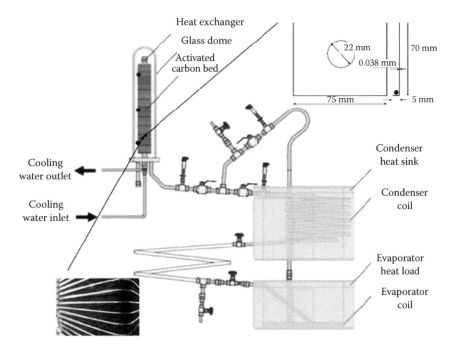

FIGURE 6.16
The experimental adsorption chiller developed by Tierney et al. (Reproduced from Tierney, M. et al., *Applied Thermal Engineering*, 110: 949–961, 2017.)

COP was 0.236, with a SCP in the range of 59–181 W/kg for a condensing temperature of 30°C, an evaporating temperature of 14°C, and a maximum generation temperature of 100°C.

6.5 Solar Cooling with Adsorption Chillers

As is shown in literature, the most recent research focuses on the exploitation of solar energy toward the design of a solar cooling unit (Fadar et al. 2009a; Berdja et al. 2014; Abu-Hamdeh et al. 2013; Du et al. 2016; El-Sharkawy et al. 2008; Fong et al. 2010; Zhai et al. 2009; Luo et al. 2006; González and Rodríguez 2007; Suleiman et al. 2012).

El Fadar et al. (2009a) modeled a continuous adsorption refrigeration unit powered by parabolic trough collectors. Water is heated in the receiver of the PTCs and is then pumped into the heat storage tank (Figure 6.17). The hot water is then used to heat the two adsorbers, which operate asynchronously (when the first is in adsorption phase, the other is in desorption phase). The cold water required for adsorption is provided by the cold water tank. The adsorption cycle includes two adsorbent beds with activated carbon-ammonia as the working pair. For an evaporating temperature of 0°C, a condensing temperature of 30°C, and a heat source temperature of 100°C, a COP of 0.43 was obtained, with a SCP of 104 W/kg.

Abu-Hamdeh et al. (2013) developed and evaluated the performance of a prototype adsorption refrigeration unit powered by a parabolic trough collector. The single-axis tracking collector is positioned such to maximize the yearly energy harvest. The adsorbent bed has four 1.8 m-long steel tubes and a perforated coaxial inner tube with a 20.1 mm inner diameter. The required desorption heat is provided by the hot water storage tank

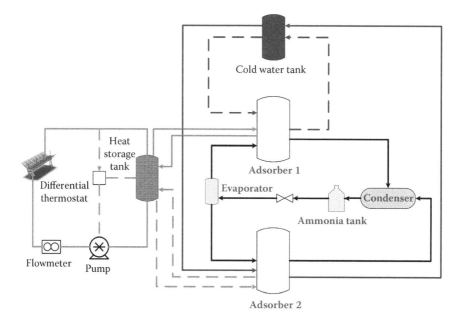

FIGURE 6.17
Schematic of the proposed solar adsorption unit proposed by El Fadar et al. (From Fadar, A. El et al., *Solar Energy*, 83 (6): 850–861, 2009a.)

of the solar collector circuit. For the adsorption phase, the cooling effect that is produced is used to cool down the surrounding space. The condenser is a simple steel tube, while the evaporator is a spirally coiled copper tube. The working pair used in this application was olive waste (adsorbent)-methanol. An overview of the system described is shown in Figure 6.18. Regarding the system's performance, for a condensing temperature of 25°C, an evaporating temperature of 8°C, and a maximum generation temperature of 120°C, the gross cycle COP was equal to 0.75, while the optimum solar COP ranged between 0.18–0.2. As was identified, increasing the tank volume and the collector area up to certain levels enhanced the system's performance.

Du et al. (2016) compared the performance of a novel solar adsorption refrigerator with SAPO-34-water and ZSM-5 zeolite-water as its working pairs. The required heat for adsorption is supplied by a parabolic trough collector, as shown in Figure 6.19. The

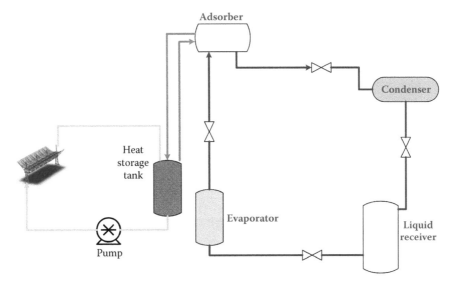

FIGURE 6.18
The experimental solar refrigeration setup proposed by Abu-Hamdeh et al. (From Abu-Hamdeh, Nidal H. et al., *Energy Conversion and Management*, 74: 162–170, 2013.)

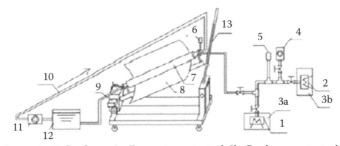

1. Evaporator 2. Condenser 3a. Evaporator water tank 3b. Condenser water tank
4. Vacuum pump 5. Condenser 6. Bed pressure sensor 7. Adsorption bed
8. Parabolic trough 9. Stepper motor 10. Cooling water loop 11. Water pump
12. Water tank 13. Angle adjust lever

FIGURE 6.19
The solar adsorption refrigeration unit proposed by Du et al. (Reproduced from Du, S. W. et al., *Solar Energy*, 138: 98–104, 2016.)

adsorber's bed was installed in the trough to enhance solar energy utilization. The cycling process starts with solar radiation heating up the adsorber. The desorption is then initiated, which increases the temperature and the pressure of the bed. When the pressure is equal to the corresponding condensation pressure, the respective valve connects the bed to the condenser. In order to cool down the stream for the adsorption phase, cooling water is supplied from the water tank. Then, the water is led into the evaporator, where it evaporates before restarting the cycle. The results of the analysis for the two working pairs are presented in Table 6.1.

Fong et al. (2010) investigated a solar hybrid system for air conditioning applications in subtropical cities. The hybrid system consists of four main sub-systems: solar energy collection, an adsorption chiller, desiccant dehumidification, and radiant ceiling cooling. The temperatures for the silica gel-water adsorption chiller at a nominal point are 80/30/18°C. A year-round analysis was carried out and the average values of the system's performance were determined. Regarding the adsorption chiller, the variation of the COP throughout the year was not very high, with values in the range 0.533–0.590.

Zhai et al. (2009) carried out an energy and exergy analysis to evaluate the performance of a solar trigeneration system, based on a helical screw expander and eight 10 kW silica gel-water adsorption chillers (Figure 6.20). For the collection of solar energy, parabolic trough collectors with cavity absorbers were implemented. The adsorption cycle is provided with the required heat from a heat exchanger that cools down the steam exiting the

TABLE 6.1

Results of the Comparison of the Working Pairs Investigated by Du et al.

Adsorbent	SAPO-34	ZSM-5 Zeolite
θ_{des} (°C)	60	100
θ_c (°C)	31–35	n/a
θ_e (°C)	15	n/a
$\dot{Q}_e$ (kW)	8.08	8.74
COP	0.122	0.060
SCP (W/kg)	169.74	91.20

Source: Du, S. W. et al., *Solar Energy*, 138: 98–104, 2016.

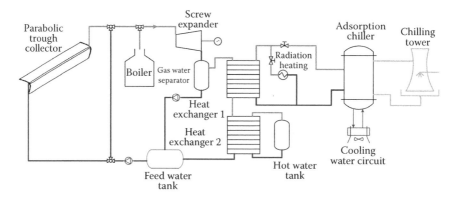

FIGURE 6.20
The solar trigeneration system investigated by Zhai et al. (From Zhai, H. et al., *Applied Energy*, 86 (9): 1395–404, 2009.)

screw expander, while cooling water for the adsorption phase is provided from a cooling tower. The energy and exergy efficiency of the system are 0.58 and 0.152, respectively. A large amount of the lost energy is a result of the poor COP of the adsorption chillers, which in the best case scenario of a 95°C regeneration temperature does not exceed 0.45 (for a condensation temperature of 30.7°C and an evaporation temperature of 20.6°C).

Luo et al. (2006) evaluated the performance of a solar adsorption chiller for a cooling grain depot. The system uses a silica gel-water adsorption chiller, which consists of two identical single-bed adsorption units and an additional stage consisting of a methanol evaporator. The solar water heating system is used to provide the required heat for the desorption phase. On the other hand, the cooling water for the adsorption phase is provided from a cooling tower. The cooling load that is produced is transferred to the grain depot through a fan coil. Furthermore, mass recovery is used in the cycle to enhance the performance of the system. Mass recovery benefits from the mismatch of pressure and temperature between the adsorption and desorption bed. By applying a tube and a valve to connect the two beds, vapor can be transferred at the end of each half-cycle of the process from the low pressure bed to the high pressure bed until the two beds reach equilibrium. The next half cycle will start with isosteric conditions, resulting in enhanced cooling production, without affecting the COP (Pons and Poyelle 1999).

Based on the experiments carried out, it was found that the adsorption chiller can operate in hot water temperatures above 65°C. Furthermore, the daily performance of the adsorption chiller was measured for specific dates and it was found that, in the best case scenario, the cycle's COP was equal to 0.331, while the refrigerating capacity was 5.29 kW ($\theta_e = 20°C$, $\theta_c = 32°C$, and $\theta_{des} = 85°C$).

Aristov et al. (2007) used simulations to evaluate and develope a solar refrigeration system based on an adsorption cycle, as shown in Figure 6.21. For the analysis, several chemisorbents were evaluated for their water adsorption characteristics. $CaCl_2$ in silica gel composite sorbent was found to be the most efficient sorbent for water adsorption, resulting in a cycle's COP equal to 0.6–0.8 ($\theta_e = 5°C$, $\theta_c = 35°C$, and $\theta_{des} = 80°C$).

Lemmini and Errougani (2005, 2007) developed a solar single-bed adsorption unit powered by a flat-plate collector, as shown in Figure 6.22. The working pair used in the experimental setup was methanol-microporous activated carbon (AC-35). Several experiments were carried out, achieving a maximum solar COP of 0.078 with a second law efficiency of 0.71 ($\theta_e = -8.4°C$, $\theta_c = 17°C$, and $\theta_{des} \approx 90°C$).

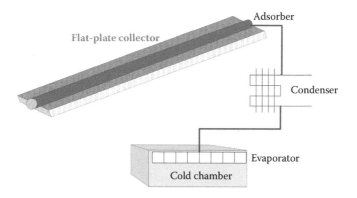

FIGURE 6.21
The solar adsorption unit developed by Aristov et al. (From Aristov, Yu I. et al., *Chemical Engineering Journal*, 134 (1–3): 58–65, 2007.)

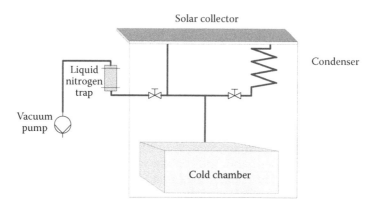

FIGURE 6.22
The solar refrigeration unit based on adsorption cycle developed by Lemmini and Errougani. (From Lemmini, Fatiha, and Abdelmoussehel Errougani, *Renewable Energy*, 32 (15): 2629–2641, 2007.)

Gonzalez and Rodriguez (2007) measured the performance of a novel solar adsorption unit with activated carbon-methanol as its working pair. For solar energy harvesting, an array of four compound parabolic concentrators was used. The adsorber was placed inside the CPC, partially exposed to solar radiation. The condenser of the adsorption chiller was a water-cooled condenser of cylindrical shape with tubes through which the water flowed, while the evaporator consisted of a grid of vertical copper tubes, as shown in Figure 6.23. The condenser was connected to the evaporator with a valve that was never shut under normal operating conditions, resulting in the methanol directly falling into the evaporator after condensation. Several daily measurements were carried out to evaluate the performance of the chiller. On a day with 19.5 MJ/m² solar irradiation (hot source temperature 38–116°C), for a condensation temperature of 20.4°C and an evaporation temperature of –1.1°C, the measured solar COP was 0.096.

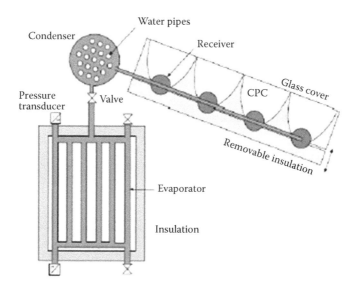

FIGURE 6.23
The solar adsorption unit tested by Gonzalez et al. (Reproduced from González, Manuel I., and Luis R. Rodríguez, *Energy Conversion and Management*, 48 (9): 2587–2594, 2007.)

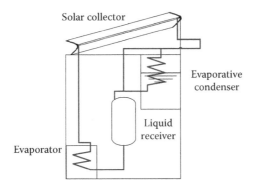

FIGURE 6.24
The single stage solar adsorption unit tested by Anyanwu and Ogueke. (From Anyanwu, E. E., and N. V. Ogueke, *Applied Thermal Engineering*, 27 (14–15): 2514–2523, 2007.)

Anyanwu and Ogueke (2007) carried out simulations to evaluate the performance of a solar adsorption unit using activated carbon-methanol as the working pair. The single-stage adsorption unit is presented in Figure 6.24. Based on the finite element method that was used as the solution for the system, a maximum solar COP of 0.025 was measured. The corresponding evaporation temperature was 0°C, while the maximum plate temperature was 100°C.

Chang et al. (2009) investigated a heat and cooling unit powered by a solar collector field. A schematic of the investigated setup is presented in Figure 6.25. The solar heating mode was able to heat two tanks of 1000 l makeup water to 50°C, while a backup gas-fired boiler was implemented to cover potential needs on cloudy days. The silica bed/water adsorption chiller has a nominal capacity of 9 kW. The experiments were carried out for a hot water temperature of 80°C, a condensation temperature of 30°C, and a chilled water inlet temperature of 14°C. The results showed that a COP of 0.37 could be achieved, with a SCP of 72 W/kg.

Baiju and Muraleedharan (2011) experimentally studied the performance of a two-bed adsorption unit powered by a parabolic trough collector, as shown in Figure 6.26.

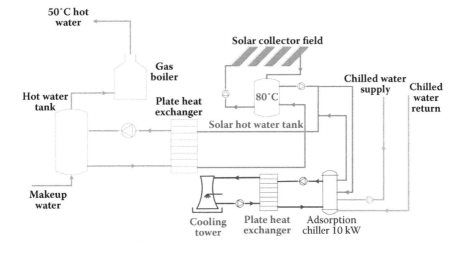

FIGURE 6.25
The solar heating and cooling unit investigated by Chang et al. (From Chang, W. S. et al., *Applied Thermal Engineering*, 29 (10): 2100–2105, 2009.)

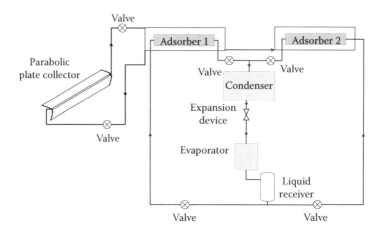

FIGURE 6.26
The experimental solar adsorption setup investigated by Baiju and Muraleedharan. (From Baiju, V., and C. Muraleedharan, *International Journal of Engineering Science and Technology*, 3 (7): 5754–5764, 2011.)

The adsorption cycle working pair was methanol activated carbon. The system's nominal cooling duty was to cool down 10 l of water from 30°C to 10°C within one hour. The average COP and the respective second law efficiency under daytime operation was 0.196 and 0.175, while the respective values for night operation were 0.335 and 0.269, for an optimum hot source temperature of 72.4°C. The corresponding values for SCP were 47.8 and 68.2 W/kg for daytime and night operation, respectively.

Suleiman et al. (2012) investigated the dynamic performance of a solar adsorption chiller with an activated carbon-methanol working pair, as shown in Figure 6.27. The adsorbent is packed into the annular space between the two co-axial pipes of the adsorption bed. The adsorber is installed inside the hot water tank that is heated up by the solar collector. When the temperature of the water tank reaches the desorption temperature, the desorption phase starts. As the heating increases the bed temperature to its maximum value, desorption continues. When the maximum temperature is reached, desorption ends and methanol is led to the condenser. After condensation, methanol is stored in the receiver before being transferred to the evaporator. By the end of the desorption phase, circulation through the collector stops, and cold water from the grid is pumped into the tank to cool down the adsorber and start the adsorption process. A year-long simulation was carried out to estimate the performance of the system. The average COP was 0.608, while the corresponding annual average solar COP was equal to 0.024, for temperatures of 0°C for the evaporation and 25°C for the condensation. The desorption temperature was at least 80°C.

Habib et al. (2013) simulated the performance of a two-stage four-bed adsorption system with silica gel-water as its working pair. A schematic of the investigated system is presented below in Figure 6.28. The heat required to drive the desorption process was provided by evacuated tube collectors. Based on the period of a year, the heat source temperature ranged between 40°C and 95°C. For the single-stage operation (hot source temperature of 80°C), a cooling water temperature of 30°C, and a chilled water inlet temperature of 14°C, the reported COP was approximately 0.48. For lower hot source temperatures, the system operated in two-stage mode, with a COP of approximately 0.27 for a hot source temperature of 50°C.

Alahmer et al. (2016) evaluated a 11 kW two-stage silica gel-water adsorption chiller powered by evacuated tube collectors using TRNSYS simulations. The investigated chiller,

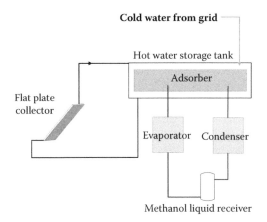

FIGURE 6.27
The solar adsorption unit proposed by Suleiman et al. (From Suleiman, R. et al., *International Journal of Renewable Energy Research*, 2 (4): 657–664, 2012.)

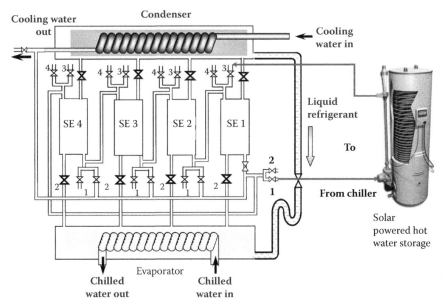

1. Hot water outlet; **2:** Coolant inlet to the absorber; **3:** Hot water inler; **4:** Coolant outlet from the adsorber

FIGURE 6.28
The two stage solar adsorption unit investigated by Habib et al. (Reproduced from Habib, Khairul et al., *Energy*, 63: 133–141, 2013.)

shown integrated in the whole system in Figure 6.29, was considered to have a COP of 0.6 for hot water temperature of 80°C, a chilled water inlet temperature of 15°C, and cooling water temperature of 28°C. Several parameter investigations were carried out for the regions of Perth, Australia, and Amman, Jordan. It was reported that the average COP during summer period was 0.491 in the case of Perth. The respective value for Amman was 0.467. The corresponding average cooling capacities were 10.30 kW and 8.46 kW for Perth and Amman, respectively.

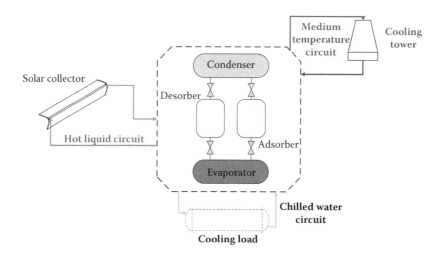

FIGURE 6.29

The solar adsorption unit investigated by Alahmer et al. (From Alahmer, Ali et al., *Applied Energy*, 175: 293–304, 2016.)

Jaiswal et al. (2016) simulated a single-stage two-bed silica gel-water adsorption chiller powered by an evacuated tube solar collector field. A parametric analysis was conducted on the considered system, shown in Figure 6.30, to identify the influence of certain parameters, including the solar collector area and the cycle time. It was found that increases in both the cycle time and the solar collector area have beneficial effects on the cooling capacity. On the other hand, an increase in the solar collector area was found to have a negative effect on the system's COP.

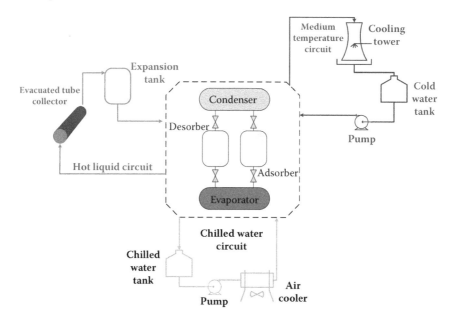

FIGURE 6.30

The theoretical solar adsorption model studied by Jaiswal et al. (From Jaiswal, Ankush Kumar et al., *Solar Energy*, 136: 450–459, 2016.)

6.6 Overview of Adsorption Systems Reported in Literature

This section presents an overview of the previously discussed systems as proposed and investigated in literature. As shown in Table 6.2 and Table 6.3, silica gel-water is the most commonly investigated working pair in both simulation and experimental studies, due to its competitive performance and availability.

The Carnot coefficient of performance for an adsorption cycle is calculated with the following expression (Sharonov and Aristov 2008; Meunier et al. 1998; San and Hsu 2009):

$$COP_{carnot} = \frac{T_e(T_{des} - T_c)}{T_{des}(T_c - T_e)} \tag{6.1}$$

Hence, the second law efficiency of an adsorption cycle is equal to

$$\eta = \frac{COP}{COP_{carnot}} \tag{6.2}$$

At this time, there are only a few manufacturers across the world that produce commercial adsorption chillers. These manufacturers include Bry-Air, Fahrenheit AG, Weatherite Manufacturing Limited, Mayekawa USA, and GBU. The aforementioned companies focus mostly on single-stage adsorption chillers with either silica gel-water or zeolite-water as the working pair. Companies that focus on the use of water as an adsorbent are also justified by Table 6.4, which shows that, according to the reports from literature, experimental water adsorption cooling systems have the highest second law efficiency.

Fahrenheit has developed two adsorption chiller product series, one with a silica gel-water working pair and one with a zeolite-water working pair. The ©Fahrenheit eZea (zeolite-water adsorption chiller), depending on the model, has a capacity range of 13 kW up to 104 kW, with a COP of 0.53, for a desorption temperature of 85°C, a condensation temperature of 25°C, and an evaporation temperature of 15°C (©Fahrenheit 2016). The cycle of this commercial model will be discussed further in a subsequent section.

6.7 Process Model

The following section presents the model that was developed to simulate the operation of an investigated zeolite-water, single–stage, two-bed adsorption chiller, as the one shown in Figure 6.31.

6.7.1 Basic Assumptions

In order to simplify the process, the following assumptions were made:

  i. The adsorbent particles are of an identical size and have the same properties, and are uniformly distributed in both beds. The specific heat and the density of the adsorbent remain constant.

TABLE 6.2

Overview of the Theoretically Investigated Adsorption Units and Their Main Parameters as Discussed in this Chapter

Refrigerant	Adsorbent	θ_{des} (°C)	θ_c (°C)	θ_e (°C)	COP	η (%)	SCP (W/kg)	Remarks	Source
Ammonia									
	Composite BaCl$_2$	61	35	15	0.598	53.34*	n/a	single bed	(Zhong et al. 2007)
	Activated carbon	200	30	15	0.345	5.00	6,500*	single bed plate type generator	(Critoph and Metcalf 2004)
	Activated carbon	200	50	3	0.55	29.53	200	four beds	(Metcalf et al. 2012)
	Activated carbon	110	25	5	0.0845 (solar)	2.74 (solar)	n/a	single stage/solar powered	(Allouhi et al. 2015)
	Activated carbon	100	30	0	0.43	25.18	104	solar powered/single stage/two bed	(Fadar et al. 2009b)
Water									
	Zeolite 13x/CaCl$_2$ (upper cycle) Silica gel (bottom cycle)	75	30	10	0.275	15.03	n/a	cascade system/two bed per cycle	(Sadeghlu et al. 2015)
	Zeolite/aluminum foam	400	40	10	0.56	11.09	190	two bed/mass recovery	(Hu et al. 2009)
	Activated carbon fiber cloth/CaCl$_2$	88.4	33.75	10	0.7	38.84	n/a	single bed	(Ye et al. 2014)
	Molecular sieves	90	30	5	0.205	11.15	854*	four bed	(San and Lin 2008)
	Composite silica gel	80	35	5	0.6–0.8	42–56	n/a	single bed	(Aristov et al. 2007)
	Silica gel	90	29.4	6.7	0.68	33.05	n/a	four bed/partial vacuum pressure	(Loh et al. 2009)
	Silica gel	86.3	31.1	14.8	0.46	16.96	n/a	single stage/two bed	(Chua et al. 1999)
	Silica gel	90	30	5	0.41	22.30	1083*	four bed	(San and Lin 2008)
	Silica gel	110	25	5	0.3843 (solar)	12.46 (solar)	n/a	single stage/solar powered	(Allouhi et al. 2015)
	Silica gel	90	30	14	0.48	16.19	n/a	two bed/solar	(Habib et al. 2014)
	Silica gel	80	30	18	0.533–0.590	15.52–17.17	n/a	solar powered (evacuated tubes)/single bed/hybrid AC system	(Fong et al. 2010)

(Continued)

TABLE 6.2 (CONTINUED)

Overview of the Theoretically Investigated Adsorption Units and Their Main Parameters as Discussed in this Chapter

Refrigerant	Adsorbent	θ_{des} (°C)	θ_c (°C)	θ_e (°C)	COP	η (%)	SCP (W/kg)	Remarks	Source
	Silica gel	95	30.7	20.6	0.45	8.86	n/a	solar (PTCs) trigeneration system with 8 single-stage adsorption chillers	(Zhai et al. 2009)
	Silica gel	80	28	15	0.60	17.60	n/a	solar powered	(Alahmer et al. 2016)
	Zeolite	110	25	5	0.1905 (solar)	6.17	n/a	single stage/solar powered	(Allouhi et al. 2015)
Methanol									
	Activated carbon	90	30	5	0.405	22.03	1333*	four bed	(San and Lin 2008)
	Activated carbon	110	25	5	0.2476 (solar)	8.03 (solar)	n/a	single stage/solar powered	(Allouhi et al. 2015)
	Activated carbon	90	30	14	0.45	15.18	n/a	two bed/solar	(Habib et al. 2014)
	Activated carbon	80	25	0	0.608 0.024(solar)	35.73	n/a	solar powered (flat-plate collector)/single bed	(Suleiman et al. 2012)
	Activated carbon	<100	n/a	0	0.025(solar)	n/a	n/a	solar powered single bed	(Anyanwu and Ogueke 2007)
	Activated carbon fibers	110	25	5	0.3287 (solar)	10.65 (solar)	n/a	single stage/solar powered	(Allouhi et al. 2015)
	Silica gel	85	30	15	0.48	16.27	86	single stage/multi-cooling tubes	(Wang and Zhang 2009)
Ethanol									
	Activated carbon	110	25	5	0.1479 (solar)	4.79 (solar)	n/a	single stage/solar powered	(Allouhi et al. 2015)
	Activated carbon fibers	90	30	14	0.55	18.55	n/a	two bed/solar	(Habib et al. 2014)
	Zeolite	110	25	5	0.0191 (solar)	0.62 (solar)	n/a	single stage/solar powered	(Allouhi et al. 2015)

* Values that are considered extremely high.

TABLE 6.3

Overview of the Experimental Adsorption Units and Their Main Parameters as Discussed in this Chapter

Refrigerant	Adsorbent	θ_{des} (°C)	θ_c (°C)	θ_e (°C)	COP	η (%)	SCP (W/kg)	Remarks	Source
Ammonia									
	BaCl$_2$/vermiculite	90	30	10	0.52–0.55	22.2–23.5	300–680	single bed	(Veselovskaya et al. 2010)
	CaCl$_2$/activated carbon	144	29	−21	0.26	18.70	474	two beds with two condensers and a single evaporator	(Lu and Wang 2013)
	CaCl$_2$/activated carbon	130	25	5	0.23	6.35	230	two beds	(Pan et al. 2014)
	CaCl$_2$/BaCl$_2$	95	30	0	0.201	12.50	220.3	trigeneration system with an ORC and a two-stage adsorption chiller	(Jiang et al. 2014)
	CaCl$_2$/expanded graphite	140.3	25	−15	0.27	15.00	422.2	two beds	(Li et al. 2010)
	CaCl$_2$/Silica gel	125	27–35	−2.5	0.2	8.86–12.26	364	two stage/cogeneration system	(Bao et al. 2014)
Water									
	SWS-8L composite	90	30	15	0.3	9.45	389	single bed	(Freni et al. 2012)
	microporous activated carbon/silica gel/CaCl$_2$	115	27	5	0.7	24.42	378	single bed	(Tso and Chao 2012)
	Activated carbon	115	27	5	0.37	12.91	65		(Du et al. 2016)
	SAPO-34 (silica gel)	60	31–35	15	0.122	7.78–11.15	169.74	solar powered (PTC)/single bed	(Chang et al. 2007)
	Silica gel	80	30	14	0.45	17.71	176	single bed/common heat exchanger for evaporation-condensation	
	Silica gel	80	30	14	0.37	13.8	72	two bed/solar	(Chang et al. 2009)
	Silica gel	82.1	31.6	12.3	0.49	23.31	n/a	two adsorption chambers	(Chen et al. 2010)
	Silica gel	85	32	20	0.331	9.16	n/a	solar adsorption chiller with two single-bed adsorption units and mass recovery	(Luo et al. 2006)
	Silica gel	80	30	15	0.53	19.49	68	single bed	(Vodianitskaia et al. 2017)
	Silica gel	84.4	30.5	16.5	0.43	13.79	104.6	two chamber/mass recovery	(Xia et al. 2009)

(Continued)

TABLE 6.3 (CONTINUED)

Overview of the Experimental Adsorption Units and Their Main Parameters as Discussed in this Chapter

Refrigerant	Adsorbent	θ_{des} (oC)	θ_c (oC)	θ_e (oC)	COP	η (%)	SCP (W/kg)	Remarks	Source
Methanol	Zeolite	80–90	41	18	0.42	24.59–30.04	n/a	Stirling engine coupled with two-bed adsorption chiller	(Flannery et al. 2017)
	ZSM-5 (zeolite)	100	n/a	n/a	0.06	n/a	91.2	solar powered (PTC)/ single bed	(Du et al. 2016)
	Silica gel/LiCl	85.2	31.4	15.2	0.41	15.34	225	solar powered/ single stage/ two bed	(Lu and Wang 2014)
		88	25	-4	0.13	8.03	n/a	two bed	
	Silica gel/LiCl	96	31	15	0.41	12.93	244	two beds	(Lu and Wang 2013)
	LiCl/SiO2	85	30	10	0.32–0.4	14.71–18.40	2,500*	single bed	(Gordeeva and Aristov 2011; Gordeeva et al. 2009a; Gordeeva et al. 2009b)
	CaCl2/silica gel	84.8	29.8	15.3	0.41	13.41	n/a	two bed/mass recovery	(Gong et al. 2012)
	Activated carbon	38–116	20.4	-1.1	0.096 (solar)	3.09 (solar)	n/a	solar powered (CPC)/ single bed	(González and Rodríguez 2007)
	Activated carbon	90	17	-8.4	0.078 (solar)	71.0	n/a	single bed	(Lemmini and Errougani 2007)
	Activated carbon	72.6 (opt.)	n/a	n/a	0.334	26.9	68.2	two stage	(Baiju and Muraleedharan 2011)
	Olive waste	120	25	8	0.75 (gross) 0.18–0.2 (solar)	18.77	n/a	solar powered (PTC)/ single bed	(Abu-Hamdeh et al. 2013)
Ethanol	SG/LiBr composite	90	30	7	0.64	31.80	n/a	single bed	(Brancato et al. 2015)
				-2	0.72	51.43	n/a		
	Activated carbon cloth	100	30	14	0.236	7.01	59–181	possible to utilize solar irradiation/single stage	(Tierney et al. 2017)
Glycol	SWS-1L composite	95–100	35	7–12	0.25	11.58–15.33	150–200	single bed	(Freni et al. 2007)

* Values questionably high in comparison to other literature.

TABLE 6.4

Overview of the Average Performance Indicators for the Systems Described
in Chapter 1 Based on the Type of Adsorbent Used

Adsorbent		Average COP	Average η (%)	Average SCP (W/kg)
Ammonia	(simulations)	0.481	19.90	152
	(experimental)	0.283	14.35	366.8
Water	(simulations)	0.478	18.99	190
	(experimental)	0.382	16.68	180.2
Methanol	(simulations)	0.486	22.30	793
	(experimental)	0.344	14.19	234.5
Ethanol	(simulations)	0.55	18.55	n/a
	(experimental)	0.532	19.41	120

FIGURE 6.31
The adsorption chiller. (From Fahrenheit, "Fahrenheit: eZea", accessed January 2017, https://fahrenheit.cool
/en/zeo/.)

   ii. The refrigerant, the adsorbent particles, and the bed's material within each bed,
       separately, are assumed to have the same temperature.
  iii. There is neither temperature nor pressure variation in the condenser.
  iv. Heat losses have been neglected, because it is considered that sufficient insulation
      has been applied.
   v. Thermal resistance between adsorbate and adsorbent is neglected.
  vi. The gaseous phase behaves as ideal gas.
 vii. The exit from the condenser is saturated liquid.
viii. The exit from the evaporator is saturated vapor.

6.7.2 Adsorption Isotherms and Kinetics

The equilibrium uptake of the water-zeolite working pair will be estimated using the Dubinin-Astakhov model (Llano-Restrepo and Mosquera 2009; Kayal et al. 2016).

$$\frac{x^*}{x_o} = \exp\left\{-\left[\frac{R \cdot T}{E_a} \ln\left(\frac{P_s}{P_w}\right)\right]^n\right\} \tag{6.3}$$

For the investigated working pair, the activation energy, E_a, is considered equal to 1192.3 kJ/kg (Mette et al. 2014), the limiting adsorbate uptake, x_o, is equal to 0.21 kg/kg, while the heterogeneity constant, n, is equal to 5 (Kayal et al. 2016).

The adsorption rate is calculated using the linear driving force (LDF) model (Liu and Leong 2008; Wu et al. 2009; Saha et al. 2009):

$$\frac{dx}{dt} = \frac{15 \cdot D_{so} \exp\left(-\dfrac{E_a}{R \cdot T}\right)}{R_p^2}(x^* - x) \tag{6.4}$$

The mean pore radius, R_p, is considered equal to 50 nm, while the pre-exponential coefficient D_{so} is considered as one of the values to be calibrated based on the experimental results (Sayılgan et al. 2016).

6.7.3 Evaporator

The energy balance in the evaporator shows that the balance is dominated mainly by the heat interaction between the adsorber and the evaporator and the heat interaction inside the evaporator between the chilled water, the heat exchanger walls, and the refrigerant. The overall energy balance equation is expressed by the following differential equation:

$$\frac{dT_e}{dt}\left(m_{cu,e}C_{p,cu} + m_{refr}C_{p,refr}\right) = -\left[h_{fg} + h_{refr}(P_e, T_{ads}) - h_{refr}(T_e)\right]m_{ze}\frac{dx_{ads}}{dt} + \dot{Q}_{ch,w} \tag{6.5}$$

where $\dot{Q}_{ch,w}$ refers to the heat flux toward the chilled water circuit, which is equal to:

$$\dot{Q}_{ch,w} = \dot{m}_{ch,w}C_{p,ch,w}(T_{ch,w,in} - T_{ch,w,out}) \tag{6.6}$$

The chiller water temperature at the outlet is calculated using the logarithmic mean temperature difference (LMTD) method, as shown below (Habib et al. 2011; Bergman and Incropera 2011):

$$T_{ch,w,out} = T_e + (T_{ch,w,in} - T_e)\exp\left[-\frac{(UA)_e}{\dot{m}_{ch,w}C_{p,ch,w}}\right] \tag{6.7}$$

6.7.4 Adsorber

Under the assumption that the refrigerant, the walls of the bed, and the adsorbent are at the same temperature, the energy balance in the adsorber is the following expression (Askalany et al. 2017; Chua et al. 1999; Rezk and Al-Dadah 2012; Yang 2009):

$$
\frac{dT_{ads}}{dt}\left(m_{cu,ads}C_{p,cu} + m_{ze}C_{p,ze} + m_{ze}C_{p,refr}\frac{dx_{ads}}{dt} \right) = \left[Q_{st} + h_{refr}(P_e, T_{ads}) - h_{refr}(T_e)\right]m_{zeol}\frac{dx_{ads}}{dt} + \dot{m}_{MT1,w}C_{p,MT1,w}(T_{MT1,w,in} - T_{MT1,w,out})
$$

(6.8)

The isosteric heat of adsorption, Q_{st}, is considered equal to $4.5 \cdot 10^5$ J/kg. The MT1 water temperature at the outlet is calculated using the LMTD method, as shown below (Habib et al. 2011; Bergman and Incropera 2011):

$$
T_{MT1,w,out} = T_{ads} + (T_{MT1,w,in} - T_{ads})\exp\left[-\frac{(UA)_{ads}}{\dot{m}_{MT1,w}C_{p,MT1,w}} \right]
$$

(6.9)

6.7.5 Desorber

The equations for desorption are equivalent to those of the adsorption, with the difference that the balance in the desorber is determined by the condenser (Askalany et al. 2017; Chua et al. 1999; Rezk and Al-Dadah 2012; Yang 2009):

$$
\frac{dT_{des}}{dt}\left(m_{cu,des}C_{p,cu} + m_{ze}C_{p,ze} + m_{ze}C_{p,refr}\frac{dx_{des}}{dt} \right)
$$
$$
= \left[Q_{st} + h_{refr}(P_c, T_{des}) - h_{refr}(T_c)\right]m_{zeol}\frac{dx_{des}}{dt}
$$
$$
+ \dot{m}_{HT,w}C_{p,HT,w}(T_{HT,w,in} - T_{HT,w,out})
$$

(6.10)

The HT water temperature at the outlet is calculated using the LMTD method, as shown below (Habib et al. 2011; Bergman and Incropera 2011):

$$
T_{HT,w,out} = T_{des} + (T_{HT,w,in} - T_{des})\exp\left[-\frac{(UA)_{des}}{\dot{m}_{HT,w}C_{p,HT,w}} \right]
$$

(6.11)

6.7.6 Condenser

The energy balance in the condenser shows that the balance is dominated mainly by the heat interaction between the desorber and the condenser and the heat interaction inside the condenser between the medium temperature (MT2) water, the heat exchanger walls, and the refrigerant. The overall energy balance equation is expressed by the following differential equation (Wang et al. 2006; Saha et al. 2009; Askalany et al. 2017):

$$\frac{dT_c}{dt}(m_{cu,c}C_{p,cu} + m_{refr}C_{p,refr}) = \left[h_{fg} + h_{refr}(P_c, T_{des}) - h_{refr}(T_c)m_{ze}\frac{dx_{des}}{dt} + \dot{Q}_{MT2,w}\right] \quad (6.12)$$

where $\dot{Q}_{MT2,w}$ refers to the heat flux toward the MT2 water circuit, which is equal to:

$$\dot{Q}_{MT2,w} = m_{MT2,w}Cp_{,MT2,w}(T_{MT2,w,in} - T_{MT2,w,out}) \quad (6.13)$$

The MT2 water temperature at the outlet is calculated using LMTD method, as shown below (Habib et al. 2011; Bergman and Incropera 2011):

$$T_{MT2,w,out} = T_c + (T_{MT2,w,in} - T_c)\exp\left[-\frac{(UA)_c}{m_{MT2,w}C_{p,MT2,w}}\right] \quad (6.14)$$

Concerning the heat capacities of the wall material, $C_{p,cu}$, and the zeolite, $C_{p,ze}$, these are taken to be equal to 510 J/kgK and 880 J/kgK, respectively.

6.7.7 Performance Indicators

The cycle average produced cooling capacity is calculated using the following expression:

$$\dot{Q}e = \frac{\int_0^{t_{cycle}} \dot{m}_{ch,w}C_{p,ch,w}(T_{ch,w,in} - T_{ch,w,out})dt}{t_{cycle}} \quad (6.15)$$

On the other side, the cycle average heat input is calculated by the following expression:

$$\dot{Q}_{des} = \frac{\int_0^{t_{cycle}} \dot{m}_{HT,w}C_{p,HT,w}(T_{HT,w,in} - T_{HT,w,out})dt}{t_{cycle}} \quad (6.16)$$

The specific cooling power of the investigated adsorption cycle is derived by the following:

$$SCP = \frac{\dot{Q}e}{m_{zeol}} \quad (6.17)$$

While the corresponding coefficient of performance is equal to

$$COP = \frac{\dot{Q}_e}{\dot{Q}_{des}} \quad (6.18)$$

The Carnot coefficient of performance for the adsorption cycle is calculated using Eq. (2.72), while the second law efficiency of the zeolite-water adsorption cycle is calculated from Eq. (2.74).

6.8 Model Solution and Results

Figure 6.32 presents a schematic flow diagram for the solution of the thermodynamic model that was already presented, showing the process for the calculation of the properties in the cycle at every moment.

In the following section, an analysis of the overall results provided by the model for a specific operating point will be presented and discussed. Table 6.5 presents the basic data for the operating point that will be examined. As shown, the point is in the range within which the model was validated in the previous sections, so as to ensure the reliability of the presented results. For the executed simulations, it is considered that the constant thermal input temperature is 90°C, which is easily achieved by an appropriate solar collector unit.

Figure 6.33 presents the inlet and outlet temperatures of the secondary water circuits. The period for each of the streams is equal to half of a cycle time and signals the switch between the two beds' operation, as was thoroughly discussed. Regarding the cooling water outlet, the average temperature can be further reduced by reducing the cooling water flow rate. Thus, for the same temperature levels, the cooling output will be consumed in order to cool

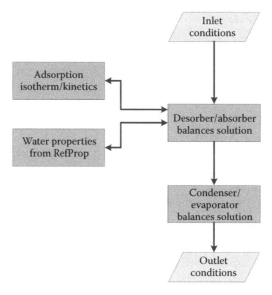

FIGURE 6.32
Cycle solution subsystem for a specific time.

TABLE 6.5

Design Point of the Analyzed Adsorber

Property	Value
HT water inlet temperature (°C)	90
MT water inlet temperature (°C)	25
Average cooling water outlet temperature (°C)	12.5
HT water flowrate (l/h)	2500
MT water flowrate (l/h)	5100
Cooling water flowrate (l/h)	2900

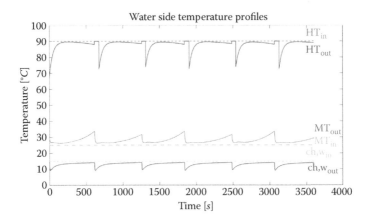

FIGURE 6.33
Temperature profiles of the water streams used for heat supply/rejection and cooling as predicted by the model.

down a smaller volume of water per second, thereby cooling the water to an even lower temperature. The MT water exiting the condenser has a profile that is typical to the case of the condenser in an adsorber and also follows the behavior of the condensing water in the primary circuit of the adsorber, as shown in Figure 6.34. Finally, the HT water outlet decreases significantly in every change phase because the three-way valves switch the HT water circuit from the hot, already desorbed, bed to the cold bed. Thus, at the beginning of each such switch, the drop in the outlet temperature is significant until the bed is heated up again.

Figure 6.34 presents the temperature profiles of the heat transfer fluid (water) in the condenser and the evaporator of the adsorber, respectively, while Figure 6.35 presents the corresponding temperature profiles in the two adsorber beds. Based on the profiles of the two figures, the cycle time is estimated to be approximately 1250 s. The temperature profiles of the condenser and the evaporator are highly dependent on the water flow rate that is cooled in each case. Regarding the temperatures of the adsorber beds, the rapid change in

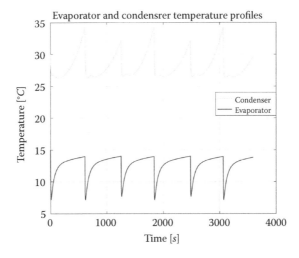

FIGURE 6.34
Temperature profiles of the evaporator and the condenser of the investigated adsorber as predicted by the model.

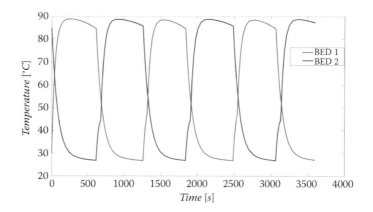

FIGURE 6.35
Temperature profiles of the two adsorber beds as predicted by the model.

their value during the adsorption and desorption phase, is due to the relatively high value of the overall heat transfer of each bed.

Figure 6.36 presents the coefficient of performance (right figure) and the second law efficiency (left figure) for the investigated operating point. The high values of the COP during the first cycle are a result of transient phenomena and do not reflect the actual COP in steady state operation, which is more visible, after two cycles of the adsorber. As shown in Figure 6.36, the steady state COP for the investigated operating point is approximately 0.51, which matches with the performance data provided by the manufacturer. This value of the COP is comparable with most of the commercially available adsorbers. Regarding the second law efficiency, the maximum value of 18.32% shows that there is significant room for improvement. The fluctuations in the second law efficiency are a result of the fluctuations of the evaporation and condensation temperatures, which, based on Eq. (6.16), results in fluctuations in the Carnot COP and, thus, in the second law efficiency. Finally, the SCP

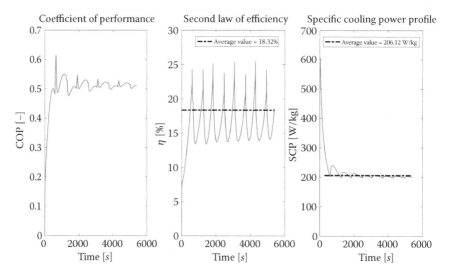

FIGURE 6.36
COP, second law efficiency, and SCP behavior based on the model predictions.

for the investigated operating point is presented. As shown, at steady state conditions, the value of the SCP is approximately 206.12 W/kg.

6.9 Adsorption Cooling Applications

6.9.1 Fahrenheit eZea Case Study

In this section, the realization of an adsorption cycle in a water-zeolite commercial domestic scale adsorption chiller is discussed. Given the fact that the Fahrenheit eZea adsorption chiller is a commercial product, only some basic technical data will be presented here. In order to enable the continuous production of cooling energy, the adsorption chiller that is reviewed in this section has a four-phase working cycle. The sequence of the phases is also presented in Figure 6.37 and Figure 6.38.

6.9.1.1 Phase 1

In the first phase of a working cycle, high temperature (HT) water flows through the adsorber (Adsorber No.1) and desorption starts. This results in a flow of water that has accumulated on the inner surface of the zeolite, into the condenser. The heat from the condensation is rejected to the medium temperature (MT) water circuit. At the same time, heat is taken from the low temperature (LT) water circuit to vaporize water in the evaporator. The water vapor is then fed into the adsorber (Adsorber No.2), where it is adsorbed. The adsorption process results in heat generation, which is rejected to the environment via the MT circuit. Phase 1 continues until the temperature, specified by the user, is achieved at the outlet of the LT water circuit.

6.9.1.2 Phase 2

After Phase 1 ends, a heat recovery step follows. The three-way valves, as seen in Figure 6.37 and Figure 6.38, are switched such that a recirculation of the MT water circuit enables the heat transfer from Adsorber No.1 to Adsorber No.2, until a certain temperature difference between the two beds is achieved. During this Phase, Adsorber No.1 is cooled down and thus starts to adsorb water, while Adsorber No.2 starts the desorption process.

6.9.1.3 Phase 3

Phase 3 is the reverse process of Phase 1. In this phase, the three-way valves are switched so that Adsorber No.2 connects to the HT circuit. Thus Adsorber No.2 is a desorber in this phase, while Adsorber No.1 connects to the MT water circuit and adsorbs water. Phase 3 continues until the temperature specified by the user is achieved at the outlet of the LT water circuit.

6.9.1.4 Phase 4

Following Phase 3, the three-way valves switch so that MT water circuit connects to previously desorbed Adsorber No.2 and exchanges heat with Adsorber No.1. Adsorber No.2

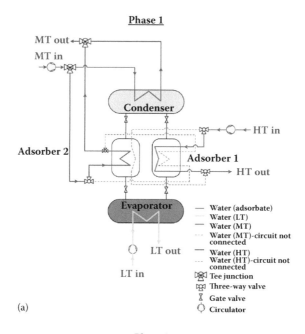

(a)

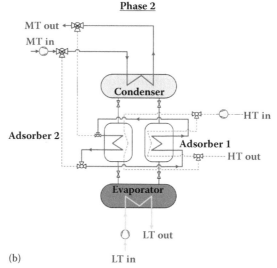

(b)

FIGURE 6.37
Sequence of phases in a working cycle of the eZea adsorption chiller: (a) Phase 1 (b) Phase 2.

adsorbs water, rejecting heat that is transferred to the desorbing Adsorber No.1, until a certain temperature difference between the two beds is achieved. After the completion of Phase 4, the cycle restarts with Phase 1.

6.9.1.5 Performance Data

As already mentioned, performance data for the discussed chiller is not publicly available, and thus only key design data was allowed to be reported and is shown in Table 6.6.

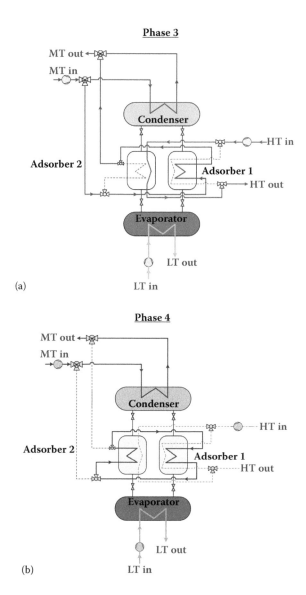

FIGURE 6.38
Sequence of phases in a working cycle of the eZea adsorption chiller: (a) Phase 3 (b) Phase 4.

6.9.2 The University of Freiburg Hospital Case

The University of Freiburg Hospital is another example of a solar adsorption installation. A lab section with a total area of 550 m² is powered by a solar adsorption system, which is equipped with two variable air flow rate ventilation systems. The adsorption chiller, model Nishiyodo NAK 20/70, has a cooling capacity of 70 kW$_c$, with a nominal COP of 0.6, while the nominal output temperature of the chilled air is 9°C. The required heat is supplied to the system by vacuum tube collectors, as shown in Figure 6.39, (model Seido 2-16) with a 167 m² net surface area and an efficiency of 32%. Typical supply temperature to the adsorption chiller is approximately 75°C. The heat rejection of the adsorption system is realized by a closed wet cooling tower.

TABLE 6.6

Technical Data for eZea 10 IPS

Parameter	Value
Basic Performance Data	
$\theta_{ch,w,in}$ (°C)	15
$\theta_{MT,w,in}$ (°C)	22–45
$\theta_{LT,w,in}$ (°C)	8–21
$\theta_{HT,w,in}$ (°C)	75–95
$\dot{m}_{LT}$ (kg/s)	0.55–0.80
$\dot{m}_{MT}$ (kg/s)	1.13–1.40
$\dot{m}_{HT}$ (kg/s)	0.44–0.69
$\dot{Q}_{hp}$ (kW)	up to 13
$\dot{Q}_{refr}$ (kW)	up to 40
CO_{Prefr}	up to 0.53
Max pressure water circuits (bar)	4
Dimensions	
W (mm) × D (mm) × H (mm)	670 × 560 × 1652
Weight (kg)	approx. 234.5

The system is also equipped with two storage tanks, a 6 m³ heat storage tank, and a 2 m³ cold storage tank. Backup heat can be supplied by district heating in order to power the adsorption chiller during the summer period. During the winter period, the district air heating is supplied directly to the lab section assisted by the solar collectors. During summer days with clear skies, the solar collectors can provide up to 90% of the required heat input in the adsorption chiller. The annual specific collector yield was 365 kWh/m² for 2003 (SOLAIR, Kalkan et al. 2012). The total cost of the system was approximately €352,000, with three-fourths of the total cost subsidized by the state. The annual maintenance costs are estimated to be about €12,000 (Wang and Oliveira 2006).

6.9.3 Fraunhofer Institute for Solar Technology—Freiburg, Germany

A small-scale solar thermal system was developed and is operated by the Institute for Solar Energy System (ISE) in Freiburg, Germany. The system uses a 5.5 kWc adsorption

FIGURE 6.39
The solar collector field at the University of Freiburg Hospital. (Reproduced from Kalkan, Naci, E. et al., *Renewable and Sustainable Energy Reviews*, 16 (8): 6352–6383, 2012.)

TABLE 6.7

Technical Specifications of the Desiccant Cooling
Installation at the Fraunhofer ISE, Freiburg, Germany

General Information	
Application	space cooling
Location	Freiburg, Germany
Total cooling area	42 m²
Start of operation	2007
Solar System	
Collector type	flat-plate collectors
Collector area	22 m²
Tilt angle	30°
Storage	
Hot storage	2 m³
Cold storage	–
Air Conditioning Unit	
Working pair	silica gel-water
Cooling capacity	5.5 kW
Auxiliary Equipment	
Backup heat from institute's CHP unit	
Performance	
Solar fraction	n/a
COP	0.43

Source: Kalkan, Naci, E. et al., *Renewable and Sustainable Energy Reviews*, 16 (8): 6352–6383, 2012.

chiller for the cooling and heating needs of the canteen kitchen in the Fraunhofer ISE, with a total area of 42 m². The heating mode capacity of the chiller is rated at 12 kW. The system has operated since 2007 and has three modes of operation:

- Cooling mode during the summer period
- Heat rejection from the chiller via three ground tubes
- Heating mode during the winter period

Additional heat can be provided by the Institute's CHP unit. Based on measurements conducted between August 2008 and July 2009, the average COP was 0.43 (Kalkan et al. 2012). A summary of the key technical specifications for the installation at the Fraunhofer ISE, Freiburg is listed in Table 6.7.

Nomenclature

A	Surface	[m²]
c_p	Specific heat capacity	[J/kg.K]
COP	Coefficient of performance	–
COP_{carnot}	Carnot's coefficient of performance	–

D_{so}	Pre-exponential constant of LDF equation	$[m^2/s]$
E_α	Activation energy	$[J/kg]$
H	Enthalpy	$[J/kg]$
h_{fg}	Heat of vaporization	$[J/kg]$
M	Total mass	$[kg]$
$\dot{m}$	Mass flow	$[kg/s]$
N	Heterogeneity constant	–
P	Pressure	$[Pa]$
P_s	Saturation pressure	$[Pa]$
P_w	Partial vapor pressure	$[Pa]$
$\dot{Q}$	Heat flux	$[W]$
Q_{st}	Isosteric heat of adsorption	$[J/kg]$
R	Gas constant	$[J/kg.K]$
R_p	Average pore radius	$[m]$
SCP	Specific cooling power	$[W/kg]$
T	Temperature	$[K]$
T	Time	$[s]$
U	Overall heat transfer coefficient	$[W/m^2.K]$
X	Adsorbate uptake	$[kg/kg]$
x_o	Limiting adsorbate uptake	$[kg/kg]$
x^*	Equilibrium adsorbate uptake	$[kg/kg]$

Greek Symbols

θ	Temperature	$[°C]$
η	Second law efficiency	–

Subscripts

ads	Adsorber
C	Condensation
Ch,w	Chilled water
Cu	Copper (wall material)
des	Desorber
E	Evaporation
Hp	Heat pump mode
HT	Hot water connected to desorber
MT1	Medium temperature water connected to adsorber beds

MT2	Medium temperature water connected to condenser
refr	Refrigerant (water in the adsorption cycle circuit)
W	Water
Ze	Zeolit

References

Abu-Hamdeh, Nidal H., Khaled A. Alnefaie, and Khalid H. Almitani. 2013. "Design and Performance Characteristics of Solar Adsorption Refrigeration System Using Parabolic Trough Collector: Experimental and Statistical Optimization Technique." *Energy Conversion and Management* 74: 162–170.

Alahmer, Ali, Xiaolin Wang, Raed Al-Rbaihat, K. C. Amanul Alam, and B. B. Saha. 2016. "Performance Evaluation of a Solar Adsorption Chiller Under Different Climatic Conditions." *Applied Energy* 175: 293–304.

Allouhi, A., T. Kousksou, A. Jamil, T. El Rhafiki, Y. Mourad, and Y. Zeraouli. 2015. "Optimal Working Pairs for Solar Adsorption Cooling Applications." *Energy* 79: 235–247.

Anyanwu, E. E., and N. V. Ogueke. 2007. "Transient Analysis and Performance Prediction of a Solid Adsorption Solar Refrigerator." *Applied Thermal Engineering* 27 (14–15): 2514–2523.

Aristov, Yu I., D. M. Chalaev, B. Dawoud, L. I. Heifets, O. S. Popel, and G. Restuccia. 2007. "Simulation and Design of a Solar Driven Thermochemical Refrigerator Using New Chemisorbents." *Chemical Engineering Journal* 134 (1–3): 58–65.

Askalany, Ahmed A., Stefan K. Henninger, Mohamed Ghazy, and Bidyut B. Saha. 2017. "Effect of Improving Thermal Conductivity of the Adsorbent on Performance of Adsorption Cooling System." *Applied Thermal Engineering* 110: 695–702.

Askalany, Ahmed A., M. Salem, I. M. Ismail, Ahmed Hamza H. Ali, and M. G. Morsy. 2012. "A Review on Adsorption Cooling Systems with Adsorbent Carbon." *Renewable and Sustainable Energy Reviews* 16 (1): 493–500.

Baerlocher, Ch., and L.B. McCusker. 2016. "Database of Zeolite Structures." Last Modified October 1. http://www.iza-structure.org/databases/.

Baiju, V., and C. Muraleedharan. 2011. "Performance Study of a Two Stage Solar Adsorption Refrigeration System." *International Journal of Engineering Science and Technology* 3 (7): 5754–5764.

Bao, Huashan, Yaodong Wang, Constantinos Charalambous, Zisheng Lu, Liwei Wang, Ruzhu Wang, and Anthony Paul Roskilly. 2014. "Chemisorption Cooling and Electric Power Cogeneration System Driven by Low Grade Heat." *Energy* 72: 590–598.

Berdja, Mohand, Brahim Abbad, Ferhat Yahi, Fateh Bouzefour, and Maamar Ouali. 2014. "Design and Realization of a Solar Adsorption Refrigeration Machine Powered by Solar Energy." *Energy Procedia* 48: 1226–1235.

Bergman, T. L., and Frank P. Incropera. 2011. *Fundamentals of Heat and Mass Transfer*. 7th ed. Hoboken, NJ: Wiley.

Brancato, V., A. Frazzica, A. Sapienza, L. Gordeeva, and A. Freni. 2015. "Ethanol Adsorption Onto Carbonaceous and Composite Adsorbents for Adsorptive Cooling System." *Energy* 84: 177–185.

Cecen, Ferhan, and Ozgur Aktas. 2011. *Activated Carbon for Water and Wastewater Treatment: Integration of Adsorption and Biological Treatment*. Weinheim: Wiley.

Cejka, Jiri. 2007. *Introduction to Zeolite Science and Practice*. Amsterdam: Elsevier.

Chang, W. S., C. C. Wang, and C. C. Shieh. 2007. "Experimental Study of a Solid Adsorption Cooling System Using Flat-Tube Heat Exchangers as Adsorption Bed." *Applied Thermal Engineering* 27 (13): 2195–2199.

Chang, W. S., C. C. Wang, and C. C. Shieh. 2009. "Design and Performance of a Solar-Powered Heating and Cooling System Using Silica Gel/Water Adsorption Chiller." *Applied Thermal Engineering* 29 (10): 2100–2105.

Chen, C. J., R. Z. Wang, Z. Z. Xia, J. K. Kiplagat, and Z. S. Lu. 2010. "Study on a Compact Silica Gel–Water Adsorption Chiller without Vacuum Valves: Design and Experimental Study." *Applied Energy* 87 (8): 2673–2681.

Chen, J. Y. 2017. "Introduction." In *Activated Carbon Fiber and Textiles*, 3–20. Oxford: Woodhead Publishing.

Chua, H. T., K. C. Ng, A. Malek, T. Kashiwagi, A. Akisawa, and B. B. Saha. 1999. "Modeling the Performance of Two-Bed, Sillica Gel-Water Adsorption Chillers." *International Journal of Refrigeration* 22 (3): 194–204.

Chua, H. T., K. C. Ng, A. Malek, T. Kashiwagi, A. Akisawa, and B. B. Saha. 2001. "Multi-Bed Regenerative Adsorption Chiller—Improving the Utilization of Waste Heat and Reducing the Chilled Water Outlet Temperature Fluctuation." *International Journal of Refrigeration* 24 (2): 124–136.

Comyns, Alan E. 2009. "Useful Zeotypes." *Focus on Catalysts* 2009 (4): 1–2.

Critoph, R. E., and S. J. Metcalf. 2004. "Specific Cooling Power Intensification Limits in Ammonia–Carbon Adsorption Refrigeration Systems." *Applied Thermal Engineering* 24 (5–6): 661–678.

Cui, Qun, Gang Tao, Haijun Chen, Xinyue Guo, and Huqing Yao. 2005. "Environmentally Benign Working Pairs for Adsorption Refrigeration." *Energy* 30 (2–4): 261–271.

Du, S. W., X. H. Li, Z. X. Yuan, C. X. Du, W. C. Wang, and Z. B. Liu. 2016. "Performance of Solar Adsorption Refrigeration in System of SAPO-34 and ZSM-5 Zeolite." *Solar Energy* 138: 98–104.

El-Sharkawy, I. I., B. B. Saha, S. Koyama, J. He, K. C. Ng, and C. Yap. 2008. "Experimental Investigation on Activated Carbon–Ethanol Pair for Solar Powered Adsorption Cooling Applications." *International Journal of Refrigeration* 31 (8): 1407–1413.

El-Sharkawy, Ibrahim I., Animesh Pal, Takahiko Miyazaki, Bidyut Baran Saha, and Shigeru Koyama. 2016. "A Study on Consolidated Composite Adsorbents for Cooling Application." *Applied Thermal Engineering* 98: 1214–1220.

Fadar, A. El, A. Mimet, and M. Pérez-García. 2009a. "Modelling and Performance Study of a Continuous Adsorption Refrigeration System Driven by Parabolic Trough Solar Collector." *Solar Energy* 83 (6): 850–861.

Fadar, A. El, A. Mimet, and M. Pérez-García. 2009b. "Study of an adsorption refrigeration system powered by parabolic trough collector and coupled with a heat pipe." *Renewable Energy* 34 (10): 2271–2279.

Fahrenheit. "Fahrenheit: eZea." Accessed January 2017. https://fahrenheit.cool/en/zeo/.

Fang, Y. T., T. Liu, Z. C. Zhang, and X. N. Gao. 2014. "Silica Gel Adsorbents Doped with Al, Ti, and Co Ions Improved Adsorption Capacity, Thermal Stability and Aging Resistance." *Renewable Energy* 63: 755–761.

Flannery, Barry, Robert Lattin, Oliver Finckh, Harald Berresheim, and Rory F. D. Monaghan. 2017. "Development and Experimental Testing of a Hybrid Stirling Engine-Adsorption Chiller Auxiliary Power Unit for Heavy Trucks." *Applied Thermal Engineering* 112: 464–471.

Fong, K. F., C. K. Lee, T. T. Chow, Z. Lin, and L. S. Chan. 2010. "Solar Hybrid Air-Conditioning System for High Temperature Cooling in Subtropical City." *Renewable Energy* 35 (11): 2439–2451.

Freni, A., Gaetano Maggio, Alessio Sapienza, Andrea Frazzica, Giovanni Restuccia, and Salvatore Vasta. 2016. "Comparative Analysis of Promising Adsorbent/Adsorbate Pairs for Adsorptive Heat Pumping, Air Conditioning and Refrigeration." *Applied Thermal Engineering* 104: 85–95.

Freni, A., F. Russo, S. Vasta, M. Tokarev, Yu I. Aristov, and G. Restuccia. 2007. "An Advanced Solid Sorption Chiller Using SWS-1L." *Applied Thermal Engineering* 27 (13): 2200–2204.

Freni, Angelo, Alessio Sapienza, Ivan S. Glaznev, Yuriy I. Aristov, and Giovanni Restuccia. 2012. "Experimental Testing of a Lab-Scale Adsorption Chiller Using a Novel Selective Water Sorbent 'Silica Modified by Calcium Nitrate.'" *International Journal of Refrigeration* 35 (3): 518–524.

Fujioka, Keiko, and Hiroshi Suzuki. 2013. "Thermophysical Properties and Reaction Rate of Composite Reactant of Calcium Chloride and Expanded Graphite." *Applied Thermal Engineering* 50 (2): 1627–1632.

Gong, L. X., R. Z. Wang, Z. Z. Xia, and Z. S. Lu. 2012. "Experimental Study on an Adsorption Chiller Employing Lithium Chloride in Silica Gel and Methanol." *International Journal of Refrigeration* 35 (7): 1950–1957.

González, Manuel I., and Luis R. Rodríguez. 2007. "Solar Powered Adsorption Refrigerator with CPC Collection System: Collector Design and Experimental Test." *Energy Conversion and Management* 48 (9): 2587–2594.

Gordeeva, Larisa G, Angelo Freni, Yuri I Aristov, and Giovanni Restuccia. 2009a. "Composite Sorbent of Methanol 'Lithium Chloride in Mesoporous Silica Gel' for Adsorption Cooling Machines: Performance and Stability Evaluation." *Industrial & Engineering Chemistry Research* 48 (13): 6197–6202.

Gordeeva, Larisa G., and Yuriy I. Aristov. 2011. "Composite sorbent of methanol "Licl in Mesoporous Silica Gel" for Adsorption Cooling: Dynamic Optimization." *Energy* 36 (2): 1273–1279.

Gordeeva, Larisa G., Alexandra D. Grekova, Tamara A. Krieger, and Yuriy I. Aristov. 2009b. "Adsorption Properties of Composite Materials (LiCl + LiBr)/silica." *Microporous and Mesoporous Materials* 126 (3): 262–267.

Habib, Khairul, Biplab Choudhury, Pradip Kumar Chatterjee, and Bidyut Baran Saha. 2013. "Study on a Solar Heat Driven Dual-Mode Adsorption Chiller." *Energy* 63: 133–141.

Habib, Khairul, Bidyut Baran Saha, Anutosh Chakraborty, Shigeru Koyama, and Kandadai Srinivasan. 2011. "Performance Evaluation of Combined Adsorption Refrigeration Cycles." *International Journal of Refrigeration* 34 (1): 129–137.

Habib, Khairul, Bidyut Baran Saha, and Shigeru Koyama. 2014. "Study of Various Adsorbent–Refrigerant Pairs for the Application of Solar Driven Adsorption Cooling in Tropical Climates." *Applied Thermal Engineering* 72 (2): 266–274.

Henninger, Stefan K, Felix Jeremias, Harry Kummer, and Christoph Janiak. 2012. "MOFs for Use in Adsorption Heat Pump Processes." *European Journal of Inorganic Chemistry* 2012 (16): 2625–2634.

Hu, Peng, Juan-Juan Yao, and Ze-Shao Chen. 2009. "Analysis for Composite Zeolite/Foam Aluminum–Water Mass Recovery Adsorption Refrigeration System Driven by Engine Exhaust Heat." *Energy Conversion and Management* 50 (2): 255–261.

Hu, Yuanyang, Liwei Wang, Lu Xu, Ruzhu Wang, Jeremiah Kiplagat, and Jian Wang. 2011. "A Two-Stage Deep Freezing Chemisorption Cycle Driven by Low-Temperature Heat Source." *Frontiers in Energy* 5 (3): 263–269.

Inagaki, Michio, and Feiyu Kang. 2014. "Fundamental Science of Carbon Materials." In *Materials Science and Engineering of Carbon: Fundamentals*. 2nd ed., 17–217. Oxford: Butterworth-Heinemann.

Jaiswal, Ankush Kumar, Sourav Mitra, Pradip Dutta, Kandadai Srinivasan, and S. Srinivasa Murthy. 2016. "Influence of Cycle Time and Collector Area on Solar Driven Adsorption Chillers." *Solar Energy* 136: 450–459.

Jiang, L., L. W. Wang, W. L. Luo, and R. Z. Wang. 2015. "Experimental Study on Working Pairs for Two-Stage Chemisorption Freezing Cycle." *Renewable Energy* 74: 287–297.

Jiang, Long, LiWei Wang, RuZhu Wang, Peng Gao, and FenPing Song. 2014. "Investigation on Cascading Cogeneration System of ORC (Organic Rankine Cycle) and Cacl2/Bacl2 Two-Stage Adsorption Freezer." *Energy* 71: 377–387.

Kalkan, Naci, E. A. Young, and Ahmet Celiktas. 2012. "Solar Thermal Air Conditioning Technology Reducing the Footprint of Solar Thermal Air Conditioning." *Renewable and Sustainable Energy Reviews* 16 (8): 6352–6383.

Kayal, Sibnath, Sun Baichuan, and Bidyut Baran Saha. 2016. "Adsorption Characteristics of AQSOA Zeolites and Water for Adsorption Chillers." *International Journal of Heat and Mass Transfer* 92: 1120–1127.

Kim, Chaehoon, Kanghee Cho, Sang Kyum Kim, Eun Kyung Lee, Jong-Nam Kim, and Minkee Choi. 2017. "Alumina-Coated Ordered Mesoporous Silica as an Efficient and Stable Water Adsorbent for Adsorption Heat Pump." *Microporous and Mesoporous Materials* 239: 310–315.

Kim, Young-Deuk, Kyaw Thu, and Kim Choon Ng. 2014. "Adsorption Characteristics of Water Vapor on Ferroaluminophosphate for Desalination Cycle." *Desalination* 344: 350–356.

Kiplagat, J. K., R. Z. Wang, R. G. Oliveira, and T. X. Li. 2010. "Lithium Chloride—Expanded Graphite Composite Sorbent for Solar Powered Ice Maker." *Solar Energy* 84 (9): 1587–1594.

Kulprathipanja, Santi, and InterScience Wiley. 2010. *Zeolites in Industrial Separation and Catalysis*. Weinheim: Wiley-VCH.

Küsgens, Pia, Marcus Rose, Irena Senkovska, Heidrun Fröde, Antje Henschel, Sven Siegle, and Stefan Kaskel. 2009. "Characterization of Metal-Organic frameworks by Water Adsorption." *Microporous and Mesoporous Materials* 120 (3): 325–330.

Lemmini, F., and A. Errougani. 2005. "Building and Experimentation of a Solar Powered Adsorption Refrigerator." *Renewable Energy* 30 (13): 1989–2003.

Lemmini, Fatiha, and Abdelmoussehel Errougani. 2007. "Experimentation of a Solar Adsorption Refrigerator in Morocco." *Renewable Energy* 32 (15): 2629–2641.

Li, S. L., J. Y. Wu, Z. Z. Xia, and R. Z. Wang. 2009. "Study on the Adsorption Performance of Composite Adsorbent of Cacl2 and Expanded Graphite with Ammonia as Adsorbate." *Energy Conversion and Management* 50 (4): 1011–1017.

Li, S. L., Z. Z. Xia, J. Y. Wu, J. Li, R. Z. Wang, and L. W. Wang. 2010. "Experimental Study of a Novel Cacl2/Expanded Graphite-NH3 Adsorption Refrigerator." *International Journal of Refrigeration* 33 (1): 61–69.

Liu, J., and J. Yu. 2016. "Toward Greener and Designed Synthesis of Zeolite Materials." In *Zeolites and Zeolite-Like Materials*, edited by Leonid M. Kustov, 1–32. Amsterdam: Elsevier.

Liu, Y., and K. C. Leong. 2008. "Numerical Modeling of a Zeolite/Water Adsorption Cooling System with Non-Constant Condensing Pressure." *International Communications in Heat and Mass Transfer* 35 (5): 618–622.

Llano-Restrepo, Mario, and Martín A. Mosquera. 2009. "Accurate Correlation, Thermochemistry, and Structural Interpretation of Equilibrium Adsorption Isotherms of Water Vapor in Zeolite 3A by Means of a Generalized Statistical Thermodynamic Adsorption Model." *Fluid Phase Equilibria* 283 (1–2): 73–88.

Loh, W. S., I. I. El-Sharkawy, K. C. Ng, and B. B. Saha. 2009. "Adsorption Cooling Cycles for Alternative Adsorbent/Adsorbate Pairs working at Partial Vacuum and Pressurized Conditions." *Applied Thermal Engineering* 29 (4): 793–798.

Lu, G. Q., and X. S. Zhao. 2004. *Nanoporous Materials: Science and Engineering*. London: Imperial College Press.

Lu, Z. S., and R. Z. Wang. 2013. "Performance Improvement and Comparison of Mass Recovery in Cacl2/Activated Carbon Adsorption Refrigerator and Silica Gel/Licl Adsorption Chiller Driven by Low Grade Waste Heat." *International Journal of Refrigeration* 36 (5): 1504–1511.

Lu, Z. S., and R. Z. Wang. 2014. "Study of the New Composite Adsorbent of Salt Licl/Silica Gel–Methanol Used in An Innovative Adsorption Cooling Machine Driven by Low Temperature Heat Source." *Renewable Energy* 63: 445–451.

Luo, H. L., Y. J. Dai, R. Z. Wang, J. Y. Wu, Y. X. Xu, and J. M. Shen. 2006. "Experimental Investigation of a Solar Adsorption Chiller Used for Grain Depot Cooling." *Applied Thermal Engineering* 26 (11–12): 1218–1225.

Maggio, G., L. G. Gordeeva, A. Freni, Yu I. Aristov, G. Santori, F. Polonara, and G. Restuccia. 2009. "Simulation of a Solid Sorption Ice-Maker Based on The Novel Composite Sorbent 'Lithium Chloride In Silica Gel Pores.'" *Applied Thermal Engineering* 29 (8–9): 1714–1720.

Martínez, C., and A. Corma. 2013. "Zeolites" In *Comprehensive Inorganic Chemistry II*, 2nd ed, edited by Kenneth Poeppelmeier, 103–131. Amsterdam: Elsevier.

Metcalf, S. J., R. E. Critoph, and Z. Tamainot-Telto. 2012. "Optimal Cycle Selection in Carbon-Ammonia Adsorption Cycles." *International Journal of Refrigeration* 35 (3): 571–580.

Mette, Barbara, Henner Kerskes, Harald Drück, and Hans Müller-Steinhagen. 2014. "Experimental and Numerical Investigations on the Water Vapor Adsorption Isotherms and Kinetics of Binderless Zeolite 13X." *International Journal of Heat and Mass Transfer* 71: 555–561.

Meunier, F., P. Neveu, and J. Castaing-Lasvignottes. 1998. "Equivalent Carnot Cycles for Sorption Refrigeration: Cycles de Carnot Équivalents pour la Production de Froid par Sorption." *International Journal of Refrigeration* 21 (6): 472–489.

Mitra, Sourav, Pramod Kumar, Kandadai Srinivasan, and Pradip Dutta. 2015. "Performance Evaluation of s Two-Stage Silica Gel + Water Adsorption Based Cooling-Cum-Desalination System." *International Journal of Refrigeration* 58: 186–198.

N'Tsoukpoe, Kokouvi Edem, Holger Urs Rammelberg, Armand Fopah Lele, Kathrin Korhammer, Beatriz Amanda Watts, Thomas Schmidt, and Wolfgang K. L. Ruck. 2015. "A Review on the Use of Calcium Chloride in Applied Thermal Engineering." *Applied Thermal Engineering* 75: 513–531.

Oliveira, R. G., and R. Z. Wang. 2007. "A Consolidated Calcium Chloride-Expanded Graphite Compound for Use in Sorption Refrigeration Systems." *Carbon* 45 (2): 390–396.

Pan, Q. W., R. Z. Wang, Z. S. Lu, and L. W. Wang. 2014. "Experimental Investigation of an Adsorption Refrigeration Prototype with the Working Pair of Composite Adsorbent-Ammonia." *Applied Thermal Engineering* 72 (2): 275–282.

Pflitsch, Christian, Benjamin Curdts, Martin Helmich, Christoph Pasel, Christian Notthoff, Dieter Bathen, and Burak Atakan. 2014. "Chemical Vapor Infiltration of Activated Carbon with Tetramethylsilane." *Carbon* 79: 28–35.

Pistocchini, Lorenzo, Silvia Garone, and Mario Motta. 2016. "Air Dehumidification by Cooled Adsorption in Silica Gel Grains. Part I: Experimental Development of a Prototype." *Applied Thermal Engineering* 107: 888–897.

Pons, M., and F. Poyelle. 1999. "Adsorptive Machines with Advanced Cycles for Heat Pumping or Cooling Applications: Cycles a Adsorption Pour Pompes a Chaleur ou Machines Frigor: Figues." *International Journal of Refrigeration* 22 (1): 27–37.

Rezk, Ahmed R. M., and Raya K. Al-Dadah. 2012. "Physical and Operating Conditions Effects on Silica Gel/Water Adsorption Chiller Performance." *Applied Energy* 89 (1): 142–149.

Sadeghlu, Alireza, Mortaza Yari, and Hossein Beidaghy Dizaji. 2015. "Simulation Study of a Combined Adsorption Refrigeration System." *Applied Thermal Engineering* 87: 185–199.

Sah, Ramesh P., Biplab Choudhury, and Ranadip K. Das. 2015. "A Review on Adsorption Cooling Systems with Silica Gel And Carbon as Adsorbents." *Renewable and Sustainable Energy Reviews* 45: 123–134.

Saha, Bidyut Baran, Anutosh Chakraborty, Shigeru Koyama, and Yu I. Aristov. 2009. "A New Generation Cooling Device Employing Cacl2-in-Silica Gel–Water System." *International Journal of Heat and Mass Transfer* 52 (1–2): 516–524.

San, Jung-Yang, and Hui-Chi Hsu. 2009. "Performance of a Multi-Bed Adsorption Heat Pump Using SWS-1L Composite Adsorbent and Water as the Working Pair." *Applied Thermal Engineering* 29 (8–9): 1606–1613.

San, Jung-Yang, and Wei-Min Lin. 2008. "Comparison Among Three Adsorption Pairs for Using as the Working Substances in a Multi-Bed Adsorption Heat Pump." *Applied Thermal Engineering* 28 (8–9): 988–997.

Sandoval, M. V., J. A. Henao, C. A. Ríos, C. D. Williams, and D. C. Apperley. 2009. "Synthesis and Characterization of Zeotype ANA Framework by Hydrothermal Reaction of Natural Clinker." *Fuel* 88 (2): 272–281.

Sayılgan, Şefika Çağla, Moghtada Mobedi, and Semra Ülkü. 2016. "Effect of Regeneration Temperature on Adsorption Equilibria and Mass Diffusivity of Zeolite 13x-Water Pair." *Microporous and Mesoporous Materials* 224: 9-16.

Shahata, Mohamed M. 2016. "Adsorption of Some Heavy Metal Ions by used Different Immobilized Substances on Silica Gel." *Arabian Journal of Chemistry* 9 (6): 755–763.

Sharonov, V. E., and Yu I. Aristov. 2008. "Chemical and Adsorption Heat Pumps: Comments on the Second Law Efficiency." *Chemical Engineering Journal* 136 (2–3): 419–424.

Sing, Kenneth S. W. 1998. "Adsorption Methods for the Characterization of Porous Materials." *Advances in Colloid and Interface Science* 76–77: 3–11.

SOLAIR. "University Hospital in Freiburg, Germany." Accessed October 2017. http://www.solair-project.eu/180.0.html.

Srivastava, N. C., and I. W. Eames. 1998. "A Review of Adsorbents and Adsorbates in Solid–Vapour Adsorption Heat Pump Systems." *Applied Thermal Engineering* 18 (9–10): 707–714.

Suleiman, R., C. Folayan, F. Anafi, and D. Kulla. 2012. "Transient Simulation of a Flat Plate Solar Collector Powered Adsorption Refrigeration System." *International Journal of Renewable Energy Research* 2 (4): 657–664.

Tangkengsirisin, Vichan, Atsushi Kanzawa, and Takayuki Watanabe. 1998. "A Solar-Powered Adsorption Cooling System Using a Silica Gel–Water Mixture." *Energy* 23 (5): 347–353.

Tatlier, Melkon. 2017. "Performances of MOF Vs. Zeolite Coatings in Adsorption Cooling Applications." *Applied Thermal Engineering* 113: 290–297.

Thu, Kyaw, Bidyut Baran Saha, Anutosh Chakraborty, Won Gee Chun, and Kim Choon Ng. 2011. "Study on an Advanced Adsorption Desalination Cycle with Evaporator–Condenser Heat Recovery Circuit." *International Journal of Heat and Mass Transfer* 54 (1–3): 43–51.

Tierney, M., L. Ketteringham, and M. Azri Mohd Nor. 2017. "Performance of a Finned Activated Carbon Cloth-Ethanol Adsorption Chiller." *Applied Thermal Engineering* 110: 949–961.

Tso, C. Y., and Christopher Y. H. Chao. 2012. "Activated Carbon, Silica-Gel and Calcium Chloride Composite Adsorbents for Energy Efficient Solar Adsorption Cooling and Dehumidification Systems." *International Journal of Refrigeration* 35 (6): 1626–1638.

Utchariyajit, Kanchana, and Sujitra Wongkasemjit. 2008. "Structural Aspects of Mesoporous Alpo4-5 (AFI) Zeotype Using Microwave Radiation and Alumatrane Precursor." *Microporous and Mesoporous Materials* 114 (1–3): 175–184.

Veselovskaya, J. V., R. E. Critoph, R. N. Thorpe, S. Metcalf, M. M. Tokarev, and Yu I. Aristov. 2010. "Novel Ammonia Sorbents 'Porous Matrix Modified by Active Salt' for Adsorptive Heat Transformation: 3. Testing of 'Bacl2/Vermiculite' Composite in a Lab-Scale Adsorption Chiller." *Applied Thermal Engineering* 30 (10): 1188–1192.

Vodianitskaia, Paulo J., José J. Soares, Herbert Melo, and José Maurício Gurgel. 2017. "Experimental Chiller with Silica Gel: Adsorption Kinetics Analysis and Performance Evaluation." *Energy Conversion and Management* 132: 172–179.

Wang, D. C., Z. Z. Xia, and J. Y. Wu. 2006. "Design and Performance Prediction of a Novel Zeolite–Water Adsorption Air Conditioner." *Energy Conversion and Management* 47 (5): 590–610.

Wang, D. C., and J. P. Zhang. 2009. "Design and Performance Prediction of an Adsorption Heat Pump with Multi-Cooling Tubes." *Energy Conversion and Management* 50 (5): 1157–1162.

Wang, L., L. Chen, H. L. Wang, and D. L. Liao. 2009a. "The Adsorption Refrigeration Characteristics of Alkaline-Earth Metal Chlorides and Its Composite Adsorbents." *Renewable Energy* 34 (4): 1016–1023.

Wang, L. W., R. Z. Wang, and R. G. Oliveira. 2009b. "A Review on Adsorption Working Pairs for Refrigeration." *Renewable and Sustainable Energy Reviews* 13 (3): 518–534.

Wang, R. Z., and R. G. Oliveira. 2006. "Adsorption Refrigeration—An Efficient Way to Make Good Use of Waste Heat and Solar Energy." *Progress in Energy and Combustion Science* 32 (4): 424–458.

Wu, J. Y., and S. Li. 2009. "Study on Cyclic Characteristics of Silica Gel–Water Adsorption Cooling System Driven by Variable Heat Source." *Energy* 34 (11): 1955–1962.

Wu, Wei-Dong, Hua Zhang, and Da-Wen Sun. 2009. "Mathematical Simulation and Experimental Study of a Modified Zeolite 13X–Water Adsorption Refrigeration Module." *Applied Thermal Engineering* 29 (4): 645–651.

Xia, Z. Z., R. Z. Wang, D. C. Wang, Y. L. Liu, J. Y. Wu, and C. J. Chen. 2009. "Development and Comparison of Two-Bed Silica Gel–Water Adsorption Chillers Driven by Low-Grade Heat Source." *International Journal of Thermal Sciences* 48 (5): 1017–1025.

Yang, Pei-zhi. 2009. "Heat and Mass Transfer in Adsorbent Bed with Consideration of Non-Equilibrium Adsorption." *Applied Thermal Engineering* 29 (14–15): 3198–3203.

Yang, R. T. 2003. *Adsorbents: Fundamentals and Applications*. Hoboken, N.J.: Wiley-Interscience.

Ye, Hong, Zhi Yuan, Shiming Li, and Lei Zhang. 2014. "Activated Carbon Fiber Cloth and Cacl2 Composite Sorbents for a Water Vapor Sorption Cooling System." *Applied Thermal Engineering* 62 (2): 690–696.

Zhai, H., Y. J. Dai, J. Y. Wu, and R. Z. Wang. 2009. "Energy and Exergy Analyses on a Novel Hybrid Solar Heating, Cooling and Power Generation System for Remote Areas." *Applied Energy* 86 (9): 1395–404.

Zheng, Weihua, Jingtian Hu, Sammuel Rappeport, Zhen Zheng, Zixing Wang, Zheshen Han, James Langer, and James Economy. 2016. "Activated Carbon Fiber Composites for Gas Phase Ammonia Adsorption." *Microporous and Mesoporous Materials* 234: 146–154.

Zhong, Y., R. E. Critoph, R. N. Thorpe, Z. Tamainot-Telto, and Yu I. Aristov. 2007. "Isothermal Sorption Characteristics of the Bacl2–NH3 Pair In A Vermiculite Host Matrix." *Applied Thermal Engineering* 27 (14–15): 2455–2462.

7

Alternative and Hybrid Cooling Systems

7.1 Alternative Cooling Systems

7.1.1 Isothermal Dehumidification

Apart from the conventional open desiccant cooling cycle already discussed briefly in Chapter 2, the replacement of the adiabatic adsorptive process with an isothermal process has been investigated as a potential method for increasing efficiency (Nóbrega and Brum 2014). Toward that principle, Ge et al. (2009) developed and experimentally investigated the performance of a two-stage desiccant rotary cooling system. The setup, shown in Figure 7.1, consists of honeycombed silica gel-haloids composite desiccant wheels, an air preconditioning unit, heat exchangers for heat recovery, and air heaters for the recovery of the desiccant. The implementation of two stages allows for, as reported by the authors, the attainment of a design absolute humidity with lower temperatures than those in a single-stage case. The system's performance was evaluated under different ambient conditions and proved to work more efficiently than the single-stage system, with reported COPs in the range 0.8–1.4 for regeneration temperatures of 50–90°C.

Using simulations, La et al. (2012) investigated the energetic and exergetic efficiency of a novel desiccant cooling cycle incorporating isothermal dehumidification and regenerative evaporative cooling. Compared to the exergy efficiency of a conventional rotary desiccant cooling cycle, which for the investigated conditions was measured to be 8.6%, the novel cycle proved to be more efficient with a value as high as 29.1%. The corresponding COP was 0.74 for space cooling with a regeneration temperature of 52.3°C for the first stage and 62.0°C for the second stage, respectively.

Rady et al. (2009) proposed the use of macro-encapsulated PCMs in dehumidifying desiccant beds to realize isothermal dehumidification. According to the authors, the phase change materials enhanced the performance during the effective first stage of the adsorption process. The overall system efficiency was found to increase with a decrease in the phase change temperature of the PCM. The implementation of PCM led the adsorbing air streams to exit at lower temperature but with a higher moisture content, which eventually resulted in a system with a potential higher cooling capacity, at the expense of a dehumidification process with less sensible heating.

7.1.2 Ejector Cooling

Ejector cooling is a technology introduced over a century ago. During the early twentieth century, this technology was used for the air conditioning of trains and large buildings (Kim et al. 2007). A conventional steam ejector refrigeration system, as shown in Figure 7.2,

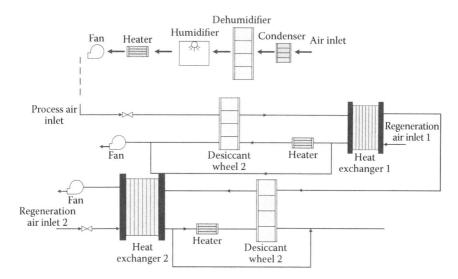

FIGURE 7.1

Schematic of the two-stage desiccant rotary cooling system. (Adapted from Ge, T. S. et al., *International Journal of Refrigeration*, 32 (3): 498–508, 2009.)

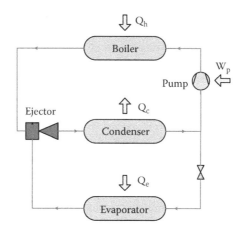

FIGURE 7.2

Schematic of a conventional steam jet ejector.

consists of an upper power cycle and a bottoming refrigeration cycle. In the upper cycle, low-grade heat, such as waste heat, is used to boil high pressure liquid refrigerant. The generated vapor is then led to the ejector, where it accelerates through the nozzle. The expansion that takes place induces vapor from the evaporator. The two streams mix in the mixing chamber before entering the diffuser section of the steam jet ejector (Chunnanond and Aphornratana 2004; Meyer et al. 2009). The mixed fluid is then led to the condenser, where it condenses, rejecting heat to the environment. A part of the stream is then led through the pump back to the boiler for the completion of the power cycle. The remainder of the liquid is expanded through a throttling device and enters the evaporator in a two-phase state. The two-phase fluid evaporates in the evaporator, producing the cooling effect, Q_e, and the resulting vapor is then led back to the ejector. Ejector cooling systems have the advantages of low initial costs and high reliability, absence of the need for maintenance, simple design, and an absence of moving parts (Balaras et al. 2007). The main drawback of the technology is the relatively poor COP. For instance, Alexis and Karayiannis (2005) reported that at a generator temperature 82–92°C, the COP ranged between 0.035–0.199 (condenser temperature 32–40°C and evaporator temperature –10–0°C).

Huang et al. (1998) investigated an ejector cooling system with R141b as its working fluid. The system was powered by a commercially available double-glazed flat-plate collector with solar surfaces, which also served as the generator of the ejector cycle. The system was modeled on a 1D gas dynamic model, modified properly to take secondary flow choking phenomena into account. The results of the model, which were experimentally validated by the authors, presented a maximum COP of 0.6 for a generation temperature of 100°C, a condensation temperature of 30°C, and an evaporation temperature of 8°C. When the system was powered by the solar collectors, the results presented in Table 7.1 were obtained.

Grazzini and Rocchetti (2002) conducted a numerical investigation and optimization of a two-stage ejector. According to the results, a compact geometrical configuration could allow for high ejector compression ratios at the expense of a low entrainment ratio. Regarding the system's COP, a maximum value of 0.529 was obtained with water as the working fluid. The corresponding second law efficiency was 14.6%, at a generation temperature of 120°C, a condenser temperature of 30°C, and an evaporation temperature of 10°C. The achieved cooling power output at this working point was 5 kW.

Khattab (2005) developed a model to design and evaluate the performance of a solar-powered steam-jet system that was used for cooling one ton of fresh fruits and vegetables located near Cairo, Egypt. The results showed that 42–45 m² of solar collectors with selective surfaces were able to provide the optimum solar fraction on a yearly basis. The maximum obtained cooling capacity of the system was around 0.6 kW, for a condensation temperature of 20°C, an evaporation temperature of 5°C, and a generation temperature of 90°C.

Shen et al. (2005) studied a gas-liquid ejector and evaluated its performance in a solar-driven bi-ejector refrigeration system. The bi-ejector system was realized by using one ejector to suck the refrigerant from the evaporator and a second ejector as a jet pump to circulate the refrigerant from the condenser toward the generator. For the simulations, a generation temperature in the range of 75–100°C, a condensation temperature between 28–40°C, and an evaporation temperature between 3–15°C were considered. Several refrigerants—R11, R12, R22, R123, R500, R717, and R718—were evaluated in terms of the cycle's performance. Among the CFCs, R11 was found to have the highest entrainment ratio, while R123 achieved the highest entrainment ratio among the HFCs. The maximum

TABLE 7.1

Overview of the Solar Ejector Cycle Performance
Investigated by Huang et al.

Parameter	Value
Capacity (kW)	10.5
Evaporator temperature (°C)	8
Condenser temperature (°C)	32
Generator temperature (°C)	95
Ejector COP (-)	0.50
Solar collector efficiency (%)	50
Overall COP (-)	0.22

Source: Huang, B. J. et al., *Solar Energy* 64 (4): 223–226, 1998.

reported solar COP was around 0.26, achieved by R717. R134a had the highest solar COP among the HFCs, with a value of 0.19. Additionally, a parametric analysis regarding the effect of the evaporation, condensation, and generator temperature on the system's performance was conducted using R134a as the working fluid. As expected, increasing the evaporation temperature and/or decreasing the condensation temperature enhances the system COP. On the other hand, there exists an optimum for the generation temperature (around 85°C) for a COP maximization at a value of 0.18 when the condensation temperature is 35°C and the evaporation takes place at 8°C.

Vidal et al. (2006) carried out an hourly simulation for a solar-powered ejector cooling cycle using R141b as the working fluid, based on meteorological data for Florianopolis, Brazil. The results of the parametric optimization for a 10.5 kW cooling capacity showed that the system could be powered by 80 m² of flat-plate solar collectors tilted 22° and a 4 m³ hot water storage tank, achieving a yearly average solar fraction of 42%, a generation temperature of 80°C, a condensation temperature of 32°C, and an evaporation temperature of 8°C.

Yu and Li (2007) reported a novel regenerative ejector refrigeration cycle. The system implemented an auxiliary jet pump and a conventional regenerator to enhance the performance of the ejector cycle. The simulations were conducted using R141b as the working fluid. The obtained COP was 9.3–12.1% higher than the conventional cycle's COP, for a generation temperature of 80–160°C, a condensation temperature of 35–45°C, and evaporation temperature of 10°C. An investigation of the pump's outlet pressure on the system performance revealed an enhancement of up to 18% for the COP at a generation temperature of 100°C and a condensation temperature of 40°C.

The performance variations of an evacuated tube solar collector's powered ejector cooling system were investigated by Ersoy et al. (2007) for several cities in Turkey. The working fluid of the cycle was R123. The simulations were carried out for the cooling season (May–October) for office hours (8:00 a.m. to 5:00 p.m.). For a generation temperature of 85°C, a condensation temperature of 30°C, and evaporation at 12°C, the maximum obtained solar COP was 0.197, with a cooling capacity of 178.26 W/(m² of solar collector), for the city of Aydin.

A solar-powered ejector refrigeration system using R600a was investigated by Pridasawas and Lundqvist (2007). The effect of the operating conditions and different solar collector types was evaluated via dynamic simulations conducted with TRNSYS software with data for Bangkok, Thailand. Three types of solar collectors were evaluated: single-glazed flat-plate collectors, double-glazed flat-plate collectors, and evacuated tube collectors. The COP of the cooling system was about 0.8 and the solar collector efficiency was approximately 47%, for evaporation at 15°C and condensation 5°C above the ambient temperature. The optimal choice was found to be the evacuated tube collectors with a total surface of 50 m², as they were able to provide a solar fraction of approximately 65%.

Nehdi et al. (2008) investigated an ejector refrigeration system powered by solar energy. Several working fluids were evaluated—R134a, R141b, R142b, R152a, R245fa, R290, R600, and R717—for the investigated cycle. According to the results of the simulations, R717 was found to demonstrate the highest COP, with a value of 0.408 at a generation temperature of 90°C, a condenser temperature of 35°C, and an evaporation temperature of 15°C. Based on meteorological data for the city of Tunis, Tunisia, the performance of three different types of solar collectors was evaluated. Evacuated tube collectors were found to have the highest performance, resulting in a solar COP in the range of 0.21–0.28 and a corresponding exergy efficiency of 0.14–0.19.

Through experimentation, Pollerberg et al. (2009) investigated a solar-powered ejector cycle used for air conditioning applications. The system, presented in Figure 7.3, was powered by parabolic trough collectors and used water as its working fluid. The maximum obtained solar

FIGURE 7.3
The experimental setup of the solar ejector system: (1) solar collector, (2) steam jet ejector, (3) evaporator, (4) steam drum, (5) condenser, and (6) convector. (Reproduced from Pollerberg, Clemens et al., *Applied Thermal Engineering*, 29 (5): 1245–1252, 2009.)

collector efficiency was in the range of 60% for a temperature difference of 25 K. The results for the system's operation were used to develop a simulation on the potential use of this setup in different locations: Essen (Germany), Toulouse (France), Genova (Italy), Safi (Morocco), and St. Katrine (Egypt). The maximum reported COP was obtained in Egypt with a mean value of 1.13, while Morocco had the lowest COP with a value of 0.83. Based on the aforementioned results and a primary economical calculation conducted by the authors, it was concluded that a specific cold price for the investigated regions using the solar ejector cycle was in the range of €0.15/kWh in the case of Egypt to €0.62/kWh in the case of Germany.

Meyer et al. (2009) developed a small-scale ejector cooling experimental setup. Three types of solar collectors were evaluated for this application: a flat-plate type, a high efficiency flat plate, and an evacuated tube collector. The evacuated tube collector achieved the highest efficiency. However, the flat-plate collectors were preferred because they were cheaper, easier to install, and were more reliable for long-term operation. For the selected type of solar collector, the reported COP was in the range of 0.253.

Ma et al. (2010) conducted experiments on a steam ejector chiller suitable for solar energy applications. Results indicated that there was an optimum in the boiler temperature to maximize the cooling capacity and the ejector's entrainment ratio. Maximum reported values were 3.7 and 0.35, respectively. The system's COP reaches its optimum at a slightly higher boiler temperature with a value of 0.33 at a boiler temperature of 92°C and an evaporation temperature of 10°C.

Zhang et al. (2012) investigated the performance of three types of solar collectors in a solar-driven ejector air conditioning unit located in a Mediterranean region (in the city of Tunis, Tunisia). The three types that were evaluated were a heat pipe solar collector with a single borosilicate glass cover (TMA600), a heat pipe collector with double-glass borosilicate glass cover (model TZ58-1800), and a direct flow solar collector with a single borosilicate glass cover (model Cortec2). They demonstrated, according to the simulation results, an efficiency of 70%, 53%, and 59%, respectively, at their design condition. When capital and maintenance costs were also taken into consideration, model TZ58-1800 was concluded to be the optimal choice, with 46.2 m² of such solar collectors being able to provide 16.7 kW heat at a temperature of 100°C, which can fully cover the heat requirements to drive the ejector cooling system for the daylight zone of the day (10 hours) during the summer in the investigated region.

Diaconu (2012) investigated the energetic performance of a solar-assisted ejector cycle used for air conditioning. Two main system configurations were evaluated. In the first module, a theoretical secondary storage unit of infinite storage capacity was considered, so that the solar ejector would continue to operate even if the cold storage unit were completely charged. In the second module, the ejector cycle would operate only during intervals with available solar radiation and would be switched off when the cold storage unit was fully charged. According to the results of the simulations, the second configuration performed better both in terms of the first law efficiency as well as the system's COP with values of 30% and 0.0965, respectively. The corresponding optimum values for the first module were 9.5% and 0.0814, respectively.

Yapıcı and Akkurt (2012) experimentally investigated an ejector cooling system driven by hot water and using R123 as the working fluid of the cycle. Based on the experimental data, a COP of 0.42 was reported for a generator temperature of 74°C, a condensation temperature of 29°C, and an evaporation temperature of 10°C. The system was evaluated for solar energy applications by installing single-glazed selective type collectors. It was found that 9.2 m² of such solar collectors could provide a cooling capacity of 1.08 kW at an evaporation temperature of 10°C.

Al-Alili et al. (2014) presented documentation of the reported values for the thermal COP and the corresponding cycle temperatures for ejector cycles found in literature before 2008. For reasons of completion, as well as to give the reader a visual overview of the operating scales in the ejector cycles, Figure 7.4 is reproduced, presenting the performance results of several ejector cycles operating with different working fluids and under different working conditions.

Guo and Shen (2009) modeled a solar-powered ejector refrigeration system used for an air conditioning application in an office building in Shanghai, China. R134a was used as the working fluid of the cycle. The model combined a lumped method with dynamic modeling to predict the system's behavior. According to the results of the simulations, an electric energy savings of up to 80% in comparison to the compressor of a conventional

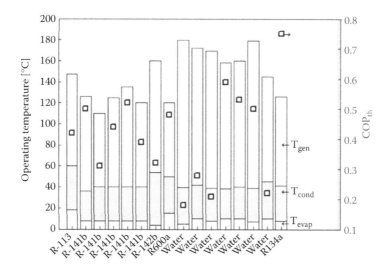

FIGURE 7.4
Overview of ejector cycle performance as reported in literature. (Reproduced from Al-Alili, Ali et al., *International Journal of Refrigeration*, 39 (Supplement C): 4–22, 2014.)

air conditioning unit could be achieved by the investigated system. During office working hours, from 9:00 a.m. to 5:00 p.m., the average COP and average solar fraction of the system were 0.48 and 82% respectively, at a generator temperature of 85°C and an evaporator temperature of 8°C, with a cooling capacity of 6 kW.

Dennis and Garzoli (2011) developed a model with TRNSYS software for a solar ejector cooling cycle employing a variable geometry ejector and a cold storage tank. The nominal cooling capacity of the modeled system was 3.5 kW at a generation temperature of 95°C, a condensation temperature of 32°C, and an evaporation temperature of 8°C. R141b was used as the refrigerant of the cycle. Meteorological data for the simulations was derived for the period of October to March in Canberra, Australia. The implementation of a variable geometry ejector enhanced the solar fraction by 8–13%. On the other hand, the use of cold storage, under the assumption of a constant solar fraction, was found to decrease the solar collector area and hence the cost of the system.

Jia and Wenjian (2012) optimized the area ratios for an air-cooled ejector cycle with a nominal capacity of 2 kW and R134a as its working fluid. The design of the ejector was realized with a 1D model. An experimental investigation was also held for six different ejector throat areas via a variable geometry ejector with a spindle under different operating conditions. According to the results of the experiments, it was concluded that the optimum area ratios have a linear dependence on the primary flow pressure. Furthermore, it was found that both the area ratios and the nozzle diameters affect the cooling capacity of the cycle. The COP was only affected by the area ratios. More specifically, an optimum value of the area ratios that maximized the system's COP and the ejector's entrainment ratio was identified. A maximum COP was reported for a primary flow pressure of 25 bar and an area ratio of 4.76 with a value of approximately 0.3. The corresponding cooling production by the system's evaporator was approximately 1.4 kW.

Tashtoush et al. (2015) evaluated the performance of a 7 kW solar-powered ejector cooling system installed in Jordan. The system used R134a as the working fluid. System modeling was developed using TRNSYS-EES software. With an optimal tilt angle of 28°, it was found that evacuated tube collectors demonstrate a better performance than flat-type collectors. Using a solar collector field of 60–70 m², a solar fraction in the range 52.0–54.2% was achieved. Under peak solar radiation and with the highest ambient temperature, the overall system efficiency had a minimum of 32%, with a solar collector efficiency between 52–92%, an ejector COP of 0.52–0.547, and a system COP of 0.32–0.47.

7.1.3 Stirling Cooling

The Stirling refrigeration cycle takes advantage of the volume change caused by the pistons, which results in compressing the flow (Kongtragool and Wongwises 2003). Solar-powered Stirling cycles offer high efficiencies at low capacities, as well as the possibility of combined power and cooling production, making this technology a potential choice when power production is also desired.

Richards and Auxer (1978) proposed a heat-activated heat pump that consisted of a natural-gas-fired Stirling engine (replacing an electric motor) driving the compressor of a vapor compression cycle (VCC). One advantage of the proposed cycle is the fact that in heating mode the waste heat of the engine can be exploited as an auxiliary heat source. Hence, the hybrid heat pump was found to have a higher capacity and a higher COP compared to conventional heating systems. With respect to cooling mode, the achieved COP of the system was reported to be around 3.2–3.5, resulting in significant energy savings, with an overall seasonal performance factor (SPF) up to twice the value of a conventional system in heating mode and around 50% better for both cooling and heating applications.

Berchowitz et al. (2008) investigated a free-piston Stirling engine directly driving the compressor of a heat pump cycle. The working gas used was CO_2. The cooling primary energy ratio (PER) was reported as a function of the heat pump temperature ratio, showing rapidly decreasing behavior with an increasing heat pump temperature ratio. The primary energy ratio was calculated by the authors based on Eq. (7.1):

$$PER_{cool} = \frac{\dot{Q}_{cool}}{\dot{W}_{in}} COP_s$$ (7.1)

where $\dot{W}_{in}$ refers to the power input of the cycle. The reported range of the PER was between 2.5 and 1.0, for a range in the heat pump temperature ratio of 1.00–1.30.

7.1.4 Electrochemical Cooling

The ability of electrochemical processes to generate heat and pressure creates the potential for utilizing such systems for cooling applications. More specifically, a reversible electrochemical cell absorbs or rejects heat when voltage is applied or reversed, respectively, allowing for implementation into a cooling cycle. The main advantage of this novel, and very young, technology is the absence, or near absence, of moving parts (Gerlach and Newell 2004). Furthermore, such systems operate with less hazardous materials, making them a more attractive choice in the direction of more environmentally friendly refrigeration systems.

Loutfy et al. (1983) patented a thermally regenerative electrochemical system. The system included an electrochemical cell with water-based electrolytes separated by an ion exchange membrane. The reported efficiencies of the system were in the range of 5–13%, proving that there is significant room for improvement before this technology becomes competitive from an energetic point of view.

Dittmar et al. (1994) investigated a novel electrochemical heat pump consisting of two identical electrochemical cells operating at different temperatures in opposite directions. The system, based on the exchange of reversible heat of reactions within the environment, had the advantage of having no electrolytic connection between the two thermal reservoirs. The heat pump concept proposed by the authors is presented in Figure 7.5. Experiments were carried out to evaluate the cells' heat of reaction in a specially designed heat flow calorimeter. Based on the results, Ni/Cd accumulators appeared to be the best candidates for electrochemical heat pumps, with a theoretical heat pump efficiency of 9.54 at a T_1 of 25°C and a T_2 equal to 60°C.

Hong et al. (1998) investigated the performance of Li-ion batteries for cooling and heating during charge and discharge. The sets were carried out at three different initial temperatures of the batteries, 35°C, 45°C, and 55°C. The cells were charged at a rate of the nominal cell capacity, and their COP was evaluated. The results of the tests are presented in Figure 7.6. As shown, the results for the COP are relatively low in all cases, with the best case scenario only reporting a COP of approximately 0.145, which is justified by the author as reasonable since such batteries are designed for power storage applications. Further obstacles that need to be addressed with the Li-ion batteries include discharge overheating and thermal run away.

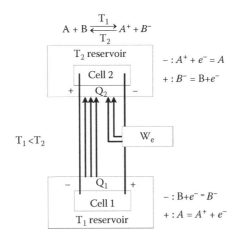

FIGURE 7.5
Schematic of the electrochemical heat pump. (Adapted from Dittmar, L. et al., "A New Concept of an Electrochemical Heat Pump System: Theoretical Consideration and Experimental Results", in *Electrochemical Engineering and Energy*, edited by F. Lapicque, A. Storck, and A. A. Wragg, 57–65, Boston, MA: Springer US, 1994.)

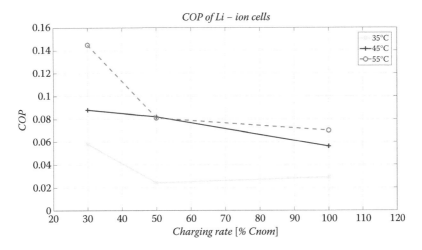

FIGURE 7.6
COP of Li-ion batteries for refrigeration applications for sets of starting temperature and charging rate. (Adapted from Hong, Jong-Sung et al., *Journal of the Electrochemical Society*, 145 (5): 1489–1501, 1998.)

7.2 Hybrid Cooling Systems

When looking to enhance the performance of cooling systems, several combined technology systems have been proposed and investigated. In this section, a brief summary of some such hybrid systems is presented.

7.2.1 Desiccant-Brayton Cascade Cycle

In a conventional Brayton cycle, air can be cooled down to an ambient temperature. The implementation of a desiccant cycle facilitates, via the evaporative cooling effect, a decrease

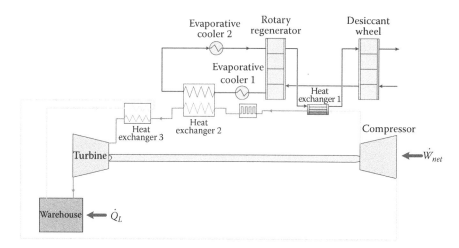

FIGURE 7.7
Schematic of the proposed Desiccant-Brayton Cascade cycle. (Adapted from Nóbrega, C. E. L., and L. A. Sphaier, *Energy and Buildings*, 55 (Supplement C): 575–584, 2012.)

in the air temperature beyond that limit, enhancing the performance of the Brayton cycle (Elsayed et al. 2006). The system, presented in Figure 7.7, was analyzed theoretically and was found to enhance the conventional Brayton cycle COP by 20% (Nóbrega and Sphaier 2012).

7.2.2 Desiccant-Vapor Compression Cycle

Peterson and Howell (1991) patented a hybrid desiccant-VCC system. In this system, the VCC was used for air conditioning, while the liquid desiccant allowed for the simultaneous cooling and dehumidification of process air. The liquid desiccants used in the system were aqueous solutions of glycol or brine. The use of circulating liquid desiccant and an adiabatic humidifier enhanced, according to the inventors, the cycle's performance, reducing the load of the compressor.

Yadav (1995) developed a thermodynamic model for a combined VCC-liquid desiccant cycle using R11 as refrigerant and H_2O-LiBr as the liquid desiccant. A parametric analysis was conducted by the author to evaluate the influence of several parameters on the system's COP and cooling capacity. In the simulations, the VCC condensation temperature was set at 50°C, the compressor efficiency was assumed to be 65%, and the level of superheating at the exit of the evaporator was set at 3 K. The air entering the condenser was considered to be 35°C with a relative humidity of 40%. The respective air inlet of the evaporator was considered to be 28.8°C with a relative humidity of 46%. The air exiting the evaporator was 17°C with a relative humidity of 72%. Results of the analysis showed that even with a lower system COP, the hybrid system was able to save more energy than the conventional system. The hybrid system's energy savings increased significantly at higher latent heat loads—a 50% increase in energy savings was calculated when the latent heat load increased from 40% to 90%. Dai et al. (2001) proposed and evaluated with experimental data a similar system (with a solution of CaCl as the liquid desiccant). To evaluate its performance, the authors defined the system's COP as the ratio of the cooling output divided by the heat input and the power consumption of the compressor with an equivalent coefficient of electric power and thermal energy of 0.3 as shown in Eq. (7.2).

$$COP_s = \frac{\dot{Q}_{cool}}{\dot{Q}_{in} + \dot{W}/0.3} \qquad (7.2)$$

The investigated system demonstrated an enhancement in cooling output and in the COP of up to 30% when compared to a similar conventional mechanical compression system. More specifically, at an ambient air inlet temperature of 12.3°C and an evaporator outlet temperature of 6.8°C, the conventional VCC COP was reported to be 0.454 with a cooling output of 0.99 kW. By implementing the desiccant cycle and additional evaporative cooling—at an ambient air inlet temperature of 14.6°C and an evaporator outlet temperature of 7.5°C—the system's COP increased to 0.638, delivering a cooling output of 1.76 kW.

Jani (2016) experimentally studied the performance of a combined solid desiccant-VCC air conditioning system installed in Roorkee, India. The system's performance was evaluated for the period of mid–March to mid-October under the hot and humid climatic conditions of the region. The results indicated that the system could operate efficiently for the specifications of the local climate, achieving a significant reduction of approximately 62% in the humidity ratio, from 18.5 g/kg of dry air down to 7.1 g/kg of dry air.

7.2.3 Absorption-Rankine Cycle

Yogi Goswami (1998) introduced a hybrid NH_3-H_2O absorption cycle and an ammonia Rankine cycle to produce power and refrigeration. A steady state simulation was carried out to evaluate the performance of the proposed cycle, enhancing, according to the authors, the conventional Rankine efficiency up to 90% when the turbine was supplied with ammonia vapor at 230°C and 27.6 bar and expanding it up to 1.4 bar. In terms of the output of the cycle, one such system with a nominal 2 MW power capacity was estimated to have a simultaneous cooling output of 175.8 kW (50 tons).

Lu and Goswami (2003) expanded upon the work of Yogi Goswami (1998) by carrying out an optimization process for the operating conditions in the cycle based on the maximization of the second law efficiency, the refrigeration output, and the power output of the cycle. The system, presented in Figure 7.8, was evaluated for variable temperature

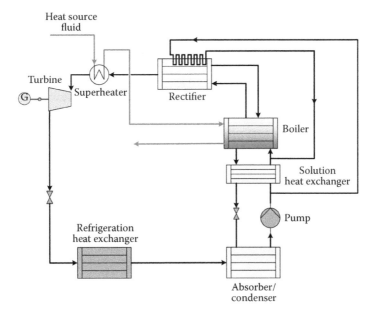

FIGURE 7.8
The novel absorption-Rankine cycle system. (Adapted from Lu, Shaoguang, and D. Yogi Goswami, *Journal of Solar Energy Engineering*, 125 (2): 212–217, 2003.)

heat sources. According to the results of the optimization, based on the working point to maximize the power output of the cycle, the obtained first law efficiency was 8.9% and the corresponding second law efficiency was 43.1%, at a heat source temperature of 360 K. The power output was equal to 107.4 kW and there was no refrigeration output at this working point. On the other hand, when the optimization's goal was the maximization of the cooling output, the performance of the cycle significantly improved: the first law efficiency was 14.1% and the corresponding second law efficiency was 53.6% at the same heat source temperature (360 K). The power output on this set point was 17.95 kW, while the refrigeration capacity was 15.4 kW.

Padilla et al. (2010) carried out a parametric analysis of a combined NH$_3$-H$_2$O absorption/ Rankine cycle similar to the cycles mentioned in this section. The sensitivity of the cycle's performance on turbine efficiency was proven by reducing the efficiency of the turbine from 100% to 50% and presenting the results of the first and second law efficiency of the cycle, which decreased from 21% down to 10% for the first law efficiency and from 92% down to 42% in the case of the second law efficiency. Furthermore, by varying the heat source temperature between 90–170°C, and with an absorber temperature of 30°C, the maximum effective first and second law efficiencies were calculated to be equal to 20% and 72%, respectively.

7.2.4 Ejector-VCC Hybrid System

Sun (1998) proposed a hybrid ejector compression refrigeration system for air conditioning applications with water as the working fluid of the ejector cycle and R21 as the fluid of the vapor compression cycle. The system was able to exploit extremely low-grade heat sources by operating with generator temperatures as low as 60–90°C. Furthermore, two additional heat exchangers were implemented within the conventional cycles, as shown in Figure 7.9,

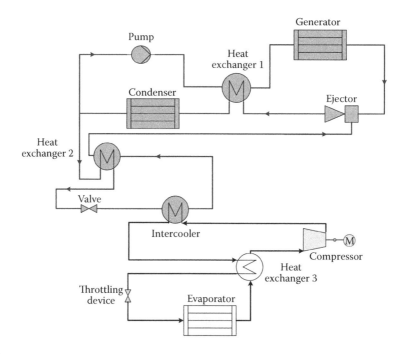

FIGURE 7.9
The hybrid ejector-vapor compression system operating with water and R21. (Adapted from Sun, Da-Wen, *International Journal of Energy Research*, 22 (4): 333–342, 1998.)

to enhance their performance. The positive influence of these heat exchangers was proven by the results of the presented simulations. A parametric study was conducted to investigate the influence of several parameters on the ejector entrainment ratio and the system's COP, including the intercooler's temperature and the condensation temperature. The maximum reported COP was around 1.25 at a generator temperature of 80°C, a condensation temperature of 40°C, an intercooler temperature (on the R21 side) of 35°C, and an evaporation temperature of 5°C.

Hernández et al. (2004) compared the performance of an ejector-vapor compression cycle with either R134a or R142b as the working fluid for ice production. The sets of simulations were created by varying the generator and condenser temperatures and the intercooler pressures for a fixed evaporation temperature of 10°C and a combined cooling capacity of 1 kW. The system's first law efficiency, coefficient of performance, and exergy efficiency, respectively, were calculated based on the equations listed below:

$$\eta s = \frac{Te \cdot (T_g - T_c)}{T_g \cdot (T_c - T_{int}) - T_c \cdot (T_{int} - T_e)} \tag{7.3}$$

$$COP_s = \frac{\dot{Q}_e}{\dot{Q}_g + \dot{Q}_{rp} + \dot{Q}_{rb}} \tag{7.4}$$

$$\eta_{ex} = \frac{\dot{Ex}_e}{\dot{Ex}_g + \dot{Ex}_{rp} + \dot{Ex}_{rb}} \tag{7.5}$$

In the case of R134a, the highest COP was reported as 0.48, at a corresponding first law efficiency of 2.2 (based on the definition of (7.3)). The exergetic efficiency was 25, the generator temperature was 85°C, the intercooler temperature was 15°C, and the condensation temperature was 30°C. The respective values for the R142b were slightly lower. However, for higher generation temperatures, the performance of R142b increased, making it the optimum fluid for medium-grade heat sources. An experimental test rig based on the simulated system was developed and installed at the same time at Centro de Investigacion en Energía de la UNAM in Temixco, Morelos, Mexico.

Elakdhar et al. (2007) developed a model with Fortran to simulate the behavior of a hybrid VCC-ejection system used for domestic purposes. Several working fluids were evaluated for their performance: R123, R124, R141b, R290, R152a, R717, R600a, and R134a. In the system, presented in Figure 7.10, the exit of the ejector was directly connected to the entrance of the compressor, without an intercooler. A parametric analysis was conducted, in which the condensation temperature varied between 28–44°C, the evaporation temperature of evaporator No.1 (see Figure 7.10) ranged between –5–10°C, and evaporator No.2's temperature ranged from –40 to –20°C. The cooling capacity of each evaporator was set to be 0.5 kW. The optimal working fluid for maximization of the COP was found to be R141b. Based on the results for the optimal fluid, the addition of the ejector was found to enhance the system's COP by up to 32%, resulting in a maximum cooling COP of around 2.3 (at a condensation temperature of 42°C, an evaporator No.1 temperature of –5°C, and an evaporator No.2 temperature of –31°C).

Using simulations, Zhu and Jiang (2012) investigated a hybrid vapor compression-ejector cooling cycle. The ejector cycle was powered by the waste heat of the compression's cycle

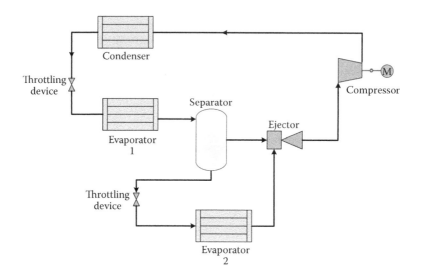

FIGURE 7.10
The investigated ejector-VCC refrigeration system. (Adapted from Elakdhar, M. et al., *Industrial & Engineering Chemistry Research*, 46 (13): 4639–4644, 2007.)

condenser. When the compressor's discharge temperature was higher than 100°C, a significant improvement occured in the system's COP through the addition of the ejector. Three working fluids were evaluated for their on- and off-design performance: R152a, R22, and R134a (the on-design condensation temperature was 50°C and the on-design evaporator temperature was –5°C). The on-design calculations showed that the three fluids had similar performances at their optimal working points, with the ejector's COP around 0.75. Furthermore, the influence of the evaporator temperature was investigated on the system's cooling output and the COP. By using R152a as the working fluid, an improvement of 5.5% in the system's average COP was found in comparison to the corresponding conventional VCC system. When R22 is used, the enhancement was 9.1%. Both fluids recorded a similar hybrid system COP in the range of 2.2–3.5 for evaporation temperatures between –10–10°C. On the other hand, when R134a was used, the improvement of the average COP was only 0.7% (the COP of the hybrid system was in the range of 1.9–2.3), mainly because of the lower compressor discharge temperature (70–90°C). In terms of the cooling output, R22 was calculated to have higher production with a cooling capacity of around 14 kW at an evaporation temperature of 0°C. At the same temperature, the hybrid R152a was delivering around 8 kW$_c$.

7.2.5 Ejector-Absorption Cycle

The use of a jet ejector has been proposed as a potential improvement in the performance of absorption cycles.

Chen (1988) investigated the implementation of an ejector-absorber in an absorption refrigeration cycle to enhance the system's performance. The system used R22-DMETEG as the working pair. A high temperature solution coming from the generator acted as the primary fluid for the ejector, and the R-22 vapor from the evaporator—after flowing through HEX1—acted as the secondary fluid (Figure 7.11). The exhaust of the ejector was connected to the inlet of the absorber, as shown in Figure 7.11.

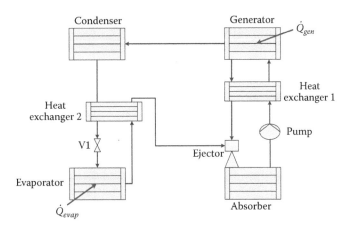

FIGURE 7.11
The investigated ejector-absorber refrigeration system. (Adapted from Chen, Li-Ting, *Applied Energy*, 30 (1): 37–51, 1988.)

Sun et al. (1996) evaluated the performance of a novel combined ejector-absorption refrigeration cycle with H2O-LiBr as the working pair of the absorption cycle Figure 7.12). By assuming nozzle and diffuser efficiencies of 85%, the results indicated that the maximum obtained COP for the proposed system was around 0.95 ($\theta_e = 5°C$, $\theta_c = 30°C$, and $\theta_g = 210°C$), which was approximately 17% higher than the respective maximum COP for a conventional H_2O-LiBr chiller (at a $\theta_e = 5°C$, $\theta_c = 30°C$, and $\theta_g = 75°C$).

Wu and Eames (1998) introduced a novel absorption cycle with an implemented ejector to enhance the system's performance. The system has a similar operation to the conventional single-effect H_2O-LiBr absorption cycle. The main difference consists of the use of a steam generator, a steam ejector, and a concentrator, as shown in Figure 7.13, which replace the high and low pressure generators used in a conventional double-effect absorption chiller. In this cycle, the steam ejector acts as a heat pump, increasing the flow of the exiting vapor and increasing the heat input at the concentrator. The main advantage of the proposed cycle is the simplicity of its design and the lower initial cost in comparison to a conventional double-effect

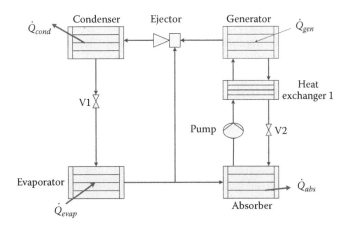

FIGURE 7.12
The combined ejector-absorption refrigeration cycle. (Adapted from Sun, Da-Wen et al., *International Journal of Refrigeration*, 19 (3): 172–180, 1996.)

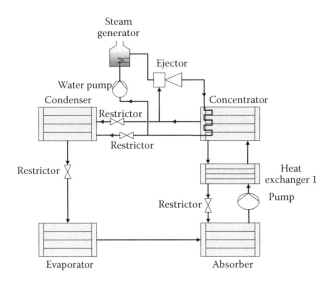

FIGURE 7.13
The ejector-absorption hybrid cycle. (Adapted from Wu, S., and I. W. Eames, *Applied Thermal Engineering*, 18 (11): 1149–1157, 1998.)

absorption cycle. Eames and Wu (2000) carried out simulations to evaluate the performance of the aforementioned cycle for different working parameters. The maximum obtained COP was equal to 1.016 for an evaporation temperature of 5°C, an absorption and condensation temperature of 30°C, and a heat source temperature of 198.3°C (saturated steam at 15 bar). The respective value for a single–effect absorption chiller was estimated to be 0.796.

Another steady state performance study of the ejector-absorption cycle was conducted by Göktun (1999). The investigated system was used to cool residences, at an ambient temperature of 307 K and a fixed evaporation temperature of 290 K. The optimization carried out revealed that an 40% increase in the COP of the conventional absorption cycle could be achieved.

Jiang et al. (2002) compared the thermo-economic performance of a hybrid absorption-ejector refrigeration system and a conventional double-effect absorption chiller. In terms of the energetic performance, the hybrid system performed less efficiently than the conventional system, with a reported COP in the range of 0.9–1.0. The annual cost of the hybrid system was lower than the conventional double-effect absorption chiller when waste heat was used to drive the two systems, and thus no cost for the heat source was considered. In a case in which the required driving heat is supplied by natural gas, the annual cost of the hybrid system increased, and in the end was slightly higher than the conventional system. The presented results indicate that there is potential for the investigated hybrid system, as long as an improvement in its efficiency can be achieved.

Alexis and Rogdakis (2002) simulated the performance of a combined ejector NH_3-H_2O absorption refrigeration cycle by evaluating two configurations. In the first configuration, shown in Figure 7.14(a), the ejector inlet was connected to the evaporator and the discharge was connected to the condenser. In the second configuration, the discharge was directed to the absorber, as shown in Figure 7.14(b). The theoretical COP for the first configuration was reported to be in the range 1.099–1.355 for a generation temperature of 237°C, a condensation temperature of 25.9–30.6°C, an absorber temperature of 48.6–59.1°C, and an evaporation temperature of –1.1–7.7°C. Based on the simulation results, the corresponding values of the COP for the second configuration were in the range of 0.274–0.382 for a generation temperature

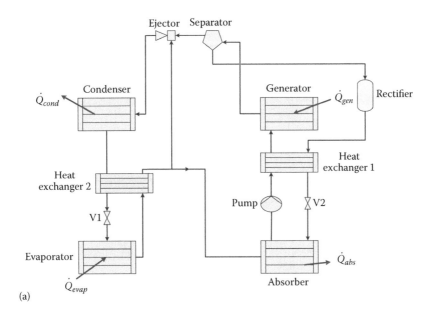

(a)

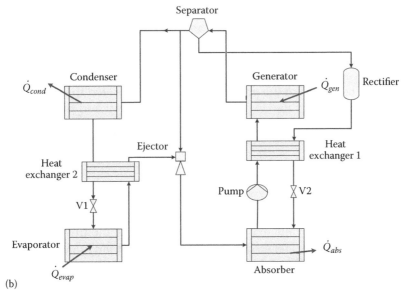

(b)

FIGURE 7.14
The two configurations of the hybrid ejector-absorption cycle. (Adapted from Alexis, G. K., and E. D. Rogdakis, *Applied Thermal Engineering*, 22 (1): 97–106, 2002.)

of 237°C, a condensation temperature of 91°C, an absorber temperature of 76.7–81°C, and an evaporation temperature of –1.1–7.7°C.

Jelinek et al. (2008) investigated the performance of a triple pressure single-stage absorption cycle with several working pairs. The absorbent in all investigated cases was dimethylethylenurea (DMEU), and the considered refrigerants were R22, R32, R124, R125, R134a, and R152a. In order to enhance the performance of the system, an ejector was implemented at the absorber inlet (Figure 7.15). Simulations were conducted to evaluate the influence of

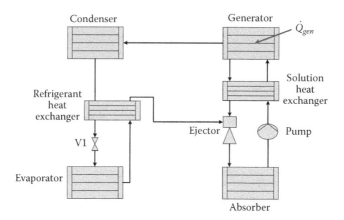

FIGURE 7.15
The triple pressure single-stage ejector-absorption cycle. (Adapted from Jelinek, M. et al., *Applied Thermal Engineering*, 42 (Supplement C): 2–5, 2012.)

generation, evaporation, and condensation temperature on the circulation ratio and the system's COP. The most favorable working pair was found to be R22-DMEU with a COP of 0.685 and a circulation ratio of 3.74 ($P_e = 4.21$ bar, $P_c = 12.55$ bar, and $\theta_g = 100°C$).

Wang et al. (2009) conducted a parametric analysis for a hybrid ejector-absorption refrigeration cycle for simultaneous power and cooling production, with NH_3-H_2O as the working pair. At the nominal point, the net power output was 612.1 kW, and the refrigeration nominal output was 246 kW, resulting in a thermal efficiency of 21% and an exergy efficiency of 35.8%. The results of the parametric analysis indicated that there is an optimum for turbine inlet pressure to maximize the net power output and the exergetic efficiency. In terms of the refrigeration output, increasing the turbine inlet pressure proved to have a negative effect on cooling production.

7.2.6 Absorption-Compressor Cycle

In order to enhance the results obtained from the simulations on a triple pressure ejector-absorption cycle, Jelinek et al. (2012) replaced the ejector with a mechanical compressor and a mixing device and evaluated the performance of the modified system using R125-DMEU as the working pair. The results showed that the system's performance was significantly improved, with a COP of 0.688 at a generator temperature of 59°C and a pressure difference between the evaporator and the absorber of 6 bar. The respective optimum values for the previous configuration with the ejector were 0.544 for the COP, at a generator temperature of 100°C and a pressure difference between the evaporator and the absorber of 0.365 bar. In both cases, the evaporation temperature was set at −5°C.

Ventas et al. (2012) conducted an experimental investigation on a thermochemical compressor implemented in a absorption-compression hybrid cycle with NH_3-$LiNO_3$ as the working pair. An adiabatic absorber was installed in the setup using fog jet injectors. The driving hot water temperatures were in the range 57–110°C, while the absorber pressures ranged approximately between 4.3–9.5 bar. The maximum obtained cycle COP ranged between 0.33 and 0.5 at generator temperatures of 95°C and 72°C, respectively. The hybridization of the cycle showed that lower grade heat sources, with driving temperatures of 57–70°C, could be exploited.

Cimsit and Ozturk (2012) analyzed the performance of cascade compression-absorption refrigeration cycles using different refrigerants in the two sections of the system. NH_3-H_2O

and H$_2$O-LiBr were considered as the potential working pairs for the absorption cycle. For the mechanical compression cycle, R134a, R410A, and R717 (NH$_3$) were evaluated and proved to have a similar behavior. R410A was concluded to be the least efficient.

Figure 7.16 presents a schematic of the investigated cascade system. The results of the simulations indicated that an electric energy saving of 48–51% could be achieved by the cascade system in comparison to a conventional VCC system. The most efficient working pair for the absorption cycle from an energetic aspect was found to be H$_2$O-LiBr, resulting in a COP of the absorption cycle of 0.75 and a COP of the cascade system of 0.592. Regarding the working fluid of the VCC, R717 and R134a presented a similar behavior. R410A was concluded to be the least efficient.

Bouaziz and Lounissi (2015) conducted a first and second law analysis of a hybrid double-stage NH$_3$-H$_2$O absorption compression cycle driven by low-grade heat sources (Figure 7.17). The results of the conducted simulations were compared with a conventional absorption chiller to evaluate the novel cycle's feasibility. The authors concluded that the novel system allowed for the exploitation of lower-grade heat sources, in the range of 60–120°C, while the conventional double-stage absorption required heat sources of 100–160°C. In terms of the system's COP, an enhancement of 25–32% was observed by the novel cycle. The minimum exergy loss obtained by the novel system was 0.8–1 kW, and the corresponding value for the conventional system was around 1.3 kW.

Xu et al. (2016) compared the performance of two hybrid absorption-VCC refrigeration cycles. In the first system (Figure 7.18 (a)), the evaporator at the top absorption cycle served as the condenser of the bottoming VCC. In the second system, the absorption cycle's evaporator was used as a subcooler for the VCC (Figure 7.18 (b)). The performance of the two configurations was evaluated for several evaporators (between −20–10°C), condensers

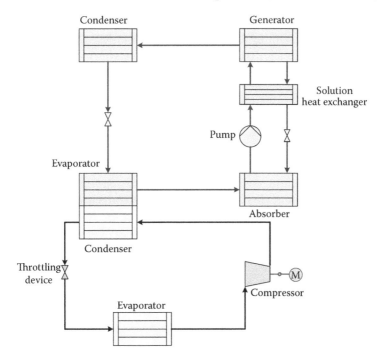

FIGURE 7.16
The cascade compression-absorption refrigeration cycle. (Adapted from Cimsit, Canan, and Ilhan Tekin Ozturk, *Applied Thermal Engineering*, 40 (Supplement C): 311–317, 2012.)

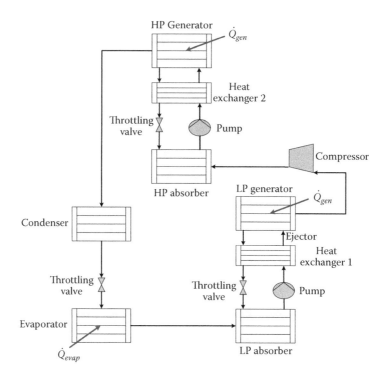

FIGURE 7.17
The double-stage absorption compression refrigeration cycle. (Adapted from Bouaziz, Nahla, and D. Lounissi, *International Journal of Hydrogen Energy*, 40 (39): 13849–13856, 2015.)

(between 35–45°C), and generator temperatures (between 90–120°C). Based on the results of the simulations, it was concluded that the first system was more efficient at lower evaporator temperatures and higher generator temperatures (the maximum obtained global COP was around 2.6). Its efficiency dropped significantly at higher evaporator temperatures and lower driving temperatures, making the evaporator-subcooler choice more preferable for such applications (the maximum obtained global COP for the second system was around 2.4).

A mathematical model of a hybrid two-stage H_2O-LiBr absorption compression chiller was developed by Dixit et al. (2017). The design of the cycle's heat exchangers was carried out to estimate the size and the cost of the proposed system. The thermodynamic performance of the proposed system was compared against a conventional two-stage absorption chiller driven by a low-grade heat source. According to the results, the proposed system was found to perform more efficiently. The optimization of the hybrid cycle led to a maximum COP of 0.435 and an exergetic efficiency of 11.83%. At the same time, the optimization reduced the overall heat transfer area of the system's heat exchangers from 79.6 m² to approximately 72 m², and thus a 5.2% reduction in the system's operational cost was also achieved (θ_g = 52°C, θ_c = 33°C, and θ_e = 9.4°C).

7.2.7 Electrochemical-Absorption Cycle

Newell (2000) proposed coupling a fuel cell and its reversible analog—an electrochemical cell—in such a way to create a refrigeration cycle. The proposed system consists mainly of four components. The base of the system is the water/hydrogen/oxygen

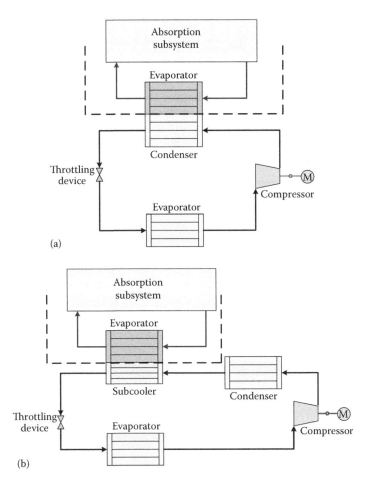

FIGURE 7.18
The two compression-absorption refrigeration cycle configurations. (Adapted from Xu, Yingjie et al., *Applied Thermal Engineering*, 106 (Supplement C): 33–41, 2016.)

fuel cell and electrochemical cell combination. The electrochemical cell serves as the evaporator of the refrigeration cycle by absorbing heat. On the other hand, the fuel cell works as the condenser equivalent by rejecting heat. A current pump has also been installed to adjust the fuel cell's voltage to a level sufficient to drive the electrochemical cell. Finally, a heat exchanger has been implemented for heat recovery between the gas streams and the water flow stream. The working pressure of the system is close to atmospheric.

The main advantages of this system, according to the author, are (1) its potential coupling with photovoltaic solar cells and (2) the fact that it is environmentally friendly, as it avoids the use of conventional refrigeration fluids with significant ODP and GWP. However, several issues have to be addressed regarding the realization of such a system, including the use of high quality electrode surfaces to limit voltage drops and reaction resistances, the cost of developing such systems, and the choice of electrolytes for the fuel and electrochemical cells. Furthermore, there are safety considerations to be addressed for such a system regarding the movement of oxygen and hydrogen.

7.2.8 Electro-Adsorption Cycle

Gordon et al. (2002) proposed a mini-scale combined adsorption thermoelectric chiller. The principle of the proposed scheme operation consists of the exploitation of the heat produced by the thermoelectric effect to drive the desorption of the adsorption cycle. Through these means, the low COP of both cycles can be significantly increased to a value of up to 0.9, according to the authors, at a minimum evaporation temperature of 11°C and a condensation temperature of 35°C. The mini-scale of the system and its high efficiency offers the potential for application in electronics cooling.

7.3 Hybrid Solar Cooling Systems

7.3.1 Solar Ejector-VCC Coupling

Sun (1997) proposed a solar-powered hybrid ejector-vapor compression cascade system, operating with water as the working fluid of the ejector cycle and R-134a as the working fluid of the VCC. The interconnection between the two cycles, as shown in Figure 7.19.

Figure 7.19, is realized by an intercooler which serves as the condenser of the bottoming vapor compression cycle. The performance analysis of the system indicated that it could enhance the COP of the conventional cycle by 50%, reaching a maximum value of almost 0.45 at a generation temperature of 80°C, a condensation temperature of 35°C, and an evaporation temperature of 5°C. Furthermore, assuming two cases of solar irradiation of 800 and 900 W/m², respectively, the solar collector area was estimated so that 5 kW of

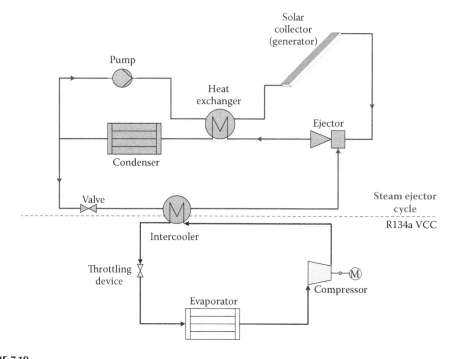

FIGURE 7.19
The solar hybrid ejector-VCC system. (Adapted from Sun, Da-Wen, *Energy Conversion and Management*, 38 (5): 479–491, 1997.)

cooling output could be achieved by the combined hybrid system. The results showed that for 900 W/m², the required collector's surface was 15 m². The respective value for solar irradiation of 800 W/m² was approximately 17 m² for maximum COP working conditions.

Using simulations, Arbel and Sokolov (2004) investigated the performance of a solar-powered combined ejector-vapor compression cycle system with R142b as the working fluid, and compared the results of the simulations with previous work carried out by Sokolov and Hershgal (1993) on the same cycle with R114 as the working fluid. The performance analysis showed an overall enhancement in efficiency (between 10–50%, depending on the working point of the cycle) because of the replacement of the working fluid.

A solar-powered hybrid ejector-compression cycle for cooling applications was evaluated thermodynamically and economically by Vidal and Colle (2010). The two cycles operated with different working fluids: R134a was used for the VCC, while the ejector stage used R141b. An overview of the proposed system is presented in Figure 7.20. An hourly simulation was conducted using meteorological data for Florianopolis, Brazil, for an air-conditioning demand of 10.5 kW. The optimized system consisted of 105 m² pf flat-plate collectors and an intercooler temperature of 19°C, which led to a solar fraction of 89% and a combined cycle COP of 0.89 (at $\theta_g = 80°C$, $\theta_c = 34°C$, and $\theta_e = 8°C$).

Dang et al. (2012) proposed the use of a hybrid solar-assisted ejector-vapor compression cycle for both heating and cooling applications. The system consists of three main circuits: the solar collector circuit, the ejector cycle, and the two-stage vapor compression cycle. The generator of the ejector cycle is powered by the heat from the solar collectors. The refrigerant, R1234ze-vapor, is then directed to the ejector, where it expands and accelerates in the ejector's nozzle. The high speed flow absorbs the flow coming from the internal heat exchanger. The mixed flow exiting the ejector is then led to the condenser before it is split into two streams: one

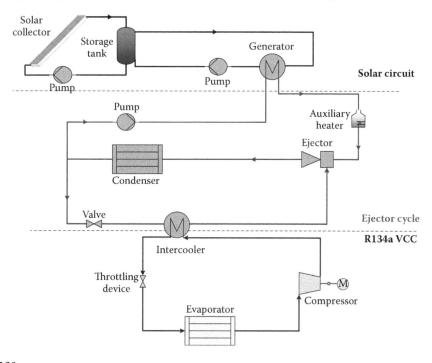

FIGURE 7.20
The solar hybrid ejector-vapor compression cycle. (Adapted from Vidal, Humberto, and Sergio Colle, *Applied Thermal Engineering*, 30 (5): 478–486, 2010.)

flowing toward the internal heat exchanger and the other one flowing toward the generator. Inside the internal heat exchanger, R1234ze cools down the working fluid, R410A, of the vapor compression cycle, hence reducing the condensation temperature and, as a result, the energy consumption of the VCC. An auxiliary condenser has been added to the ejector cycle in case the solar power is insufficient. On the other hand, when solar energy is not available, the ejector cycle is shut down and the system operates as a simple VCC. Based on the simulations, it was determined that there is an optimum value for the internal heat exchanger temperature to maximize the system's performance for each solar heat input. The system's performance was evaluated by measuring the energy savings in a 144 m² office located in Tokyo, Japan. The results showed a 50% reduction in energy consumption during heating mode and a 20% reduction of during cooling mode, with solar collectors corresponding to half of the total office area.

Chesi et al. (2012) investigated the feasibility of a solar-driven cascade ejector-vapor compression cycle system used for refrigeration applications. The performance of the system was compared with the standalone cycles for a 200 m² house at four different locations: Pantelleria (Italy), Abu Dhabi (United Arab Emirates), Caracas (Venezuela), and Singapore. Based on the meteorological data of the investigated locations, the collector efficiency was in the range of 0.50–0.61, resulting in an ejector COP of 0.25–0.30. The lowest solar fraction was reported in Abu Dhabi, with a value of 34%, while the highest value was reported in Pantelleria, with a solar fraction of 75%. Considering the energy savings, the most preferable location was determined to be Singapore with total energy savings of approximately 1 MWh on an annual basis.

7.3.2 Solar Ejector-Rankine Cycle

Nord et al. (2001) combined a thermal management and a power production cycle into a single unit powered entirely by solar energy via a concentrating solar collector (Figure 7.21). The thermal management unit was realized by the use of a jet pump to compress the flow, which reduced the vibration and weight issues and increased the system's reliability. The jet pump also allowed for the connection of the two subsystems. For the

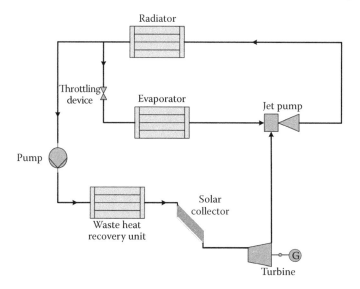

FIGURE 7.21
The combined thermal management and power production cycle. (Adapted from Nord, JW et al., *Journal of Propulsion and Power*, 17 (3): 566–570, 2001.)

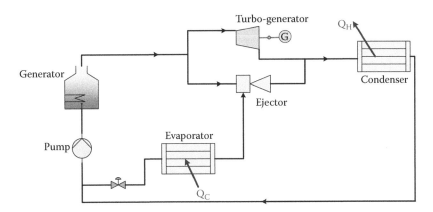

FIGURE 7.22
The hybrid solar-assisted ejector-Rankine cycle. (Adapted from Oliveira, A.C. et al., *Applied Thermal Engineering*, 22 (6): 587–593, 2002.)

simulations, the solar collector's estimated efficiency was considered to be 80%, while the rest components—the pump, the turbine and, the radiator—were assumed to each have a 95% efficiency. By varying the primary fluid inlet temperature, it was found that the system's performance improved as the temperature approached the saturation point for a given pressure. Shifting the primary fluid inlet pressure further enhanced the system's performance. Finally, an investigation of the influence of the ejector's entrainment ratio on the overall performance concluded that both the radiator and the solar collector area were highly dependent on the entrainment ratio, with values of the ratio above 2, reducing to minimum both areas.

Oliveira et al. (2002) experimentally developed and tested the feasibility of a hybrid ejector-Rankine system driven by solar collectors and a backup gas-fired burner. The system, presented in Figure 7.22, could provide power, heating, and cooling for building applications. The working fluid was R601 (n-pentane). Two prototypes were developed for the experiments. The first prototype had a cooling capacity of 2 kW and was powered by 20 m² of solar collectors, while the second had a 5 kW cooling capacity and was powered by a gas burner. Based on the results, the obtained COP of the cooling cycle was around 0.3, while the electrical efficiency of the cycle was 3–4%, at a turbine efficiency of 28%. The cost analysis that was carried out indicated that in comparison to a conventional heat pump and electricity supply from the grid, the cost of the hybrid system for the cogeneration of power and cooling was twice as costly. On the other hand, the economics improve considerably in the case of combined power, cooling, and heating, with a cost, over a 15-year period, of €0.126/kWh for the hybrid in comparison to €0.110/kWh for the conventional case.

Gupta et al. (2014) proposed and evaluated a combined ejector-Rankine cycle for power production and refrigeration. The system, driven by solar power, used duratherm 600 oil as its heat transfer fluid. A parametric analysis was conducted to investigate the influence of several parameters, including the direct normal radiation per unit area and the turbine inlet pressure on the system's first and second law efficiency. The first law efficiency of the cycle was calculated to be 14.8%, while the corresponding second law efficiency was 11.9%. According to the results, the highest exergy destruction took place in the central receiver, contributing 52.5% of the cycle's total irreversibilities, while the heliostat added another 25%.

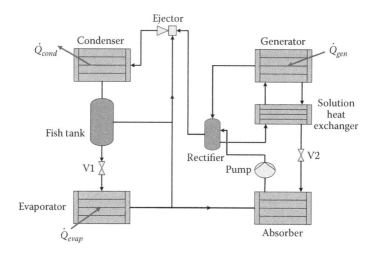

FIGURE 7.23
The hybrid solar ejector-absorption system with the implemented flash tank. (Adapted from Sirwan, Ranj et al., *Solar Energy*, 91 (Supplement C): 283–296, 2013.)

7.3.3 Solar Ejector-Absorption Cycle

Using simulations, Sirwan et al. (2013) investigated the influence of the addition of a flash tank in a solar hybrid ejector-absorption refrigeration system operating with NH_3-H_2O as the working pair. The addition of the flash tank between the condenser, the evaporator, and the ejector, as shown in Figure 7.23, aimed to enhance the ejector's entrainment ratio and thus elevate the condensation pressure. Based on the results of the simulations of several working conditions (solar collector temperature ranging from 70–120°C, a condensation temperature between 25–45°C, and an evaporation temperature ranging from –10–10°C), it was found that the modified combined cycle improved the performance of the system, with the maximum obtained COP being around 0.86 (at θ_g = 95°C, θ_c = 25°C, and θ_e = –10°C), in comparison to 0.75 (at θ_g = 120°C, θ_c = 45°C, and θ_e = 10°C) for the combined (without flash tank) cycle, and 0.575 (at θ_g = 82°C, θ_c = 25°C, and θ_e = –10°C) for the single-effect absorption cycle.

   Abdulateef et al. (2011) experimentally investigated a combined ejector absorption refrigeration system installed on the roof of the physics department lab at University Kebangsaan, in Bangi, Malaysia. The system was powered by 10 m² of evacuated tube solar collectors. The experiments conducted by the authors were for a range of the generator temperatures of 60–98°C, condenser temperatures between 23–39°C, and evaporator temperatures between 3–16°C. According to the experimental data, the COP ranged from 0.33 to 0.58. The maximum value of COP was obtained at a generation temperature of 85°C, an evaporation temperature of 16°C, a condenser temperature of 30°C, and an absorber temperature of 28°C, resulting in an average increase of 50% in comparison to a conventional absorption chiller.

7.3.4 Solar Absorption-Rankine Cycle

Using simulations, Kouremenos et al. (1991) investigated the potential of a work-producing solar-driven NH_3-H_2O absorption unit coupled with a steam Rankine cycle. The system that was designed had a theoretical efficiency that was 25% higher than that of a conventional solar-powered water-steam Rankine cycle. The proposed system obtained a 30.4% thermal efficiency in comparison to the 22.6% efficiency obtained with the respective water-steam Rankine cycle.

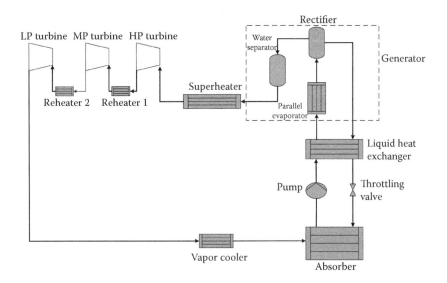

FIGURE 7.24
The solar-driven absorption-Rankine cycle. (Adapted from Kouremenos, D. A. et al., *Energy Conversion and Management*, 31 (2): 111–119, 1991.)

The system proposed by the authors, shown in Figure 7.24, replaced the conventional condensation, throttling, and evaporation process of the refrigerant in an absorption cycle with a three-stage expansion with intermediate reheaters and a superheater before the inlet of the first stage. Simulations were carried out to evaluate the performance of the proposed system during a typical year in the region of Athens, Greece. The results showed that the proposed system operated more efficiently, allowing for a 20% reduction in the required solar collector area.

Xu et al. (2000) simulated the performance of a hybrid Rankine-absorption cycle for both power and refrigeration production. The absorption cycle used NH_3-H_2O as the working pair, while the Rankine cycle operated with high concentration ammonia. The condenser of the Rankine cycle was replaced with an absorption condensation process, as shown in Figure 7.25. For a cycle's maximum temperature of 400 K, the corresponding thermal efficiency was estimated to be 23.54%. A cost calculation was also carried out, estimating that the cost for such a system powered solely by flat-plate collectors would be in the range of $2,000/kW_{th}$, while, as the authors proposed, the use of a backup natural-gas-fired heater would decrease the total cost to $1,500/kW_{th}$.

7.3.5 Solar Absorption-VCC Coupling

Chinnappa et al. (1993) developed and investigated the performance of a R-22 conventional mechanical compression air conditioning unit cascaded with a solar-assisted single-effect NH_3-H_2O absorption chiller. The overall hybrid COP was found to be 5, while the standalone COP of the mechanical compression system was equal to 2.55. The respective power consumption in the compressor of the hybrid system was 2.2 kW, in comparison to 4.36 kW for the standalone operation. Meanwhile, the absorption system's COP was in the range of 0.59–0.72, and the solar flat-plate collector efficiency was 43–50%.

Wang et al. (2012) proposed a cascade refrigeration unit combining a solar driven H_2O-LiBr absorption cycle and an electrically driven vapor compression cycle with R134a as its working fluid. The system is equipped with refrigerant and solution storage tanks for potential heat storage, as shown in Figure 7.26. The performance of the system was analyzed using meteorological

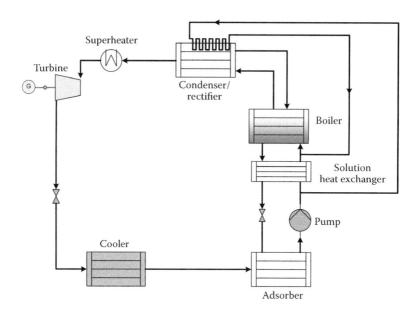

FIGURE 7.25
The hybrid absorption-Rankine cycle system. (Adapted from Xu, Feng et al., *Energy*, 25 (3): 233–246, 2000.)

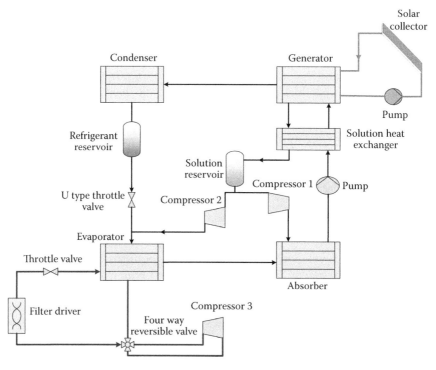

FIGURE 7.26
The cascade solar absorption-vapor compression refrigeration unit. (Adapted from Wang, Lin et al., *Energy Procedia*, 16 (Part C): 1503–1509, 2012.)

data for the July 2 in the Zhengzhou area of China. According to the results of the simulations, a COP of up to 6.1. The maximum COP of the absorption cycle was around 0.85. A solar intensity of 700 W/m² was reported at an ambient temperature of 35°C and a chilled water supply temperature of 7°C. Furthermore, the hybrid system was able to reduce power consumption by 50% in comparison to a conventional mechanical compression system in cooling mode.

Meng et al. (2013) studied a hybrid absorption-compression refrigeration cycle powered by solar energy by the means of flat-plate collectors. For the working pair of the cycle, R134a-dimethylformamide (DMF) was considered. An analysis of the energy savings was conducted and, based on the results, concluded that the proposed system has a decent potential with low-grade heat sources. The energy savings of the proposed system were estimated to be in the range of 52.7% in comparison to the conventional R134a VCC operating at a condensation temperature of 35°C, an evaporation temperature of –10°C, and a generation temperature of 90°C. The reported heat-powered COP (definition of this COP is presented in Eq. (7.6)) was 0.322, and the COP of the VCC was 3.34.

$$COP_{heat} = \frac{Q_e - W \cdot COP_{vcc}}{Q_g} \tag{7.6}$$

Boyaghchi et al. (2016) carried out an exergoeconomic analysis to evaluate the potential of a solar-powered dual-evaporator cascade absorption-vapor compression refrigeration system. The absorption system had H_2O-LiBr as its working pair, and water/copper oxide was used in the solar circuit. For the vapor compression cycle, several working fluids were evaluated: R134a, R1234ze, R1234yf, R407C, and R22. An overview of the system is presented in Figure 7.27. Based on the thermodynamic analysis, R134a was concluded to be the optimal working fluid for the VCC, with a daily COP of 0.096. From an exergoeconomic standpoint, the best fluid was found to be R1234ze with the minimum total product cost rate of $6,847/year.

7.3.6 Solar Absorption-Desiccant Cooling Cycle

Fong et al. (2011) proposed a hybrid solar cooling system that combined absorption refrigeration and desiccant dehumidification for applications in high-tech offices in a subtropical climate (Hong Kong). The proposed system was sub-divided into four main subsystems: (1) absorption refrigeration, (2) the desiccant dehumidifier, (3) radiant cooling (either passive chilled beams or active chilled beams were considered), and (4) a solar circuit consisting of flat-plate collectors (Figure 7.28). The absorption unit was a single-effect H_2O-LiBr chiller with a nominal cooling capacity of 26 kW. Simulations were carried out with TRNSYS software. According to the results, it was found that the configuration with passive chilled beams was more efficient than the one with active chilled beams, allowing for a primary energy consumption of 20% less than the active chilled beams and 36.5% less than a conventional mechanical compression system. The corresponding yearly average amounted to a solar fraction of 51%, an average COP of 0.867, and an annual solar thermal gain of approximately 38.8 MWh.

7.3.7 Solar Adsorption-Ejector

Zhang and Wang (2002) proposed a hybrid combined adsorption-ejector system for heating and cooling applications, powered by solar energy (Figure 7.29). The

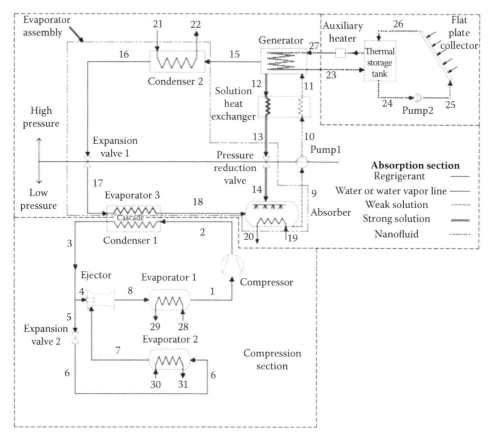

FIGURE 7.27
The cascade solar-powered dual-evaporator absorption-vapor compression refrigeration unit. (Reproduced from Boyaghchi, Fateme Ahmadi et al., *Journal of Cleaner Production*, 139 (Supplement C): 970–985, 2016.)

adsorption working pair was zeolite-water. Based on the results of the simulations, it was shown that the proposed hybrid system could reach a cooling capacity of 0.15 MJ/kg zeolite in daytime and 0.34 MJ/kg zeolite in the evening. It was able to provide 290 kg of domestic hot water at a temperature of 45°C. Additionally, the reported COP obtained by the hybrid system was 0.33 (at a θ_e = 5°C, θ_c = 30°C, and θ_{des} = 100°C). The respective value for the COP of a conventional single-stage zeolite-water adsorption chiller was 0.3.

Li et al. (2002) analyzed the performance of a solar-driven combined adsorption-ejection refrigeration system. A schematic of the investigated system is presented in Figure 7.30. A heat pipe was implemented in the system to recover the sensible heat and the heat of adsorption. The adsorption cycle provided refrigeration at night, while the ejector cycle was exploited during the daytime. More specifically, during the day, the high temperature and high pressure vapor produced in the generator of the cycle was led to the ejector, allowing for a supersonic flow at the exit of the ejector's nozzle, hence entraining the secondary flow from the evaporator. The mixed fluid, exiting the ejector after condensing in the condenser, was divided again into two streams flowing back to the evaporator and the generator, respectively. The solar energy at that time was used to drive the desorption process in the adsorber. When the pressure in the adsorber's bed reached a certain

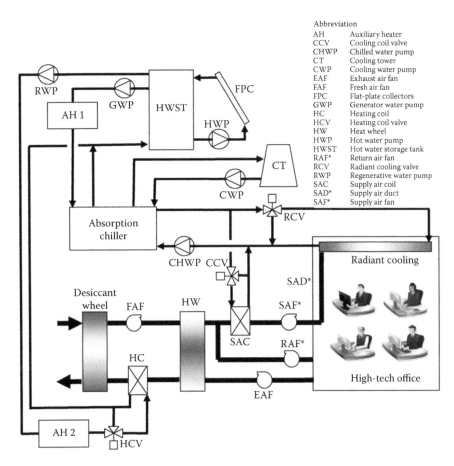

FIGURE 7.28
Schematic of the hybrid solar absorption-desiccant dehumidication system used for high-tech offices in a subtropical climate. (Reproduced from Fong, K. F. et al., *Energy Conversion and Management*, 52 (8): 2883–2894, 2011.)

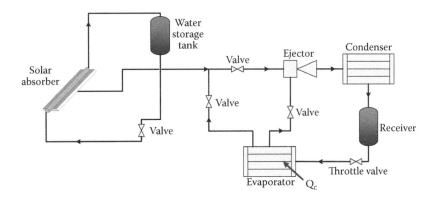

FIGURE 7.29
The hybrid solar-driven adsorption-ejector system. (Adapted from Zhang, X. J., and R. Z. Wang, *Applied Thermal Engineering*, 22 (11): 1245–1258, 2002.)

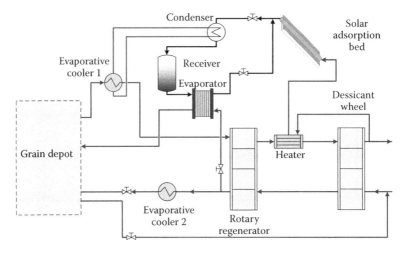

FIGURE 7.30
The hybrid solar adsorption-rotary dehumidification system. (Adapted from Dai, Y. J. et al., *Renewable Energy*, 25 (3): 417–430, 2002.)

level, the desorbed vapor was then connected to the generator as the primary fluid for the ejection cycle. During the night, when the adsorber's bed temperature decreased beyond a certain level, the adsorption process started. Based on the results of the analysis, it was concluded that the proposed system is able to provide a feasible method to overcome the intermittent behavior of a single-bed adsorption refrigeration system. By using zeolite 13X-water as the working pair of the adsorption cycle at a condensation temperature of 313 K, an evaporating temperature of 283 K, a regenerating temperature of 393 K, and a desorbing temperature of 473 K, the reported maximum COP obtained by the novel cycle was 0.4.

Sözen and Özalp (2005) developed a model to simulate the performance of a solar-powered ejector-absorption cooling system operating in Turkey. Meteorological data for sixteen cities in Turkey were used. The maximum reported COP was 0.739 (θ_e = 5°C, θ_c = 40°C, θ_{abs} = 30°C, and θ_g = 90°C), while the heat gain factor (HGF) was in the range of 1.3428K–2.85.

Bel Haj Jrad et al. (2017) carried out simulations to evaluate the performance of a solar-driven hybrid adsorption system using meteorological data for the city of Monastir, Tunisia. The hybrid system, powered by flat solar collectors, was evaluated with zeolite-water and activated carbon-methanol as the potential working pairs. Furthermore, an ejector was implemented in the system to enhance the system's performance. Results of the numerical simulations indicated that activated carbon desorbed faster than zeolite. Regarding the system's solar COP, the zeolite-water working pair reported values up to 0.082, while the activated carbon-methanol had a maximum system COP of 0.128.

7.3.8 Solar Adsorption-Desiccant Cooling Systems

Using simulations, Dai et al. (2002) investigated the performance of a combined solid adsorption-rotary desiccant dehumidification system powered by solar energy to provide cooling for a grain storage facility. The solar adsorption bed was placed on the roof of the grain depot and directly absorbed solar irradiation to drive the adsorption cycle. The working pair of the adsorption cycle was activated carbon-methanol. The grain itself was used as cold storage material during night. Moisture produced by the grain was then removed by means of the desiccant dehumidifier. The main advantages of the proposed system are that it has low operational costs and, in comparison to the standalone cycles, it allows for the removal of moisture (which a solid adsorption chiller is incapable of) and at the same time it cools down the grain to a sufficient temperature (while a standalone desiccant dehumidifier would be unable to achieve such temperature drops). The performance analysis of the proposed system showed that at an ambient temperature of 35°C, the system was able to cool down the temperature in the grain depot to 16°C at a COP of 0.42 (cooling output 6.58 kW).

Fong et al. (2010) developed a simulation model to evaluate the performance of a novel air conditioning system combining adsorption refrigeration, radiant ceiling cooling, and desiccant dehumidification (Figure 7.31). The space cooling load of the building under consideration was handled by a radiant ceiling, supplied with chilled water from the adsorption cycle. The ventilation load was covered by a desiccant dehumidifier. For the ceilings, chilled panels, passive chilled beams, and active chilled beams were all considered. The solar collectors driving the solar hybrid cooling system were evacuated tube collectors. The temperatures for the silica gel-water adsorption chiller at a nominal point were 80/30/18°C. A year-round analysis was carried out, and the average values of the system's performance were determined. Regarding the adsorption chiller, variation of the COP throughout the year was not very high, with the values in the range of 0.533–0.590.

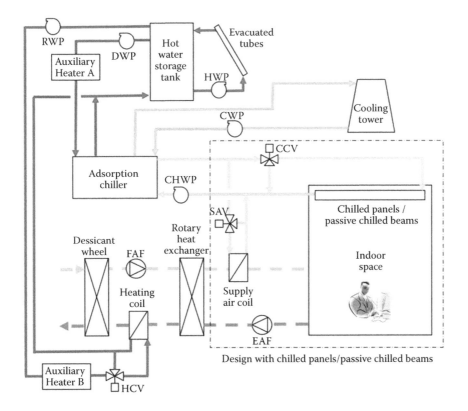

FIGURE 7.31
The solar hybrid air conditioning system. (Reproduced from Fong, K. F. et al., *Renewable Energy*, 35 (11): 2439–2451, 2010.)

Nomenclature

COP_{heat}	*Coefficient of performance for a heat-powered system*	–
$\dot{Ex}$	*Exergy flow rate*	[W]
P	*Pressure*	[Pa]
$\dot{Q}$	*Heat flux*	[W]
T	*Temperature*	[K]
$\dot{W}$	*Power*	[W]

Greek Symbols

θ	*Temperature*	[°C]
ηex	*Second law efficiency*	–
η_s	*System thermal efficiency*	–
abs	*Absorber*	
ads	*Adsorber*	
c	*Condenser*	
cool	*Cooling*	
des	*Desorbed*	
e	*Evaporator*	
g	*Generator*	
in	*Inlet*	
int	*Intercooler*	
rb	*Reversible booster*	
rp	*Reversible pump*	

Abbreviations

CFC	Chlorofluorocarbon
CPC	Compound parabolic concentrator
DMEU	Dimethylethylenurea
DMF	Dimethyl formamide
GWP	Global warming potential
HFC	Hydrofluorocarbons
HGF	Heat gain factor
ODP	Ozone depletion potential
PCM	Phase change material
PER	Primary energy ratio
PTC	Parabolic trough collectors
SPF	Seasonal performance factor
VCC	Vapor compression cycle

References

Abdulateef, JM, Nurul Muiz Murad, MA Alghoul, Azami Zaharim, and Kamaruzzaman Sopian. 2011. "Experimental Study on Combined Solar-Assisted Ejector Absorption Refrigeration System." Proceedings of the 4th WSEAS International Conference on Energy and Development-Environment-Biomedicine, July 2011.

Al-Alili, Ali, Yunho Hwang, and Reinhard Radermacher. 2014. "Review of Solar Thermal Air Conditioning Technologies." *International Journal of Refrigeration* 39 (Supplement C): 4–22.

Alexis, G. K., and E. K. Karayiannis. 2005. "A Solar Ejector Cooling System Using Refrigerant R134a in the Athens Area." *Renewable Energy* 30 (9): 1457–1469.

Alexis, G. K., and E. D. Rogdakis. 2002. "Performance Characteristics of Two Combined Ejector–Absorption Cycles." *Applied Thermal Engineering* 22 (1): 97–106.

Arbel, A., and M. Sokolov. 2004. "Revisiting Solar-Powered Ejector Air Conditioner—The Greener the Better." *Solar Energy* 77 (1): 57–66.

Balaras, Constantinos A., Gershon Grossman, Hans-Martin Henning, Carlos A. Infante Ferreira, Erich Podesser, Lei Wang, and Edo Wiemken. 2007. "Solar Air Conditioning in Europe—An Overview." *Renewable and Sustainable Energy Reviews* 11 (2): 299–314.

Bel Haj Jrad, Amal, Mohamed Bechir Ben Hamida, Rabie Ghnay, and Abdallah Mhimid. 2017. "Contribution to the Study Of Combined Adsorption–Ejection System Using Solar Energy." *Advances in Mechanical Engineering* 9 (7): 1–9.

Berchowitz, David M., Martien Janssen, and Roberto O. Pellizzari. 2008. "CO2 Stirling Heat Pump for Residential Use." International Refrigeration and Air Conditioning Conference, Purdue, USA, 2008.

Bouaziz, Nahla, and D. Lounissi. 2015. "Energy and Exergy Investigation of a Novel Double Effect Hybrid Absorption Refrigeration System for Solar Cooling." *International Journal of Hydrogen Energy* 40 (39): 13849–13856.

Boyaghchi, Fateme Ahmadi, Motahare Mahmoodnezhad, and Vajiheh Sabeti. 2016. "Exergoeconomic Analysis and Optimization of a Solar Driven Dual-Evaporator Vapor Compression-Absorption Cascade Refrigeration System Using Water/Cuo Nanofluid." *Journal of Cleaner Production* 139 (Supplement C): 970–985.

Chen, Li-Ting. 1988. "A New Ejector-Absorber Cycle to Improve the COP of an Absorption Refrigeration System." *Applied Energy* 30 (1): 37–51.

Chesi, Andrea, Giovanni Ferrara, Lorenzo Ferrari, and Fabio Tarani. 2012. "Suitability of Coupling a Solar Powered Ejection Cycle with a Vapour Compression Refrigerating Machine." *Applied Energy* 97 (Supplement C): 374–383.

Chinnappa, J. C. V., M. R. Crees, S. Srinivasa Murthy, and K. Srinivasan. 1993. "Solar-Assisted Vapor Compression/Absorption Cascaded Air-Conditioning Systems." *Solar Energy* 50 (5): 453–458.

Chum, Helena L. 1981. *Review of Thermally Regenerative Electrochemical Systems.* vol. 1. (Solar Energy Research Institute, 1981).

Chunnanond, Kanjanapon, and Satha Aphornratana. 2004. "Ejectors: Applications In Refrigeration Technology." *Renewable and Sustainable Energy Reviews* 8 (2): 129–155.

Cimsit, Canan, and Ilhan Tekin Ozturk. 2012. "Analysis of Compression–Absorption Cascade Refrigeration Cycles." *Applied Thermal Engineering* 40 (Supplement C): 311–317.

Dai, Y. J., R. Z. Wang, and Y. X. Xu. 2002. "Study of a Solar Powered Solid Adsorption–Desiccant Cooling System Used for Grain Storage." *Renewable Energy* 25 (3): 417–430.

Dai, Y. J., R. Z. Wang, H. F. Zhang, and J. D. Yu. 2001. "Use of Liquid Desiccant Cooling to Improve the Performance of Vapor Compression Air Conditioning." *Applied Thermal Engineering* 21 (12): 1185–1202.

Dang, Chaobin, Yoshitaka Nakamura, and Eiji Hihara. 2012. "Study on Ejector—Vapor Compression Hybrid Air Conditioning System Using Solar Energy." International Refrigeration and Air-Conditioning Conference, Purdue, July 16–19, 2012.

Dennis, M., and K. Garzoli. 2011. "Use of Variable Geometry Ejector with Cold Store to Achieve High Solar Fraction for Solar Cooling." *International Journal of Refrigeration* 34 (7): 1626–1632.

Diaconu, Bogdan M. 2012. "Energy Analysis of a Solar-Assisted Ejector Cycle Air Conditioning System with Low Temperature Thermal Energy Storage." *Renewable Energy* 37 (1): 266–276.

Dittmar, L., K. Jüttner, and G. Kreysa. 1994. "A New Concept of an Electrochemical Heat Pump System: Theoretical Consideration and Experimental Results." In *Electrochemical Engineering and Energy*, edited by F. Lapicque, A. Storck, and A. A. Wragg, 57–65. Boston, MA: Springer US.

Dixit, Manoj, Akhilesh Arora, and S. C. Kaushik. 2017. "Thermodynamic and Thermoeconomic Analyses of Two Stage Hybrid Absorption Compression Refrigeration System." *Applied Thermal Engineering* 113 (Supplement C): 120–131.

Eames, Ian W., and Shenyi Wu. 2000. "A Theoretical Study of an Innovative Ejector Powered Absorption–Recompression Cycle Refrigerator." *International Journal of Refrigeration* 23 (6): 475–484.

Elakdhar, M., E. Nehdi, and L. Kairouani. 2007. "Analysis of a Compression/Ejection Cycle for Domestic Refrigeration." *Industrial & Engineering Chemistry Research* 46 (13): 4639–4644.

Elsayed, S. S., Y. Hamamoto, A. Akisawa, and T. Kashiwagi. 2006. "Analysis of an Air Cycle Refrigerator Driving Air Conditioning System Integrated Desiccant System." *International Journal of Refrigeration* 29 (2): 219–228.

Ersoy, H. Kursad, Sakir Yalcin, Rafet Yapici, and Muammer Ozgoren. 2007. "Performance of a Solar Ejector Cooling-System in the Southern Region of Turkey." *Applied Energy* 84 (9): 971–983.

Fong, K. F., T. T. Chow, C. K. Lee, Z. Lin, and L. S. Chan. 2011. "Solar Hybrid Cooling System for High-Tech Offices in Subtropical Climate—Radiant Cooling by Absorption Refrigeration and Desiccant Dehumidification." *Energy Conversion and Management* 52 (8): 2883–2894.

Fong, K. F., C. K. Lee, T. T. Chow, Z. Lin, and L. S. Chan. 2010. "Solar Hybrid Air-Conditioning System for High Temperature Cooling in Subtropical City." *Renewable Energy* 35 (11): 2439–2451.

Ge, T. S., Y. Li, R. Z. Wang, and Y. J. Dai. 2009. "Experimental Study on a Two-Stage Rotary Desiccant Cooling System." *International Journal of Refrigeration* 32 (3): 498–508.

Gerlach, David W, and TA Newell. 2004. *An Investigation of Electrochemical Methods for Refrigeration.* (Air Conditioning and Refrigeration Center, College of Engineering. University of Illinois at Urbana-Champaign, 2004).

Göktun, Selahattin. 1999. "Optimal Performance of a Combined Absorption and Ejector Refrigerator." *Energy Conversion and Management* 40 (1): 51–58.

Gordon, Jeffrey M, KC Ng, HT Chua, and A Chakraborty. 2002. "The Electro-Adsorption Chiller: A Miniaturized Cooling Cycle with Applications to Micro-Electronics." *International Journal of Refrigeration* 25 (8): 1025–1033.

Grazzini, Giuseppe, and Andrea Rocchetti. 2002. "Numerical Optimisation of a Two-Stage Ejector Refrigeration Plant." *International Journal of Refrigeration* 25 (5): 621–633.

Guo, J., and H. G. Shen. 2009. "Modeling Solar-Driven Ejector Refrigeration System Offering Air Conditioning for Office Buildings." *Energy and Buildings* 41 (2): 175–181.

Gupta, Devendra Kumar, Rajesh Kumar, and Naveen Kumar. 2014. "First and Second Law Analysis of Solar Operated Combined Rankine and Ejector Refrigeration Cycle." *Applied Solar Energy* 50 (2): 113–121.

Hernández, Jorge I., Rubén J. Dorantes, Roberto Best, and Claudio A. Estrada. 2004. "The Behaviour of a Hybrid Compressor and Ejector Refrigeration System with Refrigerants 134a and 142b." *Applied Thermal Engineering* 24 (13): 1765–1783.

Hong, Jong-Sung, H Maleki, S Al Hallaj, L Redey, and JR Selman. 1998. "Electrochemical-Calorimetric Studies of Lithium-Ion Cells." *Journal of the Electrochemical Society* 145 (5): 1489–1501.

Huang, B. J., J. M. Chang, V. A. Petrenko, and K. B. Zhuk. 1998. "A Solar Ejector Cooling System Using Refrigerant R141b." *Solar Energy* 64 (4): 223–226.

Jelinek, M., A. Levy, and I. Borde. 2008. "The Performance of a Triple Pressure Level Absorption Cycle (TPLAC) with Working Fluids Based on the Absorbent DMEU and the Refrigerants R22, R32, R124, R125, R134a And R152a." *Applied Thermal Engineering* 28 (11): 1551–1555.

Jelinek, M., A. Levy, and I. Borde. 2012. "Performance of a Triple-Pressure Level Absorption/ Compression Cycle." *Applied Thermal Engineering* 42 (Supplement C): 2–5.

Jia, Yan, and Cai Wenjian. 2012. "Area Ratio Effects to the Performance of Air-Cooled Ejector Refrigeration Cycle with R134a Refrigerant." *Energy Conversion and Management* 53 (1): 240–246.

Jiang, Liben, Zhaolin Gu, Xiao Feng, and Yun Li. 2002. "Thermo-Economical Analysis Between New Absorption–Ejector Hybrid Refrigeration System and Small Double-Effect Absorption System." *Applied Thermal Engineering* 22 (9): 1027–1036.

Khattab, N. M. 2005. "Optimum Design Conditions of Farm Refrigerator Driven by Solar Steam-Jet System." *International Journal of Sustainable Energy* 24 (1): 1–17.

Kim, D. S., C. A. Infante Ferreira, and C. A. Infante Ferreira. 2007. "Solar Absorption Cooling." PhD Dissertation, Delft: 2–29.

Kongtragool, Bancha, and Somchai Wongwises. 2003. "A Review of Solar-Powered Stirling Engines and Low Temperature Differential Stirling Engines." *Renewable and Sustainable Energy Reviews* 7 (2): 131–154.

Kouremenos, D. A., E. Rogdakis, and K. A. Antonopoulos. 1991. "Anticipated Thermal Efficiency of Solar Driven NH3/H2O Absorption Work Producing Units." *Energy Conversion and Management* 31 (2): 111–119.

La, D., Y. Li, Y. J. Dai, T. S. Ge, and R. Z. Wang. 2012. "Development of a Novel Rotary Desiccant Cooling Cycle with Isothermal Dehumidification and Regenerative Evaporative Cooling Using Thermodynamic Analysis Method." *Energy* 44 (1): 778–791.

Li, C. H., R. Z. Wang, and Y. Z. Lu. 2002. "Investigation of a Novel Combined Cycle of Solar Powered Adsorption–Ejection Refrigeration System." *Renewable Energy* 26 (4): 611–622.

Loutfy, Raouf O, Alan P Brown, and Neng-Ping Yao. 1983. Low Temperature Thermally Regenerative Electrochemical System. US Patent 4,410,606, filed April 21, 1982, and issued Oct. 18, 1983. Google Patents, https://patents.google.com/patent/US4410606A/en?q=Low&q=temperature &q=thermally&q=regenerative&q=electrochemical&q=system&oq=Low+temperature+therm ally+regenerative+electrochemical+system.

Lu, Shaoguang, and D. Yogi Goswami. 2003. "Optimization of a Novel Combined Power/ Refrigeration Thermodynamic Cycle." *Journal of Solar Energy Engineering* 125 (2): 212–217.

Ma, Xiaoli, Wei Zhang, S. A. Omer, and S. B. Riffat. 2010. "Experimental Investigation of a Novel Steam Ejector Refrigerator Suitable for Solar Energy Applications." *Applied Thermal Engineering* 30 (11–12): 1320–1325.

Meng, Xuelin, Danxing Zheng, Jianzhao Wang, and Xinru Li. 2013. "Energy Saving Mechanism Analysis of the Absorption–Compression Hybrid Refrigeration Cycle." *Renewable Energy* 57 (Supplement C): 43–50.

Meyer, A. J., T. M. Harms, and R. T. Dobson. 2009. "Steam Jet Ejector Cooling Powered by Waste or Solar Heat." *Renewable Energy* 34 (1): 297–306.

Nehdi, E., L. Kairouani, and M. Elakhdar. 2008. "A Solar Ejector Air-Conditioning System Using Environment-Friendly Working Fluids." *International Journal of Energy Research* 32 (13): 1194–1201.

Newell, Ty A. 2000. "Thermodynamic Analysis of an Electrochemical Refrigeration Cycle." *International Journal of Energy Research* 24 (5): 443–453.

Nóbrega, C. E. L., and L. A. Sphaier. 2012. "Modeling and Simulation of a Desiccant–Brayton Cascade Refrigeration Cycle." *Energy and Buildings* 55 (Supplement C): 575–584.

Nóbrega, Carlos Eduardo Leme, and Nísio Carvalho Lobo Brum. 2014. "An Introduction to Solid Desiccant Cooling Technology." In *Desiccant-Assisted Cooling: Fundamentals and Applications*, edited by Carlos Eduardo Leme Nóbrega and Nisio Carvalho Lobo Brum, 1–23. London: Springer London.

Nord, JW, WE Lear, and SA Sherif. 2001. "Analysis of Heat-Driven Jet-Pumped Cooling System for Space Thermal Management." *Journal of Propulsion and Power* 17 (3): 566–570.

Oliveira, A. C., C. Afonso, J. Matos, S. Riffat, M. Nguyen, and P. Doherty. 2002. "A Combined Heat and Power System for Buildings Driven by Solar Energy and Gas." *Applied Thermal Engineering* 22 (6): 587–593.

Padilla, Ricardo Vasquez, Gökmen Demirkaya, D. Yogi Goswami, Elias Stefanakos, and Muhammad M. Rahman. 2010. "Analysis of Power and Cooling Cogeneration Using Ammonia-Water Mixture." *Energy* 35 (12): 4649–4657.

Peterson, J.L., and J.R. Howell. 1991. Hybrid Vapor-Compression/Liquid Desiccant Air Conditioner. US Patent 4,984,434, filed Sept. 12, 1989, and issued Jan. 1, 1990, Google Patents, https://patents.google.com/patent/US4984434A/en.

Pollerberg, Clemens, Ahmed Hamza H. Ali, and Christian Dötsch. 2009. "Solar Driven Steam Jet Ejector Chiller." *Applied Thermal Engineering* 29 (5): 1245–1252.

Pridasawas, Wimolsiri, and Per Lundqvist. 2007. "A Year-Round Dynamic Simulation of a Solar-Driven Ejector Refrigeration System with Iso-Butane as a Refrigerant." *International Journal of Refrigeration* 30 (5): 840–850.

Rady, M. A., A. S. Huzayyin, E. Arquis, P. Monneyron, C. Lebot, and E. Palomo. 2009. "Study of Heat and Mass Transfer in a Dehumidifying Desiccant Bed with Macro-Encapsulated Phase Change Materials." *Renewable Energy* 34 (3): 718–726.

Richards, W.D.C., and W.L Auxer. 1978. "Performance of a Stirling Engine Powered Heat Actvated Heat Pump." Intersociety Energy Conversion Engineering Conference, San Diego, CA, USA, August 20–25, 1978.

Shen, Shengqiang, Xiaoping Qu, Bo Zhang, Saffa Riffat, and Mark Gillott. 2005. "Study of a Gas–Liquid Ejector and its Application to a Solar-Powered Bi-Ejector Refrigeration System." *Applied Thermal Engineering* 25 (17): 2891–2902.

Sirwan, Ranj, M. A. Alghoul, K. Sopian, Yusoff Ali, and Jasim Abdulateef. 2013. "Evaluation of Adding Flash Tank to Solar Combined Ejector–Absorption Refrigeration System." *Solar Energy* 91 (Supplement C): 283–296.

Sokolov, M., and D. Hershgal. 1993. "Solar-Powered Compression-Enhanced Ejector Air Conditioner." *Solar Energy* 51 (3): 183–194.

Sözen, Adnan, and Mehmet Özalp. 2005. "Solar-Driven Ejector-Absorption Cooling System." *Applied Energy* 80 (1): 97–113.

Sun, Da-Wen. 1997. "Solar Powered Combined Ejector-Vapour Compression Cycle for Air Conditioning and Refrigeration." *Energy Conversion and Management* 38 (5): 479–491.

Sun, Da-Wen. 1998. "Evaluation of a Combined Ejector–Vapour-Compression Refrigeration System." *International Journal of Energy Research* 22 (4): 333–342.

Sun, Da-Wen, Ian W. Eames, and Satha Aphornratana. 1996. "Evaluation of a Novel Combined Ejector-Absorption Refrigeration Cycle—I: Computer Simulation." *International Journal of Refrigeration* 19 (3): 172–180.

Tashtoush, Bourhan, Aiman Alshare, and Saja Al-Rifai. 2015. "Hourly Dynamic Simulation of Solar Ejector Cooling System Using TRNSYS for Jordanian Climate." *Energy Conversion and Management* 100 (Supplement C): 288–299.

Ventas, R., C. Vereda, A. Lecuona, and M. Venegas. 2012. "Experimental Study of a Thermochemical Compressor for an Absorption/Compression Hybrid Cycle." *Applied Energy* 97 (Supplement C): 297–304.

Vidal, Humberto, and Sergio Colle. 2010. "Simulation and Economic Optimization of a Solar Assisted Combined Ejector–Vapor Compression Cycle for Cooling Applications." *Applied Thermal Engineering* 30 (5): 478–486.

Vidal, Humberto, Sergio Colle, and Guilherme dos Santos Pereira. 2006. "Modelling and Hourly Simulation of a Solar Ejector Cooling System." *Applied Thermal Engineering* 26 (7): 663–672.

Wang, Jiangfeng, Yiping Dai, Taiyong Zhang, and Shaolin Ma. 2009. "Parametric Analysis for a New Combined Power and Ejector–Absorption Refrigeration Cycle." *Energy* 34 (10): 1587–1593.

Wang, Lin, Aihua Ma, Yingying Tan, Xiaolong Cui, and Hongli Cui. 2012. "Study on Solar-Assisted Cascade Refrigeration System." *Energy Procedia* 16 (Part C): 1503–1509.

Wu, S., and I. W. Eames. 1998. "A Novel Absorption–Recompression Refrigeration Cycle." *Applied Thermal Engineering* 18 (11): 1149–1157.

Xu, Feng, D. Yogi Goswami, and Sunil S. Bhagwat. 2000. "A Combined Power/Cooling Cycle." *Energy* 25 (3): 233–246.

Xu, Yingjie, Ning Jiang, Qin Wang, and Guangming Chen. 2016. "Comparative Study on the Energy Performance of Two Different Absorption-Compression Refrigeration Cycles Driven by Low-Grade Heat." *Applied Thermal Engineering* 106 (Supplement C): 33–41.

Yadav, Y. K. 1995. "Vapour-Compression and Liquid-Desiccant Hybrid Solar Space-Conditioning System for Energy Conservation." *Renewable Energy* 6 (7): 719–723.

Yapıcı, R., and F. Akkurt. 2012. "Experimental Investigation on Ejector Cooling System Performance at Low Generator Temperatures and a Preliminary Study on Solar Energy." *Journal of Mechanical Science and Technology* 26 (11): 3653–3659.

Yogi Goswami, D. 1998. "Solar Thermal Power Technology: Present Status and Ideas for the Future." *Energy Sources* 20 (2): 137–145.

Yu, Jianlin, and Yanzhong Li. 2007. "A Theoretical Study of a Novel Regenerative Ejector Refrigeration Cycle." *International Journal of Refrigeration* 30 (3): 464–470.

Zhang, Wei, Xiaoli Ma, S. A. Omer, and S. B. Riffat. 2012. "Optimum Selection of Solar Collectors for a Solar-Driven Ejector Air Conditioning System by Experimental and Simulation Study." *Energy Conversion and Management* 63 (Supplement C): 106–111.

Zhang, X. J., and R. Z. Wang. 2002. "A New Combined Adsorption–Ejector Refrigeration and Heating Hybrid System Powered by Solar Energy." *Applied Thermal Engineering* 22 (11): 1245–1258.

Zhu, Yinhai, and Peixue Jiang. 2012. "Hybrid Vapor Compression Refrigeration System with an Integrated Ejector Cooling Cycle." *International Journal of Refrigeration* 35 (1): 68–78.

8

Trigeneration Systems

8.1 Introduction

In recent years, a lot of interest has focused on cogeneration (combined heat and power, or "CHP") and trigeneration (combined cooling, heat, and power, or "CCHP") systems, which aim to convert a primary energy source into the aforementioned products. This interest is justified by the fact that traditional electricity production systems have a limited efficiency (around 30–40%) (Martins, Fábrega, and d'Angelo 2012; Cho et al. 2009), so a great deal of the primary heat input is rejected to the environment. By utilizing this heat and combining it with the additional production of cooling, multigeneration systems have higher global efficiencies, thus allowing for the reduction of energy consumption and emissions.

Solar trigeneration systems are essentially trigeneration systems that utilize solar energy as their primary energy source. They thus offer the possibility of reaching high energy conversion efficiencies while simultaneously utilizing a clean energy source with no CO_2 footprint. As a result, solar trigeneration systems have the potential to significantly contribute to the decarbonization of the energy mix and the reduction of emissions.

At the heart of CHP and CCHP systems is the prime mover engine, which is used for the conversion of heat to electricity. A multitude of prime mover engines are available for multigeneration systems, which traditionally include:

- Reciprocating internal combustion engines
- Gas turbines and combined cycle plants
- Steam turbines
- Microturbines
- Organic Rankine cycle and Stirling engines (μ-CHP/CCHP)
- Fuel cells
- Hybrid PVT panels

The final selection of the most appropriate technology for a specific application is based on case-specific parameters such as:

- Scale (based on heat and power demand)
- Fuel availability
- Flexibility requirements (start-up time, part load behavior)
- Power to heat ratio

- Heating and cooling required loads and temperatures
- Maintenance requirements
- Investment costs

Most CCHP systems are essentially CHP systems with the addition of a chiller, which is integrated for the production of cooling. In principle, there are two main options for the integration of the chiller. In the first case, part of the mechanical or electrical power produced by the prime mover is used for the operation of an mechanically/electrically driven vapor compression cycle. In the second case, a portion of the generated useful heat that is produced is used for driving a thermally driven chiller, such as a sorption (absorption/adsorption) or a desiccant cooling engine.

In literature, many different configurations have been proposed for CCHP systems. These configurations can employ several types of prime mover engines (such as reciprocating engines, microturbines, ORCs, and SEs) and various cooling devices (such as sorption chillers, desiccant cooling, and heat pumps) (Sonar, Soni, and Sharma 2014) that utilize different energy sources, with solar and biomass playing a significant role among the renewable sources.

Most research on biomass-fueled trigeneration systems involves decentralized applications located in the proximity of the biomass source, with ORCs (>200 kW) (Quoilin et al. 2013; Liu, Shao, and Li 2011; Borsukiewicz-Gozdur et al. 2014) and SEs (<200 kW) considered the most popular prime movers for scales less than 2 MW (Maraver et al. 2013). Other studies focus on the development of solar-driven systems (Sarbu and Sebarchievici 2013; Kegel, Tamasauskas, and Sunye 2014a; Al-Sulaiman, Dincer, and Hamdullahpur 2013; Al-Sulaiman, Dincer, and Hamdullahpur 2011), usually with the combination of concentrating collectors coupled with ORC engines or hybrid PVT collectors as prime movers (Kalogirou and Tripanagnostopoulos 2006; Hasan and Sumathy 2010; Chen, Riffat, and Fu 2011). On most occasions, thermally activated cooling devices are considered (absorption and adsorption) (Maraver et al. 2013; Al-Sulaiman, Dincer, and Hamdullahpur 2011).

Regarding solar energy utilization, PTC systems currently constitute the most mature technology for solar thermal electricity conversion and can be used as heat input devices for CCHP systems. This is mainly because of their ability to operate under higher temperatures (up to 400°C), thus having a potential for higher efficiencies (Quoilin et al. 2013; Kalogirou 2004). In addition to solar thermal heat engines, PV panels can also function as the prime movers for multigeneration systems. In this case, the electricity produced in the cells can be used to power the compressor of a heat pump in order to generate cooling and heating. Solar cooling systems integrating PV technology have been studied by various researchers. PV technology in the residential sector is quite popular due to its numerous advantages such as ease of installation, standardization, low investment costs (ranging from €2,000/kWe to €7,000/kWe) (Villarini et al. 2014), and no need for maintenance and monitoring.

In the following section, a literature review of different studies on solar trigeneration systems including different technologies and configurations is presented.

8.2 Literature Review

Calise et al. (2016) performed a dynamic simulation and an economic assessment of a solar-geothermal trigeneration system for the production of electricity, desalinated water, space heating, and space cooling through a district heating network. The prime mover of the system was an ORC powered via geothermal and solar energy (with the installation of PTCs), while cooling was produced by an absorption chiller. The desalination was carried out by a multi-effect distillation system. The authors concluded that the main advantages of the proposed system were its high flexibility and efficiency. On the other hand, they also stressed that the system was associated with high capital costs.

Buonomano et al. (2013) investigated the design and the simulation of a solar trigeneration system using concentrating PVT dish collectors. The system was specially designed for high temperature operation and was equipped with double-effect absorption chillers. The authors concluded that their results proved the technical feasibility of the concept. However, they stressed its high costs as a result of the use of triple-junction PV cells.

Intini et al. (Intini et al. 2015; Najafi et al. 2015) investigated a trigeneration system based on a polymer electrolyte fuel cell coupled with a desiccant wheel. The authors showed that the low temperature of the produced heat (65–70°C) led to low efficiencies in the chiller, and they also remarked that the operation of the fuel processor faced flexibility issues. From annual simulations of the systems, they found that the constraints of real equipment components have significant effects on energy consumption. Thus, although it is possible to achieve energy savings in the winter period, it is impossible to do so on a yearly basis, considering the current performance of the proposed system.

Wu et al. (2018) performed an optimization of the design and the operation of trigeneration systems (Figure 8.1) integrated in commercial buildings based on ORCs by taking into account their economic and environmental performance. A solar-driven CCHP system with PTCs was compared with a biomass-driven and a natural-gas-fueled system. In all cases, cooling was generated by an absorption chiller, while an auxiliary boiler and a VCC were also incorporated. The authors concluded that the solar system exhibited the greatest potential for reducing annual costs and the biomass system had the best environmental performance.

Al-Ali and Dincer (2014) performed an energetic and exergetic study of an industrial scale multigeneration system based on the combined utilization of solar and geothermal energy (Figure 8.2). The system was able to produce a multitude of load requirements, including electric power, cooling, space heating, hot water, and industrial process heat. A low and a high temperature ORC module were used as prime mover engines, powered by geothermal and solar energy, respectively, while a single-effect LiBr-H_2O absorption chiller was used for the production of cooling. Under multigeneration operating mode, the system achieved an energetic efficiency of 78%, compared to 16.4%, which was the efficiency under standalone electricity generation. Meanwhile, the exergetic efficiency was equal to 36.6%, compared to 26.2% for single generation. Lastly, the authors identified the solar collectors as the most significant cause of irreversibilities.

Gazda and Stanek (2016) focused on the energetic and environmental assessment of a trigeneration plant powered by biogas fuel and PV panels for industrial use (Figure 8.3). An internal combustion engine was considered as the prime mover of the system. Furthermore, an adsorption chiller was used for the production of cooling, powered from the flue gases of the ICE. The authors simulated the operation of the system on an annual basis and estimated that the amount of primary energy savings and the reduction in greenhouse gas emissions were equal to 54.50% and 67.37%, respectively.

Askari et al. (2015) focused on the energy management and economic aspects of a trigeneration system including a natural gas generator and an absorption chiller by taking into account the impact of solar PVs, solar collectors, as well as fuel prices. The application of the system for covering the cooling, heating, and electricity requirements of a

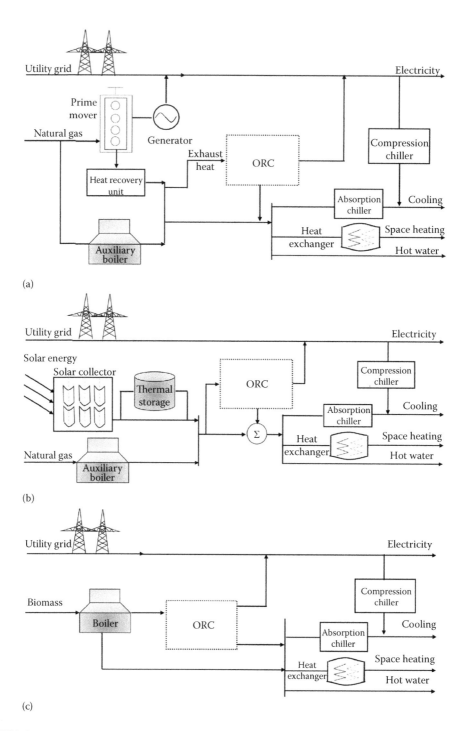

FIGURE 8.1

The trigeneration systems investigated by Wu et al. (a) CCHP—ORC. (b) Solar—ORC. (c) Biomass—ORC. (Reproduced from Wu, Qiong et al., *Energy*, 142: 666–677, 2018.)

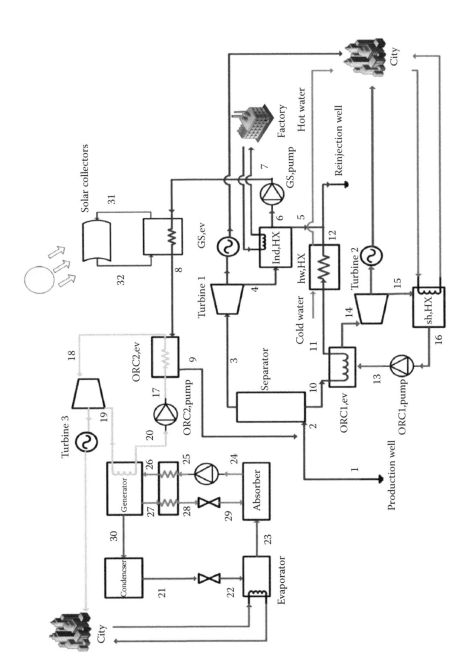

FIGURE 8.2
The trigeneration systems investigated by Al-Ali and Dincer. (Reproduced from Al-Ali, M., and I. Dincer, *Applied Thermal Engineering*, 71 (1): 16–23, 2014.)

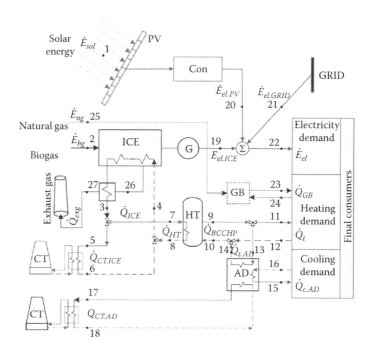

FIGURE 8.3
The trigeneration systems investigated by Gazda and Stanek. (AD: adsorption chiller; Con: Converter; CT: cooling tower; G: generator; GB: gas boiler; GRID: electric grid; HT: hot storage tank; ICE: internal combustion engine; PV: photovoltaic panels.) (Reproduced from Gazda, W., and W. Stanek, *Applied Energy*, 169 (Supplement C): 138–149, 2016.)

five-story high-rise residential building was considered. The peak electricity load of the system was equal to 48 kW, and the heating and cooling requirements were equal to 100 kW and 50 kW, respectively. Considering data corresponding to the Iranian energy market, the authors concluded that at low fuel costs ($0.1/m3), the configurations utilizing high PV capacities were not economically feasible. Meanwhile, they found that the CCHP system with an installed electrical capacity of natural gas generators equal to 44 kW had the same unit energy cost as the conventional energy systems for fuel prices below $0.1/m3. Furthermore, the authors showed that for higher fuel prices, the application of solar collectors was economically feasible.

Baghernejad at al. (2016) carried out an exergoeconomic comparison of three trigeneration systems. The first one included solid oxide fuel cells (SOFC), the second included a biomass boiler, and the third included solar collectors. The solar trigeneration system included two gas turbines and heat recovery steam generators, a steam turbine and a solar module. The systems were optimized on the basis of exergoeconomic optimization principles. The SOFC system exhibited an exergetic efficiency increase from 62.85% to 64.5%. The optimization of the biomass system led to an increase in the exergetic efficiency of 22.8%. The integrated solar trigeneration system displayed a relative exergetic efficiency increase equal to 26.34%. Lastly, the authors pointed out that the solar trigeneration system had the lowest CO_2 emissions (236.7 kg/MWh), the SOFC-trigeneration system had the highest exergy efficiency (64.5%), and the biomass-trigeneration system had the lowest unit cost of products (68.2 cents/kWh).

Xu et al. (2014) carried out an investigation of a combined cooling, heating, and power system integrating middle- and low-temperature solar thermal energy utilization and methanol decomposition. The system included a parabolic trough collector operating at temperatures between 250°C and 300°C. Solar heat was used to drive a methanol decomposition reaction, which produces syngas. The syngas was subsequently combusted in the chamber of a micro gas turbine in order to power the absorption Li-Br cooling, heating, and power systems. The total exergy input of the system was 2086.9 kW. The total exergy output corresponding to electric power, cooling, and heat was 600 kW, 35.4 kW, and 29.4 kW, respectively. The system was estimated to have an exergy efficiency of 48.81%, with a net electric-to-solar thermal efficiency of up to 22.56%. The overall energetic efficiency of the system was 76.40%.

Al-Sulaiman et al. (2011) performed an exergy modeling study of a solar-driven trigeneration system based on the integration of parabolic trough solar collectors, an ORC, and an absorption chiller (Figure 8.4). The performance of the system was evaluated for four different cases: electricity (single generation), cogeneration with heating, cogeneration with cooling, and trigeneration. Three modes of operation were considered with regard to solar energy utilization including solar energy storage and simultaneous solar energy utilization and storage. The effects of the ORC evaporator pinch point temperature, pump

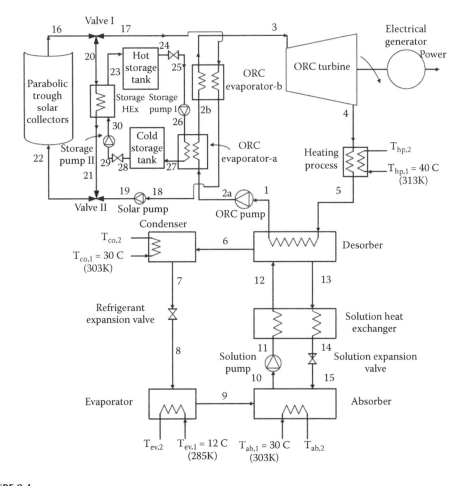

FIGURE 8.4
The trigeneration systems investigated by Al-Sulaiman et al. (Reproduced from Al-Sulaiman, Fahad A. et al., *Solar Energy*, 85 (9): 2228–2243, 2011.)

inlet temperature, and turbine inlet temperature of the ORC on the exergy efficiency of the system were evaluated in all cases. The solar utilization mode had the highest exergy efficiency. For the electricity-only generation scenario, the solar utilization exergy efficiency was 7%. The exergy efficiency of the other two modes was 3.5% (utilization and storage) and 3% (storage). Under trigeneration mode, the efficiency values increased to 20%, 8%, and 7%, respectively. The system components exhibiting the highest exergy destruction were, according to the authors, the collectors and the evaporator of the ORC. The authors also concluded that there was a very little impact from the turbine inlet pressure on the exergy efficiency of the system.

Zhai et al. (2009) considered a novel domestic-scale hybrid solar trigeneration system for remote areas. The system was based on the use of a parabolic trough collector with a cavity receiver, a screw expander, and a silica gel-water adsorption cooling module. The modeled system was able to produce 23.5 kW of electricity and 79.8 kW of cooling for a solar radiation of 600 W/m2 and an area of 600 m2. For the meteorological conditions of Dunhuang, China, the energy and exergy efficiency of the system was estimated at 58% and 15.2%, respectively.

Li et al. (2016) evaluated a hybrid trigeneration system utilizing biomass and solar energy (Figure 8.5) under exergetic and environmental criteria. The heat input to the system was derived by biomass gasification and solar collectors. The prime mover was an ICE, and cooling was generated by an absorption unit and a desiccant chiller. The solar collector was used for the generation of steam to be used for biomass gasification, and the produced syngas was fed into the ICE to produce electric power. Flue gases from the ICE were used as energy sources for the absorption chiller and for the generation of domestic hot water. The total exergy efficiency of the system was equal to 19.21%. The main constituent of exergy destruction was the gasifier, followed by the ICE.

Mohan et al. (2016) experimentally investigated a solar thermal multigeneration (not including electricity) plant including an absorption chiller located in the United Arab Emirates. The system was investigated under different modes of operation: solar cooling, cogeneration of drinking water and domestic hot water, cogeneration of cooling and desalination, and trigeneration. The system additionally involved evacuated tube solar thermal collectors, a membrane distillation unit, as well as heat exchangers, a cooling tower, and a thermal storage tank. The authors concluded that the standalone solar cooling mode was associated with a COP of 0.6 and was mostly appropriate for peak summer days. Meanwhile, under trigeneration mode, the system was more energetically efficient by 23%. In general, the plant had a payback period of 9.08 years and a net present value of $454,000.

Eisavi et al. (2017) focused on a solar trigeneration system consisting of an ORC as a prime mover engine coupled with a double-effect lithium bromide-water absorption chiller. The authors compared the performance of the system with a similar combined cycle system that included a single-effect absorption chiller. According to their results, it was determined that for the same amount of heat input, the double-effect absorption refrigeration system was associated with a 48.5% increase in the cooling capacity compared to the system including the single-effect chiller. Furthermore, the useful heat capacity was 20.5% higher. The overall performance increase led to an increase in the CHP efficiency equal to 96.0%. However, the net electrical power production was reduced by 27%. Finally, the authors determined that the primary source of exergy destruction was in the solar collectors.

Nosrat et al. (2013) investigated a hybrid trigeneration system consisting of PV technology and a fossil fueled CCHP system that included an absorption chiller to produce electricity, domestic hot water, and space heating and cooling (Figure 8.6). The authors mostly

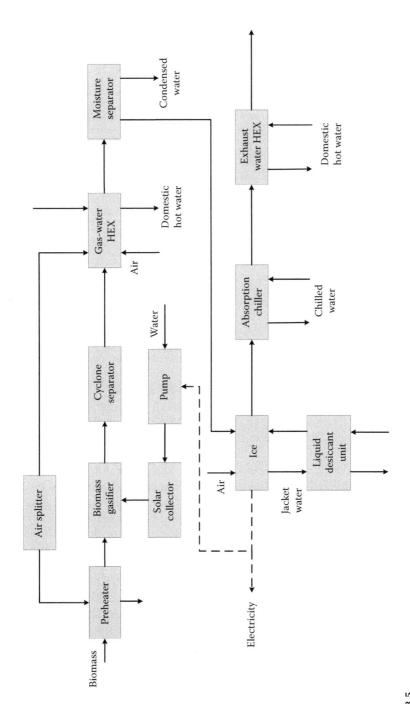

FIGURE 8.5

The trigeneration systems investigated by Li et al. (Adapted from Li, Hongqiang et al., *Applied Thermal Engineering*, 104 (Supplement C): 697–706, 2016.)

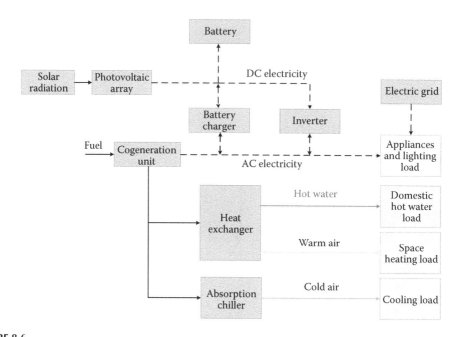

FIGURE 8.6
The trigeneration systems investigated by Nosrat et al. (Adapted from Nosrat, Amir H. et al., *Energy*, 49 (Supplement C): 366–374, 2013.)

focused on the development of a novel PV-trigeneration optimization model (PVTOM), suitable for evaluating complex PV-trigeneration systems. They compared the trigeneration system with a simple cogeneration system for domestic application, and concluded that the trigeneration system had better greenhouse gas emission performance.

Tora and El-Halwagi (2011) carried out a study on the development of a systematic procedure for the integrated design of solar trigeneration systems including absorption cooling technology. In their model, the decision-making horizon was divided into multiple steps in order to account for the seasonal fluctuation of solar radiation. Furthermore, a nonlinear programming formulation was integrated into the model. The solution of the optimization model was used to establish the optimal levels of power capacities, external heating, external cooling, heat integration, and the mix of fossil/solar energy forms to be supplied to the process as well as for the meticulous selection of the operation strategy of the system. By performing a case study, the authors estimated that a carbon credit between €5–20/tonne of CO_2 was necessary for solar trigeneration systems to be viable.

Wang and Fu (2016) carried out a thermodynamic analysis of a solar-hybrid trigeneration system integrated with methane chemical-looping combustion. The process of chemical-looping combustion occurs without the reaction of air with fuel. It is a technology that is used for capturing CO_2 at decreased energy consumption ranges. The system proposed by the authors involved CaS and $CaSO_4$ materials for the chemical-looping system. A number of parametric investigations were carried out in order to estimate the energy and exergy efficiencies at different design conditions and variable operation parameters. The authors estimated that the optimal solar heat collection temperature was equal to 900°C, and that the optimal pressure ratio of the compressor used for the chemical looping was equal to 20. The optimized energetic and exergetic efficiencies of the system were equal to 67% and 55%, respectively. Lastly, the authors remarked that additional research in the fields of CO_2 capture and utilization and solar heat storage is necessary in order to improve its feasibility and stability.

Chua et al. (2012) focused on the integration of renewable energy to trigeneration systems in commercial buildings (Figure 8.7). The investigated system comprised five subsystems: a PVT module, a solar thermal circuit, a fuel cell engine, a microturbine, and an absorption chiller-water system. A multi-criteria analysis was carried out encompassing various indexes such as operation cost reduction, energy savings, and environmental impact. Based on the results, the authors concluded that a system composed of 80% microturbines, 10% PVTs, and 10% fuel cells was the optimal one in terms of all of the above indexes.

Cioccolanti et al. (2017) assessed a residential solar trigeneration system by developing a numerical model. The system they investigated consisted of compound parabolic collectors with heat pipes, an ORC module for the production of electricity, and an absorption chiller for the production of cooling. The authors conducted dynamic and off-design simulations to evaluate the system's behavior on an annual, monthly, and daily basis. They estimated that the operational capacity of the system exceeded 2,500 hours per year. Despite the relatively decreased efficiency of the CPC collectors (due to their high temperature), the ORC had a good electrical efficiency in the winter. On the other hand, the efficiency of the ORC decreased during the summer due to the operation of the absorption chiller.

Wang et al. (2017) investigated a trigeneration system based on phosphoric acid fuel cell technology involving the integration of solar-assisted methanol reforming. The authors conducted thermodynamic simulations on the trigeneration system. They estimated that the energetic efficiencies achieved under summer and winter operating conditions were 73.7% and 51.7%, respectively, while the exergetic efficiencies were equal to 18.8% and 26.1%, respectively. The authors also estimated that when solar energy availability is different than the design point condition, the energy and exergy efficiencies in winter drop by approximately 4.7% and 2.2%, respectively, due to the decrease in the efficiency of the solar collectors.

Leiva-Illanes et al. (2017) worked on the thermoeconomic assessment of a 50 MW solar multigeneration facility for the production of electricity, fresh water via desalination, cooling, and heating while also taking into account high solar radiation intensity. The solar module consisted of concentrating solar power parabolic trough collectors, and it also

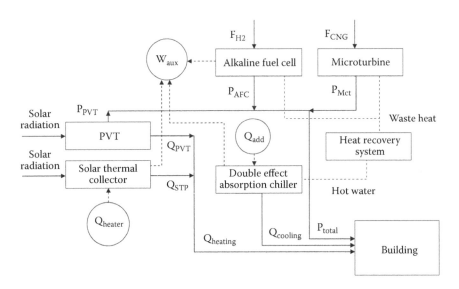

FIGURE 8.7
The trigeneration systems investigated by Chua et al. (Reproduced from Chua, K. J. et al., *Renewable Energy*, 41 (Supplement C): 358–367, 2012.)

included thermal energy storage and a backup system. The production of cooling was achieved by a single-effect absorption chiller, while a multi-effect distillation module was employed for the desalination process. By comparing the plant to standalone plants, the authors concluded that it was more efficient (in terms of exergy cost rate, unit exergy cost, and exergy efficiency) and cost competitive.

Meng et al. (2010) focused on a theoretical study of a solar/waste heat recovery trigeneration plant utilizing metal hydrides, which are special types of alloys that have the ability to absorb and desorb hydrogen reversibly. The authors optimized the system in order to increase the exergy efficiency of the metal hydride heat pump, based on the available solar input. They concluded that the proposed system was superior to a traditional CCHP system.

Kegel et al. (2014b) focused on different variants of solar thermal trigeneration systems for a multi-unit residential building in Canada. The systems included an absorption chiller. The prime mover was a natural gas CHP engine. It was estimated that the operation of the CHP engine in priority compared to the solar thermal engine led to a 21% reduction in annual costs and a 16% reduction in greenhouse gas emissions. On the other hand, the prioritization of the operation of the solar thermal collectors led to primary and secondary energy savings equal to 16% and 18%, respectively, while the greenhouse gas emission reduction was again equal to 16%. When compared to a standard CHP system, the secondary energy savings, when estimated, equal 36%. The authors also concluded that the addition of a thermally driven chiller did not produce any considerable primary or secondary energy savings, greenhouse gas emission reduction, or utility cost savings because it required the operation of an auxiliary single-stage boiler with poor part-load performance for an extended period. Finally, the authors remarked that because of relatively lower natural gas costs and higher electricity rates, the systems operating the cogeneration system in priority always resulted in the highest utility cost savings, while the addition of solar thermal and thermally driven heating/cooling led to constrained economic profits overall.

Patel et al. (2017) performed a thermoeconomic analysis of a trigeneration system based on a solar-biomass organic Rankine cycle with the integration of a cascaded vapor compression-absorption system. Different solar collector options were investigated: a paraboloid dish collector, a linear Fresnel collector, as well as parabolic trough collectors. The comparative analysis for the different collectors indicated that the dish collector had the highest solar fraction values, followed by the PTCs and the LFR. More specifically, the solar fraction and breakeven point (BEP) for the paraboloid dish system, in which n-pentane was the organic fluid and straw type biomass fuel was used, was equal to 0.254 and 7.71 years, respectively. Meanwhile, the solar fraction for the LFR-based system was estimated to be equal to 0.179. Nevertheless, the lower cost of LFR collectors as well as the lower cost of energy generation from biomass led an overall lower BEP, equal to 7.43 years. Lastly, the fully biomass-powered system had a 30% lower BEP.

Khalid et al. (2017) performed a thermoeconomic analysis of a 370 kWe solar-biomass multigeneration system designed to cover the demands of a community. The prime movers of the system were a gas turbine powered by biomass fuel and two ORC units (one at a high and one at a low temperature) powered by solar energy. Furthermore, an absorption chiller was used for covering the cooling loads. The useful heat was produced by the condensers of the ORC modules. The net present cost of the system was estimated to be equal to approximately $2,700,000 while the cost of electricity was calculated at $0.117/kWhe. Moreover, the authors estimated that the components leading to the highest exergy destruction are the combustion chamber along with the concentrating solar collectors. The exergy and energy efficiencies of the system were estimated to be equal to 34.9% and 91%, respectively, and were higher than the respective values of standalone renewable energy systems.

8.3 Case Study: The BioTRIC Trigeneration System

One of the most commonly investigated solar trigeneration concepts involves the combination of a solar-powered ORC that is used for powering a VCC that generates cooling. Meanwhile, useful heat can be extracted either from the condenser of the ORC, the condenser of the VCC, or directly from the heat transfer fluid of the solar circuit.

In the Laboratory of Steam Boilers and Thermal Plants at the National Technical University of Athens, a novel solar/biomass trigeneration system based on the combination of an ORC and a VCC has been designed and studied (Karellas and Braimakis 2016b; Braimakis, Thimo, and Karellas 2017). Furthermore, a micro-scale experimental unit has been built, including a combined natural gas/biomass boiler. The main novelty of the proposed system is the fact that the ORC and the VCC cycles are interconnected, operating with the same working fluid. Furthermore, with the installation of the ORC expander and the VCC compressor on the same shaft, it is possible to greatly reduce the electromechanical conversion losses that occur during the production and the consumption of electricity in the ORC and the VCC, respectively.

In this section of the chapter, the modeling and the operation strategy of a solar/biomass theoretical system called BioTRIC is briefly presented. Furthermore, the results and key aspects of a study—including a techno-economic comparison of the BioTRIC with a PV-heat pump system—carried out by Braimakis et al. (2017) are presented.

To begin, a general description of the system is given. The ORC-VCC system consists of a biomass boiler module, a PTC module, and an ORC interconnected to the VCC, as presented in Figure 8.8. The PV-heat pump system schematic is depicted in Figure 8.9.

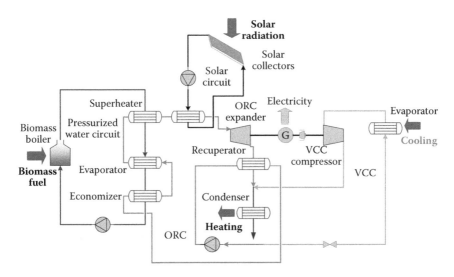

FIGURE 8.8
The BioTRIC system. (Reproduced from Karellas, S., and K. Braimakis, *Energy Conversation and Manage*, 107: 103–113, 2016a.)

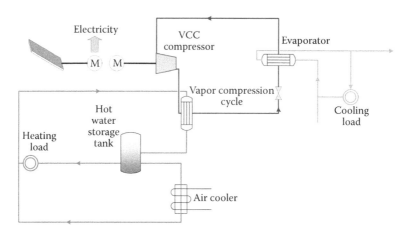

FIGURE 8.9
The PV-heat pump system.

The ORC configuration includes a pump, an expander, a condenser, and heat exchangers for the transfer of heat from the biomass boiler and the PTC circuit to the working fluid via two circuits using a heat transfer fluid. Due to the selection of dry organic fluids (positive slope of the saturated vapor curve in the T-s diagram) and the high temperature of the expander exhaust, a recuperator can be interposed between the expander and the circulation pump. The heat is mainly provided by the PTC circuit and secondarily, when solar thermal energy is inadequate, by the biomass boiler circuit. In this way, the system can achieve autonomous and continuous operation, even when solar energy is unavailable. At the same time, the system is completely independent of fossil fuels. The VCC system consists of an evaporator, a compressor, a condenser (in common with the ORC), as well as an expansion valve. The key operational parameters of the system are the pressure and the temperature of the working fluid at the inlet of the ORC expander, its condensation temperature/pressure (common with the ORC and the VCC), and the evaporation temperature/pressure of the VCC.

The condensation temperature is set at 70°C in order to cover both space heating and domestic hot water demands. The evaporation temperature of the VCC is roughly equal to 7.5°C, since this a reasonable temperature for the cooling operation of heat pumps.

One of the main parameters of the ORC-VCC system is the working fluid used. The main criteria that can influence this selection include:

- Thermodynamic parameters such as critical temperature and pressure, boiling and freezing point, vapor density, viscosity, and conductivity

- Restrictions regarding the maximum allowable pressure (corresponding to the point where the slope of the saturation curve in the temperature-entropy diagram is equal to infinity, in order to avoid the presence of liquid during the expansion process), the minimum condensing pressure, and the selection of expander.

- Other parameters such as the flammability and toxicity of the fluid, its chemical stability, environmental impact (ODP, GWP), commercial availability, and cost

Based on the diagram in Figure 8.8, the performance of the system is described by the following equations:

The total heating generated:

$$\dot{Q}_h = \dot{m}_{tot} \cdot \left(h_{11} - h_{12}\right) \tag{8.1}$$

where

$$\dot{m}_{tot} = \dot{m}_{ORC} + \dot{m}_{VCC} \tag{8.2}$$

The total cooling produced:

$$\dot{Q}_c = \dot{m}_{VCC} \cdot \left(h_1 - h_4\right) \tag{8.3}$$

The net electric output of the ORC is calculated by the following equation:

$$P_{el,ORC} = P_{el,\exp} - P_{el,pump} \tag{8.4}$$

where $P_{el,\exp}$ is the electricity produced by the generator driven by the expander and $P_{el,pump}$ is the electricity consumption of the pump.

The coefficient of performance of the VCC (in cooling mode) is given by the equation:

$$COP_c = \frac{\dot{Q}_c}{P_{comp}} = \frac{h_1 - h_4}{h_2 - h_1} \tag{8.5}$$

The electric and CHP efficiency of the PTC biomass-based ORC system are calculated as follows:

$$\eta_{el,sys} = \frac{P_{el,net}}{\dfrac{\dot{Q}_{PTC}}{\eta_{PTC}} + \dfrac{\dot{Q}_{bio,ORC}}{\eta_{bio}}} \tag{8.6}$$

$$\eta_{el,ORC} = \frac{P_{el,net}}{\dot{Q}_{PTC} + \dot{Q}_{bio,ORC}} \tag{8.7}$$

$$\eta_{CHP,sys} = \frac{P_{el,net} + \dot{Q}_{h,CHP}}{\dfrac{\dot{Q}_{PTC}}{\eta_{PTC}} + \dfrac{\dot{Q}_{bio,ORC}}{\eta_{bio}}} \tag{8.8}$$

The system operates as a cogeneration unit (heating and power) throughout the year while it additionally produces cooling during the summer. In principle, the system is heat driven, since its operation is adjusted so that there is a balance between the generated heat and the heat demand on a daily basis.

All state points of the thermodynamic cycles have fixed values, with the exception of Point 7, which corresponds to the working fluid outlet from the boiler circuit heater. The temperature of this point is determined by the heat input of the biomass boiler to the system, which in turn depends on the heat input of the PTC. The target of the control strategy of the ORC-VCC system is the operation of the system under the estimated fixed thermodynamic conditions (pressures and temperatures) at each point of the cycles. This is achieved by regulating the mass flow of the working fluid in each cycle though the pump (ORC) and the compressor (VCC) rotational speed. The flow rate of the heat transfer fluid in each one of the heat transfer loops is also adjusted accordingly.

Moreover, the mass flow rate of the working fluid is assumed to range between 80% and 100% of the nominal flow rate, and thus leads to a relatively constant isentropic efficiency for the expander, the compressor, and the pump. Therefore, efficiency penalties attributed to off-design operation of these components can be considered non-significant.

The above constraint has two implications. First, the heat output of the system has a minimum value that can possibly surpass the heat demand, and, second, during the cooling period, excess cooling can be produced. This inevitably leads either to the use of storage systems or to the rejection of excess heat or cool air to the environment, requiring additional heating and cooling rejection equipment (fan coils, cooling tower etc.).

The basic points that describe system operation are as follows:

- The system is connected to the grid.
- The ORC only operates when there is solar radiation in order to minimize biomass consumption.
- The heat required for the operation of the ORC is primarily provided by the PTC and secondarily (when solar radiation is insufficient) by the biomass boiler. When the heat produced by the PTC exceeds the heat required by the ORC, the excess heat is stored via a thermal storage tank.
- The VCC is powered from the grid while the electricity produced by the ORC is sold.
- The ORC and the VCC are sized in order to cover 50% of the maximum annual heating load (in order to reduce heat rejection) and 100% of the maximum annual cooling load respectively.

Regarding the PV-heat pump system, its design point is determined by the requirement to cover 100% of the heating and cooling load while any energy in excess is stored via a thermal storage tank. Constraints regarding the mass flow rate (between 80% and 100% of the nominal flow rate) have been taken into account. In accordance with the energy pricing policy, feed-in tariff, or net metering scheme, the compressor is powered either from the grid or the photovoltaic generator and the grid, accordingly.

One central inverter of multi-string technology (SMA-Hellas) is assumed. Taking into account its power output (10 kW) and other technical characteristics, such as minimum and maximum DC voltage (150 V and 1000 V, accordingly), maximum input current, and space availability on the roof of the building, a peak power of 9.4 kW, corresponding to an area of 49.2 m^2, has been considered.

In order to compare the BioTRIC and the PV-heat pump system, three different working fluids were investigated for the ORC-VCC system: R245fa, hexane, and cyclohexane. In each case, the evaporator pressure of the ORC was optimized in order to maximize system efficiency. The overall characteristics of the compared systems are summarized in Table 8.1.

TABLE 8.1

Design Point Data and Specifications of Both Systems

	ORC-VCC			PV-Heat Pump	
	R245fa	**Hexane**	**Cyclohexane**		
$P_{el,ORC,nom}$ (kW$_e$)	0.7	1.3	2	$P_{el,PV,nom}$ (kW$_e$)	9.4
$P_{el,com}$ (kW$_e$)	2.5–3.13	2.35–2.94	2.18–2.73	$\eta_{el,pv-HP,average}$ (%)	15.04
COP_c	2.38	2.53	2.73	$P_{el,PV}$ (kW$_e$)	4.74–7
$\eta_{el,nom,sys}$ (%)	3.71	7.62	10.05	$Q_{h,nom}$(kW$_{th}$)	20
$\eta_{el,nom,ORC}$ (%)	4.98	10.51	14.55		
$\eta_{CHP,nom,sys}$ (%)	73.5	70.7	66.7	$Q_{c,nom}$ (kW$_{th}$)	6.3
				$COP_{c,nom}$	3.36
				$COP_{h,nom}$	3.93

A detailed techno-economic comparison of the two systems was performed, taking into account Greek meteorological data. Simulations considering annual heating and cooling operation were also performed. The annual operating costs and savings that were estimated are presented in Figure 8.10.

Furthermore, the discounted payback periods for the two systems are presented in Figure 8.11.

From the results, it can be concluded that micro-scale ORCs that operate combined with biomass boilers or solar collectors can be market-competitive only if they have the potential for combined heating and power production. Moreover, if biomass is available at lower costs (waste or closely located to the unit), the savings of the system could increase. With regard to the photovoltaic-driven heat pump, although it is a more mature and common technology than ORC in residential applications, because of current national pricing policies (including net-metering for PV-generated electricity), it does not present a better economic performance.

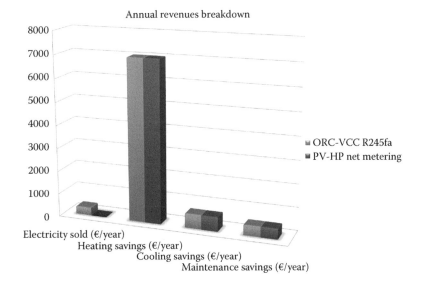

FIGURE 8.10

Annual operating costs and savings of the two systems. (Reproduced from Braimakis, K. et al., *Journal of Energy Engineering*, 143 (2): 04016048, 2017.)

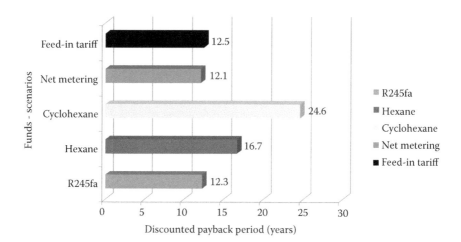

FIGURE 8.11
Discounted payback period comparison of the BioTRIC system and the PV-VCC system. (Reproduced from Braimakis, K. et al., *Journal of Energy Engineering*, 143 (2): 04016048, 2017.)

8.4 Conclusions

The first section of the chapter presented a literature review of different systems aimed at solar trigeneration. A wide variety of different configurations have been studied. The majority of the investigated concepts include absorption chillers, which utilize part of the heat that is rejected from the prime mover engines. The most commonly proposed prime mover engines include natural gas ICEs and ORCs. The second section presented a hybrid ORC-VCC system utilizing solar energy and biomass, as well as the hybrid PVT system, which also attracts the interest of a lot of researchers. The system has been technically and economically investigated by the Laboratory of Steam Boilers and Thermal Plant at the National Technical University of Athens, Greece, and could be an ideal option for decentralized applications when there is availability of biomass fuel at low cost. Furthermore, it has potential advantages in cases where net-metering schemes are applied for the pricing of PV-derived electricity.

Nomenclature

H	Specific enthalpy, kJ/kg
$\dot{m}$	Mass flow rate, kg/s
P	Power, kW
$\dot{Q}$	Geat flux, kW
T	Temperature, °C

Greek symbols

η	Efficiency, %

Abbreviations

BEP	Break even period
CCHP	Combined cooling heat and power
CHP	Combined heat and power
COP	Coefficient of performance
FPC	Flat-plate collectors
LFR	Linear Fresnel collectors
NPV	Net present value
PTC	Parabolic trough collectors
PV	Photovoltaic
SE	Stirling engine
VCC	Vapor compression cycle

Subscripts

bio	Biomass
c	Cooling
CHP	Combined heat and power
comp	Compressor
el	Electric
exp	Expander
HP	Heat pump
is	Isentropic
nom	Nominal
ORC	Organic Rankine cycle
PTC	Parabolic trough collectors
pump	Pump
PV	Photovoltaic
sys	System
VCC	Vapor compression cycle

References

Al-Ali, M., and I. Dincer. 2014. "Energetic and Exergetic Studies of a Multigenerational Solar Geothermal System." *Applied Thermal Engineering* 71 (1): 16–23. doi: https://doi.org/10.1016/j.applthermaleng.2014.06.033.

Al-Sulaiman, Fahad A., Ibrahim Dincer, and Feridun Hamdullahpur. 2013. "Thermoeconomic Optimization of Three Trigeneration Systems Using Organic Rankine Cycles: Part II—Applications." *Energy Conversion and Management* (69): 209–216.

Al-Sulaiman, Fahad A., Ibrahim Dincer, and Feridun Hamdullahpur. 2011. "Exergy Modeling of a New Solar Driven Trigeneration System." *Solar Energy* 85 (9): 2228–2243. doi: https://doi.org/10.1016/j.solener.2011.06.009.

Baghernejad, A., M. Yaghoubi, and K. Jafarpur. 2016. "Exergoeconomic Comparison of Three Novel Trigeneration Systems Using SOFC, Biomass and Solar Energies." *Applied Thermal Engineering* 104 (Supplement C): 534–555. doi: https://doi.org/10.1016/j.applthermaleng.2016.05.032.

Baniasad Askari, I., M. Oukati Sadegh, and M. Ameri. 2015. "Energy Management and Economics of a Trigeneration System Considering the Effect of Solar PV, Solar Collector and Fuel Price." *Energy for Sustainable Development* 26 (Supplement C): 43–55. doi: https://doi.org/10.1016/j.esd.2015.03.002.

Borsukiewicz-Gozdur, A., S. Wiśniewski, S. Mocarski, and M. Bańkowski. 2014. "ORC Power Plant for Electricity Production from Forest and Agriculture Biomass." *Energy Conversion and Management* 87: 1180–1185. doi: http://dx.doi.org/10.1016/j.enconman.2014.04.098.

Braimakis, Konstantinos, Antzela Thimo, and Sotirios Karellas. 2017. "Technoeconomic Analysis and Comparison of a Solar-Based Biomass ORC-VCC System and a PV Heat Pump for Domestic Trigeneration." *Journal of Energy Engineering* 143 (2): 04016048. doi: doi:10.1061/(ASCE)EY.1943-7897.0000397.

Buonomano, Annamaria, Francesco Calise, Massimo Dentice d'Accadia, and Laura Vanoli. 2013. "A Novel Solar Trigeneration System Based on Concentrating Photovoltaic/Thermal Collectors. Part 1: Design and Simulation Model." *Energy* 61 (Supplement C): 59–71. doi: https://doi.org/10.1016/j.energy.2013.02.009.

Calise, Francesco, Massimo Dentice d'Accadia, Adriano Macaluso, Laura Vanoli, and Antonio Piacentino. 2016. "A Novel Solar-Geothermal Trigeneration System Integrating Water Desalination: Design, Dynamic Simulation and Economic Assessment." *Energy* 115 (Part 3): 1533–1547. doi: https://doi.org/10.1016/j.energy.2016.07.103.

Chen, Hongbing, Saffa B. Riffat, and Yu Fu. 2011. "Experimental Study on a Hybrid Photovoltaic/Heat Pump System." *Applied Thermal Engineering* 31 (17–18): 4132–4138. doi: http://dx.doi.org/10.1016/j.applthermaleng.2011.08.027.

Cho, H., P.J. Mago, R. Luck, and L.M. Chamra. 2009. "Evaluation of CCHP Systems Performance based on Operational Cost, Primary Energy Consumption, and Carbon Dioxide Emission by Utilizing an Optimal Operation Scheme." *Applied Energy* 86: 2540–2549.

Chua, K. J., W. M. Yang, T. Z. Wong, and C. A. Ho. 2012. "Integrating Renewable Energy Technologies to Support Building Trigeneration—A Multi-Criteria Analysis." *Renewable Energy* 41 (Supplement C): 358–367. doi: https://doi.org/10.1016/j.renene.2011.11.017.

Cioccolanti, Luca, Mauro Villarini, Roberto Tascioni, and Enrico Bocci. 2017. "Performance Assessment of a Solar Trigeneration System for Residential Applications by Means of a Modelling Study." *Energy Procedia* 126 (Supplement C): 445–452. doi: https://doi.org/10.1016/j.egypro.2017.08.211.

Eisavi, Beneta, Shahram Khalilarya, Ata Chitsaz, and Marc A. Rosen. 2017. "Thermodynamic Analysis of a Novel Combined Cooling, Heating and Power System Driven by Solar Energy." *Applied Thermal Engineering* 129: 1219–1229. doi: https://doi.org/10.1016/j.applthermaleng.2017.10.132.

Gazda, Wiesław, and Wojciech Stanek. 2016. "Energy and Environmental Assessment of Integrated Biogas Trigeneration and Photovoltaic Plant as More Sustainable Industrial System." *Applied Energy* 169 (Supplement C): 138–149. doi: https://doi.org/10.1016/j.apenergy.2016.02.037.

Hasan, M. Arif, and K. Sumathy. 2010. "Photovoltaic Thermal Module Concepts and their Performance Analysis: A Review." *Renewable and Sustainable Energy Reviews* 14 (7): 1845–1859. doi: http://dx.doi.org/10.1016/j.rser.2010.03.011.

Intini, M., S. De Antonellis, C. M. Joppolo, and A. Casalegno. 2015. "A Trigeneration System Based on Polymer Electrolyte Fuel Cell and Desiccant Wheel—Part B: Overall System Design and Energy Performance Analysis." *Energy Conversion and Management* 106 (Supplement C): 1460–1470. doi: https://doi.org/10.1016/j.enconman.2015.10.005.

Kalogirou, S. A. 2004. "Solar Thermal Collectors and Applications." *Progress in Energy and Combustion Science* 30 (3): 231–295. doi: 10.1016/j.pecs.2004.02.001.

Kalogirou, S. A., and Y. Tripanagnostopoulos. 2006. "Hybrid PV/T Solar Systems for Domestic Hot Water and Electricity Production." *Energy Conversion and Management* 47 (18–19): 3368–3382. doi: http://dx.doi.org/10.1016/j.enconman.2006.01.012.

Karellas, S., and K. Braimakis. 2016a. "Energy–Exergy Analysis and Economic Investigation Oo a Cogeneration and Trigeneration ORC–VCC Hybrid System Utilizing Biomass Fuel and Solar Power." *Energy Conversation and Manage* 107: 103–113.

Karellas, Sotirios, and Konstantinos Braimakis. 2016b. "Energy–Exergy Analysis and Economic Investigation of a Cogeneration and Trigeneration ORC–VCC Hybrid System Utilizing Biomass Fuel and Solar Power." *Energy Conversion and Management* 107 (Supplement C): 103–113. doi: https://doi.org/10.1016/j.enconman.2015.06.080.

Kegel, Martin, Justin Tamasauskas, and Roberto Sunye. 2014a. "Solar Thermal Trigeneration System in a Canadian Climate Multi-Unit Residential Building." *Energy Procedia* 48: 876–887. doi: http://dx.doi.org/10.1016/j.egypro.2014.02.101.

Kegel, Martin, Justin Tamasauskas, and Roberto Sunye. 2014b. "Solar Thermal Trigeneration System in a Canadian Climate Multi-unit Residential Building." *Energy Procedia* 48 (Supplement C): 876–887. doi: https://doi.org/10.1016/j.egypro.2014.02.101.

Khalid, Farrukh, Ibrahim Dincer, and Marc A. Rosen. 2017. "Thermoeconomic Analysis of a Solar-Biomass Integrated Multigeneration System for a Community." *Applied Thermal Engineering* 120 (Supplement C): 645–653. doi: https://doi.org/10.1016/j.applthermaleng.2017.03.040.

Leiva-Illanes, Roberto, Rodrigo Escobar, José M. Cardemil, and Diego-César Alarcón-Padilla. 2017. "Thermoeconomic Assessment of a Solar Polygeneration Plant for Electricity, Water, Cooling and Heating in High Direct Normal Irradiation Conditions." *Energy Conversion and Management* 151 (Supplement C): 538–552. doi: https://doi.org/10.1016/j.enconman.2017.09.002.

Li, Hongqiang, Xiaofeng Zhang, Lifang Liu, Rong Zeng, and Guoqiang Zhang. 2016. "Exergy and Environmental Assessments of a Novel Trigeneration System Taking Biomass and Solar Energy as Co-Feeds." *Applied Thermal Engineering* 104 (Supplement C): 697–706. doi: https://doi.org/10.1016/j.applthermaleng.2016.05.081.

Liu, Hao, Yingjuan Shao, and Jinxing Li. 2011. "A Biomass-Fired Micro-Scale CHP System with Organic Rankine Cycle (ORC)—Thermodynamic Modelling Studies." *Biomass and Bioenergy* 35 (9): 3985–3994. doi: http://dx.doi.org/10.1016/j.biombioe.2011.06.025.

Maraver, Daniel, Ana Sin, Javier Royo, and Fernando Sebastián. 2013. "Assessment of CCHP Systems Based on Biomass Combustion for Small-Scale Applications through a Review of the Technology and Analysis of Energy Efficiency Parameters." *Applied Energy* 102: 1303–1313. doi: http://dx.doi.org/10.1016/j.apenergy.2012.07.012.

Martins, L.N., F.M. Fábrega, and J.V.H. d'Angelo. 2012. "Thermodynamic Performance Investigation of a Trigeneration Cycle Considering the Influence of Operational Variables." *Procedia Engineering* 42: 1879–1888.

Meng, Xiangyu, Fusheng Yang, Zewei Bao, Jianqiang Deng, Nyallang N. Serge, and Zaoxiao Zhang. 2010. "Theoretical Study of a Novel Solar Trigeneration System Based on Metal Hydrides." *Applied Energy* 87 (6): 2050–2061. doi: https://doi.org/10.1016/j.apenergy.2009.11.023.

Mohan, Gowtham, N. T. Uday Kumar, Manoj Kumar Pokhrel, and Andrew Martin. 2016. "Experimental Investigation of a Novel Solar Thermal Polygeneration Plant In United Arab Emirates." *Renewable Energy* 91 (Supplement C): 361–373. doi: https://doi.org/10.1016/j.renene.2016.01.072.

Najafi, Behzad, Stefano De Antonellis, Manuel Intini, Matteo Zago, Fabio Rinaldi, and Andrea Casalegno. 2015. "A Tri-Generation System Based on Polymer Electrolyte Fuel Cell and Desiccant Wheel—Part A: Fuel Cell System Modelling and Partial Load Analysis." *Energy Conversion and Management* 106 (Supplement C): 1450–1459. doi: https://doi.org/10.1016/j.enconman.2015.10.004.

Nosrat, Amir H., Lukas G. Swan, and Joshua M. Pearce. 2013. "Improved Performance of Hybrid Photovoltaic-Trigeneration Systems Over Photovoltaic-Cogen Systems Including Effects of Battery Storage." *Energy* 49 (Supplement C): 366–374. doi: https://doi.org/10.1016/j.energy.2012.11.005.

Patel, Bhavesh, Nishith B. Desai, and Surendra Singh Kachhwaha. 2017. "Thermo-Economic Analysis of Solar-Biomass Organic Rankine Cycle Powered Cascaded Vapor Compression-Absorption System." *Solar Energy* 157 (Supplement C): 920–933. doi: https://doi.org/10.1016/j.solener.2017.09.020.

Quoilin, Sylvain, Martijn Van Den Broek, Sébastien Declaye, Pierre Dewallef, and Vincent Lemort. 2013. "Techno-Economic Survey of Organic Rankine Cycle (ORC) Systems." *Renewable and Sustainable Energy Reviews* 22: 168–186.

Sarbu, Ioan, and Calin Sebarchievici. 2013. "Review of Solar Refrigeration and Cooling Systems." *Energy and Buildings* 67: 286–297. doi: http://dx.doi.org/10.1016/j.enbuild.2013.08.022.

Sonar, Deepesh, S. L. Soni, and Dilip Sharma. 2014. "Micro-Trigeneration for Energy Sustainability: Technologies, Tools and Trends." *Applied Thermal Engineering* 71 (2): 790–796. doi: http://dx.doi.org/10.1016/j.applthermaleng.2013.11.037.

Tora, Eman A., and Mahmoud M. El-Halwagi. 2011. "Integrated Conceptual Design of Solar-Assisted Trigeneration Systems." *Computers & Chemical Engineering* 35 (9):1807–1814. doi: https://doi.org/10.1016/j.compchemeng.2011.03.014.

Villarini, M., E. Bocci, M. Moneti, A. Di Carlo, and A. Micangeli. 2014. "State of Art of Small Scale Solar Powered ORC Systems: A Review of the Different Typologies and Technology Perspectives." *Energy Procedia* 45: 257–267.

Wang, Jiangjiang, and Chao Fu. 2016. "Thermodynamic Analysis of a Solar-Hybrid Trigeneration System Integrated with Methane Chemical-Looping Combustion." *Energy Conversion and Management* 117 (Supplement C): 241–250. doi: https://doi.org/10.1016/j.enconman.2016.03.039.

Wang, Jiangjiang, Jing Wu, Zilong Xu, and Meng Li. 2017. "Thermodynamic Performance Analysis of a Fuel Cell Trigeneration System Integrated with Solar-Assisted Methanol Reforming." *Energy Conversion and Management* 150 (Supplement C): 81–89. doi: https://doi.org/10.1016/j.enconman.2017.08.012.

Wu, Qiong, Hongbo Ren, Weijun Gao, Peifen Weng, and Jianxing Ren. 2018. "Design and Operation Optimization of Organic Rankine Cycle Coupled Trigeneration Systems." *Energy* 142: 666–677. doi: https://doi.org/10.1016/j.energy.2017.10.075.

Xu, Da, Qibin Liu, and Hongguang Jin. 2014. "Combined Cooling Heating and Power System with Integration of Middle-and-low Temperature Solar Thermal Energy and Methanol Decomposition." *Energy Procedia* 61 (Supplement C): 1364–1367. doi: https://doi.org/10.1016/j.egypro.2014.12.128.

Zhai, H., Y. J. Dai, J. Y. Wu, and R. Z. Wang. 2009. "Energy and Exergy Analyses on a Novel Hybrid Solar Heating, Cooling and Power Generation System for Remote Areas." *Applied Energy* 86 (9): 1395–1404. doi: https://doi.org/10.1016/j.apenergy.2008.11.020.

9

Solar Desiccant Cooling

Desiccant cooling is an already well-established technology with many commercially available systems. The first systems for this technology were developed in the 1930s by industries that wanted to maintain humidity at low levels. Desiccant cooling can be used in several industrial fields, including the brewing industry, to prevent sanitation problems in cellars, and the pharmaceutical industry, to avoid product contamination (Rafique et al. 2016). They are open cycle systems, using water as the refrigerant, which comes in direct contact with the ambient air. The basic principle of the desiccant cooling cycle consists of a sorption process that takes place to dehumidify the air combined with an evaporative cooling step. In most cases, evaporative cooling is realized by a direct evaporative stage. However, other configurations can also be applied, as will be presented in the section on evaporative cooling (Elsarrag et al. 2016). The sorption material in most cases is a solid desiccant, while several liquid desiccants have been used in experimental setups, as will be discussed further in this chapter. The main drawback of desiccant cooling systems is their low COP in comparison to conventional cooling systems. On the other hand, desiccant cooling cycles have several advantages, including the following:

- High energy savings
- Avoidance of environmentally harmful fluids by using water as a refrigerant
- Desiccants can simultaneously absorb/adsorb harmful substances and particles, thus serving as air filters for an air conditioned room
- Operation close to atmospheric pressure
- Low maintenance costs

9.1 Evaporative Cooling

Evaporative cooling is one of the oldest methods of cooling. The basic principle of the technology consists of cooling the air by taking advantage of water's large heat of vaporization and increasing the air's moisture content. Although conventional air conditioning has put aside this method for several years, recent energy consumption and environmental concerns have shifted scientific interest toward evaporative cooling again. Evaporative cooling systems are suitable for dry high-temperature climates.

The main challenges of evaporative cooling include the stream's increase in moisture content, which in some cases is undesirable, and evaporative cooling's inability to operate under very humid conditions, thus restricting its use in very humid climates and/ or during rainy seasons in drier climates. The combination of the dehumidifier with evaporative cooling, which is applied in most modern desiccant cooling systems, has been facilitated to feed the evaporative coolers with dry air and eliminate the aforementioned issues (Cuce and Riffat 2016). This combination allows the separate handling of sensible and latent loads. The desiccant wheel handles the latent loads, while the evaporative cooler controls the sensible loads.

9.1.1 Direct Evaporative Cooling

The working principle of evaporative cooling, presented in Figure 9.1, consists of converting sensible heat to latent heat. In its simplest form, air enters a pad, which is sprayed with water at a temperature equal to the wet bulb of the incoming air. The warm air transfers heat toward the water, which absorbs it as latent heat. Based on the amount of heat transfer, a part of the water is evaporated and embodied into the air stream, increasing its humidity levels. As a result, the exit temperature of the air stream is lower than the respective inlet temperature. On the other hand, the enthalpy decrease as a result of the temperature drop is counterbalanced by the latent heat added to the stream with moisture, and thus no enthalpy change will occur in the air stream (Porumb et al. 2016). The process is depicted in the corresponding psychometric chart in Figure 9.2. The effectiveness of a direct evaporative cooler is defined as the ratio of the dry bulb temperature drop to the theoretical maximum dry bulb temperature drop that could be achieved, which would occur if the dry

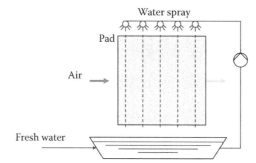

FIGURE 9.1
Working principle of direct evaporative cooling. (Adapted from Porumb, Bogdan et al., *Energy Procedia*, 85 (Supplement C): 461–471, 2016.)

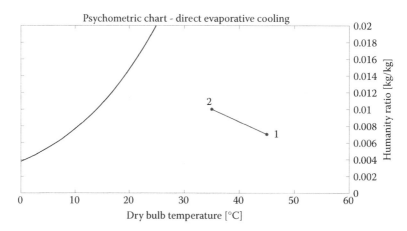

FIGURE 9.2
The evaporative cooling process in a psychometric chart.

bulb temperature could decrease down to the wet bulb temperature of the incoming air. According to Zouaoui et al. (2016), the effectiveness of a well-designed direct evaporative cooler is around 85%. Direct evaporative air conditioners are more suitable in dry climates, as they are able to produce more humid air than refrigerated air conditioners (Narayanan 2017).

9.1.2 Indirect Evaporative Cooling

In indirect evaporative cooling, the two streams do not interact directly with each other because an evaporative heat exchanger is used. The main advantage of indirect evaporative cooling is the fact that the process air is cooled without increasing its moisture content. On the other hand, the cooling process in this case is limited by the wet bulb temperature of the return air. Average efficiency of indirect evaporative cooling is around 70–80% (Zouaoui et al. 2016).

Process air flows through the dry channels (see Figure 9.3) and transfers heat via the heat exchanger walls to the return air. On the other hand, the return air flows through channels where water also flows, hence a part of the heat transferred from the process air is used to evaporate part of the water, which as a moisture is embodied in the return air. Hence,

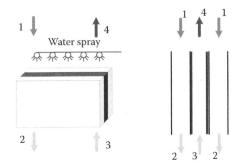

FIGURE 9.3
Working principle of indirect evaporative cooling.

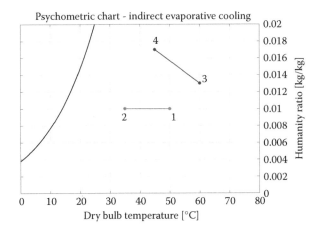

FIGURE 9.4
The indirect evaporative cooling process in a psychometric chart.

the process air is cooled without any change in its moisture content, while the return air is humidified at constant enthalpy. The process is depicted in a psychometric chart in Figure 9.4. Because the process air is not mixed with water, indirect evaporative air conditioners are more suitable in regions with moderate/humid climates (Narayanan 2017).

9.2 Dehumidifiers/Regenerators

The dehumidifier and the regenerator are among the key components of a desiccant cooling system. Solid desiccants are either fabricated in a slowly rotating wheel or packed in the form of a packed bed.

Packed bed technology is also implemented in liquid desiccant applications. Other common types of dehumidifiers and the regenerators used in liquid desiccant cooling are spray towers, falling films, and indirect contact dehumidifiers/regenerators, which mainly include the liquid-to-air membrane energy exchanger (LAMEE), the reverse-osmosis regenerator, and the electrodialysis regenerator (Abdel-Salam and Simonson 2016).

9.2.1 Desiccant Wheel

For solid desiccants, the most common dehumidifier is a slowly rotating desiccant wheel. In the desiccant wheel, mass and heat transfer take place between process and return air streams at low rotational speed. As the wheel slowly rotates, the process air is dried by the desiccant, while the hot return air drives the regeneration of the desiccant. Furthermore, due to heat transfer between the two air streams, process air exits the wheel at an increased temperature (Panaras et al. 2011). The two streams are separated by clapboard. To ensure continuous operation, sorption and regeneration take place on a periodical basis.

The desiccant wheel consists of a frame that is coated, impregnated, or fabricated with a thin layer of desiccant. The frame consists of multiple channels in the direction of the axis of the wheel's rotation (Narayanan et al. 2011). A wide variety of flow passages through the channels of the frame exist, such as honeycomb, triangular, sinusoidal (Rambhad et al. 2016). An overview of a desiccant wheel schematic is presented in Figure 9.5.

Due to the fact that the desiccant wheel is one of the key components in solid desiccant cooling systems, several studies have been conducted regarding the design, modeling, and optimization of its performance (Ahmed et al. 2005; Ali Mandegari and Pahlavanzadeh 2009; Angrisani et al. 2012; Goldsworthy and White 2012; Yamaguchi and Saito 2013; Giannetti et al. 2015; Goodarzia et al. 2017).

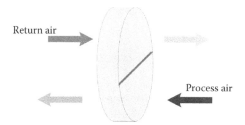

Return air

Process air

FIGURE 9.5
Schematic of a simple desiccant wheel.

One of the first studies regarding desiccant wheel modeling was presented by Zheng and Worek (1993), by combining heat and mass transfer equations into a single system. Ahmed et al. (2005) developed a mathematical model to simulate the heat and mass transfer mechanisms for the adsorption and the regeneration taking place in the desiccant wheel of a desiccant cooling system. According to a parametric analysis conducted on the results provided by the model, it was concluded that the optimum performance of the desiccant wheel was obtained at an air flow rate in the range 1–5 kg/min and a rotational speed of 0.4–1 rpm, at a regeneration temperature of 60–90°C.

9.2.2 Packed Bed

Apart from the desiccant wheel, another type of dehumidifier used for solid desiccant materials is the packed bed dehumidifier. In this case, the air flows through a bed of either granular desiccant or structured packing impregnated with desiccant. Pesaran and Mills (1987) conducted one of the first studies in the use of silica gel in packed beds for a solar-driven evaporative desiccant air conditioning system. The selection of packed beds was determined by the constraints on the pressure drop. The authors presented a transient model based on a generalized diffusion equation, and used finite difference methods for the heat and mass transfer phenomena taking place in the bed.

Hamed (2005) conducted experiments to study the performance of packed beds in solar-driven air conditioning systems. The experimental data indicated that the moisture sorption rate varies with axial distance. The desiccant layers at the packed bed's inlet adsorb moisture faster than the subsequent layers, resulting in a deterioration of the adsorptive efficiency along the bed.

Ramzy et al. (2014) proposed a heat and mass transfer model for a silica gel packed bed. The model was based on solid side resistance (SSR) and was validated against experimental data. After the validation process, the model was used to study the influence of several design parameters on the importance of axial heat conduction on the overall system's performance. It was concluded that decreasing either the particle diameter or the air flow rate, or increasing the bed length, results in an enhancement in the influence of axial heat conduction.

When a packed bed is implemented in a desiccant cooling cycle, a certain strategy has to be followed so that the bed can be directed periodically toward the regeneration stream for the regeneration of the desiccant before it is returned to the process air stream (Daou et al. 2006).

Packed beds are the most commonly applied technology in liquid desiccant cooling. In the case of liquid desiccant, the liquid solution normally enters from the top of the bed and flows over the packing, coming into direct contact with the process air. The packed bed can be either adiabatic, in which case the heat is transferred to/from the desiccant solution by the air stream, or internally heated/cooled, in which case the bed is internally cooled/heated so that the desiccant solution's temperature remains stable, enhancing its mass transfer properties (Abdel-Salam and Simonson 2016). One method to control the temperature in adiabatic packed bed dehumidifiers is to increase the solution flow rate. However, high flow rates increase the risk of desiccant carryover (Mohammad et al. 2013; Sahlot and Riffat 2016). In terms of the packing used in the bed, two types can be found: random packing and structured packing.

Bansal et al. (1997) conducted experiments to evaluate the performance of an adiabatic and an internally cooled structured packed bed dehumidifier. Results of the effectiveness of packed beds show a range of 0.38–0.55 for the adiabatic bed, while the respective range for the internally cooled packed bed was 0.55–0.706. The enhancement on the maximum moisture removal was reported to be equal to 47%, increasing from 3.4 g of moisture per kg of dry air to 5 g of moisture per kg of dry air.

9.2.3 Spray Towers

The spray tower dehumidifier/regenerator is a technology used solely for liquid desiccants. In a spray tower, the desiccant solution is sprayed in droplets at the top of the tower and comes into direct contact with the air. The already large air-to-solution contact areas achieved in spray towers allow for lower solution flow rates and air pressure drops in comparison to packed beds (Abdel-Salam and Simonson 2016).

The main drawback of spray towers is the risk for desiccant carryover, which is higher with decreasing droplet size. On the other hand, for smaller desiccant droplets, heat and mass transfer is enhanced. Hence, a compromise has to be made on each specific application between the risk for carryover and the mass and heat transfer properties of the spray tower.

9.2.4 Falling Films

In a falling film system, the desiccant solution film is fed over plates or tubes and flows by means of gravity while the process air flows over the solution film. The main advantages of falling film dehumidifiers include a lower pressure drop on the air side, low initial cost, high contact area per unit volume, and a lower risk of droplet carryover (Abdel-Salam and Simonson 2016; Jain and Bansal 2007). The main challenge of the technology involves the creation of a thin film over the entire surface, a problem which is more severe for large towers (Shah et al. 2016).

Jain et al. (2000) conducted experiments and compared the collected data against the results of a simulation model of a falling film dehumidifier and a regenerator with a liquid desiccant cooling system using a LiBr solution as desiccant. It was realized that only a fraction of the total area was involved in the mass transfer, and thus a wetness factor was introduced. The experimental data indicated that, between the top and the bottom of the dehumidifier and the regenerator, a temperature difference between 2 and 11 K was realized, while the corresponding humidity ratio difference ranged between 2–8 g/kg.

9.2.5 Indirect Contact Dehumidifiers/Regenerators

Recently, research has been conducted regarding the replacement of direct contact dehumidifiers/regenerators in liquid desiccant cooling systems with indirect ones. The main advantage of the proposed concept is the elimination of droplet carryover, a key challenge for many of the aforementioned technologies. However, due to the fact that the aim of this chapter is to present an overview of the main technologies relevant to solar cooling, this new research will not be discussed in depth because its discussion would exceed the goals of the book.

Kessling et al. (1998) experimentally investigated the performance of a liquid desiccant system coupled with an indirect evaporative cooler. The system used a LiCl solution as desiccant due to the fact that LiCl could be regenerated at temperatures lower than 80°C.

9.3 Solid Desiccant Cooling

The solid desiccant cooling cycle was discussed in detail in Chapter 2. The key component in the operation of one such system is the rotary dehumidifier, where the dehumidification of air takes place. Consecutive heat exchangers then cool the process dry air before it enters the air conditioned room. Regeneration of the desiccant is realized by the means of a low-grade heat source. For reasons of complexity, the basic process of a desiccant wheel system, as discussed in Chapter 2, is again outlined below:

- Air enters the desiccant wheel (point 1), where dehumidification and heating of the stream takes place by a return stream (point 8).
- The dehumidified air is then led through a rotary regenerator where it is cooled down.
- After the regenerator, the dry air (point 3) is further cooled down in an evaporative cooler before (point 4) it enters the room.
- At the same time, air is removed by the room to be regenerated (point 5).
- The return air is initially cooled down by an evaporative cooler before it is led to the regenerator to heat up from the hot dry air of point 2.
- Downdraft the regenerator, the return air is further heated by an external source before entering the desiccant wheel (point 8).
- In the desiccant wheel, the moisture adsorbed by the desiccant to dry the entering air is then desorbed to the return air, which eventually is exhausted to the environment.

The aforementioned process is depicted in Figure 9.6.

Solid desiccants have the advantages of a high capacity of air dehumidification, high porosity, and a high affinity for water. They are also less subject to corrosion than liquid desiccants (Zheng et al. 2014). The most commonly used solid desiccants are silica gel and zeolite. Other solid desiccant materials used in some applications are activated carbon, activated alumina, and activated clay (Nóbrega and Brum 2014).

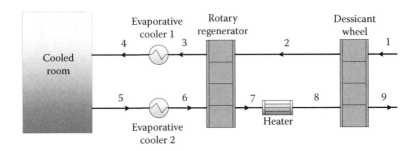

FIGURE 9.6
The basic solid desiccant cooling cycle (in ventilation mode).

9.3.1 Silica Gel

Silica gel is a non-toxic and non-corrosive desiccant derived from SiO_2. Due to its favorable properties, it is currently the most widely used desiccant material.

The regular density of silica gel has a porous size of 2 nm (Nóbrega and Brum 2014). A typical uptake of silica gel is around 0.15 g of water per g of dry silica gel. Regarding the regeneration temperature of silica gel, Zhang et al. (2003) stated that in order for the desorption phase to be maintained the nominal regeneration temperature for silica gel can be as high as 120°C. On the other hand, several studies have shown that the desorption phase can be maintained at significantly lower temperatures. Angrisani et al. (2012) experimentally estimated that the optimal regeneration temperature is approximately 65°C, while O'Connor et al. (2016) showed that under a passive ventilation mode, a silica gel desiccant wheel with regeneration temperature as low as 48.5°C could achieve a dehumidification ratio of 55%.

9.3.2 Zeolite

Zeolites are desiccant materials synthesized by sodium or calcium aluminosilicates, and they can be obtained in different pore sizes for adsorption-oriented applications (Nóbrega and Brum 2014). In comparison to silica gel, they have a smaller adsorptive uptake. However, their low cost and lower regeneration temperatures in comparison to silica gel offer a potential for the expansion of their use as adsorbents.

9.3.3 Activated Clay

Activated clay is a natural desiccant that is extensively used because of its low cost. For a relative humidity in the range of 25–30%, it has comparable performance to silica gel (Nóbrega and Brum 2014).

9.3.4 Investigations on Solid Desiccant Cooling

In 1966, Glav patented an air conditioning apparatus implementing staged regeneration in a solid desiccant dehumidifier to enhance its performance. Using simulations, Jain et al. (1995) studied the performance of solid desiccant cooling based on direct evaporative cooling in sixteen different locations in India. Different system configurations were evaluated by the model. The best performing configuration was the Dunkle cycle, presented in Figure 9.7, with a maximum obtained COP of 0.43 for the city of Trivandrum, at a dry bulb temperature of 32.9°C and a corresponding wet bulb temperature of 27.2°C.

Mazzei et al. (2002) compared the summer operating costs of a solid desiccant cooling system with the summer operating costs of a conventional cooling system for summer conditions in Italy. According to the results of the simulations for a retail store application, an interest savings of up to 35% was achieved with the desiccant cooling system, while the corresponding thermal cooling power was reduced up to 52%. Moreover, in a case in which the desiccant system is driven by waste heat, the interest savings could increase further, up to 87%. Finally, the simple payback period for a solid desiccant cooling system, based on costs in the Italian market of the period, was estimated to be 5–7 years.

Kanoğlu et al. (2004) conducted a first and second law analysis of a solid desiccant cooling system. The analysis was carried out on an experimental setup in ventilation mode using natural zeolite as the desiccant material. According to the results of the

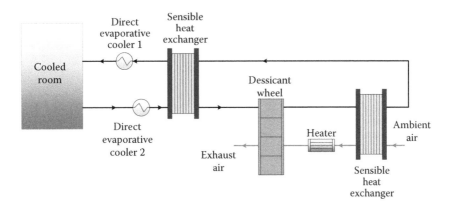

FIGURE 9.7
Schematic of the Dunkle cycle.

analysis, the obtained COP was around 0.35 and the corresponding exergetic efficiency was around 11%.

Hirunlabh et al. (2007) experimentally investigated the performance and the energy savings of a solid desiccant air conditioning system installed in Thailand. The system used silica gel as the desiccant material and was used to air condition a 76.8 m³ room. Based on experiments conducted over several days with similar ambient conditions, the system was able to record energy savings of up to 24%. A simple economic analysis showed that the system is competitive for large air conditioning applications in the humid climate of southern Thailand, with an estimated payback period of around four years.

A solid desiccant evaporative cooling system using an indirect evaporative cooler was designed by Goldsworthy and White (2011). Two separate analytical models were developed for the desiccant wheel and the indirect evaporative cooler, and subsequently they were coupled. According to the results for an ambient reference condition (a process air inlet temperature of 35°C and relative humidity of 14.3 g/kg), a regeneration temperature of 70°C and a process/regeneration air flow ratio of 0.67 maximized the system's performance, with an electrical COP exceeding 20 and a maximum cooling COP around 0.4.

Kim and Jeong (2013) developed and investigated a solid desiccant and evaporative cooling system using outdoor air exclusively. The system's thermal and energetic performance was observed. Results indicated that the investigated system could lead to potential energy savings of up to 77% compared to a conventional system.

Parmar and Hindoliya (2013) conducted a comparative study of the performance of a solid desiccant cooling system for four different climatic regions in India—Jodhpur, Mumbai, Bangalore, and New Delhi. The results of the simulations indicated that the system performed better in a warm, humid climate. The maximum reported COP was equal to 4.98 for Mumbai with an R/P ratio (regeneration air flow/process air flow) of 0.55 and a regeneration heat of 0.4 kW. The ambient conditions for the city of Mumbai were considered as follows:

- Wet bulb temperature (θ_{wb}): 23.3°C
- Dry bulb temperature (θ_{db}): 34.3°C
- Specific humidity: 13.51 g/kg of dry air

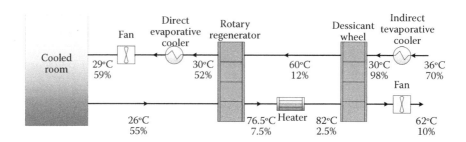

FIGURE 9.8
Schematic of the solid desiccant cooling system using direct and indirect evaporative coolers. (Adapted from Khoukhi, Maatouk, *International Journal of Energy Engineering*, 3 (4):107–111, 2013.)

Guidara et al. (2013) developed a simulation model to design and study the performance of a solid desiccant cooling system for air conditioning applications for offices in Tunisia. Three climatic conditions were considered in the model: relatively cold and humid (city of Bizerte), hot and dry (city of Remada), and moderate (city of Djerba). Three modes of operation were evaluated:

- No pre-cooling of treated air, but a final cooling by humidification
- Both pre-cooling of treated air and a final cooling by humidification
- Pre-cooling of treated air and a final cooling by means of heat exchange with a colder air stream without flowing through the humidifier

Based on the results of the simulations, each of the three aforementioned modes of operation could provide comfortable conditions for the offices of all the investigated locations. Khoukhi (2013) developed a theoretical model to simulate the performance of a solid desiccant cooling system using direct and indirect evaporative coolers under hot and humid climatic conditions. An overview of the investigated setup is presented in Figure 9.8. The results of the simulations indicated that the system could perform in a satisfactory way, delivering the air within the human comfort zone—at a temperature of 29°C and a relative humidity of 59%—for an ambient dry bulb temperature of 36°C and a relative humidity of 70%.

El Hourani et al. (2014) proposed a solid desiccant dehumidification system with an implemented two-stage evaporative cooling system. The system was designed for an air conditioning application in an office space in Beirut, Lebanon. In comparison to a single-stage evaporative cooler at the same thermal comfort condition, the proposed system allowed for water savings of around 27% and energy savings around 16%.

9.4 Liquid Desiccant Cooling

In liquid desiccant cooling, the desiccant wheel is replaced by a dehumidifier and a regenerator. More details regarding the potential options have already been presented above. In a liquid desiccant system, as with the one presented in Figure 9.9, the desiccant solution circulates between an absorber and a regenerator. Process air enters from point 1 (see Figure 9.9). Liquid desiccant solution is sprayed at point 2 above the cooling coil, absorbing moisture

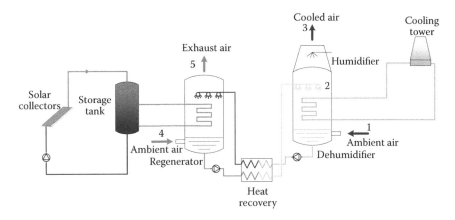

FIGURE 9.9
Schematic of a liquid desiccant system. (Adapted from Allouhi, A. et al., *Renewable and Sustainable Energy Reviews*, 44 (Supplement C): 159–181, 2015.)

from the incoming air and rejecting heat to the cooling coil. As a result, cooled, dry air exits from point 3. Simultaneously, the diluted solution coming from the dehumidifier is sprayed over a heating coil in the regenerator, which in cases of a solar-driven system as with the one presented in Figure 9.9 is connected to the solar collectors circuit. Ambient air is blown in the regenerator at point 4 and absorbs moisture from the diluted solution. At the same time, the solution is heated by the heating coils. As a result, the regenerated (concentrated) solution is ready to be sprayed back to the dehumidifier (point 2), while the hot humid air exits the regenerator at point 5. A solution heat exchanger, acting as a recuperator, is also installed in the system to preheat the cold, diluted solution exiting the dehumidifier by extracting heat from the hot concentrated solution, enhancing the system's COP.

A key advantage of liquid desiccants is their low regeneration temperature, which is approximately 50–80°C (Mohammad et al. 2013).

The earliest used liquid desiccant was triethylene glycol. However, the high viscosity of the fluid caused liquid residence issues, making the operation of the system unstable. This issue along with its low surface vapor pressure, which results in partial evaporation of triethylene glycol into the air flowing into the conditioned area, made the use of such desiccant unfavorable (Mei and Dai 2008).

The main liquid desiccant used in desiccant cooling applications is a lithium chloride (LiCl) solution due to its favorable properties. Other commonly used desiccant solutions are LiBr, $MgCl_2$, and $CaCl_2$. The main advantages of using a LiCl solution are listed below (Daou et al. 2006):

- Higher levels of air dehumidification than solid desiccants for the same driving temperature
- Lower pressure drop on the air side
- High energy storage potential by storing concentrated solutions

$MgCl_2$ has a lower price than the other common salts used in liquid desiccant solutions. However, it is less preferred because of crystallization issues. A key challenge of LiCl is the formation of crystalline hydrate, which deteriorates the adsorptive capacity of the desiccant solution—an issue that does not occurs in solid desiccants. To avoid this issue, composites made of silica gel and inorganic salts, such as halides and sulfates, have been proposed and investigated in literature (Gordeeva et al. 2008; Ge et al. 2017). The incorporation of inorganic salts in the adsorbent's structures was proven to result in composite materials with an enhanced affinity to water, ammonia, and methanol vapor, as stated by Gordeeva et al. (2008). Zhang et al. (2006) reported that fabrication of silica gel-$CaCl_2$ enhanced the moisture removal of the desiccant wheel by 10%. Jia et al. (2007) developed a composite desiccant wheel based on a composite silica gel-LiCl, and reported that the moisture removal enhancement was between 20–40% compared to the conventional silica-gel desiccant wheel.

In attempts to improve the properties of liquid desiccants, many researchers have investigated the potential of using mixtures from known liquid desiccants. Hassan and Hassan (2008) investigated a mixture with 50% w/v $H_2O/CaCl_2$ and 20% $Ca(NO_3)_2$ and concluded that the proposed mixture resulted in a significant increase in vapor pressure in comparison to conventional solutions. Xiu-Wei et al. (2008) experimentally investigated a novel mixture of $CaCl_2$ and LiCl solution and compared its performance to a conventional LiCl solution. They reported an enhancement in the dehumidification effect by the use of the proposed mixed solution equal to 20% compared to the conventional LiCl solution.

9.4.1 Investigations on Liquid Desiccant Cooling

Saman and Alizadeh (2001) proposed a coupled liquid desiccant system with an indirect evaporative cooler. A theoretical model was developed for the investigation of the cooler's performance, and was validated against experimental data under the climatic conditions of Brisbane, Australia. A parametric study was conducted using a $CaCl_2$ solution as the liquid desiccant. According to the results of the study, the optimal operation of the cooler was achieved when the solution concentration was 0.4, the solution mass flow rate was approximately 0.05 kg/s, and the process and return air velocities are around 0.7 m/s.

Al-Sulaiman et al. (2007) investigated a multi-stage evaporative cooling system using a liquid desiccant and a reverse osmosis (RO) process for the regeneration of the desiccant. The results indicated that in order to increase the desiccant solution's temperature by 22 K, the energy consumption when the regeneration was driven by RO compared to a conventional heater.

Xiong et al. (2010) developed a novel two-stage liquid desiccant dehumidification system with a $CaCl_2$ solution for the first stage and a LiCl solution for the second stage, as shown in Figure 9.10. The system was designed based on a second law analysis. Process air enters the first dehumidifier, which operates with a $CaCl_2$ solution, and then is further dehumidified in the second dehumidifier, which uses a LiCl solution. The regeneration process is reversed, with the return air regenerating the LiCl solution first and then the $CaCl_2$ solution. The staged setup of the system allows for lower regeneration temperatures, and thus reduces corrosion risks. Furthermore, to reduce further potential corrosion issues, most components were made of PVC. The results of the system's performance were compared to a conventional liquid desiccant system, showing a significant improvement due to the implementation of the novel system. More specifically, the exergetic efficiency of the two-stage system was 23.0%, whereas the respective value for the conventional liquid desiccant system was only 6.8%. Regarding the cooling COP, the two-stage system reported a COP of 0.73 in comparison to a value of 0.24 for the conventional liquid desiccant system.

She et al. (2014) proposed a novel liquid desiccant dehumidification and evaporation system, and compared its performance with a conventional VCC system under the same operating conditions. According to the results of the simulations, the proposed system could

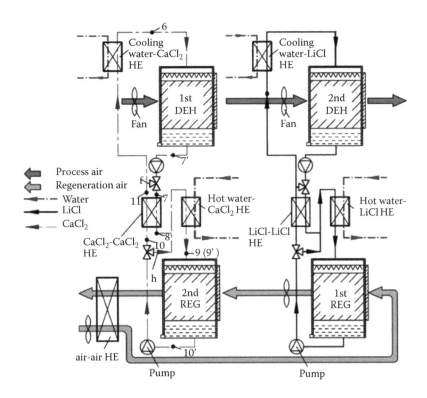

FIGURE 9.10
Schematic of the two-stage liquid desiccant dehumidification system. (Reproduced from Xiong, Z. Q. et al., *Applied Energy*, 87 (5): 1495–1504, 2010.)

achieve higher COP values than the conventional VCC. The COP of the suggested system was 16.3% higher than the conventional VCC using ambient air, while, when the air was pre-heated, the gain by the implementation of the novel cycle in the COP was as high as 18.8%.

Cihan et al. (2017) investigated a potential improvement on the liquid desiccant cycle by adding a surface additive (polycarbonate boards). The investigated system operates with a LiCl solution as the desiccant. Results indicated a significant improvement in the system's performance, with a maximum obtained thermal COP of 0.40 for the modified system, while the corresponding absorber dehumidification efficiency was around 85%.

9.5 Coupling with Solar Setups

9.5.1 Market Status

Based on data collected by Henning (2007), at the date of the survey, over 70 solar thermal cooling systems had been installed in Europe, with a total cooling capacity of 6.3 MW and a total collector area of approximately 17.5 ha. Based on the distribution presented in Figure 9.11, the most mature solar-powered thermally driven cooling technology was absorption with a 59% installed cooling capacity. By means of cooling capacity, adsorption chillers were the second most used sorption technology. Liquid desiccant cooling was

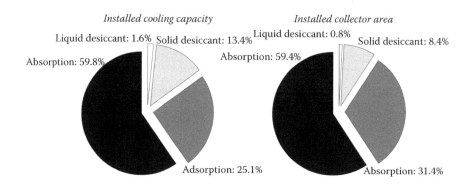

FIGURE 9.11
Distribution of solar thermal systems based on their cooling capacity and the installed collector area. (Adapted from Henning, Hans-Martin, *Applied Thermal Engineering*, 27 (10): 1734–1749, 2007.)

the least mature technology for reasons already discussed. As also shown in Figure 9.11, for the case of collector's area, the sequence of the technologies is not altered, although a significant difference in the percentages is observed mainly due to the different specific cooling capacity of each technology per unit of collector's area.

According to Mugnier and Jakob (2015), solar thermal cooling systems based on the desiccant cooling cycle have a small share in the global market of solar cooling in comparison to adsorption and absorption. More specifically, based on data reported by the authors, out of a total of 113 solar air conditioning installations, only 18 were based on desiccant cooling in 2009. The number of solar-powered desiccant cooling installations in 2012 increased to 28, with two of them using liquid desiccant.

9.5.2 Theoretical and Experimental Investigations on Solar Desiccant Cooling

Solar-assisted desiccant cooling was reported to have a potential for cooling applications as early as the 1970s. In 1973, for example, Löf pointed out the fact that the biggest advantage of solar cooling is that the peak demand for cooling matches the periods of higher solar irradiation.

Nelson et al. (1978) proposed and investigated the efficiency of two simulation models for a desiccant evaporative cooling system working on ventilation and recirculation mode, respectively, using meteorological data for Miami, Florida, in the United States. According to the simulation results, when a collector area of 45 m² is considered, the solar fraction can reach up to 92.5%, with an average overall COP of 0.67, on ventilation mode. On the other hand, the maximum obtained average overall COP was 0.79, with an aperture area of 7.5 m² and a solar fraction of 27%.

Smith et al. (1994) developed a mathematical model and simulated the performance of a solar-assisted desiccant cooling system used for an air conditioning application in residential buildings in different locations in the United States. The system was powered by 50 m² of flat-plate collectors, and a 3.75 m³ heat storage tank was also considered. The cooling capacity of the desiccant system was designed at 9.5 kW. Three locations were considered: Pittsburgh (Pennsylvania), Macon (Georgia), and Albuquerque (New Mexico). The simulations for a typical summer day showed that the proposed size of the system was able to meet the cooling loads of the house in all locations. Furthermore, it was shown that the system did not perform in a satisfying way in the Southeast of the United States, namely in Macon and Albuquerque, reporting a solar fraction of only 18%. On the other hand, the solar fraction achieved in Pittsburgh was around 73%, which was considered a competitive value by the authors.

Radhwan et al. (1999) developed a mathematical model to simulate the performance of a solar-driven liquid desiccant evaporative cooling system using a LiCl solution as desiccant. A long-term operation simulation was conducted for the meteorological conditions of Jeddah, Saudi Arabia. An auxiliary VCC chiller was considered to cover the part of the cooling load that could not be covered by the desiccant cycle. The effect of the solar collector area and the capacity of the hot storage tank, by varying the storage tank's height, were investigated based on three parameters: the desiccant replacement factor (DRF), the system thermal ratio (STR), and the solar utilization factor (SUF), defined as follows:

$$DRF = \frac{Part\ of\ cooling\ load\ provided\ by\ the\ desiccant\ system}{Cooling\ load\ using\ only\ a\ VCC} \tag{9.1}$$

$$STR = \frac{Daily\ cooling\ load\ provided\ by\ the\ desiccant\ system}{Daily\ heat\ input\ to\ the\ system} \tag{9.2}$$

$$SUF = \frac{Daily\ solar\ radiation\ utilized\ by\ the\ solar\ collectors}{Total\ heat\ input\ in\ solar\ collectors\ and\ system's\ heaters} \tag{9.3}$$

Based on the above defined parameters, on the design conditions of 50 m² of solar air heaters and a 2 m tall heat storage tank, a DRF value of around 0.85 was obtained, while the corresponding value for the STR was 29.5% and the SUF was 48.5%. Furthermore, it was concluded that increasing the solar collector's area would have no effect on the DRF. However, increasing the collector's area would result in a significant increase in the SUF and a slight decrease in the STR. Increasing the height of the heat storage tank had, as well, no influence on the DRF, while there was an optimum value above which no significant increase in either STR or SUF was observed.

Henning et al. (2001) developed a solid desiccant cooling system, powered by 20 m² flat-plate collectors. The system was installed in 1996 in a technology center in Riesa, Germany, and provided space cooling for a 330 m² seminar room. The nominal air flow rate of the desiccant cooling system was 2,700 m³/h. The system used silica gel as the desiccant material. A 2 m³ tank was implemented in the system to store hot water, as seen in Figure 9.12. According to the experimental data, a solar fraction of 7% was achieved by the system, under a collector efficiency of 0.54 and a cooling COP of 0.6.

Joudi and Dhaidan (2001) evaluated the performance of a solar-assisted heating and desiccant cooling system located in Baghdad, Iraq. During winter, space heating to the room was supplied by the solar collectors, while an auxiliary heater and a rock bed storage tank were also implemented. During summer, a solid desiccant cooling system on ventilation mode is implemented to provide the room with air conditioning, as shown in Figure 9.13. A parametric analysis was conducted using a simulation model for the aforementioned system. It was concluded that when the system was solely driven by the solar collectors, the regeneration temperature could reach up to 62°C. Furthermore, it was identified that the heat exchanger and the evaporative cooler effectiveness had a significant impact on the system's COP, while the dehumidifier did not have a big impact on the system's overall performance.

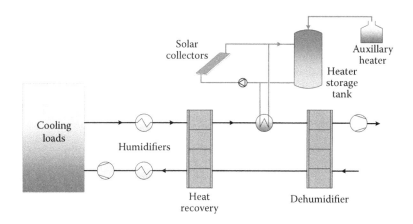

FIGURE 9.12
The solar-powered solid desiccant cooling system in Riesa, Germany. (Adapted from Henning, H. M. et al., *International Journal of Refrigeration*, 24 (3): 220–229, 2001.)

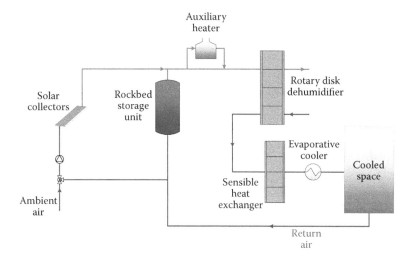

FIGURE 9.13
The solar-assisted heating and desiccant cooling system. (Adapted from Joudi, Khalid A., and Nabeel S. Dhaidan, *Energy Conversion and Management*, 42 (8): 995–1022, 2001.)

Wurtz et al. (2005) conducted a parametric analysis to maximize the performance of a liquid desiccant cooling system. The desiccant cooling that was modeled used a LiCl solution and was based on an experimental setup installed in a training room in Chambery, France. The cooling system was powered by 14.8 m² of solar liquid collectors. Increasing the air flow rate was found to enhance the system's performance, resulting in a maximum obtained system's COP of 1.862 at an air flow rate of 0.6 kg/s.

Abdalla and Abdalla (2006) presented a feasibility study of a solar-powered liquid desiccant air conditioning system using a direct evaporative cooler, located in Khartoum, Sudan. In the design case, the ambient conditions were the following:

- Wet bulb temperature: 23.4°C
- Dry bulb temperature: 43°C

The nominal return air humidity was set at 9.4 g/kg of dry air, at a dry bulb temperature of 24°C. The system was found to operate in a satisfactory way under the examined climatic conditions, supplying about 1.6 kg/s of process air.

Gommed and Grossman (2007) developed a prototype liquid desiccant cooling system, and installed it at Energy Engineering Center at the Technion, in Haifa, Israel. The system used a LiCl solution as the desiccant and was powered by 20 m² of flat-plate collectors. The system was used to air condition three offices with a total area of 35 m². The dehumidifier and the regenerator of the liquid desiccant system were both adiabatic packed towers. According to the performance results of the prototype, which has been operational since April 2003, an average dehumidification capacity of 16 kW was obtained, with an average system COP of 0.8.

Enteria et al. (2009) developed a combined solar thermal and electric desiccant cooling system, as shown in Figure 9.14. The solar circuit consisted of 10 m² of flat-plate collectors in parallel with an auxiliary 3 kW heat source for nighttime operation. A 0.322 m³ storage tank was also implemented and connected with an air compressor to remove water from the tank in cases of freezing. Experimental data over a whole day was collected, measuring a total solar collector efficiency of 53.4%, a total cooling load of 16.04 MJ, and a resulting COP of 0.25.

Hürdoğan et al. (2012) investigated the potential implementation of solar collectors in a hybrid solid desiccant-VCC cooling system. Three separate air streams were used in the system: process air, regeneration air, and return air, as shown in Figure 9.15. Based on both simulations results and experimental data from a test rig installed at Cukurova University, in Adana, Turkey, it was concluded that the implementation of solar heating to drive the system increased the regeneration temperature. Furthermore, it was shown that the exploitation of solar energy enhanced the system's COP between 50% and 120%.

Li et al. (2012) conducted a case study on a two-stage desiccant cooling/heating system installed at Himin Solar Company in China. The system was driven by 120 m² of evacuated

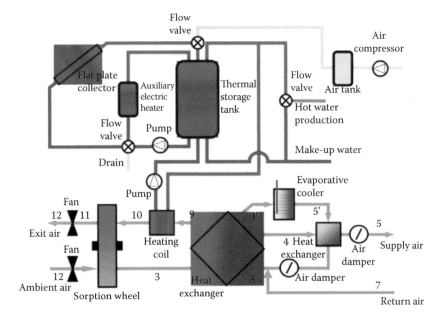

FIGURE 9.14
The combined solar thermal and electric desiccant cooling system. (Reproduced from Enteria, Napoleon et al., *Solar Energy*, 83 (8): 1300–1311, 2009.)

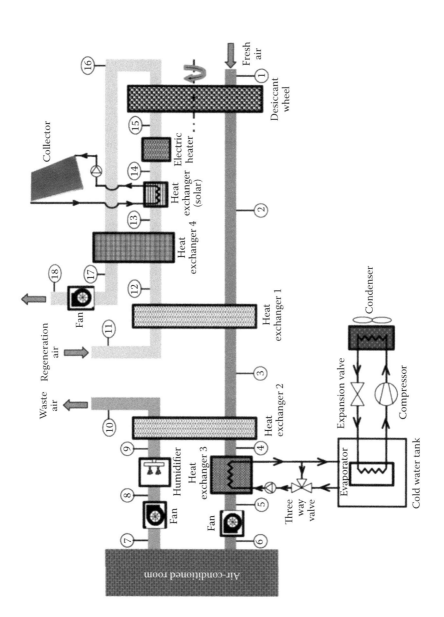

FIGURE 9.15
The hybrid solid desiccant-VCC cooling system. (Reproduced from Hürdoğan, Ertaç et al., *Energy and Buildings*, 55 (Supplement C): 757–764, 2012.)

tube collectors. The nominal cooling capacity of the desiccant system was 20 kW and it was used to air condition a space of 169 m². According to the experimental results, the system was able to perform efficiently under the investigated hot and humid climate, reaching a moisture removal level of around 14 g/kg of dry air. The average reported cooling COP was around 0.97, with a cooling capacity in the range of 16.3–25.6 kW. The solar COP in heating mode was reported to be around 0.45, with a collector efficiency of 0.50.

Infante Ferreira and Kim (2014) reported that, for direct-type evaporative coolers, the second law efficiency of a corresponding solar-driven desiccant cooling system was approximately 50%. Furthermore, based on the economic analysis conducted by the authors regarding the solar collector technology, it was concluded that flat-plate collectors were the more competitive choice when compared to concentrating parabolic trough collectors.

Finocchiaro et al. (2015) introduced the second generation of Freescoo, a novel compact solar desiccant evaporative cooling air conditioner designed at a spin-off of the University of Palermo, in Italy. The desiccant cooling cycle is based on evaporative cooling and fixed-bed adsorption, a concept that allows for the simultaneous dehumidifying and cooling of process air. The system allows for standalone operation by implementing a battery accumulator and PV cells. Two main configurations have been developed: one with an air flow rate of 500 m³/h powered by 2.4 m² of solar collectors, and the second with an air flow rate of 1000 m³/h powered by 4.8 m² of solar collectors. The corresponding cooling power of the two configurations is 2.7 and 5.5 kW, respectively. In the paper, the performance of the system with the 5.5 kW cooling output was presented. The system was used for the air conditioning of a 46 m² room. The average reported COP was around 0.9 (with a reported maximum daily average value of 1.4), while the respective average energy efficiency ratio (EER) was around 10. If real electricity derived by the grid was considered in the calculations of EER, the estimated average EER for the prototype would be as high as 30.7, according to the authors.

Elhelw (2016) conducted a performance study of a liquid desiccant cooling system powered by ETCs in the climatic conditions of Borg Al-Arab, Egypt. The influence of the ETCs' area was investigated. It was observed that varying the collectors' area does not affect the amount of water absorbed by the system. On the other hand, increasing the collectors' area enhances the desorption rate in the regenerator. Furthermore, increases in the collectors' area results in an increase in the system's COP, with a maximum value reported for July of around 0.83. The energy savings of the proposed system were quantified for an ETC area of 220 m². The introduction of solar energy as an auxiliary heat source to the thermal energy produced by a boiler results in energy savings of around 30%.

Angrisani et al. (2016) evaluated three different desiccant cooling configurations in comparison to a reference conventional HVAC system for the climatic conditions of southern Italy. The proposed systems were able to obtain primary energy savings in the range of 20–25%, while the corresponding reduction in equivalent CO_2 emissions reached values up to 40–50%.

Merabti et al. (2017) studied the performance of a solar-driven solid desiccant system, shown in Figure 9.16, used for air conditioning a building based in the climatic conditions of the Algerian coastal region. The desiccant cycle used silica gel as the desiccant material. Flat-plate collectors were considered for the solar circuit. The performance investigations were conducted for the month of July, during which the average temperature in the investigated region is 24.2°C (min/max: 18.4/30.6°C) and the average relative humidity is 68.7% (min/max: 40.8/91.2%). According to the results of the simulations, the optimal collectors' efficiency was around 0.68. Furthermore, it was concluded that correct sizing of the solar collectors' area would result in an increase in the energy savings.

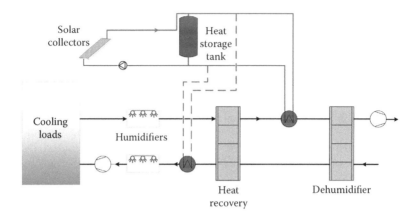

FIGURE 9.16
Schematic of the solar-driven solid desiccant system. (Adapted from Merabti, Leila et al., *International Journal of Hydrogen Energy,* 42 (48): 28997–29005, 2017.)

Frein et al. (2018) reported the experimental performance results over a two-year period of a solar-powered desiccant and evaporative cooling system coupled with a VCC heat pump. The hybrid system in question is mainly divided into three subsystems: a solar circuit, a water/water heat pump, and a desiccant cooling system. The solar circuit consists of 11 solar collectors with a total aperture area of 102 m². A flat-plate heat exchanger is used to transfer the heat from the solar subsystem to a secondary loop, and it is then stored in a 5 m³ storage tank. The heat pump has a nominal heating capacity of 10 kW and a cooling capacity of 20 kW. The performance of the system was evaluated based on the value of the primary energy ratio (PER). A reference system was also considered to quantify the primary energy savings. It was composed of two fans, a cross-flow heat recovery section, one cooling coil fed by an air-source heat pump, and a heating coil powered by a natural-gas-fired boiler.

According to the experimental data, the PER was around 20% lower than expected by the simulations. However, the system's efficiency was still considered satisfactory by the authors because it doubled the PER in comparison to the reference system.

9.6 Solar-Driven Desiccant Cooling Applications

9.6.1 Ökopark Hartberg Case

Another application of solar cooling with an open sorption cycle is the Ökopark in Hartberg, Austria. A solar-powered system is used to provide space cooling/heating and hot water for a 280 m², two-floor building. The desiccant cooling system, which uses silica gel as the desiccant material, has a nominal capacity of 30.4 kW and is powered by 12 m² of flat-plate vacuum collectors. An auxiliary 30 kW biomass boiler has also been installed. The annual average heating load of the building is approximately 85.4 MWh, while the peak cooling load is around 20 kW. A heat storage tank with 3 m³ capacity has also been installed in the system.

According to the performance results, the annual average COP of the system is approximately 0.6, with an average collector efficiency of 40%. The daily average solar gain is estimated at 60 kWh, while the corresponding daily average electricity demand is 65 kWh, according to SOLAIR (b).

The investment cost for the installation was as high as €105,000, with 60% funded by a subsidy from the government of Styria. Table 9.1 presents an overview of the main technical specifications of the solar cooling/heating installation at Ökopark.

9.6.2 Ineti Research Building

A liquid desiccant cycle was implemented in building G of the INETI Campus of the Renewable Energy Department in Lisbon, Portugal, to provide air conditioning for 12 office rooms with a total surface area of 117 m^2. The system is powered by a 16.4 kW heat pump and 24 CPC solar collectors with an aperture area of 46.1 m^2. The nominal air flow rate capacity of the desiccant cooling system is 5,000 m^3/h and its cooling capacity is 36 kW. The desiccant used in the cycle is LiCl. For the heating needs of the building, a 23 kW backup gas-fired boiler is also installed. A heat storage tank with a volume of 2 m^3 has also been implemented in the system. The peak cooling load of the building is estimated to be 39 kW, while the respective heating load, during the winter period, is 20 kW (SOLAIR a). Table 9.2 presents a summary of the key technical features for the desiccant cooling setup at the INETI building.

The system was built as part of a European research program, and thus several measurement devices have been implemented. The total cost for its replication is estimated to be around €75,000.

TABLE 9.1

Technical Specifications of the Desiccant Cooling Installation at Ökopark in Hartberg, Austria

General Information	
Application	Space cooling/heating and hot water
Location	Hartberg, Austria
Total cooling area	280 m^2
Start of operation	1999
Solar System	
Collector type	Flat-plate collectors
Collector area	12 m^2
Tilt angle	70°
Storage	
Hot storage	3 m^3
Cold storage	–
Air Conditioning Unit	
Desiccant type	Silica gel
Cooling capacity	30.4
Auxiliary Equipment	
Auxiliary biomass boiler 30 kW	
Performance	
Solar fraction	n/a
COP	0.6

TABLE 9.2

Technical Specifications of the Desiccant Cooling Installation at the INETI
Research Building in Lisbon, Portugal

General Information	
Application	Space cooling/heating
Location	Lisbon, Portugal
Total cooling area	117 m²
Start of operation	2000
Solar System	
Collector type	Compound parabolic collectors
Collector area	46.1 m²
Tilt angle	30°
Storage	
Hot storage	2 m³
Cold storage	–
Air Conditioning Unit	
Desiccant type	Lithium chloride
Cooling capacity	36
Auxiliary Equipment	
Auxiliary gas-fired boiler 23 kW	
Heat pump to drive the desiccant cycle (16.4 kW$_c$)	
Performance	
Solar fraction	n/a
COP	n/a

TABLE 9.3

Technical Specifications of the Desiccant Cooling Installation
at the Solar Info Center in Freiburg, Germany

General Information	
Application	Space cooling/heating
Location	Freiburg, Germany
Total cooling area	300 m²
Start of operation	2004
Solar System	
Collector type	Flat-plate collectors
Collector area	16.8 m²
Tilt angle	30°
Storage	
Hot storage	1.5 m³
Cold storage	Solution storages
Air Conditioning Unit	
Desiccant type	Lithium chloride
Cooling capacity	10 kW
Auxiliary Equipment	
Auxiliary heating from district heating network	
Performance	
Solar fraction	n/a
COP	1.0

9.6.3 Solar Info Center, Freiburg, Germany

An open cycle liquid desiccant cooling system, using a LiCl solution as desiccant, has been installed in the Solar Info Center (SIC) in Freiburg, Germany, to provide space cooling/heating for a seminar room and some offices, with a total area of 300 m². Flat-plate collectors with an aperture area of 16.8 m² are used to drive the cycle. The nominal air flow capacity of the system is 1,500 m³/h, and the nominal cooling capacity is equal to 10 kW. A heat storage tank of 1.5 m³ is also installed. Auxiliary heating during the winter period is supplied by a district heat network. The peak cooling load of the air conditioned area was measured to be 6 kW (SOLAIR c).

According to Kalkan et al. (2012), the average COP of the system is equal to 1.0, while the annual collector yield is approximately 270 kWh/m². The technical aspects of the desiccant cooling installation at the Solar Info Center are listed in Table 9.3.

Nomenclature

COP	Coefficient of performance for heat-powered system	–
DRF	Desiccant replacement factor	–
R/P	Regeneration air flow/process air flow ratio	–
STR	System thermal ratio	–
SUF	Solar utilization factor	–
T	Temperature	K

Greek Symbols

θ	Temperature	[°C]
η	Efficiency	–

Subscripts

Db	Dry Bulb
Wb	Wet Bulb

Abbreviations

CPC	Compound parabolic concentrator
EER	Energy efficiency ratio
ETC	Evacuated tube collector

HVAC	Heating ventilation and air conditioning
LAMEE	Liquid-to-air membrane energy exchanger
PER	Primary energy ratio
PTC	Parabolic trough collectors
PVC	Polyvinyl chloride
RO	Reverse osmosis
SSR	Solid side resistance
VCC	Vapor compression cycle

References

Abdalla, S.A., and Kamal Nasreldin Abdalla. 2006. "A Radiant Air-Conditioning System Using Solar-Driven Liquid Desiccant Evaporative Water Cooler." *Journal of Engineering Science and Technology* 1 (2): 139–157.

Abdel-Salam, Ahmed H., and Carey J. Simonson. 2016. "State-of-the-Art in Liquid Desiccant Air Conditioning Equipment and Systems." *Renewable and Sustainable Energy Reviews* 58: 1152–1183.

Ahmed, M. H., N. M. Kattab, and M. Fouad. 2005. "Evaluation and Optimization of Solar Desiccant Wheel Performance." *Renewable Energy* 30 (3): 305–325.

Al-Sulaiman, Faleh A., P. Gandhidasan, and Syed M. Zubair. 2007. "Liquid Desiccant Based Two-Stage Evaporative Cooling System Using Reverse Osmosis (RO) Process for Regeneration." *Applied Thermal Engineering* 27 (14): 2449–2454.

Ali Mandegari, M., and H. Pahlavanzadeh. 2009. "Introduction of a New Definition for Effectiveness of Desiccant Wheels." *Energy* 34 (6): 797–803.

Allouhi, A., T. Kousksou, A. Jamil, P. Bruel, Y. Mourad, and Y. Zeraouli. 2015. "Solar Driven Cooling Systems: An Updated Review." *Renewable and Sustainable Energy Reviews* 44 (Supplement C): 159–181.

Angrisani, G., F. Minichiello, and M. Sasso. 2016. "Improvements of an Unconventional Desiccant Air Conditioning System Based on Experimental Investigations." *Energy Conversion and Management* 112 (Supplement C): 423–434.

Angrisani, Giovanni, Francesco Minichiello, Carlo Roselli, and Maurizio Sasso. 2012. "Experimental Analysis on the Dehumidification and Thermal Performance of a Desiccant Wheel." *Applied Energy* 92 (Supplement C): 563–572.

Bansal, N. K., J. Blumenberg, H. J. Kavasch, and T. Roettinger. 1997. "Performance Testing and Evaluation of Solid Absorption Solar Cooling Unit." *Solar Energy* 61 (2): 127–140.

Cihan, Ertuğrul, Barış Kavasoğulları, and Hasan Demir. 2017. "Enhancement of Performance of Open Liquid Desiccant System with Surface Additive." *Renewable Energy* 114 (Part B): 1101–1112.

Cuce, Pinar Mert, and Saffa Riffat. 2016. "A State of the Art Review of Evaporative Cooling Systems for Building Applications." *Renewable and Sustainable Energy Reviews* 54: 1240–1249.

Daou, K., R. Z. Wang, and Z. Z. Xia. 2006. "Desiccant Cooling Air Conditioning: A Review." *Renewable and Sustainable Energy Reviews* 10 (2): 55–77.

El Hourani, Mario, Kamel Ghali, and Nesreen Ghaddar. 2014. "Effective Desiccant Dehumidification System with Two-Stage Evaporative Cooling for Hot And Humid Climates." *Energy and Buildings* 68 (Part A): 329–338.

Elhelw, Mohamed. 2016. "Performance Evaluation for Solar Liquid Desiccant Air Dehumidification System." *Alexandria Engineering Journal* 55 (2): 933–940.

Elsarrag, Esam, Opubo N. Igobo, Yousef Alhorr, and Philip A. Davies. 2016. "Solar Pond Powered Liquid Desiccant Evaporative Cooling." *Renewable and Sustainable Energy Reviews* 58: 124–140.

Enteria, Napoleon, Hiroshi Yoshino, Akashi Mochida, Rie Takaki, Akira Satake, Ryuichiro Yoshie, Teruaki Mitamura, and Seizo Baba. 2009. "Construction and Initial Operation of the Combined Solar Thermal and Electric Desiccant Cooling System." *Solar Energy* 83 (8): 1300–1311.

Finocchiaro, Pietro, Marco Beccali, Andrea Calabrese, and Edoardo Moreci. 2015. "Second Generation of Freescoo Solar DEC Prototypes for Residential Applications." *Energy Procedia* 70 (Supplement C): 427–434.

Frein, A., M. Muscherà, R. Scoccia, M. Aprile, and M. Motta. 2018. "Field Testing of a Novel Hybrid Solar Assisted Desiccant Evaporative Cooling System Coupled with a Vapour Compression Heat Pump." *Applied Thermal Engineering* 130: 830–846.

Ge, T. S., J. Y. Zhang, Y. J. Dai, and R. Z. Wang. 2017. "Experimental Study on Performance of Silica Gel and Potassium Formate Composite Desiccant Coated Heat Exchanger." *Energy* 141 (Supplement C): 149-158.

Giannetti, Niccolò, Andrea Rocchetti, Kiyoshi Saito, and Seiichi Yamaguchi. 2015. "Entropy Parameters for Desiccant Wheel Design." *Applied Thermal Engineering* 75 (Supplement C): 826–838.

Goldsworthy, M., and S. White. 2011. "Optimisation of a Desiccant Cooling System Design with Indirect Evaporative Cooler." *International Journal of Refrigeration* 34 (1): 148–158.

Goldsworthy, M., and S. D. White. 2012. "Limiting Performance Mechanisms in Desiccant Wheel Dehumidification." *Applied Thermal Engineering* 44 (Supplement C): 21–28.

Gommed, K., and G. Grossman. 2007. "Experimental Investigation of a Liquid Desiccant System for Solar Cooling and Dehumidification." *Solar Energy* 81 (1): 131–138.

Goodarzia, Gholamreza, Neelesh Thirukonda, Shahin Heidari, Aliakbar Akbarzadeh, and Abhijit Date. 2017. "Performance Evaluation of Solid Desiccant Wheel Regenerated by Waste Heat or Renewable Energy." *Energy Procedia* 110 (Supplement C): 434–439.

Gordeeva, Larisa G., Angelo Freni, Tamara A. Krieger, Giovanni Restuccia, and Yuri I. Aristov. 2008. "Composites 'Lithium Halides in Silica Gel Pores': Methanol Sorption Equilibrium." *Microporous and Mesoporous Materials* 112 (1): 254–261.

Guidara, Zied, Mounir Elleuch, and Habib Ben Bacha. 2013. "New Solid Desiccant Solar Air Conditioning Unit in Tunisia: Design and Simulation Study." *Applied Thermal Engineering* 58 (1): 656–663.

Hamed, Ahmed M. 2005. "Experimental Investigation on the Adsorption/Desorption Processes Using Solid Desiccant in an Inclined-Fluidized Bed." *Renewable Energy* 30 (12): 1913–1921.

Hassan, A. A. M., and M. Salah Hassan. 2008. "Dehumidification of Air with a Newly Suggested Liquid Desiccant." *Renewable Energy* 33 (9): 1989–1997.

Henning, H. M., T. Erpenbeck, C. Hindenburg, and I. S. Santamaria. 2001. "The Potential of Solar Energy Use in Desiccant Cooling Cycles." *International Journal of Refrigeration* 24 (3): 220–229.

Henning, Hans-Martin. 2007. "Solar Assisted Air Conditioning of Buildings—An Overview." *Applied Thermal Engineering* 27 (10): 1734–1749.

Hirunlabh, J., R. Charoenwat, J. Khedari, and Sombat Teekasap. 2007. "Feasibility Study of Desiccant Air-Conditioning System in Thailand." *Building and Environment* 42 (2): 572–577.

Hürdoğan, Ertaç, Orhan Büyükalaca, Tuncay Yılmaz, Arif Hepbasli, and İrfan Uçkan. 2012. "Investigation of Solar Energy Utilization in a Novel Desiccant Based Air Conditioning System." *Energy and Buildings* 55 (Supplement C): 757–764.

Infante Ferreira, Carlos, and Dong-Seon Kim. 2014. "Techno-Economic Review of Solar Cooling Technologies based on Location-Specific Data." *International Journal of Refrigeration* 39 (Supplement C): 23–37.

Jain, S., P. L. Dhar, and S. C. Kaushik. 1995. "Evaluation of Solid-Desiccant-Based Evaporative Cooling Cycles for Typical Hot and Humid Climates." *International Journal of Refrigeration* 18 (5): 287–296.

Jain, Sanjeev, and P. K. Bansal. 2007. "Performance Analysis of Liquid Desiccant Dehumidification Systems." *International Journal of Refrigeration* 30 (5): 861–872.

Jain, Sanjeev, P. L. Dhar, and S. C. Kaushik. 2000. "Experimental Studies on the Dehumidifier and Regenerator of a Liquid Desiccant Cooling System." *Applied Thermal Engineering* 20 (3): 253–267.

Jia, C. X., Y. J. Dai, J. Y. Wu, and R. Z. Wang. 2007. "Use of Compound Desiccant to Develop High Performance Desiccant Cooling System." *International Journal of Refrigeration* 30 (2): 34–353.

Joudi, Khalid A., and Nabeel S. Dhaidan. 2001. "Application of Solar Assisted Heating and Desiccant Cooling Systems for a Domestic Building." *Energy Conversion and Management* 42 (8): 995–1022.

Kalkan, Naci, E. A. Young, and Ahmet Celiktas. 2012. "Solar Thermal Air Conditioning Technology Reducing the Footprint of Solar Thermal Air Conditioning." *Renewable and Sustainable Energy Reviews* 16 (8): 6352–6383.

Kanoğlu, Mehmet, Melda Özdinç Çarpınlıoğlu, and Murtaza Yıldırım. 2004. "Energy and Exergy Analyses of an Experimental fDesiccant Cooling System." *Applied Thermal Engineering* 24 (5): 919–932.

Kessling, W., E. Laevemann, and C. Kapfhammer. 1998. "Energy Storage for Desiccant Cooling Systems Component Development." *Solar Energy* 64 (4): 209–221.

Khoukhi, Maatouk. 2013. "The Use of Desiccant Cooling System with IEC and DEC in Hot-Humid Climates." *International Journal of Energy Engineering* 3 (4): 107–111.

Kim, Min-Hwi, and Jae-Weon Jeong. 2013. "Development of Desiccant and Evaporative Cooling Based 100% Outdoor System." In *AEI 2013: Building Solutions for Architectural Engineering,* 506–515.

Li, H., Y. J. Dai, Y. Li, D. La, and R. Z. Wang. 2012. "Case Study of a Two-Stage Rotary Desiccant Cooling/Heating System Driven by Evacuated Glass Tube Solar Air Collectors." *Energy and Buildings* 47 (Supplement C): 107–112.

Mazzei, P., F. Minichiello, and D. Palma. 2002. "Desiccant HVAC Systems for Commercial Buildings." *Applied Thermal Engineering* 22 (5): 545–560.

Mei, L., and Y. J. Dai. 2008. "A Technical Review on Use of Liquid-Desiccant Dehumidification for Air-Conditioning Application." *Renewable and Sustainable Energy Reviews* 12 (3): 662–689.

Merabti, Leila, Mustapha Merzouk, Nachida Kasbadji Merzouk, and Walid Taane. 2017. "Performance Study of Solar Driven Solid Desiccant Cooling System under Algerian Coastal Climate." *International Journal of Hydrogen Energy* 42 (48): 28997–29005.

Mohammad, Abdulrahman Th, Sohif Bin Mat, M. Y. Sulaiman, K. Sopian, and Abduljalil A. Al-abidi. 2013. "Survey of Hybrid Liquid Desiccant Air Conditioning Systems." *Renewable and Sustainable Energy Reviews* 20 (Supplement C): 186–200.

Mugnier, Daniel, and Uli Jakob. 2015. "Status of Solar Cooling in the World: Markets and Available Products." *Wiley Interdisciplinary Reviews: Energy and Environment* 4 (3): 229–234.

Narayanan, R. 2017. "Heat-Driven Cooling Technologies." In *Clean Energy for Sustainable Development,* 191–212. Amsterdam Academic Press.

Narayanan, R., W. Y. Saman, S. D. White, and M. Goldsworthy. 2011. "Comparative Study of Different Desiccant Wheel Designs." *Applied Thermal Engineering* 31 (10): 1613–1620.

Nelson, J. S., W. A. Beckman, J. W. Mitchell, and D. J. Close. 1978. "Simulations of the Performance of Open Cycle Desiccant Systems Using Solar Energy." *Solar Energy* 21 (4): 273–278.

Nóbrega, Carlos Eduardo Leme, and Nísio Carvalho Lobo Brum. 2014. "An Introduction to Solid Desiccant Cooling Technology." In *Desiccant-Assisted Cooling: Fundamentals and Applications,* edited by Carlos Eduardo Leme Nóbrega and Nisio Carvalho Lobo Brum, 1–23. London: Springer London.

O'Connor, Dominic, John Kaiser Calautit, and Ben Richard Hughes. 2016. "A Novel Design of a Desiccant Rotary Wheel for Passive Ventilation Applications." *Applied Energy* 179 (Supplement C): 99–109.

Panaras, G., E. Mathioulakis, and V. Belessiotis. 2011. "Solid Desiccant Air-Conditioning Systems—Design Parameters." *Energy* 36 (5): 2399–2406.

Parmar, H., and D. A. Hindoliya. 2013. "Performance of Solid Desiccant-Based Evaporative Cooling System Undert The Climatic Zones of India." *International Journal of Low-Carbon Technologies* 8 (1): 52–57.

Pesaran, Ahmad A., and Anthony F. Mills. 1987. "Moisture Transport in Silica Gel Packed Beds—I. Theoretical Study." *International Journal of Heat and Mass Transfer* 30 (6): 1037–1049.

Porumb, Bogdan, Paula Ungureşan, Lucian Fechete Tutunaru, Alexandru Şerban, and Mugur Bălan. 2016. "A Review of Indirect Evaporative Cooling Technology." *Energy Procedia* 85 (Supplement C): 461–471.

Radhwan, A.M., M.M. Elsayed, and H.N. Gari. 1999. "Mathematical Modeling of Solar Operated Liquid Desiccant, Evaporative Air Conditioning System." *Engineering Sciences* 11 (1): 119–141.

Rafique, M. Mujahid, P. Gandhidasan, and Haitham M. S. Bahaidarah. 2016. "Liquid Desiccant Materials and Dehumidifiers—A Review." *Renewable and Sustainable Energy Reviews* 56: 179–195.

Rambhad, Kishor S., Pramod V. Walke, and D. J. Tidke. 2016. "Solid Desiccant Dehumidification and Regeneration Methods—A Review." *Renewable and Sustainable Energy Reviews* 59: 73–83.

Ramzy, K. A., Ravikiran Kadoli, and T. P. Ashok Babu. 2014. "Significance of Axial Heat Conduction in Non-Isothermal Adsorption Process in a Desiccant Packed Bed." *International Journal of Thermal Sciences* 76 (Supplement C): 68–81.

Sahlot, Minaal, and Saffa B. Riffat. 2016. "Desiccant Cooling Systems: A Review." *International Journal of Low-Carbon Technologies* 11 (4): 489–505.

Saman, W. Y., and S. Alizadeh. 2001. "Modelling and Performance Analysis of a Cross-Flow Type Plate Heat Exchanger for Dehumidification/Cooling." *Solar Energy* 70 (4): 361–372.

Shah, Niyati M. , Krunal N. Patel, and Jignesh R. Mehta. 2016. "A Review of Packed Bed and Falling Film Dehumidifiers and Regenerators for Liquid Desiccant Air Conditioning Systems." National Conference on Design, Analysis and Optimization in Mechanical Engineering, Vadodara, India.

She, Xiaohui, Yonggao Yin, and Xiaosong Zhang. 2014. "Thermodynamic Analysis of a Novel Energy-Efficient Refrigeration System Subcooled by Liquid Desiccant Dehumidification and Evaporation." *Energy Conversion and Management* 78 (Supplement C): 286–296.

Smith, R. R., C. C. Hwang, and R. S. Dougall. 1994. "Modeling of a Solar-Assisted Desiccant Air Conditioner for a Residential Building." *Energy* 19 (6): 679–691.

SOLAIR. a. "INETI Building in Lisbon, Portugal." Accessed November 2017. http://www.solair -project.eu/189.0.html.

SOLAIR. b. "Ökopark Hartberg, Austria." Accessed November 2017. http://www.solair-project .eu/177.0.html.

SOLAIR. c. "Solar Info Center SIC in Freiburg, Germany." Accessed November 2017. http://www .solair-project.eu/183.0.html.

Wurtz, Etienne, Chadi Maalouf, Laurent Mora, and Francis Allard. 2005. "Parametric Analysis of a Solar Desiccant Cooling System Using the Simspark Environment." Ninth International IBPSA Conference Montréal, Canada.

Xiong, Z. Q., Y. J. Dai, and R. Z. Wang. 2010. "Development of a Novel Two-Stage Liquid Desiccant Dehumidification System Assisted by Cacl2 Solution Using Exergy Analysis Method." *Applied Energy* 87 (5): 1495–1504.

Xiu-Wei, Li, Zhang Xiao-Song, Wang Geng, and Cao Rong-Quan. 2008. "Research on Ratio Selection of a Mixed Liquid Desiccant: Mixed Licl–Cacl2 Solution." *Solar Energy* 82 (12): 1161–1171.

Yamaguchi, Seiichi, and Kiyoshi Saito. 2013. "Numerical and Experimental Performance Analysis of Rotary Desiccant Wheels." *International Journal of Heat and Mass Transfer* 60: 51–60.

Zhang, X. J., Y. J. Dai, and R. Z. Wang. 2003. "A Simulation Study of Heat and Mass Transfer in a Honeycombed Rotary Desiccant Dehumidifier." *Applied Thermal Engineering* 23 (8): 989–1003.

Zhang, X. J., K. Sumathy, Y. J. Dai, and R. Z. Wang. 2006. "Dynamic Hygroscopic Effect of the Composite Material Used in Desiccant Rotary Wheel." *Solar Energy* 80 (8): 1058–1061.

Zheng, W, and W M Worek. 1993. "Numerical Simulation of Combined Heat and Mass Transfer Processes in a Rotary Dehumidifier." *Numerical Heat Transfer, Part A: Applications* 23 (2): 211–232.

Zheng, X., T. S. Ge, and R. Z. Wang. 2014. "Recent Progress on Desiccant Materials for Solid Desiccant Cooling Systems." *Energy* 74 (Supplement C): 280–294.

Zouaoui, Ahlem, Leila Zili-Ghedira, and Sassi Ben Nasrallah. 2016. "Open Solid Desiccant Cooling Air Systems: A Review And Comparative Study." *Renewable and Sustainable Energy Reviews* 54: 889–917.

10

Thermal Energy Storage

While the sun is the most powerful source of energy in our solar system, exploitation of solar energy is restricted to daytime hours. Under clear sky conditions, solar energy production peaks at noon and during summer. Meanwhile, electricity and heating demand is typically higher during the winter, especially in the early hours of the morning and at night. On the other hand, the peak cooling demand in the buildings sector coincides with the hours of maximum solar radiation during a day, which makes solar-driven air conditioning a very attractive candidate to replace conventional electrically powered systems, and thus reduce electricity demand during peak-load periods. However, even for cooling applications, the high intermittency of solar radiation due to changing weather and atmospheric conditions induces a temporal mismatch between energy supply and demand. Integration of short- and long-term energy storage technologies will enable more efficient use and conversion of solar energy by compensating for the diurnal and seasonal offset between energy supply and demand. Energy storage solutions will enhance predictability and increase the hours of operation of solar systems, improve energy supply security, and, especially for solar thermal electric power plants, will alleviate the problem of power grid stability.

While battery banks are the most widespread technology for storing solar-generated electricity by photovoltaics, solar radiation captured by thermal collectors is stored in thermal energy storage (TES) systems in the form of heat which can be discharged at a later stage when solar irradiation is not available. Important advantages of TES systems are their low energy storage losses, long lifetime, and relatively low costs. Typically, TES systems are classified by storage mechanism and by method of integration into the solar energy system.

Regarding integration to the system, classification refers mainly to how the charging and discharging of solar thermal energy into/from the storage unit proceeds. Two types of systems can be distinguished: active and passive systems. Passive storage systems make use of storage media that do not circulate through the system and a separate heat transfer fluid (HTF) carries thermal energy to/from the storage medium. The most common storage media for this type of system are solids, phase change materials, and compounds that serve as the chemical reactants in thermochemical storage systems. On the other hand, in active systems, the storage medium itself circulates by forced convection through a heat exchanger, which might be either the solar thermal receiver (direct systems) or a heat exchanger (indirect systems), to gain or release stored thermal energy. In direct-active systems the storage medium serves also as the heat transfer fluid of the thermal collector and thus circulates between the heat source and the heat storage tank, whereas indirect-active systems use different media to collect and store solar heat. A schematic representation of an indirect-active and a passive TES system is shown in Figure 10.1.

Thermal storage energy systems are classified according to their storage mechanism into technologies that utilize thermophysical or thermochemical processes. Thermophysical systems can be subdivided into sensible (STES) and latent (LTES) thermal energy storage systems. Sensible TES systems involve the storage of thermal energy by affecting a temperature increase in the storage medium. In latent TES, thermal energy is stored by inducing a phase

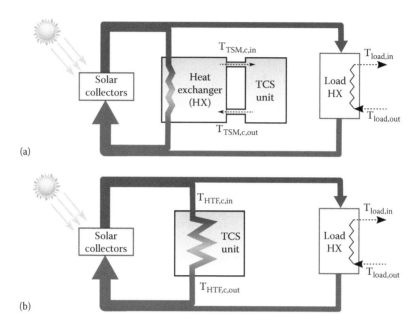

FIGURE 10.1
Schematic representation of (a) an indirect-active and (b) a passive TES system during charging.

change of the storage medium. On the other hand, thermochemical storage processes make use of the energy captured by solar collectors to drive reversible chemical reactions or sorption processes, and thus efficiently store heat in the form of high-energy-density chemical bonds.

10.1 Sensible Thermal Energy Storage

Sensible thermal energy storage is the most mature technology, and there are a large number of well-demonstrated and reliable concepts that are widely used because they offer simplicity of design and ease of operation. Assuming an incompressible thermal storage medium (TSM), the thermal energy stored in the TSM during charging is given by:

$$E_{STES} = \rho_{TSM} \cdot V_{TSM} \int_{T_c}^{T_h} c_{p,TSM}(T)\,dT \tag{10.1}$$

where ρ, V, and c_p are the density, volume, and the specific heat capacity of the TSM, respectively, and T_h–T_c is the temperature difference over which the storage operates—referred to as "temperature swing." As seen from Eq. (10.1) the amount of heat stored and, thus, the selection of the storage medium and the size of the storage unit depend directly on its volumetric heat capacity $\rho \cdot c_p$ and the temperature swing. The volumetric heat capacity of the most common STES media is relatively low and it varies in the range of 1.11–4.68 MJ·m^{-3}·K^{-1} (Tian and Zhao 2013), resulting in systems with low energy storage density that require large amounts of storage media to deliver the amount of energy necessary for a specific application.

Although selection of materials with high c_p allows for a reduction in the storage unit size, the specific heat capacity of a material should always be considered in conjunction with its specific price so as to avoid excessive capital costs. Apart from the volumetric heat capacity and material costs, other properties, design criteria, and operational practices important for a STES system include:

- Construction of storage units with a relatively low surface-to-volume ratio to enable reduction of its thermal losses to the surroundings. Higher storage efficiency implies lower solar collector areas required to compensate for these losses.

- High thermal diffusivity or heat transfer coefficient of the solid or liquid storage medium, respectively, that will enhance the heat transfer rates between the heat transfer fluid and the medium and thus enable sufficient charge and discharge rates of thermal energy.

- High quality of stored thermal energy, which can be achieved by enhancing thermal stratification within the storage tank. That is, the development of a thermal gradient across the storage tank as shown in Figure 10.2. Such temperature gradients across the storage unit are desirable since the existence of a hotter and colder zone will enable maintaining a high driving force for heat transfer at all states of charge. Specifically, a low T region at the bottom of the storage unit will ensure a high difference between the inlet temperature of the HTF during charging and the temperature of the storage medium, $T_{HTF,c,in} - T_{TSM}$, even at high states of charge, and thus facilitate efficient storage of the heat captured by the solar collectors also under demanding conditions. On the other hand, thermal stratification during discharge is important because it allows for the extraction of high-quality heat from even an almost empty store.

- High thermal, mechanical, and chemical stability of the storage medium at the temperature range of operation to avoid material degradation over a certain number of charging/discharging cycles.

- Compatibility of the storage medium with the heat transfer fluid or any heat exchangers present in the system.

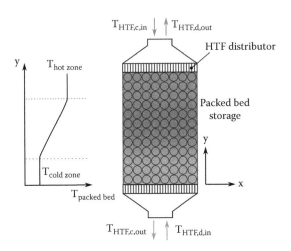

FIGURE 10.2
Thermal stratification within a storage tank using solid particles in a packed bed arrangement as the storage medium. Also indicated is the temperature distribution across the thermal storage unit.

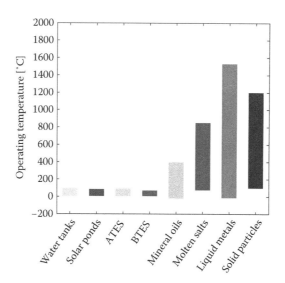

FIGURE 10.3
Operating temperature ranges of sensible thermal energy storage systems.

Overall, the selection of the appropriate storage concept and material is a complicated procedure during which technical, environmental, and financial aspects should be considered. Therefore, a multi-objective optimization methodology was developed by Fernandez et al. (2010) that allows for identifying materials with the highest performance for a specific STES application. Sensible heat storage can be either liquid (such as water, mineral oils, molten salts, etc.) or solid (rocks, concrete, metals, etc.). Figure 10.3 provides an overview of the most common STES media and systems along with their potential operating temperature ranges.

10.1.1 Liquid Media

Liquid-state STES media exhibit higher specific heat capacities (c_p), which typically vary in the range of 1.1–4.18 kJ·kg^{-1}·K^{-1}, vis-à-vis common solid-state materials with rather low c_p in the range of 0.56–1.3 kJ·kg^{-1}·K^{-1} (Tian and Zhao 2013; Siegel 2012). However, they are more difficult to contain because their vapor pressure increases with temperature and leak tightness issues may arise.

10.1.1.1 Water

Water is a very promising candidate for sensible thermal energy storage applications at temperatures below 100°C due to its high specific heat capacity (4.18 kJ·kg^{-1}·K^{-1}), wide availability, low cost, and chemical stability. Therefore, it is the most commonly used STES medium in water, space heating, and cooling systems of residential buildings.

In the majority of residential applications, water is stored in tanks that are typically made of steel, aluminum, or concrete. Depending on the storage requirements, the size of a tank can vary between a few hundred of liters for diurnal storage systems to some thousands of cubic meters for seasonal storage for district heating systems. Tanks can be located either above ground or buried underground in order to reduce thermal losses to the surroundings. Buried tanks have higher capital costs, but the higher storage efficiency enables them to acquire the same thermal storage capacity as above-ground tanks with smaller storage units.

In indirect-active water storage systems, heat is carried by a HTF through the solar collectors and a heat exchanger to the storage tanks. The most commonly used heat exchanger configurations are immersed coils, mantle, and external heat exchangers (Han, Wang, and Dai 2009), as illustrated in Figure 10.4. In all the configurations, cold water coming from the load side is supplied to the lower part, and hot water is extracted from the top of the tank during discharge in order to prevent mixing of hot and cold water streams inside the tank, with adverse effects on the degree of thermal stratification. In immersed coil heat exchanger configurations, coils are positioned inside the water storage tank, and therefore only one pump responsible for circulating the HTF throughout the system is required. Coils are typically immersed at the bottom part of the tanks to exploit the high temperature difference between the incoming flows of HTF and water, $T_{HTF,c,in} - T_{water,in}$. An important drawback of this configuration is their relatively low degree of thermal stratification. In mantle heat exchanger configurations, the HTF flow passes through an annular gap between the inner water storage tank wall and a second outer wall, thus increasing the surface area over which solar heat is transported to the storage medium. This design enhances thermal stratification and increases the system performance, but at the expense of designing more advanced storage tanks. Finally, heat transport to the water storage can proceed via external heat exchangers, as schematically shown in Figure 10.4c. This arrangement is advantageous in terms of structural requirements vis-à-vis mantle and immersed coil heat exchangers as it allows for the use of standard, low cost storage tanks while also achieving a high degree of thermal stratification.

Overall, the degree of thermal stratification in all water storage tank configurations depends on the tank geometry, water flow rates, position of water inlets and outlets, and the heat exchanger unit. Several water storage tanks are equipped with mechanisms that reduce mixing between the hot and cold layers of water stored inside the tank and thus achieve higher heat transfer rates during charging and discharging.

Apart from storage tanks, solar pond technology provides an alternative solution for storing hot water, especially in regions with limited snowfall and low mean wind speeds. In this concept, naturally existing or artificial water ponds are used to collect and store solar thermal energy in the form of sensible heat. The most representative solar pond configuration is referred to as a "salinity-gradient solar pond." In this system, the bottom of a naturally existing or an artificial water-containing pond is darkened to absorb solar

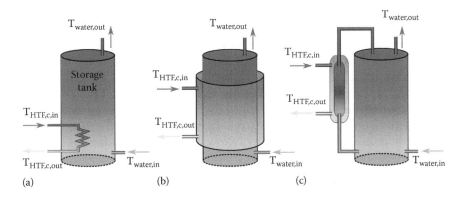

FIGURE 10.4
Schematic layout of (a) immersed coil, (b) mantle, and (c) external heat exchanger designs used to transport heat to the storage tank of indirect-active water TES systems.

radiation and transport the heat to the water, which would normally induce a natural convection flow of the warm water toward the upper layers of the pond. To prevent natural convection inside the pond from occurring, salt is added to the water and three zones of different salinity are developed across the pond: i) the upper convective zone, which is located just below the water surface and exhibits a low salinity level; ii) the non-convective zone, in which salt concentration increases with depth; and iii) the lower convective zone, where a high-concentration solution of salt in water is contained. As the brine density increases with higher salt concentrations, a density gradient also develops across the pond due to the varying salt concentrations. This concentration-driven density gradient is what ultimately counteracts temperature-driven buoyancy forces and prevents warm water at the bottom of the tank from flowing to the upper layers of the pond. Thus, the lower layer of the solar pond serves as a sensible TES system, where water is stored at temperatures typically close to ~ 70–90°C. During discharging, heat is extracted from the pond by an immersed heat exchanger. The most commonly used salts in solar ponds are $MgCl_2$, KNO_3, NH_4NO_3, $NaNO_3$, and $NH_2CO \cdot NH_2$. Salinity-gradient ponds are cost-effective solutions for STES, but complex in operation because salt concentration in the upper convective zone has to be kept at a minimum in order to reduce water reflectivity and thus maximize the solar energy input to the system while the required high salt concentration in the lower convective zone can only be achieved by continuously supplying salt to the pond. A common practice adopted for preventing mixing between the different zones involves the use of floating barriers on the free surface of the pond which prevent formation of wind-generated wakes and simultaneously reduce the convective exchange with the surroundings. Typical pond designs that were developed in order to alleviate the problems of wind mixing and operational complexity are the so-called "gel-stabilized" and "honeycomb" solar ponds. In the former, a transparent polymer gel layer floating on the free surface replaces the upper convective and non-convective zones of a conventional salinity-gradient pond, and thus eliminates the problem of wind mixing. Honeycomb solar pond configurations make use of a floating, air-filled honeycomb structure to insulate the upper surface of the pond and reduce its thermal losses to the surroundings. An important disadvantage inherent in all types of solar ponds is the high variability in their performance throughout a year. Relatively low efficiencies are obtained during the winter, and especially in high-latitude regions, since ponds, unlike flat-plate collectors, cannot be tilted relative to the ground and are thus accompanied by higher cosine losses.

Finally, hot water storage using aquifer thermal energy storage (ATES) systems has attracted considerable attention because it represents a cost-effective solution for seasonal thermal energy storage. It relies on storage of groundwater in an aquifer, such as an underground layer of water-bearing permeable rocks or unconsolidated materials. At least two thermal wells—holes drilled into the ground to penetrate an aquifer—are used in ATES systems to extract or inject groundwater from or into the ground via pumping. With this arrangement and during the summer months, excess thermal energy is pumped into the injection well. During the winter, warm groundwater stored in the aquifer can be pumped from the extraction well through a heat exchanger to satisfy the heating load of a building directly or to provide low quality energy to a heat pump. Combined ATES-heat pump systems have been shown to improve the coefficient of performance (COP) conventional heat pumps operated with ambient air by 65%. Storage temperatures in ATES systems vary typically in the range of 10–40°C, but successful operation has been demonstrated at temperatures up to 90°C (Ghaebi, Bahadori, and Saidi 2014).

10.1.1.2 *Mineral Oil Hydrocarbons, Molten Salts, and Liquid Metals*

One of the main inconveniencies of using water as a sensible thermal storage medium is its incompatibility for applications above temperatures of ~ 95°C. For such applications, water is replaced by liquid media that exhibit lower vapor pressures over the temperature range of interest. The most common liquid-state STES media for high-temperature applications are mineral oil hydrocarbons, inorganic molten salts, and liquid metals.

Hydrocarbon oils are most commonly synthetically produced oils and exhibit favorable heat transfer properties and low viscosity. Owing to the low pumping energy required for their circulation, oils are the most widely used medium in parabolic trough concentrated solar power systems. However, they undergo fast degradation at temperatures over ~ 400°C, exhibit high vapor pressures and flammability, and are relatively expensive valued at ~ \$5/kg (Tian and Zhao 2013).

Owing to their non-flammable and non-toxic character as well as to their excellent thermal stability and low vapor pressures and costs (\$1.5/kg), molten salts offer an alternative solution to hydrocarbon oils for high-temperature applications (Tian and Zhao 2013). Potassium and sodium nitrate salts are the most commonly used storage media in central tower receiver systems because they are able to operate at temperatures up to ~ 550–600°C. However, they exhibit higher solidification points (~ 150–250°C) compared to synthetic oils, thus posing a significant limitation to their use because this implies higher maintenance costs and power consumption during hours of solar unavailability to prevent them from solidifying. In order to overcome the solidification problem and enable molten salts to be used also in parabolic trough plants that are operated at lower temperatures vis-à-vis central tower receivers, research activities have been focusing on the development of novel low-melting-point molten salt materials. Recently, Zhao and Wu (2011) developed ternary mixtures of KNO_3, $LiNO_3$, and $Ca(NO_3)_2$ with melting temperatures in the range of 76–80°C, whereas Bradshaw, Cordaro, and Siegel (2009) reported the development of quaternary nitrates with a solidification point of ~ 90°C.

Finally, liquid metals have been considered for use as STES in applications with temperatures exceeding 550°C. Liquid sodium is an attractive material due to its outstanding thermal properties and low melting point (97.8°C). However, its high chemical reactivity and flammability urge the need for high safety measures and pose a significant limitation on its deployment as a STES medium.

10.1.2 Solid Media

10.1.2.1 *Packed Bed Storage*

Besides liquid media, the suitability of a broad range of solids as storage media in STES applications has been demonstrated. These materials are relatively cheap (\$0.05–5/kg), non-toxic, easy to contain, and exhibit high thermal conductivities (1–37 $W \cdot m^{-1} \cdot K^{-1}$) (Tian and Zhao 2013), thus contributing to attaining high charging/discharging rates. Due to their extended range of operating temperatures—typically between 100°C and 1,200°C— solid materials are used as STES media in a number of residential, industrial, and power generation applications in tandem with both non-concentrating and concentrating solar thermal collectors. However, due to the typically low heat capacity values of solid TES media, large volumes of storage material are required for capturing the desired amount of thermal energy with an adverse effect on the investment costs of the system.

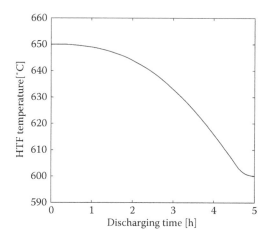

FIGURE 10.5
Variation of the outflow heat transfer fluid temperature of a 6.5 MW$_{th}$ packed bed TES unit as a function of time during discharging. (Adapted from Zanganeh, G. et al., *Applied Thermal Engineering*, 70 (1): 316–320, 2014.)

Typical solid-state materials used for STES applications are rocks, sand, concrete, and cast ceramics. These are most commonly utilized in loosely-packed beds. In packed bed TES systems—as opposed to TES units using liquid media—the storage medium cannot circulate through the system; therefore, an HTF is required to pass through the bed volume in order to recover/store thermal energy from/into it. If a gas with low specific heat capacity (such as air) is used as the HTF, it does not contribute to the storage and the system is considered passive, whereas systems using liquid HTFs with high c_p (such as water) are referred to as hybrid active-passive systems. An important disadvantage of packed bed storage systems, which applies to all STES systems, is that they are not capable of maintaining the outflow HTF temperature during discharging constant. Figure 10.5 illustrates the temporal variation in the outflow HTF temperature of a 6.5 MW$_{th}$ packed bed TES unit during discharging (Zanganeh et al. 2014), which reveals that heat is delivered at monotonically decreasing temperatures as thermal energy is extracted from the packed bed. Besides packed beds, fluidized bed storage systems have been proposed by (Hasnain 1998) in order to achieve higher heat transfer rates.

10.1.2.2 Borehole Thermal Energy Storage

Borehole thermal energy storage (BTES) systems have gained considerable interest lately because they provide a cost-effective solution for seasonal thermal energy storage in residential applications. These systems make use of the ground as the solid-state TES medium. A BTES system consists of an array of boreholes drilled in the ground, into which U-shaped pipes are inserted. During the summer months or a period of solar radiation availability, a heat transfer fluid—typically an antifreeze/water mixture—flowing through the U-shaped pipes transports excess heat from solar collectors and stores it in the ground. During the winter, heat is extracted by passing low-temperature HTF through the ground and is used either to satisfy the heating load of a building directly or to provide low quality energy to a heat pump. The entire BTES field is surrounded by layers of insulation to achieve high storage efficiency.

10.1.2.3 Particle Suspensions and Storage

Packed bed arrangements represent the most common STES technology when solid particles serve as the storage media. However, a very promising concept that is extensively investigated for its applicability in concentrated solar power (CSP) plants involves the use of particle flows to absorb and store solar energy. The use of solid particles as the heat transfer and storage media would help to overcome the limitations set on the operating conditions of CSP plants by existing liquid-state media (low solidification point and thermal decomposition of molten salts at temperatures exceeding 600°C, safety risks posed by the use of liquid metals), and thus enable exploration of more efficient thermodynamic cycles such as the supercritical Rankine or the supercritical CO_2 (s-CO_2) Brayton cycles that require working fluid temperatures above 600°C. Particles made of silicon carbide (SiC), silica (SiO_2), alumina (Al_2O_3), and zircon ($ZrSiO_4$) are the most commonly used due to their high stability at temperatures up to 1,000°C and high specific heat capacity values (Tan and Chen 2010).

A typical realization of the technology involves the use of an upward-flowing or free-falling dense gas-particle suspension as the heat transfer medium in a central receiver tower. After absorbing concentrated solar radiation, particles are transported to a high-temperature particle storage tank. During discharging, particles are passed through a heat exchanger, where the stored heat is transferred to the working fluid of a thermodynamic cycle. The low-temperature particles then flow into a low-temperature storage tank that supplies the central receiver tower, thus closing the particle circulation loop.

10.2 Latent Energy Storage (LTES)

Latent thermal energy storage systems are capable of storing large amounts of heat by changing the phase of the thermal storage medium. During charging, the storage medium initially behaves as a STES medium. It absorbs sensible heat until its temperature rises to the phase-transition point, at which large amounts of thermal energy are stored at a nearly constant temperature. For materials experiencing a solid-to-liquid phase transition, the thermal energy stored during the process is given by:

$$E_{LTES} = \rho_{s,TSM} \cdot V_{s,TSM} \int_{T_c}^{T_{mp}} c_{p,s,TSM}(T)dT + m_{TSM} \cdot a_m \cdot (\Delta h_f)_{T=T_{mp}}$$

$$+ \rho_{l,TSM} \cdot V_{l,TSM} \int_{T_{mp}}^{T_h} c_{p,l,TSM}(T)dT \tag{10.2}$$

where s and l denote the solid and liquid states of the thermal storage medium, respectively, T_{mp}, Δh_f, and m are the melting temperature, the specific heat of fusion, and the mass of the TSM, respectively, and α_m is the mass fraction of the TSM that underwent a phase change. Due to the large enthalpy changes that accompany phase transitions, latent energy storage offers a higher energy storage density vis-à-vis STES systems, thus leading to more compact storage units. During discharging, the phase change material is cooled down to its freezing temperature, where the heat of solidification is released at a nearly constant temperature. The fact that TES systems using phase change materials are capable of releasing heat nearly isothermally represents an important advantage compared to STES systems, especially for applications with strictly defined working temperature limits.

As seen from Eq. (10.2), the amount of heat stored in phase change materials depends largely on the specific heat of fusion, Δh_f, while materials with high volumetric heat capacity, $\rho \cdot c_p$, provide the option of storing considerable amounts of additional energy in the form of sensible heat within temperature intervals close to the phase transition temperature. Apart from the specific heat of fusion and volumetric heat capacity, other material properties and operational criteria that govern materials selection and the design of a LTES system are:

- Good matching of the PCM phase-transition temperature with the operating temperature of the solar thermal system under consideration.

- Selection of PCMs with preferably high thermal conductivity in order to increase the heat transfer exchange between the PCM and HTF and achieve high charging/discharging rates of the storage unit.

- Long thermal, mechanical, and chemical stability of the PCM at the operating temperatures of the system to prevent a possible degradation of its thermal properties or a decrease in the fraction of the material undergoing phase change (α_m) over its life cycle.

- The subcooling effect of the PCM—defined as the decrease of the PCM temperature during discharging of the storage unit below its phase transition temperature without initiation of the phase change phenomena—should be kept to a minimum to ensure full recovery of the thermal energy stored and avoid temperature variations in the outflow temperature of the HTF that would pose difficulties in controlling the system. For solid-liquid PCMs, where the solidification process commences with the formation of crystals, selection of materials exhibiting high rates of crystal formation rates is critical for avoiding subcooling of the liquid phase. Further techniques investigated for preventing solidification below the melting temperature involve direct contact of the PCM and an immiscible HTF (Fouda et al. 1984; Farid and Yacoub 1989), as well as the use of nucleation agents (Telkes 1952; Ryu et al. 1992).

- Low vapor pressures and small volume changes of the PCM during phase transition in order to avoid pressure build-up within the PCM container that would impose special design requirements on the PCM container.

- Selection of a non-toxic and non-flammable PCM that is chemically compatible with the construction material of the PCM container.

- Wide availability and cost-effectiveness of the PCM.

10.2.1 Phase Change Materials Classification and Properties

A wide variety of phase change materials has been developed (Lane 1986; Zalba et al. 2003; L. Cabeza, Heinz, and Streicher 2005; Sharma et al. 2009; Gil et al. 2014). They can be classified according to the type of phase change transition to: i) solid-liquid, ii) solid-gas, iii) liquid-gas, and iv) solid-solid PCMs. Solid-solid PCMs exhibit generally low phase transition enthalpy changes and slow phase transformation kinetics that would pose a limitation on the charging and discharging rates of the storage unit. On the other hand, solid-gas and liquid-gas PCMs exhibit the highest phase transition enthalpy changes, but are not favorable for thermal storage applications because storage of gaseous media typically requires high-pressure gas containers. As solid-liquid PCMs are considerably easier to contain, they are considered the most promising candidates, although they exhibit lower latent heat values than phase transitions involving gases.

Solid-liquid PCMs can be classified into three main categories according to their chemical composition: organic, inorganic, and eutectic, as shown in Figure 10.6.

Organic PCMs are subdivided into paraffin and non-paraffin (fatty acids, esters, alcohols, glycols) materials and are suitable for low-temperature applications because their melting temperatures typically vary in the range of 5–150°C. Despite their relatively low specific heat of fusion (95–260 kJ·kg^{-1}), organic PCMs exhibit high thermal and chemical stability because no phase segregation occurs during melting/solidification, high chemical compatibility with conventional container materials, whereas low or no subcooling occurs during phase transition. However, organic PCMs are materials with lower thermal conductivity (0.1–0.7 W·m^{-1}·K^{-1}) compared to inorganic materials, thus posing an important limitation on the rate of heat recovery from the TES unit during discharging. Relatively large volume changes that occur during phase transitions, high flammability, and higher costs vis-à-vis inorganic compounds are further characteristics restricting the use of organics in thermal storage applications.

Inorganic substances are classified into hydrated salts and metallic compounds, and can be used as storage media in both low- and high-temperature applications because they exhibit a very broad range of phase transition temperatures (0–900°C). Inorganic compounds are non-flammable, low-cost materials with superior thermal properties compared to organic PCMs. Their specific heat of fusion and thermal conductivity vary in the ranges of 116–492 kJ·kg^{-1} and 0.5–5.0 W·m^{-1}·K^{-1} (Zalba et al. 2003), respectively, thus leading to units with higher storage density and charging/discharging rates. However, as they are typically prone to phase segregation during melting/solidification, the available amount of phase change material as well as the initially high storage density decrease upon thermal cycling. Furthermore, the high degree of subcooling experienced by several inorganic PCMs, and especially by salt hydrates, poses additional problems to recovering the total amount energy stored, whereas high corrosiveness also hinders their use as storage media.

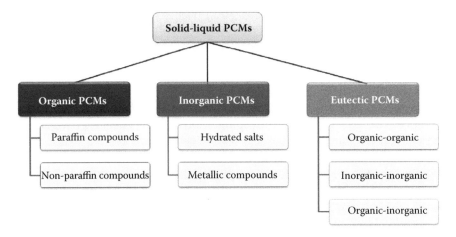

FIGURE 10.6
Classification of phase change materials.

Eutectic phase change materials are mixtures composed of two or more chemical substances. Depending on the nature of the mixture constituents, eutectic PCMs can be classified into organic-organic, inorganic-inorganic, and organic-inorganic mixtures. Their composition is determined so as to acquire a melting temperature that matches the temperature of the specific solar thermal application. Melting temperatures of typical eutectic PCMs vary in the range of 25–832°C (Zalba et al. 2003), and their heat of fusion is in the same order of magnitude as that of organic materials (123–226 kJ·kg^{-1}).

Despite the wide range of phase change materials, none of them satisfy all the requirements set for an ideal thermal storage medium. Organic materials exhibit relatively low energy storage densities, whereas inorganic compounds face serious issues relating to phase segregation and subcooling. However, the most important disadvantage of phase change materials, which is common to all material categories, is their inherently low thermal conductivities when compared to sensible TES media. Low heat transfer rates between the HTF and the PCM result not only in long response times of the TES unit and low peak power during discharging but can also induce damages to a solar thermal collector at hours of high solar irradiation if the excess energy cannot be efficiently absorbed by the TSM.

10.2.2 Containment of Phase Change Materials

In latent thermal energy storage systems a HTF—usually water or air—is used to store/recover thermal energy to/from the storage medium which is typically placed in a container. Due to the inherently low thermal conductivity of PCMs, the geometrical configuration of the PCM container has a direct impact on the rate of heat transfer between the HTF and the PCM as well as on the efficiency of the charging/discharging processes. Latent TES systems can be generally subdivided into two main types according to the configuration of the PCM container:

- Compact LTES systems
- LTES systems with encapsulated phase change materials

Figure 10.7 shows the schematic configuration of typical compact and encapsulated-PCM systems. In compact LTES systems, the phase change material is inserted into a cylindrical or rectangular container tank into which a heat exchanger is embedded. The heat exchanger in a compact LTES system could be a long heat pipe extending through the volume of the phase change material (Figure 10.7a), an annular pipe surrounding the PCM (Figure 10.7b), or multiple parallel tubes embedded into the PCM material in a so-called "shell-and-tube" system (Figure 10.7c). The latter is the most frequently investigated configuration in the literature due to its simple design, high thermal performance, and the limited pressure losses through the piping. More complex heat exchanger units can be used in systems with charge/discharge power below ~ 15–20 kW.

In encapsulated-PCM systems, the phase change material is contained in an array of small polymer or inorganic shells -referred to as "capsules"- inserted in a storage tank, as shown in Figure 10.7d. Shells may contain one or multiple PCM cores and can be either irregular or regular—typically of cylindrical or spherical—in shape. Containment of the phase change material in capsules eliminates any issues of chemical incompatibility between the HTF and the PCM, and prevents leakage of the PCM when it is in the liquid phase. Void spaces should be provided inside the storage container or the PCM capsules for compact and encapsulated PCM systems, respectively, in order to allow for the thermal

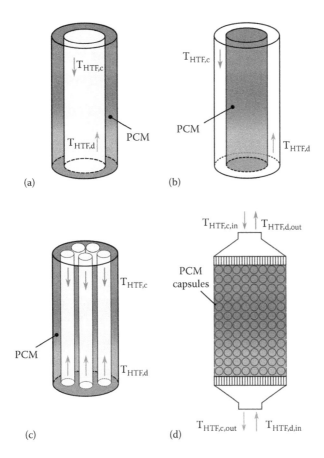

FIGURE 10.7
Schematic illustration of (a) cylindrical-pipe-flow, (b) annular-pipe-flow, (c) shell-and-tube, and (d) encapsulated PCM packed bed LTES systems.

expansion of the PCM during melting without a pressure build-up. Considerably higher PCM volume fractions and thus energy storage density can be achieved with compact LTES vis-à-vis encapsulated-PCM systems. However, this leads to lower surface-to-volume ratios and has an adverse effect on heat transfer between the PCM and the HTF and, consequently, on the charging/discharging rates of the TES unit.

10.2.3 Heat Transfer Enhancement Techniques

In order to enhance the heat absorption/release rates within the storage unit various other approaches. Besides shell-and-tube LTES systems and conventional macro-encapsulation of PCMs (i.e. containment of PCMs in large—in the range of several centimetres—shells), have been investigated.

10.2.3.1 Micro- and Nano-Encapsulation

Microencapsulation of PCMs involves engulfing individual PCM particles or droplets with sizes in the range of 1–1,000 μm in a chemically compatible thin solid wall. The use of smaller particle sizes compared to the technique of macroencapsulation leads to improved heat transfer between the PCM and the HTF due to the increased surface area available for heat exchange. However, the fabrication of such microcapsules is associated with relatively high costs compared to other TES methods and, therefore, the usage of microencapsulated PCMs is currently limited to thermal control applications. Depending on the PCM material and the technique used for the fabrication of a microcapsule, this may be formed either by applying a polymer or inorganic shell coating to PCM particles comprising the core or by embedding them in a homogeneous or heterogeneous matrix. Microcapsules of the PCM core-surrounding shell type may be used either in a powder form or can be inserted into a carrier fluid to form a microencapsulated PCM slurry that can serve both as the HTF in the solar field and storage medium in the TES unit. Based on the underlying phenomena, the most common techniques used for the production of microcapsules are classified into physical, chemical, and physicochemical methods. A comprehensive review of these fabrication techniques along with their advantages and disadvantages is provided by Jamekhorshid, Sadrameli, and Farid (2014).

An important limitation posed when using microencapsulated PCM slurries is the low mechanical stability of the microcapsules when being pumped through the closed solar collector-thermal storage circulation loop. Owing to the higher mechanical stability of smaller PCM capsules, encapsulation of PCMs at nanoscale (particle sizes in the range of 1–1,000 nm) has gained considerable attention lately. Nano-encapsulation techniques enable the production of capsules with higher mechanical strength that are suitable for long-term circulation through the solar field piping system.

10.2.3.2 Insertion of Extended Heat-Exchange Surfaces

One of the most widely applied heat transfer enhancement methods in compact LTES systems involves the insertion of extended surfaces such as metal fins into the phase change material. The wide applicability of this technique is attributed to its simplicity of design and fabrication as well as its low construction costs. Fins are usually attached to the heat exchanger tubes in order to increase the effective heat transfer area between the HTF and the PCM. A schematic configuration of the most common fin geometries is provided in Figure 10.8. Tubes equipped with longitudinal fins enhance heat transfer and lead to a decrease in the solidification time of PCMs (Castell et al. 2008). For internally finned tubes, it has been shown that the amount of PCM that undergoes a phase change increases with the number of fins embedded in the material as well as with increasing fin thickness and height (Zhang and Faghri 1996). Experimental and numerical investigations of the performance of radially finned tubes (Erek, İlken, and Acar 2005) revealed that the amount of energy stored in such a LTES system increases with increasing fin radius and decreasing spacing between the fins. A review of the geometric design parameters of fins employed for heat transfer enhancement in thermal energy storage systems is provided by Abdulateef et al. (2018).

10.2.3.3 Insertion of High-Conductivity Materials

Incorporation of high-conductivity materials such as metal powders and beads into PCMs represents a simple technique that has been shown to improve the heat transfer within the storage unit by increasing the effective thermal conductivity k_{eff} of the material. Besides increasing the heat transfer rates in the storage tank, particle insertions can serve as further nucleation points within the PCM, thus leading to higher crystallization rates during solidification. Particles impregnated into PCMs are commonly made of aluminum, copper, or silver in order to take advantage of the excellent thermal conductivities (239.3 W·m^{-1}·K^{-1}, 401.2 W·m^{-1}·K^{-1}, and

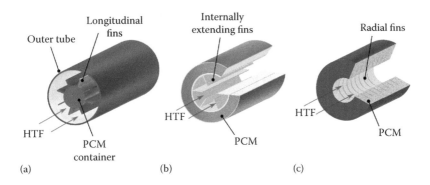

FIGURE 10.8
3D schematic view of tubes equipped with (a) longitudinal, (b) internally-extending, and (c) radial fins to enhance heat transfer in compact LTES systems.

429 W·m^{-1}·K^{-1}, respectively) of these materials. However, the overall impact of this heat transfer enhancement technique is rather limited, whereas metal particles add considerable weight to the storage system and may sink to the bottom of the storage container when the PCM is in the molten state due to their high density. Issues of chemical compatibility between the metal particles and the PCMs may also arise, because copper and aluminum are not completely inert against paraffin PCMs and some salt hydrates, respectively (Heine 1981; L. F. Cabeza et al. 2001).

In an attempt to overcome the shortcomings of metal particles, low-density materials with high conductivity have been investigated for their suitability as PCM additives. Carbon represents one of the most promising candidates because it exhibits a thermal conductivity comparable to that of aluminum and silver, and shows high chemical compatibility with the majority of PCMs. Impregnation of carbon fiber cloths and brushes into paraffin wax was shown to induce a twofold increase in the effective thermal conductivity of the thermal storage medium (Nakaso et al. 2008), thus improving the heat exchange rate in the storage tank. Carbon fiber cloths stretched among heat transfer tubes exhibited a better thermal performance vis-à-vis brush structures due to the structural discontinuity of the latter, which impaired an effective heat transfer throughout the TSM. The spatial arrangement of carbon fibers was identified to significantly influence the heat transfer properties of the material. In particular, materials impregnated with unidirectional carbon fibers exhibited an effective thermal conductivity which was approximately two times higher than that of materials using randomly oriented fibers (Fukai et al. 2000).

10.2.3.4 Impregnation of High-Conductivity Porous Structures

An alternative technique for improving the thermal performance of LTES systems is by embedding phase change materials with graphite contents in the range of 5–35 wt.%, into high-conductivity porous structures to form stable composite materials. Crucial design parameters for the performance of PCM composites are the porosity and mean pore size of the porous matrices. High porosity values imply higher volume percentages occupied by the PCM but limited heat transfer enhancement effects, whereas low porosities lead to higher effective thermal conductivities at the expense of impairing the movement of liquid PCM through the porous structure (Elgafy and Lafdi 2005). A correlation for the estimation of the effective thermal conductivity of PCMs embedded into porous structures has been developed by Mesalhy et al. (2005). Porous structures can be either graphite or metal matrices that are commonly made of aluminum, copper, or nickel.

Paraffin/compressed-expanded-natural graphite (CENG) composite materials have been characterized by (Py et al.) and effective thermal conductivities in the range of 4–70 $W \cdot m^{-1} \cdot K^{-1}$ were obtained which represents a considerable increase when compared to the thermal conductivity of pure paraffin (0.24 $W \cdot m^{-1} \cdot K^{-1}$). However, the anisotropic nature of paraffin/CENG composites leads to undesired spatial variations in their heat transfer performance. To mitigate this shortcoming, exfoliated graphite (EG) has recently attracted increasing attention. Palmitic acid/exfoliated graphite composites were synthesized by Sarı and Karaipekli (2009) and effective thermal conductivities higher than that of pure palmitic acid (0.17 $W \cdot m^{-1} \cdot K^{-1}$) were obtained. In particular, the effective thermal condictivity of Palmitic acid/EG composites varied in the range of 0.18–0.6 $W \cdot m^{-1} \cdot K^{-1}$ for mass fractions of graphite between 5 wt.% and 20 wt%, respectively. Impregnation of higher amounts of exfoliated graphite into the PCM resulted in leakage of the palmitic acid when it was in the molten state. The thermal behavior of paraffin/exfoliated graphite composites was also investigated by Sarı and Karaipekli (2007) using graphite mass fractions in the range of 2–10 wt.%. Composites with a graphite content of 10 wt.% were identified as the most promising candidates for thermal energy storage applications because they exhibited an effective thermal conductivity of 0.82 $W \cdot m^{-1} \cdot K^{-1}$ while avoiding leakage of melted paraffin during the solid-liquid phase transition. Further increase of k_{eff} is hindered by stability issues as well as by the structural discontinuity of exfoliated graphite. In an attempt to further increase the heat transfer rates in LTES systems, continuous metallic porous structures with isotropic thermal properties offer a great potential. Paraffin wax RT 27 and calcium chloride hexahydrate were used as PCMs by Zhou and Zhao (2011) to experimentally investigate their heat transfer characteristics when embedded in open-cell exfoliated graphite and metal matrices. Metal-based composites showed a better heat transfer performance as lower temperature differences between the PCM composite and the heat source were obtained compared to EG composite materials during charging.

10.2.3.5 Cascaded PCM Storage Systems

Cascaded storage units make use of multiple PCMs with different phase transition temperatures to overcome an important problem faced during charging and discharging of single-PCM systems, which is the decrease in the temperature difference between the PCM and HTF in the flow direction. During charging, the HTF in a single-PCM system transfers its heat, and its temperature decreases, thus leading to poor heat transfer at the regions close to the HTF outlet where $T_{HTF,c} - T_{PCM}$ is rather low. During discharging, extraction of heat from the PCM layers located at the end of the storage unit could become difficult as the temperature of the HTF might have reached a temperature close to the solidification temperature of the PCM. This limitation can be circumvented by using cascaded storage systems, in which multiple PCMs with different melting points are arranged in order of decreasing melting temperatures so as to achieve a constant temperature difference $T_{HTF,c} - T_{PCM}$ throughout the storage unit during the charging process. When reserving the HTF flow direction for discharging, the HTF extracts heat first from the PCM with the lowest solidification temperature and then passes through the high-temperature PCM layers. Using this technique, the temperature difference $|T_{HTF,c} - T_{PCM}|$ is stabilized during both charging and discharging and an almost constant heat flux to/from the PCM is obtained. Besides purely PCM-based systems, cascaded hybrid systems combining sensible and latent thermal storage media have also been proposed with the aim of stabilizing the HTF outflow temperature at low costs (Zanganeh et al. 2015).

10.3 Thermochemical Energy Storage (TCS)

Even higher energy storage densities vis-à-vis sensible and latent TES systems can be attained by thermochemical storage systems. Such systems use excess solar heat to drive reversible chemical reactions or chemical sorption processes, and thus efficiently store it in the form of high-energy-density chemical bonds. During charging, solar heat is absorbed by a chemical compound A, which is converted into the products B and C in an endothermic process according to:

$$A + solar\ heat\ \rightarrow B + C \tag{10.3}$$

During the endothermic step, solar heat heats up compound A to the process temperature, and then is stored in the form of chemical energy as the chemical reaction or sorption proceeds. The amount of thermal energy stored during the charging process can be obtained by:

$$E_{TCS} = n \cdot \alpha \cdot \Delta H_m \big|_{T=T_{process}} \tag{10.4}$$

where n is the molar amount of the compound A, α is the chemical conversion defined as the ratio of the molar amount of compound A reacted over the initial molar amount of A, and ΔH_m is the molar enthalpy change of sorption/reaction. Reaction products B and C are stored separately and usually at ambient temperature in order to enable thermal energy storage for long periods of time with minimal thermal losses. This characteristic represents an important advantage of thermochemical processes in comparison to STES and LTES systems, and qualifies them as very promising candidates for long-term thermal storage applications. Thermal energy losses during the storage period are related solely to the sensible heat rejected during cooling of the products B and C from the desorption/reaction temperature to the ambient temperature T_{amb}. Any other losses that might occur are attributed to degradation of the thermal properties and energy storage density of the materials due to thermal cycling. During discharging, the process is reversed and chemical compounds B and C are recombined to regenerate component A in an exothermic step where heat is released at a nearly constant temperature:

$$B + C \rightarrow A + heat \tag{10.5}$$

An important difference of thermochemical energy storage vis-à-vis STES and LTES systems is that the temperature at which charging and discharging proceeds may significantly differ from each other since the endothermic reaction steps typically occurs at higher temperatures. In this way, considerable energy and exergy losses are introduced to the system.

As seen in Eq. (10.4), the amount of heat stored in TCS systems depends largely on the amount of storage material, the molar enthalpy change of sorption/reaction ΔH_m, and the chemical conversion α. Since enthalpy changes of typical thermochemical material pairs lie in the order of MJ·kg^{-1} or GJ·kg^{-1} and are thus approximately by one 1–2 orders of magnitude higher than the enthalpy of fusion of solid-liquid PCMs, thermochemical systems exhibit clearly the highest energy density among all TES technologies. In order to exploit the advantage provided by the high ΔH_m values and maximize the amount of energy stored, the following criteria are considered important for the design of a TCS system and the selection of the thermochemical storage media:

- Selection of a reversible chemical system with an equilibrium temperature, i.e. a temperature at which the endothermic and exothermic reactions proceed at the same rate, which matches well to the operating temperature of the solar thermal system.
- Excellent process reversibility.
- High cycling stability of the material pair in order to avoid degradation of the material properties and consequently of the energy storage density of the system with increasing numbers of thermal cycles.
- High reaction rates in both process steps that would enable fast charging and discharging of the storage unit.
- Non-complex reaction conditions that would enable both process steps to proceed in simple, low cost reactors.
- Selection of low-cost, widely available, and non-toxic TCS media with favorable thermal properties and high chemical compatibility with the construction materials of the storage container.

Despite the significant advantage of TCS systems in terms of energy density, the thermochemical energy storage technology is still in an early development stage. Enhancing the heat and mass transfer performance in the TCS media by improving reactor design or by the use of composite materials, as well as reducing the total costs of a TCS system, represent the most important challenges at the present development stage.

10.3.1 Chemical Sorption Processes

In chemical sorption systems, excess solar heat is used to drive a desorption process. During desorption, one of the two substances comprising compound A (Eq. 10.3) evaporates and gets separated from the other. The product vapor is then either exhausted (open-loop system) or gets condensed and is separately stored for future use in the reverse sorption process (closed-loop system). During discharging, the condensed substance (sorbate) is re-evaporated using a low-grade heat source and gets captured by the other substance (sorbent) to regenerate compound A in an exothermic process. A schematic representation of the operating principle of a closed-loop sorption storage system is provided in Figure 10.9.

Depending on the state of the chemical compounds and the underlying phenomena, sorption processes can be classified into three subcategories: (i) gas adsorption on the surface of a microporous solid substance, (ii) gas absorption by a liquid absorbent, and (iii) solid-gas chemical reactions, which refer to a special form of adsorption that also involves a chemical reaction between the sorbate and the sorbent—referred to as chemisorption. A wide range of materials has been proposed and experimentally investigated for their suitability as working pairs in chemical sorption processes. Water-zeolites, water-silica gels, methanol-activated carbon, and ammonia-activated carbon are the most extensively studied materials for gas adsorption systems (Meunier 1986; Yu, Wang, and Wang 2013; Lu et al. 2006). For absorption processes, H_2O-LiBr, NH_3-H_2O, H_2O-NaOH, and metal alloys-hydrogen (metal hydrides) show great potential as working pairs (N'Tsoukpoe, Le Pierrès, and Luo 2013; Y. T. Kang, Chen, and Christensen 1997; B. H. Kang, Park, and Lee 1996;

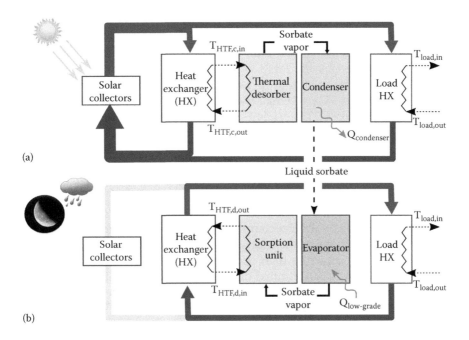

FIGURE 10.9
Schematic layout of a closed-loop sorption TCS system, illustrating the operating principles during (a) charging and (b) discharging.

Klein and Groll 2002; Marcriss, Gutraj, and Zawacki 1988; Srikhirin, Aphornratana, and Chungpaibulpatana 2001), while the NH_3- and H_2O-inorganic metal salt pairs have been identified as promising candidates for chemisorption systems (Fadhel, Sopian, and Daud 2010; Stitou, Mazet, and Bonnissel 2004; Lahmidi, Mauran, and Goetz 2006).

Sorption processes are used to store low grade solar heat at temperatures below 100°C and medium- to high-temperature heat within the range of 100–400°C. The main disadvantage of sorption systems is their low efficiency due to poor heat and mass transfer in the sorption reactor. Therefore, research is currently focused on implementing heat transfer enhancement techniques similar to those applied in LTES systems as described in Section 10.2.3. To overcome the limitations imposed by the inherently low thermal conductivity of TCS media, which for typical inorganic metal salts and hydrides varies in the range of 0.1–1 $W{\cdot}m^{-1}{\cdot}K^{-1}$, the technique of impregnating sorption materials into high-conductivity porous structures to form stable composite materials has been widely used. Besides improving the thermal performance of TCS materials, research activities are also directed toward minimizing the negative effects of the temperature swing between sorption and desorption on system efficiency. The main focus lies in novel sorption cycles that are designed to optimize thermal management (Wang and Oliveira 2006) or to lower the regeneration temperature of the sorbent (Saha, Boelman, and Kashiwagi 1995; Saha, Akisawa, and Kashiwagi 2001; Saha et al. 2003).

10.3.2 Chemical Reaction Processes

Chemical reaction processes make use of solar heat to drive reversible chemical reactions. During charging, high-grade heat is converted into chemical energy by driving the endothermic reaction step and the reaction products are stored separately for use during hours of solar unavailability. During discharging the reaction products are recombined in the reverse exothermic reaction, thus releasing high-grade heat. A schematic depiction of the operating principle of a typical chemical-reaction-based TCS system is provided in Figure 10.10.

Chemical reaction systems can be utilized to store solar heat at a very wide range of temperatures as indicated by the large variation in the equilibrium temperature of the most common material pairs shown in Table 10.1 (HSC Chemistry for Windows 1997). Similar to sorption-based TCS systems, chemical reaction systems also suffer from poor heat transfer properties of TCS materials as well as a large temperature swing between the endothermic and exothermic reaction steps, which leads to limited roundtrip efficiencies that typically lie within the range of 20–50% (Siegel 2012). Impregnating thermochemical reaction materials into high-conductivity porous structures represents also in chemical reaction systems the most widely deployed heat transfer enhancement technique.

Numerous material pairs have been investigated for their suitability as TCS media (Cot-Gores, Castell, and Cabeza 2012; Aydin, Casey, and Riffat 2015). A comprehensive list of the most extensively researched chemical compounds for TCS applications is provided in Table 10.1. Among these material pairs, the MgH_2/Mg and $NH_3(g)/N_2(g)/H_2(g)$ systems have the inherent disadvantage of requiring high operating pressures and large storage units, since gaseous products (N_2 and H_2) are formed during the dissociation of MgH_2 and NH_3. In the $SO_3(g)/SO_2(g)/O_2(g)$ system, sulfur trioxide is stored in the liquid state and has to be heated up to vaporization before its thermal dissociation reaction proceeds. A V_2O_5 catalyst is typically used to enhance the otherwise slow reaction kinetics of the dissociation reaction while the relatively high corrosiveness and toxicity of $SO_3(g)$ and $SO_2(g)$ represent further disadvantages of the system. Carbonate-based TCS systems have attracted considerable scientific attention over the past few years because they are simple chemical systems with no by-products and high enthalpy of reaction, whereas none of the forward or reverse reactions need to be catalyzed. Since metal carbonates dissociate into the respective solid-state metal oxide and CO_2, product separation is easy, but the amount of CO_2 released needs to be collected and stored for future

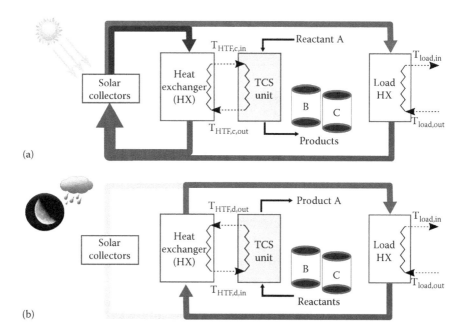

FIGURE 10.10
Operating principle of a chemical-reaction-based TCS system during (a) charging and (b) discharging.

TABLE 10.1

Properties of Common Thermochemical Reaction Material Pairs (HSC Chemistry for Windows 1997)

	Eq. Temperature T_{eq} (°C) (@ $p = 1$ bar)	Enthalpy of Reaction ΔH_{rxn} (kJ × mol^{-1})
Hydride Systems		
$MgH_2(s) \leftrightarrow Mg(s) + H_2(g)$	288	79.8
Hydroxide Systems		
$Mg(OH)_2(s) \leftrightarrow MgO(s) + H_2O(g)$	265	77.7
$Ca(OH)_2(s) \leftrightarrow CaO(s) + H_2O(g)$	518	100.1
$Fe(OH)_2(s) \leftrightarrow FeO(s) + H_2O(g)$	139	61.1
Carbonate Systems		
$MgCO_3(s) \leftrightarrow MgO(s) + CO_2(g)$	304	98.9
$CaCO_3(s) \leftrightarrow CaO(s) + CO_2(g)$	886	165.8
$FeCO_3(s) \leftrightarrow FeO(s) + CO_2(g)$	173	79.9
Redox Systems		
$2\,Co_3O_4(s) \leftrightarrow 6\,CoO(s) + O_2(g)$	936	196.5
$6\,Mn_2O_3(s) \leftrightarrow 4\,Mn_3O_4(s) + O_2(g)$	915	30.1
$Mn_2O_3(s) \leftrightarrow 2\,MnO(s) + 1/2\,O_2(g)$	1,464	169.6
Ammonia Systems		
$NH_3(g) \leftrightarrow {}^1/_2\,N_2(g) + {}^3/_2\,H_2(g)$	183	49
Organic Systems		
$CH_4(g) + CO_2(g) \leftrightarrow 2\,CO(g) + 2\,H_2(g)$	642	259.4
$CH_4(g) + H_2O(g) \leftrightarrow 3\,H_2(g) + CO(g)$	620	223.4
SO$_2$ System		
$SO_3(g) \leftrightarrow SO_2(g) + {}^1/_2\,O_2(g)$	781	97.3

use during hours of solar energy unavailability. Storage of high amounts of CO_2, however, implies potential negative environmental effects, as well as safety risks in case of a CO_2 leakage from the storage unit. Such limitations do not apply to hydroxide and redox systems that make use of H_2O vapor and O_2, respectively, as reactants in the reverse exothermic reaction step, and are therefore preferred over carbonate systems. Additionally, due to the wide availability of the H_2O vapor and O_2, hydroxide and redox systems can be operated in an open-loop cycle, i.e. without storing the gaseous products released during the charging process. The Co_3O_4/ CoO and Mn_2O_3/Mn_3O_4 systems have been identified as the most attractive redox material pairs for large-scale deployment in concentrated solar power plants since endothermic reduction for both material takes place at about 900°C, a temperature that can be achieved with the new volumetric receivers and heat transfer fluids (see Section 10.1.2.3) for solar tower systems (Tescari et al. 2014; Ströhle et al. 2016).

Numerous studies have been conducted to investigate the potential of these material pairs for TCS applications, and reactor prototypes have been built at both laboratory and demonstration scales to gain further insight into the reaction conditions and adjust the reactor design accordingly. A comprehensive review of the thermodynamic, kinetic, process simulation, reactor design, and techno-economic feasibility studies for thermochemical storage systems is provided by Pardo et al. (2014).

Nomenclature

Symbols

c_p	Specific heat capacity	$[kJ \cdot kg^{-1} \cdot K^{-1}]$
E	Energy	$[kJ]$
Δh_f	Specific heat of fusion	$[kJ \cdot kg^{-1}]$
ΔH_m	Molar enthalpy change of desorption or chemical reaction	$[kJ \cdot mol^{-1}]$
k	Thermal conductivity	$[W \cdot m^{-1} \cdot K^{-1}]$
m	Mass	$[kg]$
n	Molar amount	$[mol]$
T	Temperature	$[°C, K]$
V	Volume	$[m^3]$

Greek symbols

α	Chemical conversion	$[mol \cdot mol^{-1}]$
α_m	Mass fraction	$[m \cdot m^{-1}]$
ρ	Density	$[kg \cdot m^{-3}]$

Subscripts

amb	Ambient
c	Charging
c	Cold
eff	Effective
eq	Chemical equilibrium
h	Hot
in	Inlet
l	Liquid
mp	Melting point
rxn	Chemical reaction
s	Solid

Acronyms

ATES	Aquifer thermal energy storage
BTES	Borehole thermal energy storage
COP	Coefficient of performance

CSP	Concentrated solar power
HTF	Heat transfer fluid
LTES	Latent thermal energy storage
PCM	Phase change material
STES	Sensible thermal energy storage
TCS	Thermochemical storage
TES	Thermal energy storage
TSM	Thermal storage medium

References

Abdulateef, A. M., S. Mat, J. Abdulateef, K. Sopian, and A. A. Al-Abidi. 2018. "Geometric and Design Parameters of Fins Employed for Enhancing Thermal Energy Storage Systems: A Review." *Renewable and Sustainable Energy Reviews* 82: 1620–1635.

Aydin, D., S. P. Casey, and S. Riffat. 2015. "The Latest Advancements on Thermochemical Heat Storage Systems." *Renewable and Sustainable Energy Reviews* 41: 356–367.

Bradshaw, R. W., J. G. Cordaro, and N. P. Siegel. 2009. "Molten Nitrate Salt Development for Thermal Energy Storage in Parabolic Trough Solar Power Systems." In *ASME 2009 3rd International Conference on Energy Sustainability Collocated with the Heat Transfer and InterPACK09 Conferences*, 615–624. American Society of Mechanical Engineers.

Cabeza, L. F., J. Illa, J. Roca, F. Badia, H. Mehling, S. Hiebler, and F. Ziegler. 2001. "Middle Term Immersion Corrosion Tests on Metal-Salt Hydrate Pairs Used for Latent Heat Storage in the 32 to 36°C Temperature Range." *Materials and Corrosion* 52 (10): 748–754.

Cabeza, Luisa, Andreas Heinz, and Wolfgang Streicher. 2005. *Inventory of Phase Change Materials (PCM)*. Report C2 –IEA Solar Heating and Cooling Task 32. International Energy Association.

Castell, A., C. Solé, M. Medrano, J. Roca, L. F. Cabeza, and D. García. 2008. "Natural Convection Heat Transfer Coefficients in Phase Change Material (PCM) Modules with External Vertical Fins." *Applied Thermal Engineering* 28 (13): 1676–1686.

Cot-Gores, J., A. Castell, and L. F. Cabeza. 2012. "Thermochemical Energy Storage and Conversion: A-State-of-the-Art Review of the Experimental Research under Practical Conditions." *Renewable and Sustainable Energy Reviews* 16 (7): 5207–5224.

Elgafy, A., and K. Lafdi. 2005. "Effect of Carbon Nanofiber Additives on Thermal Behavior of Phase Change Materials." *Carbon* 43 (15): 3067–3074.

Erek, A., Z. İlken, and M. A. Acar. 2005. "Experimental and Numerical Investigation of Thermal Energy Storage with a Finned Tube." *International Journal of Energy Research* 29 (4): 283–301.

Fadhel, M. I., K. Sopian, and W. R. W. Daud. 2010. "Performance Analysis of Solar-Assisted Chemical Heat-Pump Dryer." *Solar Energy* 84 (11): 1920–1928.

Farid, M., and K. Yacoub. 1989. "Performance of Direct Contact Latent Heat Storage Unit." *Solar Energy* 43 (4): 237–251.

Fernandez, A. I., M. Martínez, M. Segarra, I. Martorell, and L. F. Cabeza. 2010. "Selection of Materials with Potential in Sensible Thermal Energy Storage." *Solar Energy Materials and Solar Cells* 94 (10): 1723–1729.

Fouda, A. E., G. J. G. Despault, J. B Taylor, and C. E. Capes. 1984. "Solar Storage Systems Using Salt Hydrate Latent Heat and Direct Contact Heat Exchange–II. Characteristics of Pilot System Operating with Sodium Sulphate Solution." *Solar Energy* 32 (1): 57–65.

Fukai, J., M. Kanou, Y. Kodama, and O. Miyatake. 2000. "Thermal conductivity enhancement of energy storage media using carbon fibers." *Energy Conversion and Management* 41: 1543–1556.

Ghaebi, H., M. N. Bahadori, and M. H. Saidi. 2014. "Performance Analysis and Parametric Study of Thermal Energy Storage in an Aquifer Coupled with a Heat Pump and Solar Collectors, for a Residential Complex in Tehran, Iran." *Applied Thermal Engineering* 62 (1): 156–170.

Gil, A., E. Oró, L. Miró, G. Peiró, Á. Ruiz, J. M. Salmerón, and L. F. Cabeza. 2014. "Experimental Analysis of Hydroquinone Used as Phase Change Material (PCM) to Be Applied in Solar Cooling Refrigeration." *International Journal of Refrigeration* 39: 95–103.

Han, Y. M., R. Z. Wang, and Y. J. Dai. 2009. "Thermal Stratification within the Water Tank." *Renewable and Sustainable Energy Reviews* 13 (5): 1014–1026.

Hasnain, S. M. 1998. "Review on Sustainable Thermal Energy Storage Technologies, Part I: Heat Storage Materials and Techniques." *Energy Conversion and Management* 39 (11): 1127–1138.

Heine, D. 1981. "The Chemical Compatibility of Construction Materials with Latent Heat Storage Materials." In *International Conference on Energy Storage*. Brighton, UK.

HSC Chemistry for Windows. 1997. Pori, Finland: RA Outokumpu.

Jamekhorshid, A., S. M. Sadrameli, and M. Farid. 2014. "A Review of Microencapsulation Methods of Phase Change Materials (PCMs) as a Thermal Energy Storage (TES) Medium." *Renewable and Sustainable Energy Reviews* 31: 531–542.

Kang, B. H., C. W. Park, and C. S. Lee. 1996. "Dynamic Behavior of Heat and Hydrogen Transfer in a Metal Hydride Cooling System." *International Journal of Hydrogen Energy* 21 (9): 769–774.

Kang, Yong Tae, Weibo Chen, and Richard N. Christensen. 1997. *A Generalized Component Design Model by Combined Heat and Mass Transfer Analysis in NH_3-H_2O Absorption Heat Pump Systems.* American Society of Heating, Refrigerating and Air-Conditioning Engineers, Inc., Atlanta, GA (United States).

Klein, H.-P., and M. Groll. 2002. "Development of a Two-Stage Metal Hydride System as Topping Cycle in Cascading Sorption Systems for Cold Generation." *Applied Thermal Engineering* 22 (6): 631–639.

Lahmidi, H., S. Mauran, and V. Goetz. 2006. "Definition, Test and Simulation of a Thermochemical Storage Process Adapted to Solar Thermal Systems." *Solar Energy* 80 (7): 883–893.

Lane, G. A. 1986. *Solar Heat Storage: Latent Heat Materials.* Vol. II: Technology. CRC.

Lu, Z. S., R. Z. Wang, L. W. Wang, and C. J. Chen. 2006. "Performance Analysis of an Adsorption Refrigerator Using Activated Carbon in a Compound Adsorbent." *Carbon* 44 (4): 747–752.

Marcriss, R. A., J. M. Gutraj, and T. S. Zawacki. 1988. *Absorption Fluid Data Survey: Final Report on Worldwide Data, ORLN/sub/8447989/3, Inst.* Inst. Gas. Tech.

Mesalhy, O., K. Lafdi, A. Elgafy, and K. Bowman. 2005. "Numerical Study for Enhancing the Thermal Conductivity of Phase Change Material (PCM) Storage Using High Thermal Conductivity Porous Matrix." *Energy Conversion and Management* 46 (6): 847–867.

Meunier, F. 1986. "Theoretical Performances of Solid Adsorbent Cascading Cycles Using the Zeolite-Water and Active Carbon-Methanol Pairs: Four Case Studies." *Journal of Heat Recovery Systems* 6 (6): 491–498.

Nakaso K, H. Teshima, A. Yoshimura, S. Nogami, S. Hamada, and J. Fukai. 2008. "Extension of heat transfer area using carbon fiber cloths in latent heat thermal energy storage tanks." *Chem Engineering and Processing: Process Intensification* 47: 879–885.

N'Tsoukpoe, K. E., N. Le Pierrès, and L. Luo. 2013. "Experimentation of a LiBr–H_2O Absorption Process for Long-Term Solar Thermal Storage: Prototype Design and First Results." *Energy* 53: 179–198.

Pardo, P., A. Deydier, Z. Anxionnaz-Minvielle, S. Rougé, M. Cabassud, and P. Cognet. 2014. "A Review on High Temperature Thermochemical Heat Energy Storage." *Renewable and Sustainable Energy Reviews* 32: 591–610.

Ryu, H. W., S. W. Woo, B. C. Shin, and S. D. Kim. 1992. "Prevention of Supercooling and Stabilization of Inorganic Salt Hydrates as Latent Heat Storage Materials." *Solar Energy Materials and Solar Cells* 27 (2): 161–172.

Saha, B. B., A. Akisawa, and T. Kashiwagi. 2001. "Solar/Waste Heat Driven Two-Stage Adsorption Chiller: The Prototype." *Renewable Energy* 23 (1): 93–101.

Saha, B. B., E. C. Boelman, and T. Kashiwagi. 1995. "Computational Analysis of an Advanced Adsorption-Refrigeration Cycle." *Energy* 20 (10): 983–994.

Saha, B. B., S. Koyama, T. Kashiwagi, A. Akisawa, K. C. Ng, and H. T. Chua. 2003. "Waste Heat Driven Dual-Mode, Multi-Stage, Multi-Bed Regenerative Adsorption System." *International Journal of Refrigeration* 26 (7): 749–757.

Sarı, A., and A. Karaipekli. 2007. "Thermal Conductivity and Latent Heat Thermal Energy Storage Characteristics of Paraffin/Expanded Graphite Composite as Phase Change Material." *Applied Thermal Engineering* 27 (8): 1271–1277.

Sarı, A., and A. Karaipekli. 2009. "Preparation, Thermal Properties and Thermal Reliability of Palmitic Acid/Expanded Graphite Composite as Form-Stable PCM for Thermal Energy Storage." *Solar Energy Materials and Solar Cells* 93 (5): 571–576.

Sharma, A., V. V. Tyagi, C. R. Chen, and D. Buddhi. 2009. "Review on Thermal Energy Storage with Phase Change Materials and Applications." *Renewable and Sustainable Energy Reviews* 13 (2): 318–345.

Siegel, N. P. 2012. "Thermal Energy Storage for Solar Power Production." *Wiley Interdisciplinary Reviews: Energy and Environment* 1 (2): 119–131.

Srikhirin, P., S. Aphornratana, and S. Chungpaibulpatana. 2001. "A Review of Absorption Refrigeration Technologies." *Renewable and Sustainable Energy Reviews* 5 (4): 343–372.

Stitou, D., N. Mazet, and M. Bonnissel. 2004. "Performance of a High Temperature Hydrate Solid/gas Sorption Heat Pump Used as Topping Cycle for Cascaded Sorption Chillers." *Energy* 29 (2): 267–285.

Ströhle, S., A. Haselbacher, Z. R. Jovanovic, and A. Steinfeld. 2016. "The Effect of the Gas–Solid Contacting Pattern in a High-Temperature Thermochemical Energy Storage on the Performance of a Concentrated Solar Power Plant." *Energy & Environmental Science* 9 (4): 1375–1389.

Tan, T., and Y. Chen. 2010. "Review of Study on Solid Particle Solar Receivers." *Renewable and Sustainable Energy Reviews* 14 (1): 265–276.

Telkes, M. 1952. "Nucleation of Supersaturated Inorganic Salt Solutions." *Industrial & Engineering Chemistry* 44 (6): 1308–1310.

Tescari, S., C. Agrafiotis, S. Breuer, L. de Oliveira, M. Neises-von Puttkamer, M. Roeb, and C. Sattler. 2014. "Thermochemical Solar Energy Storage via Redox Oxides: Materials and Reactor/heat Exchanger Concepts." *Energy Procedia* 49: 1034–1043.

Tian, Y., and C.-Y. Zhao. 2013. "A Review of Solar Collectors and Thermal Energy Storage in Solar Thermal Applications." *Applied Energy* 104: 538–553.

Wang, R. Z., and R. G. Oliveira. 2006. "Adsorption Refrigeration—An Efficient Way to Make Good Use of Waste Heat and Solar Energy." *Progress in Energy and Combustion Science* 32 (4): 424–458.

Yu, N., R. Z. Wang, and L. W. Wang. 2013. "Sorption Thermal Storage for Solar Energy." *Progress in Energy and Combustion Science* 39 (5): 489–514.

Zalba, B., J. M. Marín, L. F. Cabeza, and H. Mehling. 2003. "Review on Thermal Energy Storage with Phase Change: Materials, Heat Transfer Analysis and Applications." *Applied Thermal Engineering* 23 (3): 251–283.

Zanganeh, G., M. Commerford, A. Haselbacher, A. Pedretti, and A. Steinfeld. 2014. "Stabilization of the Outflow Temperature of a Packed-Bed Thermal Energy Storage by Combining Rocks with Phase Change Materials." *Applied Thermal Engineering* 70 (1): 316–320.

Zanganeh G., R. Khanna, C. Walser, A. Pedretti, A. Haselbacher and A. Steinfeld, 2015. "Experimental and numerical investigation of combined sensible-latent heat for thermal energy storage at 575°C and above." *Solar Energy* 114: 77–90.

Zhang, Y., and A. Faghri. 1996. "Heat Transfer Enhancement in Latent Heat Thermal Energy Storage System by Using the Internally Finned Tube." *International Journal of Heat and Mass Transfer* 39 (15): 3165–3173.

Zhao, C.-Y., and Z. G. Wu. 2011. "Thermal Property Characterization of a Low Melting-Temperature Ternary Nitrate Salt Mixture for Thermal Energy Storage Systems." *Solar Energy Materials and Solar Cells* 95 (12): 3341–3346.

Zhou, D., and C.-Y. Zhao. 2011. "Experimental Investigations on Heat Transfer in Phase Change Materials (PCMs) Embedded in Porous Materials." *Applied Thermal Engineering* 31 (5): 970–977.

11

Economic Evaluation of Solar Cooling Technologies

11.1 Introduction

Chapter 11 focuses on the economic aspects of different solar cooling technologies. More specifically, the first section gives a general outline of different possible solar cooling options, along with a brief presentation of their main technical and practical advantages and disadvantages.

The second section includes a literature review encompassing a large number of studies on the economic assessment of solar cooling systems. The purpose of the review is to inform the reader about different approaches followed by various researchers for investigating the economic competitiveness of solar cooling concepts. Furthermore, it aims to highlight the main qualitative and quantitative results that have been derived from these studies. Although these results are not, in most cases, directly comparable to one another due to the highly diverse assumptions and methodologies applied in each study, they can give valuable insights into the key issues associated with the economic performance of solar cooling systems.

The following section of the chapter compiles different equipment cost data regarding solar collectors and chillers as reported in the literature. This compilation aims to provide information regarding the costs of the most important equipment components that need to be combined for the deployment of integrated solar cooling solutions. These include solar thermal collectors, photovoltaic panels, and different types of thermally driven chillers (absorption, adsorption, and desiccant). It should be noted that although additional components are required for the construction of solar cooling installations (i.e. storage tanks, cooling towers, pumps, control systems), these are considered to be the most crucial and influential on investment costs.

The last section investigates the economic feasibility of sorption solar cooling systems via a case study. The goal of the study is to provide insights into the competitiveness of these systems by considering their annual operation in different building types (single houses, apartment blocks, offices) located in two countries (Italy and Germany), taking into account the solar radiation potential and the cooling loads in each case. Furthermore, a series of parametric investigations are performed to determine the influence of key system parameters (area of solar collectors, volume of thermal storage tank, nominal solar collector temperature) on the economic performance of the systems.

11.2 Overview of Solar Cooling Technologies

There are two principle energy conversion pathways for the production of cooling from solar radiation. The first one involves the conversion of solar power to electricity via the use of PV panels and the subsequent utilization of the generated electricity for powering a mechanically driven chiller. Solar electric cooling systems include vapor compression cycle engines, thermoelectric devices, Stirling engines, and, finally, thermoacoustic and magnetic chillers. The second pathway is based on the conversion of the solar radiation to heat, which is subsequently used to power the chiller. Solar thermal cooling systems include sorption (absorption and adsorption) chillers, desiccant cooling systems, and ejector cooling cycle engines. The main advantages and disadvantages of solar electric and solar thermal cooling technologies are listed in Table 11.1 and Table 11.2, respectively.

An overview of the energy conversion pathways employed by different solar cooling technologies, along with the corresponding efficiencies and capital costs, was provided by Kim (2007) and is reproduced in Figure 11.1. It should be noted that the efficiencies and capital costs presented in this figure are typical only for the smallest machines available, which can be comparable to one another. Existing chillers based on these technologies can, however, exhibit a wide variation of cooling capacities, from a few tens to several megawatts.

By observing the efficiency and cost values presented in Figure 11.1, and by taking into account the fact that a conventional electrically powered vapor compression cycle has a very low specific investment cost of around €200–300/kW_{th} and a COP of at least 3, it can be seen that solar cooling systems are in principle much more expensive than conventional systems, with their costs being higher as a result of multiple factors. As can also be observed, the cost of solar conversion equipment (PV panels and collectors) constitutes a substantial component of solar cooling systems. As a result, the economic competitiveness of solar cooling applications is even worse when there is no pre-existing solar system, since the cost of

TABLE 11.1

Advantages and Disadvantages of Solar Electric Cooling with PV Panels

Chiller Type	Advantages	Disadvantages
Vapor compression cycle	High reliability Simpler capacity control mechanism Easier implementation Technological maturity	Necessary to include electricity storage (battery) to cope with varying electricity production rate for autonomous operation
Thermoelectric cooling	No moving parts No refrigerant Can be made small and portable	Very low COP (0.3 to 0.6)
Stirling engine	In theory, efficiency equal to that of Carnot cycle Competitive for small sizes with relatively large surface-to-volume ratio	Low COP Limited power density due to poor heat transfer between working fluids and the environment
Thermoacoustic	Simple construction with no moving parts High reliability	Efficiency lower than vapor compression systems Low power density So far, very small capacities
Magnetic cooling	High COP (comparable and even higher than that of conventional systems)	Cost of magnetic material is too high

Source: Kim, D.S., "Solar Absorption Cooling", PhD diss., Mechanical Maritime and Materials Engineering (TU Delft), 2007.

TABLE 11.2

Advantages and Disadvantages of Solar Thermal Cooling Technologies

Chiller Type	Advantages	Disadvantages
Absorption system	Operate silently High reliability No auxiliary energy for operation of the small system Simpler capacity control mechanism Easier implementation Low-temperature heat supply	High installation cost and large installation area in case of continuous system Quite complicated system and requires advanced knowledge for maintenance High heat release to the ambient.
Adsorption system	Low maintenance costs No moving parts Low heat source temperatures	Poor thermal conductivity of the adsorbent Very sensitive to low temperature during nighttime Low COP Intermittent in basic systems Bulky machine
Desiccant system	Uses water as a working fluid, which is environmentally safe Can be integrated with a ventilation and heating system Low heat release to the ambient (in the case of liquid desiccant systems)	Difficult design for small applications and complex control strategy, especially in humid areas Crystallization risk in liquid desiccant systems Requires dehumidifier Rotating elements (desiccant wheel, sensible heat regenerators) need maintenance
Ejector system	Low operating temperature heat can be used Low operating costs	Low COP Complex design of the ejector Specific ambient temperature ranges are required

Source: Allouhi, Amine et al., *Renewable and Sustainable Energy Reviews*, 50 (Supplement C): 770–781, doi: https://doi.org/10.1016/j.rser.2015.05.044, 2015.

installing new collectors or panels greatly overwhelms capital investment costs. This is also showcased in Table 11.3, in which the cost distribution among the different components of small- and large-scale cooling systems is presented, as reported by Allouhi et al. (2015).

In a presentation by Daniel Mugnier of TECSOL, the economic feasibility of solar thermal cooling systems was discussed, among other issues (Mugnier 2012). Regarding the cost reduction potential of solar cooling kits, the figures summarized in Table 11.4 were reported for the different cost components.

It can be seen that the highest cost reduction potential concerns the sorption chillers themselves, as well as the recooler and control systems.

Despite higher capital costs, the main advantage of solar cooling rests in the fact that, being based on the utilization of freely available solar energy, it can lead to primary energy savings. Meanwhile, apart from the environmental benefits related to the reduction of CO_2 and pollutant emissions due to the avoidance of fossil fuel consumption, solar cooling can also help level off peak electricity demand on very hot summer days, thus enhancing grid stability. From an economic competitiveness standpoint, the actual reduction of operating fuel costs that can be achieved by substituting conventional cooling engines with solar cooling systems depends on many factors, such as energy policy legislation that determines renewable energy investment subsidies and electricity pricing schemes, the price of energy resources such as natural gas, oil, and electricity, and, last but not least, the cost of the technology itself. Other factors that affect the economics of solar cooling systems include the

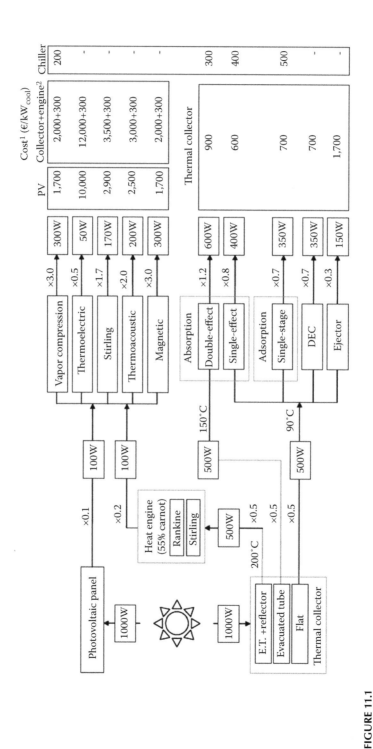

FIGURE 11.1
Overview of the efficiencies and capital costs of different solar cooling schemes. (Reproduced from Kim, D.S., "Solar Absorption Cooling," PhD diss., Mechanical Maritime and Materials Engineering (TU Delft), 2007.)

TABLE 11.3

Component Cost Distribution of Solar Cooling Systems

	Small Scale	Large Scale
Collectors	35%	35%
Chiller	15%	27%
Control	10%	6%
Auxiliary equipment	35%	20%
Other	5%	12%

Source: Allouhi, Amine et al., *Renewable and Sustainable Energy Reviews*, 50 (Supplement C): 770–781, doi: https://doi.org/10.1016/j.rser.2015.05.044, 2015.

TABLE 11.4

Cost Reduction Potential of Solar Cooling Kits as of 2012

Component	Cost Reduction Potential
Solar collectors and storage	Maximum 10% reduction
Small-scale sorption chillers	Up to 50% cost reduction potential for serial mass-produced kits (higher than 500 units)
Recooler	Cost reduction potential between 40–50%
Control	A minimum 60% cost reduction potential combined with an increase in system performance
Installation	10–30% cost reduction potential through standardized solar cooling kits

Source: Mugnier, Daniel, "Solar Thermal Energy For Cooling and Refrigeration: Status and Perspectives", Solar Cooling International Conference, Intersolar Fair, Munchen, 2012.

magnitude of cooling loads, solar radiation intensity, the hybridization of the technology, and also the integration of existing equipment (such as solar collectors and auxiliary heat sources and heating and cooling units). In practice, the high capital costs that are associated with the high cost of solar cooling cannot be compensated by the profits from energy savings. As a result, so far, solar cooling is hardly attractive if financial support schemes and incentives are not provided for its implementation. In this respect, political and financial support from the government plays an important role in its promotion (Kim 2007).

Of course, the economics of solar cooling systems can be potentially improved if the chillers are coupled to pre-existing solar infrastructure. This could be the case mainly for low driving heat temperature solar thermal technologies, considering the wide proliferation of flat-plate and evacuated tube collectors. On the other hand, PV panels are comparatively less diffused, and thus the economic performance and market penetration of solar electric cooling applications could be restrained by this fact.

Indeed, solar collectors are one of the most critical components of solar cooling systems, because one of the main differences among solar thermal cooling systems concerns the required temperature of the driving heat. An overview of the different sorption technologies, the driving heat temperatures, the COP values, and the suitable collector types is given in Table 11.5. For most of the technologies, the driving heat temperature is lower than 100°C, and thus flat-plate collectors can be used. Nevertheless, it can be seen that for certain applications, temperatures exceeding 100°C are required. In this case, evacuated tube or even parabolic trough collectors are necessary. The selection of a solar collector type is also dependent on the consideration of thermal energy storage systems, which typically operate at different temperatures. Meanwhile, the selection of different collector

TABLE 11.5

Marketed Solar Sorption Technologies for Cooling

Type of System	Water Chillers (Closed Thermodynamic Cycles)				Direct Air Treatment (Open Thermodynamic Cycles)			
Physical Phase of Sorption Material	Liquid		Solid		Liquid		Solid	
Sorption material and refrigerant	Water Ammonia	Lithium-bromide Water	Zeolite Water	Silica gel Water	Lithium-chloride Water	Lithium-chloride Water	Lithium-chloride Water	Silica gel (or zeolite) Water
Type of cycle	1-effect	1-effect	2-effect	1-effect	1-effect	1-effect	Cooled sorption process	Desiccant rotor
COP thermal range	0.5–0.75	0.65–0.8	1.1–1.4	0.5–0.75	0.5–0.75	0.5–0.75	0.7–1.1	0.6–0.8
Driving temperature range (1C)	70–100 120–180	70–100	140–180	65–90	65–90	65–90	60–85	60–80
Solar collector technology	FPC, ETC, SAT	FPC, ETC	SAT	FPC, ETC	FPC, ETC	FPC, ETC	FPC, ETC, SAHC	FPC, ETC, SAHC

Source: Allouhi, Amine et al., *Renewable and Sustainable Energy Reviews*, 50 (Supplement C): 770–781, doi: https://doi.org/10.1016/j.rser.2015.05.044, 2015.
Abbreviations: FPC: flat-plate collector; ETC: evacuated tube collector; SAT: single-axis tracking collector; SAHC: solar air heating collector.

types is also subject to techno-economic criteria, because there exists a trade-off between the purchase costs and the performance of different types of collectors.

Based on the previous discussion, and as can also be observed in Table 11.6, the economic benefits that can be gained when installing thermally driven chillers and coupling them with existing solar collectors are highly dependent on the market distribution of the different collector types. Interestingly, in all the regions of the world, flat-plate collectors significantly outnumber evacuated tube collectors. However, the situation is completely reversed in China, where evacuated tube collectors are by far the most popular type. This is a fact that should be taken into account when evaluating the economic potential of retrofitting solar chillers to existing applications.

In any case, the number of solar cooling installations is currently increasing at a high rate each year, mostly in Europe, as can be observed in Figure 11.2, indicating the interest of the market in solar cooling.

More specifically, according to the IEA (Mauthner, Weiss, and Spörk-Dür 2015), by the end of 2014, an estimated 1,175 solar cooling systems were installed worldwide. The market showed a positive trend between 2004 and 2014, but the growth rates decreased from 32% in 2007–2008 to 12% in 2013–2014. Approximately three-quarters of solar cooling installations worldwide are installed in Europe, most notably in Spain, Germany, and Italy. The

TABLE 11.6

Distribution of Solar Collectors by Category as of 2015

Collector Type	USA	Australia	Sub-Sahara Africa	Latin America	Europe	Asia w/o China	China
Flat unglazed collector	88%	59%	52%	33%	4%	0%	0%
Flat glazed collector	11%	39%	37%	61%	83%	92%	8%
Evacuated tube	1%	2%	11%	6%	13%	8%	92%

Source: Mauthner, F., W. Weiss, and M. Spörk-Dür, *Solar Heat Worldwide: Markets and Contribution to the Energy Supply 2013*, 2015 edition, AEE-Institute for Sustainable Technologies A8200 Gleisdorf, Austria, 2015.

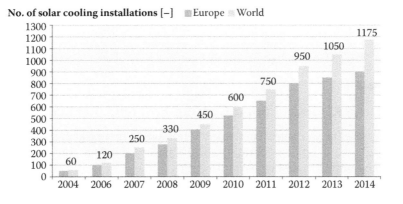

FIGURE 11.2

The market development of solar air conditioning and cooling systems from 2004–2014 in Europe and worldwide as reported by the IEA. Data compiled from Climasol, EURAC, Fraunhofer ISE, Green Chiller, Rococo, Solem Consulting, and Tecsol. (From Mauthner, F., W. Weiss, and M. Spörk-Dür, *Solar Heat Worldwide: Markets and Contribution to the Energy Supply 2013*, 2015 edition, AEE-Institute for Sustainable Technologies A8200 Gleisdorf, Austria, 2015.)

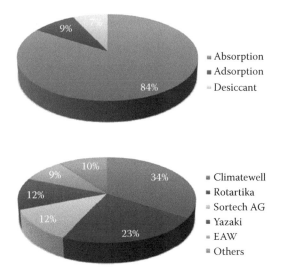

FIGURE 11.3
Technology and market share of solar cooling systems internationally as of 2015. (Reproduced from Allouhi, Amine et al., *Renewable and Sustainable Energy Reviews*, 50 (Supplement C): 770–781, doi: https://doi.org/10.1016/j .rser.2015.05.044, 2015.)

majority of solar air conditioning and cooling applications installed are equipped with high performance flat-plate or evacuated tube collectors. By contrast, some examples for thermal cooling machines driven by concentrating solar thermal collectors (e.g. parabolic troughs or Fresnel collectors) were reported in India, Australia, and Turkey.

In Figure 11.3, the market share of solar cooling systems as of 2015 on an international level is presented, along with the main manufacturers (Allouhi, Kousksou, Jamil, Bruel et al. 2015).

Absorption technology takes up the great majority of market shares (about 82%), followed by adsorption systems (11%). Absorption machines, which are marketed by Climatewell and Rotartica, are adopted, respectively, in 34% and 23% of cases. Sortech AG and Invensor are the only manufacturers of adsorption chillers present at the small scale for domestic applications (Allouhi, Kousksou, Jamil, El Rhafiki et al. 2015).

11.3 Literature Review of Solar Cooling Economic Evaluation Studies

This section presents an extensive overview of studies focused on investigating the economic performance of solar cooling applications. In each case, the studied systems are reported, along with the most important results.

Allouhi et al. (2015) performed an economic and environmental comparison between a conventional vapor compression cooling system and a solar absorption and a solar adsorption chiller in Morocco. The authors took into account the costs of multiple equipment components involved in the implementation of each system, including the solar collectors, the storage tank, the cooling tower, and the cost of the chillers. The payback period of the absorption and the adsorption system for the base case scenario was estimated to be equal to 24 and 25 years, respectively. These high values are attributed by the authors to the high investment costs of the solar cooling technologies, with the cost of the solar field being the most significant cost component, followed by the absorption and adsorption modules. Furthermore, the authors stressed the importance of increasing the performance of solar air conditioning systems. The COP of the chillers ranged from 0.65 to 1.4 and 0.5 to 0.75 for the LiBr-H_2O and the silica gel-water systems, respectively, being substantially lower than the respective values of vapor compression systems. The authors also reported that a major barrier concerning the market growth of solar cooling systems is the fact that solar systems are generally complex and require additional features such as storage tanks, pumps, and auxiliary components. They are also associated with surface limitations that impose constraints on their capacity.

Lazzarin (2014) conducted a techno-economic comparison between solar PV and solar thermal absorption and adsorption cooling schemes. The PV modules were used to power a conventional vapor compression chiller, while the thermal sorption systems included single- and double-effect absorption chillers and an adsorption chiller coupled with flat-plate, evacuated tube, and parabolic trough collectors. The authors simulated the systems for daily operation and found that although the specific cost of PV is much higher than solar thermal, the investment costs are very similar for PV solar cooling systems and FPC-, ETC-, and PTC-driven absorption, for a daily production of 10 kWh of cooling. The competitiveness of solar PV cooling technologies is due to the decreasing cost of PVs and the improvement in the efficiency of the technology. Furthermore, solar thermal chillers also require the installation of storage tanks, while their COP may decrease when they operate under ON-OFF cycles. Similar to other studies, the authors reported that the most important cost element of solar cooling systems is the cost of the solar circuit. Of course, a major advantage for solar thermal versus PV cooling systems is the ability of the former to produce useful heat in the months when cooling is not required.

Al-Ugla et al. (2016) carried out a techno-economic analysis of solar cooling systems for a commercial building in Saudi Arabia. The authors investigated the economic competitiveness between a conventional vapor compression system, a solar LiBr-H_2O system, and a PV-vapor compression system. They concluded that the solar absorption system was more economically viable than the solar PV-vapor compression system, also noting an effect of the economy of scale on the competitiveness of the systems. The PBP of the solar absorption and the PV system were estimated to equal 111.5 and 23.9 years, respectively, compared to a conventional system.

Lambert and Beyene (2007) performed a thermo-economic analysis of a solar-powered adsorption heat pump over a number of different locations in the United States. The adsorption heat pump consisted of evacuated flat-panel solar thermal collectors and an adsorption chiller. This system was compared to a solar electric cooling system consisting of PV arrays with battery storage coupled with an electrically powered vapor compression chiller and a conventional vapor compression chiller powered by an electrical grid. The authors concluded that the solar-driven cooling systems were profitable only when there are moderate or high electricity prices. The authors also reported that, based on their findings, the best location for solar thermal cooling systems is Hawaii.

Gabbrielli et al. (2016) carried out a performance and economic comparison of three different solar LiBr absorption cooling systems. Two different locations were considered, Brindisi and Gela (Italy). The first system included evacuated tube solar collectors coupled with a single-effect chiller, while the second and third systems consisted of compact linear concentrating Fresnel collectors coupled with a single and a medium temperature double-effect chiller, respectively. The single-effect chiller operated at a driving temperature of 88°C, while the single-effect chiller operated at 170°C. The nominal cooling capacities of the three systems were in the range of 109–308 kW$_{th}$. The systems were evaluated on an annual operation basis considering combined heating and cooling generation. The selection of Fresnel type collectors was justified by the fact that they are the cheapest solar thermal concentrating collectors, and they also exhibit the lowest land occupancy and easiness of mirror cleaning. The authors conducted sensitivity analyses to investigate the impact of the specific cost of the solar field on the levelized cost of cooling (LCOC) production. They estimated that because the specific solar field cost varied from €125–300/m², the LCOC approximately varied in the range of €0.030–0.080/kW$_c$ in Brindisi, and from €0.020–0.065/kWh$_c$ in Gela. The lowest costs were achieved in the case of the double-effect absorption chiller. The authors concluded that further technological improvements are necessary in order to reduce the solar field cost to the range of €150–200/m². Meanwhile, they stressed that it is important to pursue further innovations in the performance of mirrors, receiver technology (as far as materials, coatings, and production techniques are concerned), and the production of collector fields with different materials in different field segments to improve the economics of solar cooling.

Tsoutsos et al. (2003) carried out a techno-economic comparison between a solar absorption and an adsorption system in 2003. Using contemporary economic data, they concluded that the payback period exceeded the lifetime of the solar cooling systems, and highlighted the necessity of providing subsidies for instigating market penetration.

Ferreira and Kim (2014) published a techno-economic review of solar cooling technologies based on location-specific data. In their study, they considered solar electric (PV) cooling as well as solar thermal technologies, including absorption, adsorption, and desiccant cooling. The authors used a variety of data to determine the costs of different solar collector types (flat plate, evacuated tube, parabolic trough, compound parabolic dish) and their efficiencies. Meanwhile, they took into account the specific operational and performance characteristics (temperatures, COP) for different types of chillers and evaluated their economic competitiveness with a case study. The authors highlighted the barrier of high investment costs that inhibit the market expansion of small capacity (<20 kW) solar air conditioning systems. They concluded that for regions in Central Europe, vapor compression cycles powered by PV panels are the best option, due to the recent extreme reductions in the investment costs of PV technology. The systems are followed by vapor compression cycles driven by electricity delivered by parabolic dish collectors coupled with Stirling engines. Among thermally driven systems, the best performance was exhibited by double-effect absorption technology equipped with concentrating trough collectors closely followed by desiccant systems equipped with flat-plate solar collectors. Adsorption applications were shown to be the most expensive. As far as Mediterranean regions are concerned, the authors concluded that, again, the PV-VCC systems are the best option, followed by parabolic dish collectors coupled with Stirling engines. However, the authors remarked that double-effect LiBr-H$_2$O chillers cannot operate in these climate conditions, due to the crystallization line limit. As a result, single-effect absorption chillers should be used, which are followed in competitiveness by desiccant systems coupled with flat-plate collectors.

Noro and Lazzarin (2014) also compared the economic profitability of solar thermal and photovoltaic systems for Mediterranean conditions. Similar to other studies, parabolic trough, evacuated tube and flat-plate collectors were considered. As far as the chillers are concerned, a single- and double-effect LiBr absorption chiller, a silica gel adsorption chiller (water tower cooled), and a GAX ammonia-water absorption chiller were considered. Furthermore, mono-crystalline and amorphous PV modules coupled to air- and water-cooled vapor compression heat pumps were investigated. The analysis was carried out considering a typical office building. In accordance with other studies, the authors found that PV panels with silicon cells performed better than solar thermal technologies. However, parabolic trough collectors coupled with double-effect LiBr absorption chillers were able to exhibit comparable performance. The authors estimated that under the base case assumptions, the PV cooling systems exhibited payback periods ranging from 4.8 years (PV aSi with water-cooled VCC) to 14.8 years (PV mSi with air-cooled VCC). On the other hand, the authors found that when a solar fraction of 70% is considered, the solar thermal cooling systems had a very low performance compared to the standard VCC cycle, and thus the annual savings were negative. As a result, investment did not have a payback period. Similar to other researchers, the authors of this study also noted that a major factor that results in the superior economic performance of PV solutions is the fast reduction of the cost of the technology in recent years. Nevertheless, the authors also mentioned that a complete analysis considering both cooling and heating could reverse the results of economic profitability.

Otanicar et al. (2012) investigated the prospects of solar cooling by carrying out an economic and environmental assessment of different technologies. More specifically, they compared solar PV systems with solar thermal systems, including desiccant, absorption, and adsorption chillers. The authors concluded that for the PV prices of 2012, the cost of solar electric cooling systems was highly tied to the COP. However, for lower prices, the impact of the COP becomes less substantial. On the other hand, although solar thermal systems have lower costs regarding the solar system, chillers have higher costs. Moreover, the authors pointed out that the costs of solar thermal cooling systems are not projected to drop as much as the cost of PV cooling applications over the next 20 years due to the relative stability exhibited by the cost of solar collectors and heat storage equipment. In fact, according to the authors, solar thermal cooling systems could compete with solar electric cooling systems only if equipment costs decrease and the COP of thermal refrigeration increases to surpass the value of 1. As far as the environmental aspects are considered, the study indicated that even considering the associated impact of global warming caused by refrigerants, solar electric systems have lower projected emission values compared to thermal technologies, mostly because of the significantly higher COP of vapor compression cycles and the comparatively increased collector area footprint of thermal systems.

Mokhtar et al. (2010) also focused on the techno-economic comparison of different solar cooling technologies. The authors focused on single-, double-, and triple-stage absorption chillers coupled with solar thermal collectors (ETC, FPC, Fresnel, and parabolic), vapor compression machines powered by PV modules, and large-scale parabolic trough and solar towers coupled with VCC cycles. A total of 25 different solar cooling technologies were considered based on the climatic conditions and cooling demand time series of Abu Dhabi. The authors evaluated the performance of these technologies from a thermodynamic and an economic perspective. The primary index used was the cooling generation cost, which represents the price-life-cycle cost of the solar cooling systems. Based on the results of the study, as expected, large-scale plants proved to have a better economic performance. For these large scales, the most competitive technologies included concentrating solar thermal power plants coupled with vapor compression cycles. For smaller scales, the authors reported that Fresnel concentrators and thin film PV cells were the most cost competitive options. However, in terms of efficiency, multicrystaline PV cells exhibited the best performance. The authors also stressed the strong impact of the solar radiation intensity and time distribution throughout the cooling period on the assessment of the systems, as well as the high influence of the heat rejection mechanism (dry versus wet) on the economics of the systems. Furthermore, they added that absorption chillers experience a severe performance degradation in hot conditions, and as a result their application in such climates may be problematic.

Eicker and Pietruschka (2009) focused on the investigation of the optimization and economics of solar closed cycle absorption and desiccant cooling systems driven by heat at temperatures below 100°C. The authors concluded that while absorption chillers can be used in a wide range of applications with cold distribution based on water or air systems, desiccant cooling systems are recommended if the need for fresh air humidity is high in a building. For all solar thermal cooling technologies, the reduction of auxiliary electrical energy consumption is a major goal, because otherwise primary energy savings are not significant. This suggests an optimized control strategy, especially for partial load operation, and a good design of the heat source and heat sink circuits. Based on the full economic analysis, taking into account capital and operating costs, the total system costs for commercially available absorption solar cooling systems in southern Europe were estimated equal to €180–320/MWh$_c$. The authors stressed the dependence of this value on the cooling load profile and the control strategy. For example, in Germany, where there is a significantly lower cooling energy demand, these costs rise to €680/MWh. According to the authors, the most significant cost components of solar absorption cooling systems include the solar thermal system as well as the chiller machine. As far as desiccant cooling systems are concerned, the authors estimated that about two-thirds of the investment costs are due to the desiccant air conditioning and distribution system, with solar collectors taking up 10–15% of the investment. The total cooling cost was calculated in the range from €300–900/MWh$_c$, again depending on the load profile of the application.

In another study, the same authors (Eicker et al. 2014) investigated energetic and economic aspects of the performance of solar single-effect absorption cooling systems for different locations worldwide. The cooling capacities of the chillers ranged from 106 kW$_c$ to 229 kW$_c$, and a cooling water temperature of 7°C supply and 14°C return was considered. Based on the results of the economic investigation, it was determined that solar thermal cooling is more viable in hot climates than in moderate European climates, with annual operating costs being strongly dependent on the locations, varying from €0.25/kWh$_c$ to €1.01/kWh$_c$ in Germany and from €0.13/kWh$_c$ to €0.30/kWh$_c$ in Spain. Accordingly, the costs for regions with even hotter climates, such as Jakarta, Indonesia, and Riyadh, Saudi Arabia, can be as low as €0.09/kWh$_c$ to €0.15/kWh$_c$. Regarding the optimal sizing of absorption chillers, the authors remarked that while it is difficult to establish specific quantitative guidelines for selecting the nominal capacity of the chiller in relation to the peak cooling load of buildings, undersizing the chiller is beneficial from both energetic and economic perspectives. In fact, the authors reported that it is possible to reduce the nominal cooling capacity of the chiller to about 60% of the maximum cooling load without any decrease in the solar fraction. Furthermore, according to the authors, due to the relatively reduced cooling requirements in most European locations, careful system design including key parameters such as machine power, collector surface, and storage tank size should be followed.

Blackman et al. (2015) conducted a techno-economic evaluation of solar-assisted heating and cooling systems based on the integration of sorption chillers and solar collectors. The solar cooling systems included a PV-VCC system, a solar thermal sorption system to produce heating and cooling, a hybrid system based on the combination of a PV and a solar thermal system, and a reference system consisting of a conventional VCC and a natural gas boiler for cooling and heating, respectively. The systems were evaluated by considering climate data and cooling loads for a building in Madrid, Spain. Compared to the reference case, the annual cost savings amounted to €173–346 for the PV system, €153–386 for the solar absorption system, and €205–615 for the hybrid system.

Huang et al. (2011) also investigated the thermodynamic performance and the economic competitiveness of solar thermal heating and cooling systems. However, in this study, an ejector cooling cycle was considered, coupled with a conventional heat pump as an auxiliary cooling device. The nominal cooling capacity of the system was equal to 3.5 kW_c. Two different installation locations were examined, Taipei and Tainan. A daily cooling load of 35 kWh_c was considered, dispersed at a daily operation time of ten hours. The authors carried out a detailed cost analysis for the system, taking into account the cost of each individual component, such as the heat exchangers (generator, condenser, evaporator), the refrigerant pump, receiver, cooling tower, piping, control system, and the frame. Interestingly, the cost of the ejector was estimated to equal approximately $95/$kW_{th}$. From the results of the economic analysis, the authors concluded that higher installed cooling capacities led to increased net present values (NPV) and shorter payback periods, leading, thus, to overall better economic performance. The payback periods were estimated equal to 4.8 years in Tainan and 6.2 years in Taipei. Lastly, the authors pointed out that although the COP of ejector cooling systems is low (considered equal to 0.2 in their techno-economic analysis), it can be further improved upon. Meanwhile, the ejector chiller has a relatively low manufacturing cost at small scales.

Allouhi et al. (2015) conducted a review of different solar cooling systems. Several performance, technical, market, and economic criteria were discussed for the different options, along with their advantages and disadvantages. Regarding the economics of solar cooling, the authors reported that adsorption chillers have very low maintenance costs, despite their low efficiency. Meanwhile, desiccant systems have higher investment costs due to their complex design. Lastly, thermo-mechanical technologies based on the implementation of Rankine and Stirling cycles were reported as having the highest investment costs along with high maintenance costs and are, according to the authors, unsuitable for small-scale applications. The authors provided data to show that the investment cost of solar absorption and adsorption cooling systems is around 267% and 336% of the cost of conventional vapor compression cycles, respectively. Overall, the authors concluded that despite the undeniable potential of solar cooling processes, several barriers need to be overcome for their worldwide implementation. The major obstacles are their high installation cost and low performance. Therefore, subsidies and other payment schemes must be undertaken by policy makers. It is also necessary to achieve performance enhancement in order to compete with conventional solutions. This is possible through the research of optimal designs and the development of novel, highly efficient options eventually based on hybrid configurations.

Shirazi et al. (2016) performed an energetic, economic, and environmental assessment of different solar-assisted heating and cooling systems based on absorption technology coupled with evacuated tube collectors, considering the transient operation of the system. More specifically, they considered four different configurations. In the first configuration, a gas-fired heater was used for backup. In the second configuration, a vapor compression chiller served as backup. The other two configurations were similar to the second one, the only difference being that the nominal design capacity of the absorption chiller was reduced to 50% and 20%. The authors concluded that the first system had very low primary energy savings, and suggested that gas-fired backup systems should only be considered for multi-effect absorption chillers. Meanwhile, the second system was the most energy efficient of all. Finally, the authors showed that the systems examined were not economically profitable due to the high capital costs of the collectors and the chillers.

More specifically, the payback periods of the four systems were very high, equal to 93.6, 94.5, 49.4, and 36.2 years, respectively. However, they highlighted that if financial support schemes are implemented, such as a 50% government subsidy of the capital cost, a payback period of 4.1 years can be attained with the fourth system.

Rowe and White (2014) performed a survey of international incentive schemes for solar cooling systems. The survey was conducted with the help of representatives and contributors from different countries, including Australia, Austria, France, Germany, Italy, and the United States. They identified 65 support mechanisms, which they put into different categories and evaluated. Among the direct measures, subsidies on capital cost are by far the most popular, followed by tax deductions, competitive grants, and a provision of access to capital. Furthermore, they found that the majority of reported incentives were in the United States, followed by Australia and Italy. Finally, the authors reported that most of the applications that were supported were residential.

Desideri et al. (2009) compared the technical and economic aspects of different solar cooling systems for industrial refrigeration (meat manufacturing) and air conditioning (hotel in a tourist town) in Italy. In the first case, the investigated system consisted of an absorption chiller powered by flat-plate collectors. In the second case, a hybrid trigeneration (space heating, domestic hot water production, and space cooling) plant was proposed. The estimated payback periods for the two cases were equal to 7 and 12 years, respectively.

Calise (2012) focused on high temperature solar heating and cooling systems based on the combination of parabolic trough collectors with a double-stage LiBr-H_2O absorption chiller. A biomass heater was considered as an auxiliary heating and cooling source. Climatic conditions of the Mediterranean were taken into account, and more specifically the regions of Italy, Spain, Egypt, France, Greece, and Turkey. The primary energy savings that were achieved were up to 80%. Meanwhile, the author found that the payback periods, even without considering any financial support scheme, are significantly shorter than the system's operating life. As expected, they become even shorter when funding is considered. According to the author, the results highlight the superiority of double-effect high temperature adsorption chillers against low temperature cooling systems from an economic standpoint, which is attributed to the substantially higher COP of the technology, despite the lower efficiency of parabolic trough collectors, combined with the lower cost of parabolic collectors compared to evacuated tube collectors. As expected, the economic results improved for hotter climates. In general, the author stressed the fact that high temperature solar heating and cooling systems are very attractive for the cases examined, also remarking that the only limitation for further deployment of the technology is the insufficient commercial availability of high temperature concentrating solar collectors.

In another study, Calise et al. (2011) performed a thermo-economic comparison of solar heating and cooling systems installed in an office building located in Naples, Italy. These included a single-stage LiBr absorption chiller at different capacities coupled with evacuated tube solar collectors and an auxiliary electrical heat pump or a gas-fired heater. The authors developed a zero-dimensional transient simulation model in order to analyze the performance of the system, and they also employed cost correlations to evaluate the capital and operating costs of each configuration. The authors stressed the effect of suitable selection of the solar collector area and the volume of the storage tank on the economic profitability of the systems they examined. Among the different financing schemes that they considered, they proposed one based on a combination of feed-in tariffs along with a slight capital cost subsidy.

Baitaneh and Taamneh (2016) presented the recent (as of 2016) developments in solar cooling systems based on sorption (absorption and adsorption) technology. Regarding absorption technology, the authors reviewed a large number of studies and reported that they all showed that the performance of solar cooling applications is a strongly nonlinear function of various parameters such as the size and the type of the collector, the geographical location, as well as the operating conditions and the auxiliary energy source that is considered. A second conclusion drawn by this review is the fact that absorption systems have the potential to become economically feasible under specific conditions. On one hand, increasing the COP (by decreasing the operating temperature of the condenser or increasing the operating temperature of the evaporator, or both) leads to a reduction of the payback period. As far as adsorption chillers are concerned, the authors reported that several studies have indicated that they have a significantly higher capital cost and lower operational cost compared to conventional systems, since they do not involve the running cost of fossil fuels, electricity transmission costs, or energy conversion costs, and their maintenance is also less expensive. Meanwhile, the high cost of adsorption pairs like zeolite-water and activated carbon-methanol is another factor that restricts their market growth despite their technical success (Bataineh and Alrifai 2015). In general, the authors concluded that for the time being the costs of solar sorption cooling systems make them not competitive, mostly due to the increased prices of the solar collectors. However, the authors pointed out that integrating thermal energy storage with double-purpose systems (heating and cooling) could improve the feasibility of solar cooling technologies.

Gebreslassie et al. (2010) focused on the development of a systematic approach to minimize the economic and environmental life-cycle impact of solar-assisted absorption cooling applications installed in Tarragona and Barcelona. More specifically, the authors proposed a novel mixed-integer nonlinear programming (MINLP) model to optimize the design point and operating conditions of the systems and combined it with an LCA analysis to account for environmental sustainability. In their model, a NH_4-water absorption chiller was considered, while the driving heat was produced by natural-gas-fired heaters, steam boilers, waste heat, and solar thermal (FPC, ETC, and CPC) collectors. The system was investigated on the basis of fixed cooling demands, which typically correspond to industrial, but not domestic, applications. From the results, the authors noted that there is a conflict between the economic and the environmental performance of the system. Meanwhile, they reported that the environmental impact of such systems can be greatly mitigated if an investment focus on the solar subsystem is made, for example by increasing the number of solar collectors and thus the solar fraction of the application. Lastly, the authors reported that their optimization method led to reduced computation times.

Montagnino (2017) carried out a review of existing projects on solar cooling technologies, examining various aspects, such as the design as well as the performance and the economic competitiveness. The author provided data from a project focused on solar cooling technologies (Navarro-Rivero and Ehrismann 2012), which included 57 solar cooling installations in Europe, revealing a widely dispersed specific cost for the technology, ranging from €1,200–27,000/kW$_c$, with most of the systems in the range of €2,000–6,000/kW$_c$ for small cooling capacities below 300 kW$_c$ and mostly from a few kW$_c$ to 50 kW$_c$. Citing data from another study by the Solar Heating and Cooling Programme of the International Energy Agency (SHC-IEA 2016a), the author reported that the prices of pre-engineered small (lower than 35 kW$_c$) systems, excluding installation costs, have recently dropped from €6,000/kW$_c$ in 2007 to €4,500/kW$_c$ in 2013. In accordance with other researchers, the author highlighted the fact that the payback time of solar cooling installations still remains too long. The author also added that the implementation of such systems is still dependent on a sophisticated design phase, which is required in order to optimize the primary energy savings and reduce the energy consumption of the auxiliary system, thus extending its

TABLE 11.7

Costs of Solar Cooling Systems Compiled by SHC, IEA

Scale	Specific Cost ($€/kW_{th}$)
Small capacity (<10 kW_{th})	7,302
Medium capacity (<50 kW_{th})	4,472, 3,968, 2,012
Large capacity (>50 kW_{th})	2,653, 2,611, 1,866, 1,320

Source: SHC-IEA, "Technology and Quality Assurance for Solar Thermal Cooling Systems", *IEA Solar Heating and Cooling Task 48*, http://task411.iea-shc.org/data/sites /1/publications/Task48%20D3%20brochure%20final.pdf, accessed on 30 March 2018, 2016b.

operating hours and improving its economic competitiveness. He noted, however, that some nations have already started to support the market proliferation of solar cooling systems by providing incentive policies and appreciating the positive effect of reducing the cost of adapting the electricity infrastructure to summer peak demand, thus promoting grid stability. The author also pointed out that the recent significant decrease in PV technology combined with the increased COP of conventional vapor compression chillers have made the solar electric PV option exceptionally attractive compared to solar thermal cooling pathways. Lastly, the author highlighted the importance of hybridization with conventional thermal fuels, polygeneration, and seasonal switchers in plant operation as ways to reduce the payback period of solar cooling investments.

In the context of Solar Heating and Cooling Task 48 that was carried out by SHC-IEA (2016b), a total of 12 solar cooling projects were selected for different applications like office buildings, school/institute buildings, commercial buildings, and residential buildings in North America, Europe, and mostly Southeast Asia. Based on the evaluation of these systems, the specific installed system costs that are presented in Table 11.7 were reported.

11.4 Compilation of Cost Data for Solar Cooling Technologies

In this section, a compilation of cost data regarding costs of solar collectors and different chillers is given. The data was retrieved from an extensive literature review and can be used to provide the typical values of the different components, which is an essential step for the economic evaluation of solar cooling applications.

Table 11.8 presents the reported costs of different solar collector types and PV panels from the literature. Note that each source is based on a different reference year, so care must be taken to convert the reported values to present day values.

Table 11.9 summarizes the reported costs of absorption chillers from different literature studies.

Table 11.10 summarizes the reported costs of desiccant cooling chillers. It is notable that there is a general scarcity of data regarding the costs of such applications.

Finally, Table 11.11 summarizes the cost of various adsorption chillers as retrieved from literature studies.

TABLE 11.8

Reported Costs of Different Types of Solar Collectors and PV Panels

Cost (€/m²) unless Stated Otherwise	Characteristics	Year	Reference
–	Evacuated tube	2016	Gabbrielli et al. (Gabbrielli, Castrataro, and Del Medico 2016)
350–450	Compact linear Fresnel		
171	Flat plate	2003	Tsoutsos et al. (Tsoutsos et al. 2003)
250	Evacuated tube	2015	Allouhi et al. (Allouhi, Kousksou, Jamil, El Rhafiki et al. 2015)
350	Flat plate	2014	Lazzarin (Noro and Lazzarin 2014)
650	Evacuated tube		
450	Parabolic trough		
650	PV		
230	Flat-plate collectors	2016	Al-Ugla et al. (Al-Ugla et al. 2016)
1,776 (€/kW$_e$)	PV		
616–740 (5,500–6,500 €/kW$_e$)	PV single crystal or polycrystalline, not thin film	2007	Lambert and Beyene (Lambert and Beyene 2007)
175	Flat plate, single glazed, black paint		
200	Flat plate, double glazed, black paint		
200	Flat plate, double glazed, black paint		
225	Flat plate, double glazed, solar selective		
250	Evacuated flat plate		
375	Compound parabolic		
250–400	Residential building PV	2014	Ferreira and Kim (Infante Ferreira and Kim 2014)
150–250	Utility building PV		
750	Evacuated tube		
370	Flat plate		
540	Parabolic trough		
350	Flat plate	2014	Noro and Lazzarin (Noro and Lazzarin 2014)
650	Evacuated tube		
450	Parabolic trough		
330	PV mSi		
130	PV asi		
300	Evacuated tube	2017	Bellos et al. (Bellos, Tzivanidis, and Tsifis 2017)
472.5	Evacuated tube	2016	Shirazi et al. (Shirazi et al. 2016)
200	Parabolic trough	2012	Calise (Calise 2012)
771–783	Evacuated tube	2010	Gebreslassie et al. (Gebreslassie et al. 2010)
196–271	Flat plate		
377	Compound parabolic concentrating		

TABLE 11.9

Reported Costs of Absorption Chillers

Capacity (kW$_{th}$)	Capital Cost (€/kW$_{th}$)	Characteristics	Year	Reference
109–174	400	Single effect	2016	Gabbrielli et al. (Gabbrielli,
253–308	320	Double effect		Castrataro, and Del Medico 2016)
176	1,433	Single effect/hot water	2017	U.S. Department of Energy
1,547	549	Double effect/hot water		(Energy 2017)
	352		2015	Allouhi et al. (Allouhi, Kousksou, Jamil, El Rhafiki et al. 2015)
1	400	Single effect	2014	Lazzarin (Noro and Lazzarin
1	700	Double effect		2014)
1,500	516			Al-Ugla et al. (Al-Ugla et al. 2016)
	400 (Henning, Heating, and Programme 2004)	Single effect	2014	Compiled by Ferreira and Kim (Infante Ferreira and Kim 2014)
	400 (Kim and Infante Ferreira 2008)			
	210 (Mokhtar et al. 2010)			
	250			
	300 (Kim and Infante Ferreira 2008)	Double effect		
	550 (Mokhtar et al. 2010)			
	855 (Otanicar, Taylor, and Phelan 2012)			
	700			
	400	LiBr single effect	2014	Noro and Lazzarin (Noro and
	700	LiBr double effect		Lazzarin 2014)
106–229	700	Single effect	2013	Eicker et al. (Eicker et al. 2014)
70	600	LiBr single effect	2017	Bellos et al. (Bellos, Tzivanidis, and Tsifis 2017)
209–1,023	585	Single effect	2016	Shirazi et al. (Shirazi et al. 2016)

TABLE 11.10

Reported Costs of Desiccant Chillers

Capacity (kW$_{th}$)	Capital Cost (€/kW$_{th}$)	Year	Reference
–	370	2014	Compiled by Ferreira and
–	1,065		Kim (Infante Ferreira and
–	700 (Otanicar, Taylor, and Phelan 2012)		Kim 2014)

TABLE 11.11

Reported Costs of Adsorption Chillers

Capacity (kW$_{th}$)	Cost (€/kW$_{th}$)	Year	Reference
–	1,000	2003	Tsoutsos et al. (Tsoutsos et al. 2003)
	520	2015	Allouhi et al. (Allouhi, Kousksou, Jamil, El Rhafiki et al. 2015)
1	600	2014	Lazzarin (Noro and Lazzarin 2014)
633 MJ/day	4,237 (€)	2007	Lambert and Beyene (Lambert and Beyene 2007)
	850 (Henning, Heating, and Program 2004)	2014	Compiled by Ferreira and Kim (Infante Ferreira and Kim 2014)
	500 (Kim and Infante Ferreira 2008)		
	855 (Otanicar, Taylor, and Phelan 2012)		
	700		
	600	2014	Noro and Lazzarin (Noro and Lazzarin 2014)

11.5 Economic Evaluation Case Studies

In this section, an economic evaluation of some of the solar cooling technologies that were previously presented is carried out. The installation of solar cooling technology in order to cover the cooling demand is investigated for different building types and geographical locations corresponding to different cooling loads and meteorological data (solar radiation intensity and ambient temperature variation) via the calculation of the payback period of each case. The evaluation is carried out by comparison to a conventional VCC system, which is currently the standard and most popular cooling technology.

The solar cooling systems examined include sorption (absorption and adsorption) chillers. In each case, two solar collector types are investigated (flat plate and evacuated tube), with different efficiency coefficients and specific costs. Meanwhile, the installation of PV panels (instead of the implementation of solar thermal cooling systems) in order to reduce the electricity consumption of a standard VCC chiller is considered. In each case, cost correlations from the literature are employed in order to determine the capital investment costs. It is assumed that the sorption cooling systems are combined with a natural gas boiler, which plays the role of an auxiliary heat source.

A series of sensitivity analyses are carried out in order to assess the impact of the total area of the solar collectors, the size of the storage tank, and the nominal temperature of the solar collectors, which are the most important design variables of solar thermal cooling systems on the attained energy savings and the economic performance.

11.5.1 System Description and Modeling

A schematic depicting the configuration of the solar cooling technologies based on sorption is given in Figure 11.4.

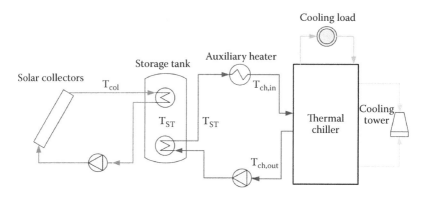

FIGURE 11.4
Schematic of the solar cooling concept.

11.5.1.1 Step 1. Selection of Geographical Location and Building Types, Calculation of Cooling Loads

The first step for modeling the systems includes the selection of the geographical location of the study and the types of buildings to be considered. The locations selected include: (a) Rome, Italy and (b) Berlin, Germany. These locations correspond to high/low cooling loads and solar radiation intensity, respectively. For each location, three different building types are considered: (1) a single house, (2) an apartment block, and (3) an office building, each of which correspond to different loads and use hours. The characteristics and the use hours of the buildings are summarized in Table 11.12 and Table 11.13. These different building types are considered in order to examine the impact of different loads and operating hours on the economic performance of the solar cooling systems.

The ambient temperature and the solar radiation intensity are retrieved on an hourly basis for the average day of each month for each location by the Photovoltaic Geographical Information System of the European Commission (Commission 2017). In order to estimate the hourly variation of the cooling loads, the web tool MIT Design Advisor is employed, which was developed by the Massachusetts Institute of Technology (Technology 2000–2009).

The peak cooling loads and annual cooling demand for the investigated cases are summarized in Table 11.14.

TABLE 11.12

Common Assumptions for All Building Types

Orientation	N–S/E–W
Primary façade orientation	east
Indoor air temperature	20–26°C
Maximum relative humidity	60%
Window area	25% of exterior wall area
Floor height	3 m
Window type	double glazed with blinds
Wall and roof U-Valued	2 W/m^{2}K

TABLE 11.13

Specific Assumptions for Each Building Type

	Single House	Apartment Block	Office Building
Floor area	12 × 12 m	18 × 18 m	18 × 18 m
No of floors	1	5	5
Occupancy schedule	4:00 p.m.–12:00 a.m.	4:00 p.m.–12:00 a.m.	8:00 a.m.–6:00 p.m.
Person density	0.025	0.025	0.75
Lighting	200 lux	200 lux	400 lux
Equipment load	5 W/m²	5 W/m²	15 W/m²

TABLE 11.14

Peak Cooling Loads and Annual Cooling Demand for the Examined Cases

		Single House	Apartment Block	Office Building
Peak cooling load (kW$_c$)	Italy	4.9	26.4	192.6
	Germany	1.2	2.1	53.3
Annual cooling demand (kWh$_c$)	Italy	2,234	9,279	146,364
	Germany	177	191	20,221

11.5.1.2 Step 2. Modeling and Sizing of the Solar Cooling Systems

The cooling load at each hour i is expressed by the variable $Q_{cool,i}$. After the cooling loads are calculated, the heat flux that is required for the operation of the thermally driven chiller is calculated by the equation:

$$Q_{h,ch,i} = \frac{Q_{cool,i}}{COP_{ch}}$$ (11.1)

In the above equation, COP_{ch} is the coefficient of performance of the chiller. The COP is a significant performance parameter of the chillers. Its value depends on the exact temperatures at which the processes (adsorption/desorption, generation/evaporation, condensation) of the cooling cycle occur. As a result, in reality, the actual COP of the chillers is highly variable depending on the operating conditions of the system, which are directly dependent on the temperature and the mass flow rate of the heat transfer fluid that provides heat to the chiller, the temperature and mass flow rate of the cooled water, and the condensation temperature at which heat is rejected into the environment. In turn, the aforementioned temperatures are affected by the solar radiation intensity, the size of the collectors and the storage tank, and the ambient temperature and the cooling load, which constantly varies during the operation of the chiller. In order to reduce the complexity of the interdependence of the above variables and derive as much as possible a simplified but accurate estimation of the COP of the chiller, the following assumptions are followed:

- The upper temperature of the chiller is constant, depending on the chiller category.
- The temperature of the heat transfer fluid at the chiller inlet ($T_{ch,in}$) is constant, 5 K higher than the upper chiller temperature.

- The temperature drop of the heat transfer fluid as it passes through the chiller is constant, equal to 5 K.
- The evaporation and condensation temperatures of the chiller are constant.
- The COP for each chiller category is constant.

In each case, to optimize the system, the maximum (nominal) temperature of the solar collectors is varied within a range of 30 K. The assumptions for each chiller type are summarized in Table 11.15.

As previously stated, temperatures and the COP are not constant during the operation of the chiller. However, for the purpose of the present techno-economic analysis, which does not aim to focus on the detailed technical aspects governing the operation of the chiller, the values presented in Table 11.15 can be considered constant and representative of average values without a significant loss in accuracy of the economic evaluation.

The storage tank is also a critical component of solar cooling systems. For its modeling, it is assumed that full mixing occurs inside it, and hence no stratification effects are taken into account. Therefore, at each time-step, the storage tank has a uniform temperature, equal to $T_{st,i}$. It is also assumed that the temperature of the heat transfer fluid exiting the storage tank is equal to $T_{st,i}$. Subsequently, the heat transfer fluid is driven through the auxiliary heater, where its temperature increases to the temperature that is required for the operation of the thermal chiller ($T_{ch,in}$). Therefore, at each time step, the heat provided by the auxiliary heater is equal to:

$$Q_{h,aux,i} = \dot{m}_{ch,i} c_{p,htf} \left(T_{ch,in} - T_{st} \right) \tag{11.2}$$

Obviously, if the storage tank temperature is higher than the required chiller inlet temperature, the auxiliary heater does not operate. Furthermore, as previously stated, it is assumed that the outlet temperature of the heat transfer fluid exiting the heat exchanger of the chiller is also always fixed at a value corresponding to:

$$T_{ch,out} = T_{ch,in} - 5 \tag{11.3}$$

Thus, the mass flow rate of the fluid entering the chiller heat exchanger $\dot{m}_{ch}$ is calculated by the equation:

$$\dot{m}_{ch,i} = \frac{Q_{h,ch,i}}{c_{p,htf} \left(T_{ch,in} - T_{ch,out} \right)} \tag{11.4}$$

TABLE 11.15

Boundary Conditions for the Operation of the Solar Thermal Cooling Systems

	Absorption Chiller		Adsorption Chiller	
	Li-Br Single Effect	Li-Br Double Effect	Zeolite-Water	Silica Gel-Water
Upper chiller temperature (°C)	90	120	75	85
Heat transfer fluid Temperature at chiller inlet (°C)-$T_{ch,in}$	95	125	80	90
Maximum solar collector temperature range (°C)	95–125	125–155	80–110	120
COP	0.75	1.35	0.50	0.60

The fuel thermal input from the auxiliary heater is given by the equation:

$$Q_{f,i} = \frac{Q_{h,aux,i}}{\eta_{aux}} \tag{11.5}$$

In the above equation, η_{aux} is the thermal efficiency of the auxiliary heater.

The next step in the simulation and sizing is the selection of the total area of the solar collectors. This variable is expressed via the surface parameter, which is defined as:

$$A_p = \frac{A_{col}}{A_{b,base}} \tag{11.6}$$

In the above equation, A_{col} is the area of the solar collectors and $A_{b,base}$ is the area of the land covered by the building (equal to the area of each floor). Obviously, in the case that the collectors are installed on the roof of the building, A_p must in theory be necessarily lower than one. Note that in practice this value must be even lower (at around 0.5–0.75). If A_p is higher than one, solar collectors must be installed on additional building surfaces or on additional land areas adjacent to the building, with repercussions to the cost of the applications. In the present evaluation, it is considered that for all cases $A_p \leq 1$, which is the most common case for solar thermal applications in buildings in urban areas.

At each time-step, the useful heat that is provided to the heat transfer fluid by the solar collectors is estimated by the equation:

$$Q_{sol,heat,i} = A_{col}\eta_{sol,i}G_i \tag{11.7}$$

In the above equation, G is the solar radiation intensity and η_{sol} is the efficiency of the solar collectors. The efficiency is typically described by the equation:

$$\eta_{sol,i} = c_0 - c_1\left(\frac{T_{col} - T_{amb,i}}{G_i}\right) - c_2 G_i\left(\frac{T_{col} - T_{amb,i}}{G_i}\right)^2 \tag{11.8}$$

The coefficients c_0, c_1, and c_2 depend on the type of the collector. For the present evaluation, their values for flat-plate and evacuated tube collectors are retrieved from the study of Ferreira and Kim (2014). They are summarized in Table 11.16.

The second parameter that is used for sizing a solar cooling system is the size of the storage tank, which is expressed by its volume V_{tank}. The sensible heat that can be stored in the

TABLE 11.16

Efficiency Coefficient for Flat-Plate and Evacuated Tube Solar Collectors

	c_0	c_1	c_2
Flat plate	0.80	3.02	0.0113
Evacuated tube	0.84	2.02	0.0046

storage tank, considering the maximum temperature of the heat transfer fluid of the solar collectors T_{col} and a reference temperature T_{ref}, is described by the equation:

$$Q_{st} = M_{ST} c_{p,htf} \left(T_{col} - T_{ref} \right) = \rho_{htf} V_{htf} c_{p,htf} \left(T_{col} - T_{ref} \right) \tag{11.9}$$

In the above equation, M_{st} is the total mass of the heat transfer fluid that can be filled inside the tank, and is equal to the density of the heat transfer fluid (ρ_{htf} multiplied by the volume of the storage tank V_{htf}). The density and the heat capacity of the heat transfer fluid are evaluated at the reference temperature.

Of course, the size of the storage tank is subject to specific spatial and practical limitations. In order to avoid an infeasible or unreasonable design, it is assumed that the tank's diameter D ranges from 0.2 to 2 meters.

The temperature of the heat transfer fluid inside the storage tank at each hour *i+1* is calculated by considering the temperature of the previous hour *i* through the energy balance equation:

$$T_{st,i+1} = T_{st,i} + \frac{Q_{stored,i}}{M_{st} c_{p,htf}} \tag{11.10}$$

Obviously, the temperature of the storage tank cannot exceed the maximum design temperature of the solar collectors or drop below the ambient temperature (it is assumed that the storage tank is located at an external location). If the temperature of the heat transfer fluid inside the tank becomes too high, then the collector's circuit is disconnected from the tank and heat is rejected to the environment in order to maintain the temperature of the tank at the design temperature of the system (T_{col}). Therefore, in this case, $Q_{stored} = 0$. When this is not the case, the heat flux that is stored in the storage tank at any given hour is given by the equation:

$$Q_{stored,i} = Q_{sol,heat,i} - Q_{loss,i} - Q_{u,i} \tag{11.11}$$

In the above equation, Q_{loss} are the heat losses from the storage tank to the environment. The heat losses are described by the equation:

$$Q_{loss,i} = U_{loss} A_{st,ex} \left(T_{st,i} - T_{amb,i} \right) \tag{11.12}$$

where U_{loss} is the heat loss coefficient of the storage tank. In the present analysis, it is assumed to be equal to 0.5 W/m²K, and $A_{st,ex}$ is the external surface of the cylindrical storage tank. Lastly Q_u is the heat flux provided from the heat transfer fluid to the chiller and the auxiliary heater. Its value depends on the operation mode of the system, the temperature of the storage tank, and the ambient temperature.

At each time-step, the calculation of the amount of heat that is delivered from the solar circuit to the adsorption chiller is carried out based on the methodology presented in Table 11.17.

Naturally, when the cooling load is zero, the chiller is non-operational and it is not provided with heat from the solar circuit ($Q_u = 0$). When the cooling load is higher than zero, two cases are distinguished. If the temperature in the storage tank is lower than the temperature of the heat transfer fluid at the chiller outlet ($T_{ch,out}$), the solar circuit is bypassed, and, again, no heat is delivered for the production of cooling ($Q_u = 0$). If the temperature in the storage

TABLE 11.17

Delivered Heat to the Chiller from the Solar Collectors
Circuit (Q_u) at Each Hour

Conditions	Q_u
Cooling load = 0	0
Cooling load > 0	
If $T_{st} < T_{ch,out}$	0
If $T_{ch,out} < T_{st} < T_{ch,in}$	
$Q_{st,max} = M_{st}c_{p,htf}(T_{st}-T_{ch,out})$	Min $(Q_{st,max}, Q_{htf,max})$
$Q_{htf,max} = m_{htf}c_{p,htf}(T_{st}-T_{ch,out})$	
If $T_{st} > T_{ch,in}$	
$Q_{st,max} = M_{st}c_{p,htf}(T_{st}-T_{ch,out})$	
$Q_{htf,max} = m_{htf}c_{p,htf}(T_{ch,in}-T_{ch,out})$	
if $Q_{h,ch} > Q_{st,max}$	$Q_{st,max}$

tank is higher than $T_{ch,out}$, heat is provided to the heat transfer fluid, and its temperature is determined by the energy balance equations, taking into account the second law of thermodynamics and ensuring that no temperature crossovers occur. The temperature of the heat transfer fluid at the storage tank outlet can be either equal to the nominal driving heat temperature of the chiller ($T_{ch,in}$) or lower. In the second case, the auxiliary heater needs to provide additional heat to the heat transfer fluid after it exits the storage tank.

The above analysis concerns solar thermal cooling technologies. In addition to these technologies, an additional scenario considering the installation of PV panels to produce electricity instead of solar thermal collectors is considered. The produced electrical energy for each month can be retrieved using the web tool found in the Photovoltaic Geographical Information Systems of the European Commission (Commission 2017), by considering a crystalline silicon panel and a nominal efficiency of 12%.

Lastly, all the above cases are compared to a reference scenario involving the installation of a conventional heat pump based on the implementation of a VCC for covering the cooling loads. The COP of the conventional system is considered fixed, corresponding to an average value for the cooling period equal to 3. Furthermore, the total annual electricity consumption for its operation is calculated via the equation:

$$E_{e,vcc,t} = \frac{Q_{cool,t}}{COP_{vcc}} \qquad (11.13)$$

11.5.2 Economic Evaluation Methodology

The index used for the economic evaluation of the solar cooling systems is the payback period. The PBP is the amount of years that are required until the cumulative savings that are derived from the investment are equal to the initial total capital investment (TCI) cost. If a fixed annual nominal net profit is assumed, the simplified PBP (using a zero value of interest rate) is calculated by the following equation:

$$PBP = \frac{TCI}{C_t} \qquad (11.14)$$

TABLE 11.18

Capital Cost Estimation Data and Correlations

Component	Specific Cost
Solar collectors (Eicker and Pietruschka 2009)	
Flat-plate collectors	$C_{sp,fpc,col} = 297.79 A_{col}^{-0.0426}$ (€/m²)
Evacuated tube collectors	$C_{sp,etc,col} = 829.49 A_{col}^{-0.1668}$ (€/m²)
Storage tank (Eicker and Pietruschka 2009)	$C_{sp,tank} = 16.011 V_{tank}^{-0.2403}$ (€/m³)
Chillers (Neyer et al. 2016)	
Absorption chiller	$3,700 Q_{th}^{-0.45}$ (€/kW$_{th}$)
Adsorption chiller	$1,680\ Q_{th}^{-0.17}$ (€/kW$_{th}$)
Vapor compression chiller (air cooled) (Neyer et al. 2016)	$1,219.8 Q_{th}^{-0.292}$ (€/kW$_{th}$)
Cooling tower (dry) (Neyer et al. 2016)	$C_{ct} = 46.8 Q_{th} + 26,311.6$ (€)
Auxiliary heater (natural gas boiler) (Neyer et al. 2016)	$C_{sp,aux} = 600 Q_{nom}^{-0.289}$ (€/kW$_{th}$)
PV panels (Infante Ferreira and Kim 2014)	$C_{sp,pv} = 300$ €/m²

The first step for the economic analysis is the calculation of the TCI that corresponds to each scenario. The TCI is calculated based on the cost of the individual equipment components that are included in each system. These are summarized in Table 11.18.

The specific investment costs of the different solar cooling technologies (as presented in Table 11.18) as a function of the cooling capacity are plotted in Figure 11.5. It can be observed that while for higher cooling capacities adsorption chillers are more expensive than single-effect absorption chillers, for cooling capacities below 20 kW$_{th}$, they have a significantly lower specific cost (lower than half of absorption chillers) and constitute the least expensive sorption technology.

The total equipment cost for the solar thermal cooling systems is the sum of the cost of the solar collectors, the storage tank, the chiller, the cooling tower, and the auxiliary heater, and is determined by the equation:

$$C_{eq,sc} = C_{col} + C_{tank} + C_{ch} + C_{ct} + C_{aux} \tag{11.15}$$

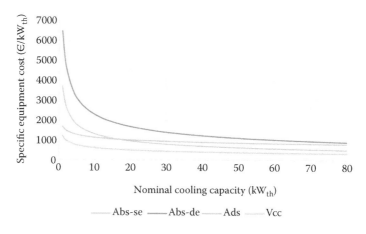

FIGURE 11.5
Specific investment cost correlations for different types of cooling technologies used in the present evaluation.

The total investment cost of solar cooling systems is derived by accounting for the following additional costs (Neyer et al. 2016):

- Control, electricity, and monitoring (10% of equipment cost)
- Design, planning, and commissioning (20% of equipment cost)
- Labor cost (30% of equipment cost)
- Indirect and other costs (5% of equipment cost)

As a result, the total capital investment cost is given by the equation:

$$C_{tc\ i,sc} = 1.65 C_{eq,sc} \tag{11.16}$$

For the PV system, the equipment cost only involves the solar panels. An additional 30% labor cost is considered for the estimation of the total investment cost.

For the conventional VCC system scenario, the equipment cost only includes the VCC module and the cooling tower:

$$C_{eq,vcc} = C_{vcc} + C_{ct} \tag{11.17}$$

The total capital investment cost of the conventional system is estimated by also accounting for the following cost components (Neyer et al. 2016):

- Auxiliaries (50% of VCC cost)
- Control, electricity, and monitoring (7% of VCC)
- Design, planning, and commissioning (20% of VCC)
- Labor cost (30% of VCC)
- Indirect and other costs (5% of VCC)

Based on the above assumptions, the total capital investment cost of the VCC system is determined by the equation:

$$C_{tc\ i,vcc} = 2.12 C_{eq,vcc} \tag{11.18}$$

The operational costs of the solar cooling systems include the running fuel costs of the auxiliary heater. These are determined by the equation:

$$C_{f,sc} = Q_{f,t} C_{sp,f} \tag{11.19}$$

In the above equation, $Q_{f,t}$ is the annual thermal energy of the fuel combusted in the auxiliary heater, while $C_{sp,f}$ is the fuel price.

The electricity consumption costs of the VCC system are equal to:

$$C_{e,vcc} = E_{e,vcc,t} C_{sp,e} \tag{11.20}$$

where Ee,vcc,t is the annual electricity consumption of the system and $C_{sp,e}$ the electricity price.

Meanwhile, the revenues from the electricity produced by the PV panels are equal to the electricity produced by the panels multiplied by the electricity price:

$$C_{e,pv} = E_{e,pv}C_{sp,e} \tag{11.21}$$

Excluding the material and maintenance costs, the annual savings achieved by the operation of solar cooling systems are equal to the difference between the electricity consumption costs of the conventional VCC system minus the fuel costs of the auxiliary heater of the solar cooling system. They are thus determined by the equation:

$$C_{s,sc-vcc} = E_{e,vcc,t}C_{e,sp} - Q_{f,t}C_{f,sp} \tag{11.22}$$

Meanwhile, the solar fraction (SF) of the solar cooling is defined as the ratio of the annual cooling energy that is produced by solar radiation divided by the total annual cooling energy. It is expressed by the equation:

$$SF = \frac{Q_{ch,t}}{Q_{cool,t}} \tag{11.23}$$

Given the previous analysis, the solar fraction can also be expressed by the equation:

$$SF = \frac{COP_{ch}\sum_{i} Q_{u,i}}{\sum_{i} Q_{cool,i}} \tag{11.24}$$

Considering that the non-solar-derived cooling is produced through the operation of the auxiliary heater, the annual fuel costs of the solar cooling system are expressed by the equation:

$$C_f = \frac{(1-SF)Q_{cool,t}}{\eta_{aux}COP_{ch}}C_{f,sp} \tag{11.25}$$

The electricity costs corresponding to the VCC system are given by the equation:

$$C_{e,vcc} = \frac{Q_{cool,t}}{COP_{VCC}}C_{e,sp} \tag{11.26}$$

A summary of the values of the general economic parameters that are used for the economic evaluation is given in Table 11.19.

TABLE 11.19

General Economic Parameters

Fuel cost (natural gas)	€0.061/kWh$_{th}$
Electricity cost (purchase and selling price grid)	€0.15/kWh$_e$

The payback period for the solar thermal cooling systems is given by the equation:

$$PBP_{sc-vcc} = \frac{C_{tci,sc} - C_{tci,vcc}}{Q_{cool,t}\left(\dfrac{C_{e,sp}}{COP_{vcc}} - \dfrac{C_{f,sp}(1-SF)}{h_{aux}COP_{ch}}\right)} \tag{11.27}$$

Meanwhile, the payback period for the PV-VCC technology is given by the equation:

$$PBP_{sc-vcc} = \frac{C_{TCI,pv}}{C_{e,sp}\sum_i Q_{sol,i}\eta_{pv,i}} \tag{11.28}$$

Obviously, for the solar cooling system to be theoretically viable (which means having a positive payback period), the annual savings (i.e. the denominator of the fraction in the above equation should be positive). This means that the following inequality should be satisfied:

$$SF > 1 - \frac{\eta_{aux}COP_{ch}C_{e,sp}}{COP_{vcc}C_{f,sp}} \tag{11.29}$$

Note that material and maintenance costs have been excluded from the above analysis. In practice, solar cooling systems are expected to have higher maintenance costs than conventional VCC units, since they include a far larger number of components (solar collectors, storage tank, auxiliary heater, extra piping accompanied by more sophisticated control, and automations), while VCC heat pumps constitute a far more mature technology. In practice, this means that the actual minimum SF value should be even lower than the one given in in order to ensure the theoretical economic feasibility of the investment.

Considering all other parameters fixed, increasing the solar fraction of the solar cooling system can be achieved in three ways: by increasing (1) the surface area and (2) the efficiency of the solar collectors (i.e. using ETC collectors instead of FPC collectors), and (3) the thermal storage capacity of the storage tank by increasing its volume and the temperature of the solar collectors. Of course, the above measures all lead to an increase in capital costs. A trade-off between the initial costs of the solar cooling systems and the annual savings can hence be observed. Evidently, further increasing these parameters (such as the area of the solar collectors and the storage tank) after an SF value of 1 has been achieved has a negative impact on the economics of the system, since the capital costs increase while the savings remain constant. As a result, it is important to optimize the solar cooling system in order to reach the most cost-efficient combination of capital costs and annual savings.

In the present study, for each case, three independent optimization variables are considered:

- The total area of the solar collectors (expressed by the parameter A_p)
- The volume of the storage tank (expressed by the parameter V_{tank})
- The temperature of the heat transfer fluid at the outlet of the collectors (T_{col})

The reason for the selection of these variables is the fact that they constitute straightforward key design parameters that highly effect the performance of the system and its cost. While the area of the collectors largely determines the useful heat that is available for use by the chiller, both the storage tank volume and the temperature of the collectors influence the solar thermal storage capacity and the solar fraction of the solar cooling system. Therefore, in each case, the performance of the system is optimized with respect to these parameters. The search bounds for the variables are summarized in Table 11.20.

Before proceeding to the results section, some additional considerations regarding the economic assessment carried out in the present study should be discussed. It is clear that the assessment only takes into account the cooling loads, which is an unrealistic assumption for domestic and commercial buildings. In reality, it makes no sense to install expensive solar collectors without using them from the production of heating during the winter period, unless no heating is required (which is mostly typical in industrial environments). In fact, the economic competitiveness of solar cooling systems can be significantly improved if the energy savings from the production of heating during the winter period are taken into account.

Two different options can be distinguished regarding the combined heating and cooling production of solar cooling systems. The first one involves retrofitting the thermally driven chiller into an existing solar heating installation. Although this option could lead to significantly reduced capital costs compared to the installation of a whole solar cooling system, it should be noted that it may require many equipment modifications for the effective and efficient integration of the thermal chiller to the solar heating circuit. These modifications involve possibly increasing the solar collector area and the storage tank and the installation of additional auxiliary heaters, circulation pumps, piping, and control systems. Furthermore, since thermally driven chillers typically require higher temperatures than heating equipment, it is possible that the existing collectors may need to be entirely substituted if their nominal operating temperature is too low. The second option is to design and install a solar heating and cooling system from scratch. In this case, it is necessary to optimize the components of the system by considering its annual operation, including the winter and summer periods.

For the purpose of the present analysis, the operation of a solar cooling system exclusively for the production of cooling is evaluated. First, this is because the analysis of integrated solar heating and cooling systems is substantially different in many aspects and

TABLE 11.20

Optimization Variables and Search Bounds

Optimization Variable	Min	Max
A_p	0.01	0.80
V_{tank} (m³)	0.5	25
T_{col} (°C) ($T_{ch,in}$ depending on chiller technology)	$T_{ch,in} + 5$	$T_{ch,in} + 30$

extends beyond the scope of the present study. Second, considering both the energy savings from heating and cooling in the economic evaluation can be misleading, since it may conceal the contribution of the solar cooling system in the results. It should be highlighted, of course, that better economic results can be obtained by combined solar heating and cooling systems.

11.5.3 Economic Evaluation Results

11.5.3.1 Optimization Results

The results are organized into two sections. In the first section, the optimal cases (payback periods and optimization variables) for each system are presented for the different locations and building types. In the second section, a series of sensitivity analyses are performed in order to provide an insight into the influence of the independent optimization variables on the economic performance of the system.

The optimal PBP values for the solar cooling systems for Italy are plotted in Figure 11.6. The results include the different collector types (FPC and ETC) and buildings cases (single house, apartment block, and offices).

It should be highlighted that for Germany, the solar cooling systems are highly unfavorable, since the payback period in all cases exceeds 100 years. This is because in Germany the solar radiation intensity is decreased compared to Italy, and the cooling loads are also very low. As a result, it can be deduced that the application of solar cooling systems in this central European region currently faces significant economic challenges.

As expected, the results for Italy are significantly improved compared to those corresponding to Germany. However, even in this case, the payback periods are generally high, even in the best case scenarios (office buildings equipped with absorption chillers), ranging from approximately 30 to 60 years. In all the other cases, the payback periods are very high, since they exceed 100 years. As a result, as a general conclusion, it is apparent that either the integration of solar heating in addition to cooling should be considered in order to further increase the annual energy savings and costs, or financial support schemes must be introduced to promote these solar cooling technologies.

The installation of PV panels combined with VCC units is, in most cases, profitable compared to solar thermal systems since the payback period of this scenario is equal to 41 and 67 years in Italy and Germany, respectively. The only cases in which the economic performance of PV-VCC is comparable to that of the sorption technologies are for Italy (Mediterranean climate), office building types, and absorption chillers. This is due to the

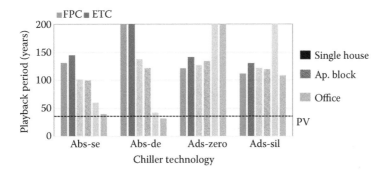

FIGURE 11.6
Optimal PBP values for the solar cooling technologies for Italy.

reduced investment costs of the PV panels, which constitute a less complex installation involving far fewer equipment components. It is important to highlight that the specific costs of PV technology exhibit a continuous declining trend in recent years, accompanied by a simultaneous increase in attainable efficiencies. In this study, the same price has been assumed for selling and buying electricity. In reality, this is not always so, since in many instances (according to specific national legislations), feed-in tariff schemes are implemented to support the market of renewable energy electricity, which are associated with significantly higher electricity selling prices. On the other hand, it should be also stated that in the last few years, there have been trends to regress these support measures, via the introduction of less remunerating schemes for PV technology, such as net-metering.

The economic results regarding sorption chillers can be explained with the help of Eq. (11.28). A crucial difference regarding the solar fraction variation of the three building types must be first discussed. Obviously, office buildings exhibit the highest cooling loads (due to their longer operating hours), followed by apartment blocks and single houses. On the other hand, the maximum area of solar collectors that can be installed in each building type has an upper bound, which in this evaluation has been assumed to equal 80% of the rooftop surface. Given this limitation, it is apparent that it is far easier to reach high SF values in single houses than it is for office buildings, with apartment blocks covering a middle ground between the two. As a result, the relative significance of the annual fuel cost reduction that can be achieved by the solar cooling systems is higher for single house buildings and lower for apartment blocks and office buildings. On the other hand, the smaller the cooling capacity of the application, the higher the specific investment costs, due to the effect of the economy of scale. Therefore, two conflicting trends coexist and determine the economic profitability of the solar cooling technologies for each building type. In summary:

- For single house buildings, more relative energy savings are possible but at higher specific capital costs.
- For apartment blocks and offices, less relative energy savings are possible but at lower specific capital costs.

In the case of single houses, the payback period is over 100 years for all types of chillers. Apart from the high specific investment costs, another factor that inhibits the economic performance of solar cooling systems in single houses concerns their limited operating hours, which range from 4:00 p.m. to 12:00 a.m. Within this time frame, the cooling loads are in principle lower compared to midday hours, during which the ambient temperature is maximized. Furthermore, there is a temporal discrepancy between the solar radiation availability (which is higher during the previous hours of the day) and the cooling load. The ultimate result is a combination of high specific capital investment costs and a decreased solar energy utilization, which inhibits the economics of solar cooling systems in this case.

The situation is somewhat better for apartment blocks. The main difference between these buildings and single houses is the higher required cooling capacities, which lead to lower specific capital investment costs for the equipment components. Meanwhile, despite the higher annual cooling loads, the larger available area for solar collectors enables solar coolers to achieve high SF values. The result is an overall improved economic profitability, which is mirrored in the comparatively (to the single house buildings) lower payback periods. However, even in this case, the payback periods in all cases exceed 100 years (being equal to almost 100 years for single-stage absorption chillers, which constitute the most cost-effective option).

As far as the office building case is concerned, the results vary greatly between absorption and adsorption chillers. On one hand, the economic performance of absorption chillers is highly improved in office buildings compared to the other building types. On the other hand, in all cases of adsorption chillers, with the exception of silica gel chillers coupled with ETCs, the economic performance for office buildings deteriorates compared to apartment blocks. This significant difference in the economic profitability between absorption and adsorption chillers is due to the very high cooling loads of office buildings, which make higher COP values necessary for achieving sufficient energy savings. As a result, adsorption chillers exhibit a significantly inferior performance compared to absorption-type chillers. In fact, only with silica gel-water adsorption chillers (which have a higher COP than zeolite-water chillers), the investment leads to energy savings, and only ETCs are employed, which can lead to higher SF values due to their enhanced efficiency.

For the same reason, for single house buildings, the use of FPCs leads to better economic results for all chiller technologies, since the low efficiency of these collectors does not inhibit the attainment of high SF values due to the decreased overall cooling loads of these buildings. Accordingly, in the case of apartment block buildings, which constitute the average case between single houses and office buildings, both collector types lead to a similar economic performance.

From the above results, under the assumptions this study has followed, it can be concluded that:

- Solar cooling technology is economically feasible for Mediterranean regions with high solar energy availability and cooling loads.
- Solar absorption chillers are preferable in high capacity applications (services and utility buildings), where they have favorable economy of scale and low COP values are necessary to achieve high SF values.
- Solar adsorption chillers are preferable for small-scale applications (single house buildings), where they have favorable economy of scale and low COP values are not necessary to achieve high SF values.
- FPCs are preferable for small-scale applications, where high collector efficiency is not necessary to achieve high SF values.
- ETCs are preferable for larger-scale applications, where high collector efficiency is necessary to achieve high SF values.
- PV-VCC solar cooling systems are substantially more cost effective than sorption technologies, with the exception of office buildings, where absorption chiller exhibit comparable performance.

In order to explore the potential of retrofitting solar cooling systems into existing solar heating infrastructures, the payback period without taking into account the cost of the solar circuit has been estimated. The results for the case of Italy are plotted in Figure 11.7.

For single houses located in Italy, both types of adsorption chillers are associated with payback periods of around 10 years. The performance of absorption chillers remains significantly lower. For apartment blocks, single-effect absorption chillers exhibit the best performance, with payback periods of around 17 years, followed by adsorption chillers, with payback periods slightly lower than 30 years. Lastly, the payback periods achieved

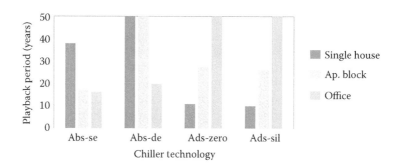

FIGURE 11.7
Optimal PBP values for the solar cooling technologies excluding the cost of the solar circuit for Italy.

for office buildings are minimized for absorption chillers and are between 15 and 20 years for single- and double-effect chillers, respectively.

In Germany, the economic performance of absorption chillers in all cases is not viable, with payback periods higher than 90 years, despite the reduction of the capital costs owing to the exclusion of the solar collector circuit. On the other hand, it is interesting that in the case of adsorption chillers in single house buildings and apartment blocks, the capital cost of the solar cooling systems is lower than that of conventional systems, and thus the investment is inherently very profitable. This occurs due to the fact that cooling loads are very low in Germany, and hence the capital costs of installing adsorption chillers only are greatly reduced.

Next, the optimal values of the design variables in the case of Italy are presented in the diagrams of Figure 11.8.

First, it can be seen that the area parameter of the collectors and the storage tank volume tends to have lower values for single houses and apartment blocks, and adopts significantly higher values for office buildings. This is because more collectors and larger storage tanks are necessary to achieve the SF values that are necessary for economically feasible performance when the loads are higher (as is the case for office buildings). Notably, there is not a significant influence on the values of these parameters for different chiller technologies. Regarding the type of the collectors, FPCs are associated with higher optimal collector surfaces compared to ETCs. This can be explained by the lower efficiency of these collectors that results in an increase of the surface area that is required for achieving higher solar fractions. A similar effect can be observed regarding the influence of the solar collector types on the storage tank volume. ETCs offer the capability of reducing the required storage tank volume that is required for ensuring the economic feasibility of the investment.

As far as the optimal collector temperature is concerned, it is highly dependent on the chiller type. Of course, this is in part due to the fact that each chiller has a different driving heat temperature, as is also presented in Table 11.15. On the other hand, there is clearly an influence between the building and the collector type on the optimal temperature of the collectors. In general, higher collector temperatures are selected when ETCs are considered. This is because of the higher efficiencies of these collectors, which permit operation at higher temperatures without a significant decrease in the useful heat produced by the solar circuit. In fact, the highest possible collector temperatures (within the range investigated) are almost always optimal for these types of collectors. Finally, the building type also affects the optimal collector temperature, even considering the same collector and chiller technology. In principle, for single houses and apartment blocks, higher temperatures are more favorable.

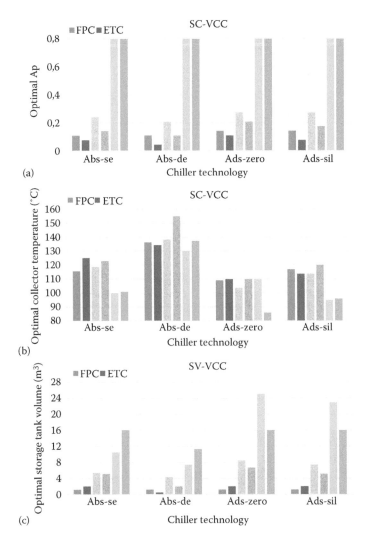

FIGURE 11.8
Optimal design variables for the solar cooling technologies: (a) A_p, (b) storage tank volume, and (c) collector temperature for Italy.

11.5.3.2 Parametric Analyses

In the qualitative diagrams presented in Figure 11.9, the variation in the payback period of the solar cooling systems as a function of the different optimization parameters is presented. As can be observed, a common pattern governs the optimization of these variables. More specifically, for each variable, there is theoretically an intermediate optimal value that minimizes the payback period.

In Figure 11.9a, the variation of the payback period as a function of the area of the solar collectors and the storage tank volume is plotted. The optimal values of these variables are closely related to their contribution to the solar fraction of the system and its variation. Starting from their lower bounds, as they increase, the solar fraction increases, leading to a subsequent increase in energy savings. However, after they reach a specific value, the solar fraction is either maximized and reaches unity or its increase rate may become

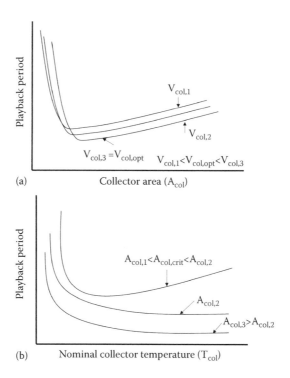

FIGURE 11.9
Variation in payback period as a function of (a) the collector area and (b) the nominal collector temperature.

very slow. The latter can occur when one of the variables imposes a bottleneck on the solar fraction of the system. For example, if the area of the solar collectors is too small, the absorbed solar radiation is limited, and thus increasing the storage volume does not lead to a significant increase in the solar fraction, or vice versa. Beyond this point, there are no more economic benefits to be achieved by increasing the area of the solar collectors or the volume of the storage tank since the energy savings are either fixed (SF = 1) or increasing at a lower rate than the increase rate of the capital cost, thus causing the payback period to increase. It should also be noted that the payback period is significantly more sensitive to the area of the collectors than to the volume of the storage tank. This is because the cost of the former is significantly higher than that of the latter, while its influence on the solar fraction is more important, since it directly determines the capacity of the solar energy that is absorbed by the system.

Regarding the nominal temperature of the solar collectors (Figure 11.9b), its influence on the payback period is more complicated. A higher collector temperature leads to an increase of the storage capacity of the tank (in the form of sensible heat). However, it simultaneously leads to a decrease in the efficiency of the solar collectors, and thus to a decrease of the useful heat than can be delivered to the storage tank. Consequently, two different cases can be distinguished regarding the overall influence of the temperature of the collectors on the payback period. In both cases, an initial increase in the temperature starting from its lower bound leads to a drop in the payback period, since the positive effect of the storage capacity increase is predominant over the decrease in the efficiency of the collectors. However, for higher values, if the area of the collectors or the volume of the storage tank are very low, further increasing the collector temperature has a negative influence on the payback period, since the increasing heat losses in the collectors are more significant than the energy benefits of the increasing storage capacity. This leads to a global minimum of the payback period for an intermediate value of the collector temperature. Note that in Figure 11.9, the depicted qualitative relation between the payback period and different A_{col} values is similar for different V_{tank} values. On the other hand, if the A_{col} and V_{tank} are sufficiently high, then a solar fraction of unity is achieved. In this case, further increasing the T_{col} does not lead to a variation in the SF value and energy savings. Meanwhile, the capital cost of the system does not change (in the present study, it has been assumed that the impact of the collector temperature on the cost of the equipment components is negligible). Therefore, the payback period in this case assumes a fixed value as the temperature of the collectors increases.

11.6 Conclusions

The first purpose of Chapter 11 was to provide a broad literature review of different studies on the economic feasibility of solar cooling systems and their constituent components along with a summary of different key economic values and parameters (capital costs and payback periods). First, it is evident that there is a wide variation in reported cost values among different investigations, especially those concerning the costs of adsorption chillers. This is to be expected since this technology is comparatively novel and has yet to be fully commercialized. Second, there is also a large variation regarding the values of the economic indexes (NPV, PBP, cost of cooling energy) that are calculated in different studies. This is also to be anticipated given the fact that the economic performance of solar cooling systems is highly affected by geographic location, which is closely tied to solar radiation intensity and cooling requirements. Furthermore, the methodologies followed also differ, sometimes in crucial aspects, such as the scope of the system (solar heating and cooling versus exclusive solar cooling).

The second goal of this chapter was to present a consistent and detailed methodology for the evaluation of solar cooling systems. A techno-economic evaluation of sorption technologies (absorption, adsorption) and photovoltaic systems was carried out for three different building types (single houses, apartment blocks, and office buildings) for two regions: Italy and Germany. It was shown that new dedicated cooling systems (i.e. that are not used for heating) are economically infeasible in Germany. On the other hand, there is potential to achieve economic feasibility in Italy by installing solar absorption systems in large utilities services of service buildings with high cooling loads, or adsorption chillers in single house buildings, if financial support schemes are introduced or these systems are also used for the production of heating. If the retrofitting of chillers into existing solar circuits is considered, the results are significantly improved, especially in the case of Italy and in the application of adsorption chillers in single houses and apartment blocks in Germany. Regarding the optimal design of these systems, it was shown that it is necessary to properly select the area of the collectors and storage tank in order to maximize the solar fraction, while, at the same time, avoid oversizing the equipment components. Furthermore, it is important to avoid bottlenecks that could lead to increased costs without an effective increase in energy savings.

Nomenclature

A	Area, m^2
C	Cost/cash flow
COP	Coefficient of performance
c$_p$	Specific heat capacity, kJ/kgK
E	Electrical energy
G	Solar radiation, W/m^2
h	Specific enthalpy, kJ/kg
$\dot{m}$	Mass flow rate, kg/s
M	Mass, kg
P	Power, kW
p	Pressure, bar
PBP	Payback period
Q	Heat, kW
SF	Solar fraction
T	Temperature, °C
U	Heat transfer coefficient, W/m^{2}K
V	Volume, m^3

Greek symbols

η	Efficiency, %
ρ	Density, kg/m^3

Subscripts

amb	Ambient
aux	Auxiliary heat source
ch	Chiller
col	Solar collector
cond	Condenser
cool	Cooling
ct	Cooling tower
e	Electric
e	Electricity
eq	Equipment
etc	Evacuated tube collectors
f	Fuel

fpc	Flat-plate collectors
htf	Heat transfer fluid
in	Inlet
is	Isentropic
max	Maximum
min	Minimum
nom	Nominal
out	Outlet
pv	Photovoltaic
ref	Reference
s	Savings
sc	Solar cooling system
sol	Solar
sp	Specific
st	Storage
tank	Tank
th	Thermal
t	Total (annual)
u	Useful heat
vcc	Vapor compression cycle

Abbreviations

COP	Coefficient of performance
ETC	Evacuated tube collectors
FPC	Flat-plate collectors
NPV	Net present value
PTC	Parabolic trough collectors
TCI	Total capital investment cost
VCC	Vapor compression cycle

References

Al-Ugla, A. A., M. A. I. El-Shaarawi, S. A. M. Said, and A. M. Al-Qutub. 2016. "Techno-Economic Analysis of Solar-Assisted Air-Conditioning Systems for Commercial Buildings in Saudi Arabia." *Renewable and Sustainable Energy Reviews* 54 (Supplement C): 1301–1310. doi: https://doi.org/10.1016/j.rser.2015.10.047.

Allouhi, A., T. Kousksou, A. Jamil, P. Bruel, Y. Mourad, and Y. Zeraouli. 2015. "Solar Driven Cooling Systems: An Updated Review." *Renewable and Sustainable Energy Reviews* 44 (Supplement C): 159–181. doi: https://doi.org/10.1016/j.rser.2014.12.014.

Allouhi, Amine, Tarik Kousksou, Abdelmajid Jamil, Tarik El Rhafiki, Youssef Mourad, and Youssef Zeraouli. 2015. "Economic And Environmental Assessment of Solar Air-Conditioning Systems in Morocco." *Renewable and Sustainable Energy Reviews* 50 (Supplement C): 770–781. doi: https://doi.org/10.1016/j.rser.2015.05.044.

Bataineh, K. M., and S. Alrifai. 2015. "Recent Trends In Solar Thermal Sorption Cooling System Technology." *Advances in Mechanical Engineering* 7 (5): 1–20. doi: 10.1177/1687814015586120.

Bataineh, Khaled, and Yazan Taamneh. 2016. "Review and Recent Improvements of Solar Sorption Cooling Systems." *Energy and Buildings* 128 (Supplement C): 22–37. doi: https://doi.org/10.1016/j.enbuild.2016.06.075.

Bellos, Evangelos, Christos Tzivanidis, and Georgios Tsifis. 2017. "Energetic, Exergetic, Economic and Environmental (4E) Analysis of a Solar Assisted Refrigeration System for Various Operating Scenarios." *Energy Conversion and Management* 148 (Supplement C): 1055–1069. doi: https://doi.org/10.1016/j.enconman.2017.06.063.

Blackman, Corey, Chris Bales, and Eva Thorin. 2015. "Techno-economic Evaluation of Solar-Assisted Heating and Cooling Systems with Sorption Module Integrated Solar Collectors." *Energy Procedia* 70 (Supplement C): 409–417. doi: https://doi.org/10.1016/j.egypro.2015.02.142.

Calise, F., M. Dentice d'Accadia, and L. Vanoli. 2011. "Thermoeconomic Optimization of Solar Heating and Cooling Systems." *Energy Conversion and Management* 52 (2): 1562–1573. doi: https://doi.org/10.1016/j.enconman.2010.10.025.

Calise, Francesco. 2012. "High Temperature Solar Heating and Cooling Systems for Different Mediterranean Climates: Dynamic Simulation and Economic Assessment." *Applied Thermal Engineering* 32 (Supplement C): 108–124. doi: https://doi.org/10.1016/j.applthermaleng.2011.011.037.

Commission, European. 2017. "Photovoltaic Geographical Information System (PVGIS)—Interactive Maps." *http://re.jrc.ec.europa.eu/pvgis/apps4/pvest.php.* Accessed on 30 March 2018.

Desideri, Umberto, Stefania Proietti, and Paolo Sdringola. 2009. "Solar-powered cooling systems: Technical and Economic Analysis on Industrial Refrigeration and Air-Conditioning Applications." *Applied Energy* 86 (9): 1376–1386. doi: https://doi.org/10.1016/j.apenergy.2009.01.011.

Eicker, Ursula, and Dirk Pietruschka. 2009. "Optimization and Economics of Solar Cooling Systems." *Advances in Building Energy Research* 3 (1): 45–81. doi: 10.3763/aber.2009.0303.

Eicker, Ursula, Dirk Pietruschka, Maximilian Haag, and Andreas Schmitt. 2014. "Energy and Economic Performance of Solar Cooling Systems World Wide." *Energy Procedia* 57 (Supplement C): 2581–2589. doi: https://doi.org/10.1016/j.egypro.2014.10.269.

Energy, US Department of. 2017. "Asorbtion Chillers for CHP Chillers." https://energy.gov/sites/prod/files/2017/06/f35/CHP-Absorption%20Chiller-compliant.pdf.

Gabbrielli, Roberto, Piero Castrataro, and Francesco Del Medico. 2016. "Performance and Economic Comparison of Solar Cooling Configurations." *Energy Procedia* 91 (Supplement C): 759–766. doi: https://doi.org/10.1016/j.egypro.2016.06.241.

Gebreslassie, Berhane H., Gonzalo Guillén-Gosálbez, Laureano Jiménez, and Dieter Boer. 2010. "A Systematic Tool for the Minimization of the Life Cycle Impact of Solar Assisted Absorption Cooling Systems." *Energy* 35 (9): 3849–3862. doi: https://doi.org/10.1016/j.energy.2010.05.039.

Henning, H. M. 2004. *Solar-Assisted Air-Conditioning in Buildings: A Handbook for Planners.* Springer.

Huang, B. J., J. H. Wu, R. H. Yen, J. H. Wang, H. Y. Hsu, C. J. Hsia, C. W. Yen, and J. M. Chang. 2011. "System Performance and Economic Analysis of Solar-Assisted Cooling/Heating System." *Solar Energy* 85 (11): 2802–2810. doi: https://doi.org/10.1016/j.solener.2011.011.011.

Infante Ferreira, Carlos, and Dong-Seon Kim. 2014. "Techno-Economic Review of Solar Cooling Technologies Based on Location-Specific Data." *International Journal of Refrigeration* 39 (Supplement C): 23–37. doi: https://doi.org/10.1016/j.ijrefrig.2013.09.033.

Kim, D. S., and C. A. Infante Ferreira. 2011. "Solar Refrigeration Options—A State-Of-The-Art Review." *International Journal of Refrigeration* 31 (1): 3–15. doi: https://doi.org/10.1016/j.ijrefrig.2007.07.011.

Kim, D.S. 2007. "Solar Absorption Cooling." PhD diss. Mechanical Maritime and Materials Engineering (TU Delft).

Lambert, Michael A., and Asfaw Beyene. 2007. "Thermo-Economic Analysis of Solar Powered Adsorption Heat Pump." *Applied Thermal Engineering* 27 (8): 1593–1611. doi: https://doi.org /10.1016/j.applthermaleng.2006.09.005.

Lazzarin, Renato M. 2014. "Solar Cooling: PV or Thermal? A Thermodynamic and Economical Analysis." *International Journal of Refrigeration* 39 (Supplement C): 38–47. doi: https://doi.org /10.1016/j.ijrefrig.2013.05.012.

Mauthner, F., W. Weiss, and M. Spörk-Dür. 2015. *Solar Heat Worldwide: Markets and Contribution to the Energy Supply 2013.* 2015 edition. AEE-Institute for Sustainable Technologies A8200 Gleisdorf, Austria.

Mokhtar, Marwan, Muhammad Tauha Ali, Simon Bräuniger, Afshin Afshari, Sgouris Sgouridis, Peter Armstrong, and Matteo Chiesa. 2010. "Systematic Comprehensive Techno-Economic Assessment of Solar Cooling Technologies Using Location-Specific Climate Data." *Applied Energy* 87 (12): 3766–37711. doi: https://doi.org/10.1016/j.apenergy.2010.06.026.

Montagnino, Fabio Maria. 2017. "Solar Cooling Technologies. Design, Application and Performance of Existing Projects." *Solar Energy* 154 (Supplement C): 144–157. doi: https://doi.org/10.1016/j .solener.2017.01.033.

Mugnier, Daniel. 2012. "Solar Thermal Energy For Cooling and Refrigeration: Status and Perspectives." Solar Cooling International Conference. Intersolar Fair, Munchen.

Navarro-Rivero, Pilar, and Bjorn Ehrismann. 2012. "Durability Issues, Maintenance and Costs of Solar Cooling Systems." *Task Report 5.3.2.* Deliverable D5.3 on QAiST project.

Neyer, D., J. Neyer, A. Thür, R. Fedrizzi, A. Vittoriosi, S. White, and H. Focke. 2016. "Collection of Criteria to Quantify the Quality and Cost Competitiveness for Solar Cooling Systems." *IEA Solar Heating and Cooling Task 48.* http://task411.iea-shc.org/data/sites/1/publications/Task%20 48%20Activity%20B7%20Final%20report%20May%202015.pdf.

Noro, M., and R. M. Lazzarin. 2014. "Solar Cooling between Thermal and Photovoltaic: An Energy and Economic Comparative Study in the Mediterranean Conditions." *Energy* 73 (Supplement C): 453–464. doi: https://doi.org/10.1016/j.energy.2014.06.035.

Otanicar, Todd, Robert A. Taylor, and Patrick E. Phelan. 2012. "Prospects for Solar Cooling – An Economic and Environmental Assessment." *Solar Energy* 86 (5): 1287–1299. doi: https://doi.org /10.1016/j.solener.2012.01.020.

Rowe, Daniel, and Stephen White. 2014. "Review of International Solar Cooling Incentive Schemes." *Energy Procedia* 57 (Supplement C): 3160–3170. doi: https://doi.org/10.1016/j .egypro.2015.06.065.

SHC-IEA. 2016a. "The Future of Solar Cooling." *Shc Solar Update.* http://task53.iea-shc.org/data/sites /1/publications/2016-05-Task53 The%20Future%20of%20Solar%20Cooling.pdf. Accessed on 30 March 2018.

SHC-IEA. 2016b. "Technology and Quality Assurance for Solar Thermal Cooling Systems." *IEA Solar Heating and Cooling Task 48.* http://task411.iea-shc.org/data/sites/1/publications/Task48%20 D3%20brochure%20final.pdf. Accessed on 30 March 2018.

Shirazi, Ali, Robert A. Taylor, Stephen D. White, and Graham L. Morrison. 2016. "Transient Simulation and Parametric Study of Solar-Assisted Heating and Cooling Absorption Systems: An Energetic, Economic, and Environmental (3E) Assessment." *Renewable Energy* 86 (Supplement C): 955–971. doi: https://doi.org/10.1016/j.renene.2015.09.014.

Technology, Massachusetts Institute of. 2000–2009. "The MIT Design Advisor: Building Simulation in Minutes." http://designadvisor.mit.edu/design/.

Tsoutsos, Theocharis, Joanna Anagnostou, Colin Pritchard, Michalis Karagiorgas, and Dimosthenis Agoris. 2003. "Solar Cooling Technologies in Greece. an Economic Viability Analysis." *Applied Thermal Engineering* 23 (11): 1427–1439. doi: https://doi.org/10.1016/S1359 -4311(03)00089-9.

Index

Page numbers followed by f and t indicate figures and tables, respectively.

T - #0472 - 071024 - C27 - 254/178/20 - PB - 9780367733179 - Gloss Lamination